Hydrogen-Bonded Liquids

NATO ASI Series

Advanced Science Institutes Series

A Series presenting the results of activities sponsored by the NATO Science Committee, which aims at the dissemination of advanced scientific and technological knowledge, with a view to strengthening links between scientific communities.

The Series is published by an international board of publishers in conjunction with the NATO Scientific Affairs Division

A Life Sciences **B Physics**	Plenum Publishing Corporation London and New York
C Mathematical and Physical Sciences **D Behavioural and Social Sciences** **E Applied Sciences**	Kluwer Academic Publishers Dordrecht, Boston and London
F Computer and Systems Sciences **G Ecological Sciences** **H Cell Biology**	Springer-Verlag Berlin, Heidelberg, New York, London, Paris and Tokyo

Series C: Mathematical and Physical Sciences - Vol. 329

Hydrogen-Bonded Liquids

edited by

John C. Dore
Physics Department,
University of Kent at Canterbury,
Canterbury, U.K.

and

José Teixeira
Laboratoire Leon Brillouin,
C.E.N.,
Saclay, France

Kluwer Academic Publishers

Dordrecht / Boston / London

Published in cooperation with NATO Scientific Affairs Division

Proceedings of the NATO Advanced Study Institute on
Hydrogen-Bonded Liquids
Cargèse, Corsica, France
April 3–15, 1989

Library of Congress Cataloging-in-Publication Data

NATO Advanced Study Institute on Hydrogen-bonded Liquids (1989 : Institut scientifique de Cargèse)
Hydrogen-bonded liquids : proceedings of the NATO Advanced Study Institute on Hydrogen-bonded Liquids, held at the Institut scientifique de Cargèse, Corsica, April 3-15, 1989 / edited by John C. Dore and José Teixeira.
p. cm. -- (NATO ASI series. Series C, Mathematical and physical sciences ; vol. 329)
Includes indexes.
ISBN 0-7923-1055-1 (alk. paper)
1. Hydrogen bonding--Congresses. 2. Liquids--Congresses. I. Dore, John C. II. Teixeira, José. III. Title. IV. Series: NATO ASI series. Series C, Mathematical and physical sciences ; no. 329.
QD461.N355 1989
541'.042--dc20 90-20716

ISBN 0–7923–1055–1

Published by Kluwer Academic Publishers,
P.O. Box 17, 3300 AA Dordrecht, The Netherlands.

Kluwer Academic Publishers incorporates the publishing programmes of D. Reidel, Martinus Nijhoff, Dr W. Junk and MTP Press.

Sold and distributed in the U.S.A. and Canada
by Kluwer Academic Publishers,
101 Philip Drive, Norwell, MA 02061, U.S.A.

In all other countries, sold and distributed
by Kluwer Academic Publishers Group,
P.O. Box 322, 3300 AH Dordrecht, The Netherlands.

Printed on acid-free paper

Printed in the Netherlands

CONTENTS

Preface

The study of liquids covers a wide range of scientific disciplines, primarily in physics and chemistry. As a result of this disparate activity the links between new developments in remote fields are seldom co-ordinated into a single conference. The objective of the present meeting was to gather together people with different forms of expertise. Previous ASI meetings on the liquid state have been held over an extended period and have occurred on a three-yearly basis. The first meeting in this series was on 'Structure and Dynamics of Liquids' in 1980 and was held on the island of Corsica. The next meeting on 'Molecular liquids: Dynamics and Interactions' was held in Florence in 1983 and was followed by 'Aqueous Solutions' at the Institut d'Etudes Scientifiques de Cargèse in 1986. It therefore seemed a natural choice to select Cargèse for the next meeting in 1989 and to choose a topic which emphasised a particular area of liquid state studies. Due to our own involvement in collaborative research we considered that 'Hydrogen-bonded liquids' would be an appropriate topic. One of its attractions, was that there was much new material coming from widely disparate investigations and it would be a convenient time to draw together the different strands. The particular interest in water was clearly central to this topic but it was thought desirable to set this development in the wider context of other systems in which hydrogen-bonding plays a significant role. The objective was therefore to develop a multi-disciplinary programme in which current activities could be represented with a range of contributions from physicists, chemists and biologists in one sense and from theorists, simulators and experimentalists in another sense.

In the planning stages it became apparent that the achievement of a satisfactory balance between the different subject areas would be difficult but we were pleased to find that our enthusiasm was frequently echoed by people on our priority list of invited speakers The original intention was to hold the meeting in June but booking arrangements proved difficult for a two-week course and we eventually decided to move the date forward to April. The significance of the choice of date and venue was to have an important influence on the conduct of the meeting but we did not realise its significance at the time.

The start of 1989, produced an increase in unrest on Corsica due to regional issues. Being an island in the Mediterranean, meant that the most effective weapon in the dispute was disruption of travel services to and from the island. For the ASI meeting it was convenient to follow usual arrangements and book a group flight from Paris to Ajaccio and as a precaution a back-up flight on the following day. When Sunday 2nd April arrived, about half the conference delegates, some with wives and children met at Orly airport. All looked well but suddenly it was announced that the control tower at Ajaccio airport had been occupied by strikers and the flight was cancelled half-an-hour before the scheduled take off! Worse was to follow!! It had been assumed by the agency that the flight had been cleared and the back-up reservation had been cancelled - nothing could now be guaranteed until Thursday! With half the conference delegates already on the island and the other half stranded in Paris, everything seemed in complete disarray and no obvious solution was available.

The group stayed overnight in the airport. The weather for April was particularly severe and heavy rain was replaced by snow during the following morning. With little to do but wait for further developments it was decided to start the talks in the hotel but the only suitable area was an open space, close to the restaurant. Furthermore, there was building work inside the hotel and pneumatic drills were in action for part of the afternoon when it was assumed very few people would be in the hotel. Despite the background noise, which later changed to the clatter of knives and forks amidst a rising level of general conversation, several excellent introductory talks were given; it is doubtful that the speakers will ever have to perform again in such adverse conditions! As a finale to the evening there was a remote chance of a flight so the group reassembled in the departure area. Unfortunately the original 170-seater aircraft had already been reduced to 115 places due to some technical difficulties and the group of ~ 50 was given a priority stand-by allocation. Amazingly, everyone was able to board with only a few seats left. It was a routine flight to Ajaccio with a large number of happy people who did not appear to mind missing their evening meal. The coach was fortunately waiting at the airport and the group boarded it through torrential rain. A brief stop at a roadside cafe to take on some food turned into a rather lengthy delay and it was long after midnight when the coach descended the steep and winding track to the Institute to unite happily with the other members of the meeting. The heavy rain persisted as various groups were shown to their hotels and other residences in the village of Cargèse. The following morning provided a chance to establish a revised programme of talks and to commence the meeting proper. However, there were a few delegates who had been unable to make the journey due to other travel problems or who considered the further uncertainties on how to get off the island in time for other commitments were simply too great a risk to take. It did not take too long for the 'island' group and the 'Paris' group to intermingle and some repeat performances of the lectures given in the hotel were organised. This produced several interesting and revealing variants for a few talks since they were not always given by the original speaker!

Throughout the two week meeting the strike situation and travel conditions remained unpredictable. One of the lecturers who was involved with experimental work to test 'cold fusion' claims gave a short presentation on current developments before leaving early in case his team discovered something important and had written a paper without his name on it. His bus to Ajaccio did not run so he took a taxi but then had to wait for a day or two to catch an operational flight off the island.

One of the attractions of the Institute is its proximity to the beach but the weather remained poor for the time of year and the sea was much too cold for swimming. One fortunate exception was on the day of the excursion to Capo Rosso and the magnificent coastal scenery in the north part of the island, leading to an energetic walk over the headland in hot and sunny conditions. A few days later there was a violent storm with lightning and heavy winds, perhaps a suitable backdrop to a meeting on hydrogen-bonding but far from the expectations of this Mediterranean island.

As the meeting progressed into its second week, thoughts began to focus on the return group journey to Paris. Uncertainties over flight arrangements grew and it was decided that the only sure way was to bring forward the Friday programme and made a sea crossing. The group therefore made the $2^1/_2$ hour coach journey across the island to the port of Propriano to board the boat to Marseille. There was a magnificent sunset over the seas as the boat left the port. A few hours later as people had settled into their berths in the bows of the ship the tranquility was broken by a violent storm which occurs somewhat infrequently in the Mediterranean. About half the party were sea-sick and not all the rest felt like an early morning breakfast as the boat slid calmly past the white rocks tinged with red from the rising sun as the boat entered the harbour.

The rest of the journey involving the Metro across town, a coach trip to the airport and the flight back to Paris could possible be described as relatively uneventful but it was not entirely without incident. Arrival at Charles de Gaulle airport mean some people had to trek back to Orly to collect left luggage before resuming their journeys. There was a sense of relief that the travel problems were finally settled as everyone dispersed to return to their own countries. It had certainly been a meeting which would remain long in the memory, particularly for those who suffered on the sea-crossing. Hopefully, some of the science of hydrogen-bonding in the liquid state will eventually emerge as the main features of the meeting but it seemed as if water in its more elemental forms (e.g. rain, snow, sea etc. had impressed itself on all delegates in some form or other.

We wish to thank the Scientific Committee of the North Atlantic Treaty Organisation (NATO/OTAN) for providing financial support for the Advanced Study Institute. We also wish to acknowledge additional sponsorship from C.N.R.S. and C.E.A. On the personal side, we are particularly indebted to Marie France Hanseler who dealt with all the local organisation. The problems outlined above cannot fully represent the continual pressure which she had to endure. It must have been an administrative nightmare but without her persistent optimism and skilled negotiating the whole enterprise could have collapsed into disorder and chaos. Several later conferences at the Institute were cancelled, so, in some respects, we were lucky to be able to meet at all. The friendly and tolerant atmosphere (while under extreme provocation?) was a direct result of the collective response of all participants and we wish to express our appreciation for making a difficult organisational job as pleasant as it could be under these circumstances. Finally, we hope that the book will be a useful reference source for future work. It has taken longer to compile and publish than had been intended but this particular field is not changing as rapidly as some other areas of science. The disruption of the programme has inevitably led to some shortcomings in the final presentation but we hope and anticipate that it will have provided a broad survey of current activities and given a focal point to all the varied approaches to the difficult task of understanding hydrogen-bonded liquids, particularly water. There were some occasions when one might have attributed some malevolence to 'water', considering that we were subjected to such a dramatic exposure to its influences.

J.C.Dore and Jose Teixeira

List of Participants

a) Directors:

Dr.John C Dore,
Physics Laboratory,
University of Kent,
Canterbury,
Kent. CT2 7NR. UK

Dr.José Teixeira,
Laboratoire Leon Brillouin,
CEN-Saclay,
91191 Gif/Yvette Cedex,
FRANCE.

b) Lecturers (invited and contributed)

Prof.C.A.Angell,
Arizona State University, Tempe,
USA.

Dr.M-C. Bellissent-Funel,
Lab.Leon Brillouin, Saclay, France.

Prof.H.J.C.Berendsen,
Lab.Physical Chemistry,
Univ.Groningen, Holland.

Dr.L.Bosio,
E.S.P.C.I, Paris, France.

Prof.D.Bratko,
Chem.Dept., Univ. of Lyubliana,
Yugoslavia.

Prof.S.Bratos,
Physique Theorique des Liquides,
Univ. Paris VI, France.

Prof.S.H.Chen,
M.I.T, Cambridge, USA.

Dr.G.D'Arrigo,
Dipart.di Fisica, Univ.di Roma, Italy.

Dr.E.Del Giudice,
Dipart. di Fisica, Milan, Italy.

Prof.A.Geiger,
Physical Chemie, Univ.Dortmund,
W.Germany

Prof.J.P.Hansen,
ENS de Lyon, France.

Prof.M.Klein,
Dept.of Chemistry, Univ of
Pennsylvania, USA.

Prof.H.D.Ludemann,
Universitat Regensburg, W.Germany

Dr.E.Lang,
Inst.Biophysik, Univ.Regensburg,
W.Germany

Prof.A.Luzar,
Inst.of Microbio, Lyubliana,
Yugoslavia.

Dr.V.Mazzacuratti,
Dipart.di Fisica, Univ.di Roma, Italy.

Dr.Y.Marechal,
CEN-Grenoble, France.

Dr.H.D.Middendorf,
Clarendon Lab., Univ. of Oxford, UK.

Prof.D.G.Montague,
Phys.Dept., Willamette Univ.,
Oregon,USA.

Prof.G.Palinkas,
Hung.Acad.Sci.Pusztaszeri, Budapest,
Hungary.

Dr.D.K.Ross,
School of Phys. & Space Res.,
Birmingham, UK.

Prof.P.Rossky,
Dept.of Chem., Univ.of Texas, Austin,
USA.

Dr.G.Salvetti,
IFAM-CNR, Pisa, Italy.

Dr.A.K.Soper,
Rutherford Appleton Lab., Oxon, UK.

Dr.A.Stone,
Chem.Lab. Univ.of Cambridge, UK.

Prof.G.E.Walrafen,
Dept.of Chem., Howard Univ,
Washington, USA.

c) Participants

Mme. C.Alba-Simionesco,
Lab.de Chimie Phys., Paris, France.

Mr. Alonso,
Inst.Estructura de la Materia, Madrid, Spain.

Mme. Alvarez,
Inst.Estructura de la Materia, Madrid, Spain.

Mr.J.W.N.Bieze,
Gorlaeus Lab., State.Univ. Leiden, Holland.

Mr.K.Bradley,
Argonne Nat.Lab., Argonne, USA.

Mme. del C. Bujan Nunez,
Universidad de Santiago, Spain.

Prof.A.A.Da Costa,
Dept. of Chem., University, Coimbra, Portugal.

Mr.E.Davies,
Phys.Lab., Univ. of Kent, UK.

Mme.F.Floris,
Dipart.di Chimica E chimica Ind., Pisa, Italy.

Mr.A.Grasso,
Dip.di Fisica Univ.di Messina, Italy.

Mrs.J.Green,
Chem.Dept., Arizona State Univ., Tempe, USA.

Mme.E.Guardia,
Fisica,Eng.Nucl.Univ.Politec, Barcelona, Spain.

Dr.A.Guner,
Dept.of Chem., Hacettepe Univ., Ankara, Turkey.

Mr.M.Has,
Univ.Regensburg, Regensburg, W.Germany

Mr.K.Hoffer,
Inst.Anorganische, Univ.Innsbruck, Austria.

Mr.D.W.Huxley,
Physic Lab., Univ.of Kent, UK.

Mr.M.Ilczyczyn,
Inst.of Chemistry,Wroclaw, Poland.

Dr.J.C.Li,
School of Phys. & Space Res, Birmingham, UK.

Dr.C-K.Loong,
IPNS, Argonne Nat.Lab., USA.

Mr.S.Magazu,
Dipart.di Fisica, Messina, Italy.

Mr.D.Majolino,
Dipart.di Fisica, Messina, Italy.

Mr.J.Mitchell,
Chem.Lab., Univ.of Cambridge, UK.

Mr.J.A.Padro,
Dep.Fisica Fondamental, Univ.Barcelona, Spain.

Dr.L.Pusztai,
Theo.Chem.Lab., L,Eotvos Univ., Budapest, Hungary.

L.F.D.Santos,
Dept.Chem., Univ.Porto, Portugal.

Mr.G.I.Swan,
Phys.Lab., Univ.of Kent, UK.

Dr.Y.Tamura,
Max-Planck-Institut fur Chemie, Mainz, W.Germany

Dr.A.Tani,
Dipart.di Chimica, Pisa, Italy.

Mme.C.Tojo,
Dpto.Quimica Fisica, Universidad de Santiago, Spain.

Dr.E.Tombari,
Ist.Fisica Atomica & Mol.del., Pisa, Italy.

Mr.Z.H.Tromp,
Gorlaeus Lab.Dept.Phys.,Chem.,State Univ.Leiden, Holland.

Dr.J.Turner,
Dept.Crystallography,Birkbeck College, London, UK.

Prof.U.Wanderlingh,
Messina, Italy.

Mr.T. Welton,
School of Chem.& Mol.Science, Univ.of Sussex, Brighton, UK.

Mme.F.Yigit,
Dept. of Chem., Hacettepe Univ. , Ankara, Turkey.

INTRODUCTION

The liquid state has a familiar presence in everyday experience and on a simple basis is easily distinguishable from solids and gases by its physical properties. At a molecular level the behaviour of liquids still poses many fundamental problems to our understanding. Any satisfactory scientific theory requires a conceptual basis in which the main features governing the observed phenomena can be put into a mathematical framework to give predictive power. The process of developing an acceptable description therefore rests on the assumption that the necessary approximations built into any model representation are relatively simple and effective. The foundations of solid state theory are well known and have been understood for many years, particularly in the case of crystalline materials with a well-defined periodicity to the lattice structure. Defective structures are also capable of treatment when the concentration remains small and the properties can be investigated in terms of departure from the ideal periodic model. As increasing disorder is introduced, the problem of finding a suitable basis for quantitative evaluation becomes more pronounced and the approximations/assumptions become much more difficult to justify. The structural disorder and complex dynamics of even the simplest systems therefore poses significant questions, particularly for the development of a comprehensive theory to describe the liquid state.

The phenomenon of hydrogen-bonding has been recognised for many years as an important influence on thermodynamic properties. The most obvious example, frequently quoted, is that of water, which in the absence of hydrogen-bonding would be a gas at room-temperature. While it is easy to observe the influence of hydrogen-bonding on bulk properties, a detailed representation at the molecular level requires a quantitative treatment of the interaction forces. Although a full-scale calculation of molecular properties can often be made via *ab initio* methods the influence of hydrogen-bonding on extensive molecular assemblies is not particularly well developed. The

J.C. Dore and J. Teixeira (eds.), Hydrogen-Bonded Liquids, 1–3.

simplifying basis again starts with either the ordered structures of crystalline materials or the properties of small molecular clusters. Thus, there is considerable information about hydrogen-bonding geometries in the crystalline state and the tendency to form linear structures of the X-H - - - Y type. Structures may be classified in terms of the XY separation and the effects of symmetric or non-symmetric placement of the hydrogen atom on the bond can be investigated by studying correlations of structure with spectroscopic information The results give a valuable basis for understanding the average properties within specific configurations. The opposite approach involves the study of small clusters of molecules where the effects on the local configurations can be explored through direct calculation of energies. It is in these conditions that the presence of co-operativity is noted and its importance in the stabilisation of particular configurations becomes apparent. Experimental studies by molecular beam techniques provide a link between the theoretical concepts and practical reality. The topic of hydrogen-bonding in the liquid phase again occupies a middle ground where the ordered structure of crystalline systems is inapplicable yet the features cannot be readily generalised from a simple increase in the size of a small molecular assembly.

In the context of the previous remarks, it can be seen that hydrogen-bonded liquids pose a particular problem in which many of the established techniques of chemical physics will seem to be relatively crude and inadequate. It seems that the difficulties could be so severe that an effective description encompassing all observed phenomena may not be feasible. It is, indeed, possible that this area would not have attracted so much scientific interest apart from the single factor that water belongs to this group of materials. The all-pervading presence of water in our physical and biological environment indicates the importance of hydrogen-bonding and requires that we attempt to understand the essential principles despite the complexity. The properties of water also appear, at the superficial level, to be very different from other liquids such that it needs to be treated as a special case. However, much recent work has shown that the specific features can be understood on a microscopic scale and that they can be related to other hydrogen-bonded systems. Present research therefore seeks to place the phenomenon of hydrogen-bonding, which has particular effects on water, into the general context of liquid state science and to work towards a broad understanding of the most important features required to establish a quantitative description. This task provides a fascinating challenge and progress is necessarily slow but the pieces of this complex jigsaw puzzle are gradually being put into place as shown by the following papers.

J.C.Dore and J.Teixeira

BASIC CONCEPTS AND TOOLS OF LIQUID STATE THEORY

J.P. HANSEN
Laboratoire de Physique ; Ecole Normale Supérieure de Lyon
69364 Lyon Cedex 07, France

ABSTRACT. Some of the basic concepts and theoretical tools of the Statistical Mechanics of Liquids are summarized and illustrated by a few applications to the microscopic structure and dynamics of hydrogen-bonded liquids.

1. Microscopic Structure and Static Properties of Liquids

1.1 Introduction : the liquid state

The purpose of these two lectures is to introduce some of the concepts and tools of Statistical Mechanics of Liquids and to show, on the basis of a few illustrations, how this theoretical framework may be used to analyse the static and dynamical properties of hydrogen-bonded liquids. Historically, liquid state theory has been developed for the study of "simple" liquids, like rare gases, alkali metals or the simplest, homonuclear, diatomics, but the basic tools may readily be generalized to analyse more complex liquids, including associated liquids, ionic solutions or even macromolecular systems.

Liquids constitute one of the condensed phases of matter. They share with the solid state (crystalline or amorphous) a high degree of geometrical packing, wich leads to a low compressibility (typically a few percent of the gas compressibility) and well-defined short-range order, which extends over several molecular diameters. In fact the liquid and solid phases of a given substance have comparable densities (typically such that the reduced number density $\rho^* = N\sigma^3/V \simeq 1$, if N/V is the number of molecules per unit volume, and σ their diameter), since the relative volume change on melting, $\Delta V/V = (V_l - V_s)/V_s$ rarely exceds 10% ; note that liquids are generally slightly less dense than solids ($\Delta V > 0$), with the notable exception of water.

The most notable features which distinguish liquids from solids are :

a) the absence of long-range order characteristic of crystalline solids.

b) the absence of resistance to shear : if a low-frequency shear stress is applied, a solid will respond elastically, while a liquid will flow. However at sufficiently high frequencies, of the order of the inverse Maxwell relaxation time $\tau = \eta/G_\infty$ (where η is the shear viscosity, and G_∞ the instantaneous shear modulus) even a liquid will respond elastically (appearance of visco-elastic shear waves).

J.C. Dore and J. Teixeira (eds.), Hydrogen-Bonded Liquids, 5–24.

c) a much higher self-diffusion constant D in the liquid ; in most liquids near their triple point, D is of the order of 10^{-5} cm^2/sec, typically three orders of magnitude larger than in the solid, where it is mediated by crystal imperfections (defects, interstitials etc.).

For these (and other) reasons past attempts to describe a liquid as an "imperfect crystal" (the so-called cell-theories of the liquid state) have been abandoned since the early sixties, giving rise to the development of modern statistical theories of liquids and disordered media.

1.2 Classes of liquids and intermolecular forces

A cc of liquid contains of the order of $N = 10^{22}$ molecules which, in turn, are each made up of n nuclei and electrons. In the Born-Oppenheimer approximation the latter follow instantaneously the motions of the nuclei and determine intra-molecular and inter-molecular forces. Each molecule is then characterized by 3n degrees of freedom, of which 3 correspond to center-of-mass (CM) translations, 2 or 3 are associated with rotations (2 polar angles for linear molecules, and 3 Euler angles for non-linear molecules), and the remaining 3(n-1) - 2(3) correspond to intramolecular vibrations. If one is not directly interested in these vibrations, which are generally much faster than the rotations and translations, one may make an additional Born-Oppenheimer approximation and consider the molecules to be rigid assemblies of atoms. Let $\underline{q}_i = (\underline{R}_i, \underline{\Omega}_i)$ denote the coordinates of the ith rigid molecule ($\underline{R}_i$ is the CM position, and $\underline{\Omega}_i$ the set of 2 or 3 angular coordinates). The total potential energy for a given configuration $\underline{q}^N = (\underline{q}_1, \ldots \underline{q}_N)$ of the N molecules can then be expanded in a sum of successive terms involving an increasing number of molecules : pair interaction, triplet interaction etc :

$$V_N(\underline{q}^N) = \sum_{(i,\, j)} v_2(\underline{q}_i, \underline{q}_j) + \sum_{(i,\, j,\, k)} v_3(\underline{q}_i, \underline{q}_j, \underline{q}_k) + \ldots \tag{1.1}$$

In many applications neglecting all terms beyond the pair-wise additive term is an acceptable approximation, and theories of the liquid state are generally tractable only in that approximation. However much recent research deals with the effects of triplet forces, which become important for highly polarizable atoms and molecules or at high pressures. In the following we shall restrict ourselves to pair-wise additive models.

According to the nature of inter-molecular forces, one may attempt an (arbitrary) classification of liquids, in increasing order of complexity.

a) Atomic liquids, made up of mono-atomic molecules (rare gases). The pair interactions have spherical symmetry and contain a short-range repulsion (the atoms are impenetrable) and long-range van der Waals attractions.

b) Simple liquid metals (with simple electronic band structures, like alkali metals) which are made up of positive ions and electrons ; second-order perturbation theory and the adiabatic approximation reduce the system to an assembly of pseudo-atoms, very similar to the first class.

c) Simple non-polar molecular liquids, made up of molecules without a permanent dipole moment ; the main difference compared to atomic liquids is the anisotropy of the short-range repulsion (non-spherical shape).

d) Polar liquids containing molecules with permanent dipoles $\underline{\mu}_i$. The resulting dipolar interaction :

$$v_2^d(i, j) = \underline{\mu}_i \cdot \underline{\underline{T}}(\underline{R}_i - \underline{R}_j) . \underline{\mu}_j \tag{1.2a}$$

where

$$T^{\alpha\beta}(\underline{R}) = \nabla^{\alpha} \nabla^{\beta} (1/R) \tag{1.2b}$$

is the dipolar tensor, is long-ranged ($\sim R^{-3}$), leading to the usual importance of boundary contributions to the dielectric properties.

e) Hydrogen-bonded liquids composed of strongly polar molecules with intermolecular bonds involving protons "shared" between donors and acceptors. These bonds lead to strongly directional forces and a tendency towards bond-network formation.

f) Ionic liquids involving positive and negative ions and a polar solvent (electrolyte solutions). The special case where the solvent is absent corresponds to ionic melts (like molten Na Cl) which are simpler liquids and relatively well understood. Ionic liquids are of course dominated by long-range Coulomb forces and molecular polarizability.

More generally, much recent effort is going into the study of models involving polarizable (rather than rigid) molecules. In principle the anisotropic interaction potential between two molecules may be expanded in the complete set of "rotational invariants" (Blum and Torruela, 1972) according to :

$$v_2(i, j) = \sum_{\substack{mnl \\ \mu\nu}} v_{\mu\nu}(mnl \; ; \; R_{ij}) \, \Phi_{\mu\nu}^{mnl} (i, j) \tag{1.3}$$

The $\Phi_{\mu\nu}^{mnl}$ are combinations of products of generalized spherical harmoncics (Wigner matrix elements) depending on the angular coordinates $\underline{\Omega}_i$ and $\underline{\Omega}_j$, and on the orientation of the vector $\underline{R}_{ij} = \underline{R}_i - \underline{R}_j$ respectively. For axially symmetric molecules the indices $\mu = \nu = 0$ are dropped. For example the dipole-dipole interaction (2) may be expressed as :

$$v_2^d(i, j) = -\frac{\mu^2}{R_{ij}^3} \Phi^{112}(\underline{\Omega}_i, \underline{\Omega}_j, \underline{R}_{ij}^o) \tag{1.4}$$

More generally the representation in terms of rotational invariants is well adapted to the multipolar interactions between molecular charge distributions. However it is less suitable for representing the rapidly varying short-range anisotropic repulsion associated with non-spherical molecular shapes, because too many terms would then be needed in the slowly convergent sum (1.3). In that case it is preferable to resort to <u>interaction-site</u> models, in which a molecule is represented by a set of discrete interaction sites that are commonly located at the sites of the atomic nuclei. The total potential energy of two interaction-site molecules is obtained as the sum of spherically symmetric interaction-site potential :

$$v_2(i, j) = \sum_{\alpha} \sum_{\beta} v_{\alpha\beta} (| \underline{r}_{i\alpha} - \underline{r}_{j\beta} |) \tag{1.5}$$

where $\underline{r}_{i\alpha}$ denotes the position of site α on molecule i, $v_{\alpha\beta}$ is a site-site potential, and the sums on α and β run over all interaction sites in the respective molecules. Interaction-site models may also be used to account for the permanent molecular dipoles, quadrupoles etc. by assigning fractional charges to the sites. A widely used example is the ST2 model for water. An H_2O molecule is, in this model, a rigid assembly of five sites : a central Lennard-Jones sphere and four fractional charges distributed in tetrahedral directions around the centre ; two positive charges ($q_+ = 0.23e$) at a distance of 1Å representing the hydrogen atoms, and two negative charges ($q_- = -0.23e$) at a distance of 0.8Å representing the electron orbitals. This charge distribution reproduces correctly the known dipole moment $\mu = 1.85$ Debye along the symmetry axis of the molecule.

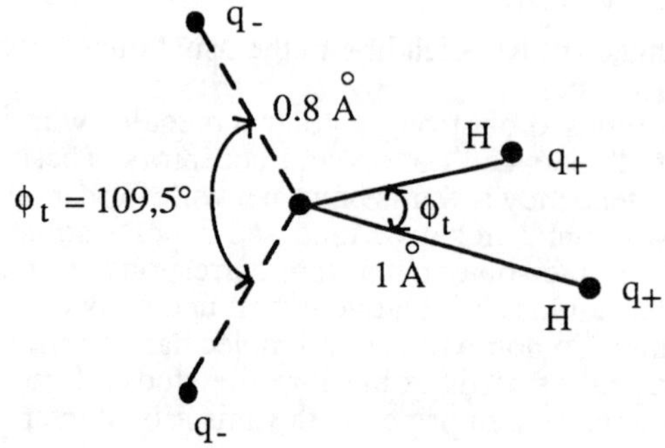

Figure 1

Thus the total potential energy of two H_2O molecules is made up of 5^2 site-site interactions, which lead to highly directional forces, the energy going through a pronounced minimum in the O - H ...O hydrogen-bond configuration. Rigid molecular models, like ST2, generally lead to an underestimation of the static dielectric constant ε of water. The instantaneous dipole moment of molecule i is the sum of its permanent dipole $\underline{\mu}_i$ and of the dipole induced by the local electric field $\underline{E}_i$ acting on that molecule :

$$\underline{m}_i = \underline{\mu}_i + \underline{\underline{\alpha}} \cdot \underline{E}_i \tag{1.6a}$$

$$\underline{E}_i = \sum_{j(\neq i)} \underline{\underline{T}}(\underline{R}_i - \underline{R}_j).\underline{m}_j \tag{1.6b}$$

Clearly molecular polarizability introduce a complicated many-body interaction which must be handled by iteration in a computer simulation of polarizable molecules (Levesque et al. 1984). Alternatively the fluctuating dipoles $\underline{m}_i$ may be regarded as additional degrees of freedom in a dynamical simulation (Sprik and Klein, 1988). To a good approximation, the properties of a polarizable dipolar fluid can be related to those of a fluid with an <u>effective</u> permanent dipole $\mu^2_{eff} = < |\underline{m}_i|^2 >$ (Carnie and Patey, 1982). For water, μ_{eff} turns out to be substantially larger than the bare permanent dipole μ, thus leading to a considerable enhancement of ε.

1.3 Pair distribution functions

Liquids are characterized on the molecular level by strong spatial correlations linked to steric (excluded volume) effects. These correlations give rise to short-range order which extends over several molecular diameters and which may be quantitavely described by pair distribution functions. The presentation will be restricted here to homogeneous liquids, for which translational invariance implies that the one-particle density :

$$\rho(\underline{R}) = <\sum_{i=1}^{N} \delta(\underline{R} - \underline{R}_i)> \tag{1.7}$$

is constant and equal to the macroscopic number density $\rho = N/V$.

The CM pair distribution function $g_c(R)$ is defined such that the mean number of molecules at a distance R, within dR, from the CM of a given molecule is :

$$n(R)\,dR = 4\Pi\,\rho\,R^2\,g_c(R)\,dR \tag{1.8}$$

For a uniform distribution of molecules (i.e. in the absence of any positional correlations, as in an ideal gas) $g_c(R) \equiv 1$. This is the limiting value of $g_c(R)$ in a liquid, whenever $R \gg \sigma$. On the other hand, $g_c(R) = 0$ for $R \leq \sigma$ because molecules are inpenetrable. A formal Statistical Mechanics definition of $g_c(R)$ is (Hansen and Mc Donald, 1986) :

$$\rho\, g_c(R) = \frac{1}{N} < \sum_i \sum_{j \neq i} \delta(\underline{R} + \underline{R}_i - \underline{R}_j) > \tag{1.9}$$

For atomic liquids, $g_c(R)$ completely specifies the micoscopic pair structure, as determined e.g. by X-ray or neutron diffraction experiments which measure directly the static structure factor, S(k), the Fourier transform of $g_c(R)$:

$$\begin{aligned} S(k) &= \frac{1}{N} < \rho_{\underline{k}}\, \rho_{-\underline{k}} > \\ &= 1 + \rho \int e^{i\underline{k}\cdot\underline{R}}\, g_c(R)\, d\underline{R} \end{aligned} \tag{1.10}$$

where $\rho_{\underline{k}}$ denotes a Fourier component of the microscopic density :

$$\rho_{\underline{k}} = \sum_{i=1}^{N} e^{i\underline{k}\cdot\underline{R}_i} \tag{1.11}$$

For molecular liquids the spatial distribution of neighbouring molecules around a given molecule depends not only on the relative distance of the CM, but also on the relative orientation of the non-sperical molecules. The corresponding molecular distribution function for ahomogeneousliquid may then be defined by a generalization of equ. (1.9) :

$$\frac{\rho}{\Omega^2}\, g(\underline{R}, \underline{\Omega}, \underline{\Omega}') = < \sum_i \sum_{j \neq i} \delta(\underline{R} + \underline{R}_i - \underline{R}_j)\, \delta(\underline{\Omega}_i - \underline{\Omega})\, \delta(\underline{\Omega}_j - \underline{\Omega}')> \tag{1.12}$$

The CM pair distribution function is the angular average of g :

$$g_c(R) = \frac{1}{\Omega^2} \int\!\!\int g(\underline{R}, \underline{\Omega}, \underline{\Omega}')\, d\underline{\Omega}\, d\underline{\Omega}' \tag{1.13}$$

In equs. (12) and (13), $\Omega = \int d\underline{\Omega} = 4\Pi$ for linear molecules and 8Π for non-linear ones. To simplify the notation, it is convenient to use the symbol $i \equiv (\underline{R}_i, \underline{\Omega}_i)$ to denote both the CM and angular coordinates of molecule i ; under these conditions :

$$g_c(R) = < g(1, 2) >_{\underline{\Omega}_1, \underline{\Omega}_2} \tag{1.14}$$

where angular brackets now denote an unweighted average over the specified angles, and $R = |\underline{R}_1 - \underline{R}_2|$

The static angular correlations in a molecular liquid may be characterized by a set of "order parameters" defined by :

$$G_l = \rho \int < g(1, 2)\, P_l(\underline{u}_1 \cdot \underline{u}_2) >_{\underline{\Omega}_1, \underline{\Omega}_2} d\underline{R} \tag{1.15}$$

where P_l denotes the l - th order Legendre polynomial and $\underline{u}_i$ is a unit vector along a symmetry axis or the dipole moment of molecule i. The value of the first-rank order parameter G_1 determines the dielectric constant of a polar fluid, while G_2 is related to the integrated intensity of the spectrum observed in deplolarized light - scattering experiments.

Like the intermolecular pair potential (1.3) the molecular pair distribution function may be expanded, within a laboratory - fixed frame of reference, in an infinite series of rotational invariants (Gray and Gubbins, 1984 ; Hansen and Mc Donald, 1986) :

$$g(1, 2) = \sum_{\substack{m\,n\,l \\ \mu\nu}} g_{\mu\nu}(m\,n\,l\,;\,R)\,\Phi_{\mu\nu}^{m\,n\,l}(1, 2) \tag{1.16}$$

The coefficients $\hat{g}_{\mu\nu}(m\ n\ l\ ;\ k)$ of the corresponding expansion of the Fourier transform $\hat{g}$ of g with respect to $\underline{R} = \underline{R}_1 - \underline{R}_2$ are related to the $g_{\mu\nu}(m\ n\ l\ ;\ R)$ by Hankel transforms.

An auxiliary function which plays an important role in the formulation of approximations for the molecular pair ditribution function is the <u>direct correlation function</u> c(1, 2) which is related to the <u>total correlation function</u> h(1, 2) = g(1, 2) - 1 by the Ornstein-Zernike convolution relation :

$$h(1, 2) = c(1, 2) + \frac{\rho}{\Omega}\int c(1, 3)\, h(3, 2)\, d3 \tag{1.17}$$

In the simplest models of molecular liquids, a molecule is regarded as a hard sphere with a point dipole, a point quadrupole, etc. fixed at its centre. The intermolecular potential reduces then to a spherically symmetric excluded volume and the pair-wise additive electrostatic interactions between the multipoles, as e.g. in equ. (1.4). For such models a widely used theory is the "Mean Spherical Approximation" (MSA) which is summarized by the "closure relations" :

$$g(1, 2) = 0\ ,\quad R_{12} < \sigma \tag{1.18a}$$

$$c(1, 2) = -v(1, 2) / k_B T\ ,\quad R_{12} > \sigma \tag{1.18b}$$

which, together with the OZ relation, form a closed set which can be solved analytically (Wertheim, 1971 ; Blum, 1972). The MSA has played an important role in the recent discrete solvent models of ionic solutions (Hansen, 1984). More elaborate closures, like the HNC equation, lead to considerably more reliable predictions for the pair structure of molecular liquids (Fries and Patey, 1985 ; Fries and Salamito, 1988).

When an interaction-site model is used to represent the intermolecular potential, the natural way to describe the structure of the fluid is in terms of site-site distribution functions. If the coordinates of site α on molecule i are denoted by $\underline{r}_{i\alpha}$ and those of site β on molecule j (j ≠ i) by $\underline{r}_{j\beta}$, then the site-site radial distribution fucntion is defined in a manner similar to (1.9) :

$$\rho\, g_{\alpha\beta}(r) = \frac{1}{N} < \sum_i \sum_{\neq j} \delta(\underline{r} + \underline{r}_{i\alpha} - \underline{r}_{j\beta}) > \tag{1.19}$$

Comparison of equs (1.12) and (1.19) shows that the site-site functions are constrained angular contractions of the full molecular distribution function, namely :

$$g_{\alpha\beta}(r) = \int d\underline{R}_{12} < g(1, 2)\, \delta[\underline{R}_{12} + \underline{l}_{2\beta}(\underline{\Omega}_2) - \underline{l}_{1\alpha}(\underline{\Omega}_1) - \underline{r}] >_{\underline{\Omega}_1, \underline{\Omega}_2} \tag{1.20}$$

where $\underline{l}_{i\alpha} = \underline{r}_{i\alpha} - \underline{R}_i$ is the vector displacement of site α on molecule i from the molecular centre. Conversely, the full molecular distribution function connot be reconstructed from any finite set of $g_{\alpha\beta}(r)$. The latter have however the advantage of being easily visualized, since they are radial functions. They are also sufficient to determine the X - ray or neutron diffraction patterns. Since neutrons are scattered by the atomic nuclei in the molecules, the measured structure factor $S(\underline{k})$ is entirely determined by an intramolecular form factor and by the Fourier transforms of the site-site pair distribution functions. Thus, for the neutron diffraction pattern :

$$S^N(k) = S^N_{intra}(k) + S^N_{inter}(k) \qquad (1.21a)$$

$$S^N_{intra}(k) \sim \sum_{\alpha} b_\alpha^2 + \sum_{\alpha} \sum_{\neq \beta} b_\alpha b_\beta \frac{\sin kl_{\alpha\beta}}{kl_{\alpha\beta}} \qquad (1.21b)$$

$$S^N_{inter}(k) \sim \sum_{\alpha} \sum_{\beta} b_\alpha b_\beta [S_{\alpha\beta}(k) - 1] \qquad (1.21c)$$

$$S_{\alpha\beta}(k) = \rho \int [g_{\alpha\beta}(r) - 1] e^{-i\underline{k}\cdot\underline{r}} \underline{dr} \qquad (1.21d)$$

where the sums run over all nuclei in the molecule, b_α is the coherent neutron scattering length of nucleus α, and $l_{\alpha\beta}$ is the separation of nuclei α and β. $S^N{}_{intra}$ is easily calculated from the known molecular structure, and hence Fourier inversion of the measured $S^N(k)$ immediately yields a weighted sum of the site-site distribution functions. In the case of D_2O the resulting combination is:

$$g^N(r) = 0.09\, g_{OO}(r) + 0.42\, g_{OD}(r) + 0.49\, g_{DD}(r) \qquad (1.22)$$

The X-ray diffraction pattern from H_2O or D_2O, on the other hand, is dominated by the contribution from $g_{OO}(r)$. Thus different linear combinations of the $g_{\alpha\beta}(r)$ are measured by the various radiation scattering techniques. Moreover the technique of isotopic substitution in neutron diffraction experiments also leads to different linear combinations of the $g_{\alpha\beta}(r)$ which may thus be resolved, at least in principle. Important structural information on hydrogen-bonded liquids has thus been obtained, e.g. in the case of liquid alcohols (Montague et al., 1986).

Site-site pair distibution functions are also easily "measured" in Monte Carlo (MC) or Molecular Dynamics (MD) computer simulations of models of molecular and hydrogen-bonded liquids. A good example is taken from the work of Klein and Mc Donald (1979) on HF, which they modelled by a three-site intermolecular potential with fractional charges distributed such as to reproduce the known dipole and quadrupole moments of the molecule. Results for the three partial distribution functions g_{HH}, g_{HF} and g_{FF} at T = 278 K are shown in Fig. 2. Examination of the peak positions, and of the running coordination numbers :

$$n_{\alpha\beta}(r) = 4\Pi\rho \int_0^r g_{\alpha\beta}(s) s^2\, ds \qquad (1.23)$$

allows to conclude that the microscopic structure of liquid HF may be interpreted as a tangled network of winding chains made up of a succession of hydrogen-bonded monomers.

The integral equations (MSA, HNC etc.) alluded to earlier have been generalized to calculate the site-site pair distribution functions (e.g. the RISM approximation of Chandler and Andersen (1972) ; see also Chandler, 1976 ; Hansen and Mc Donald, 1986).

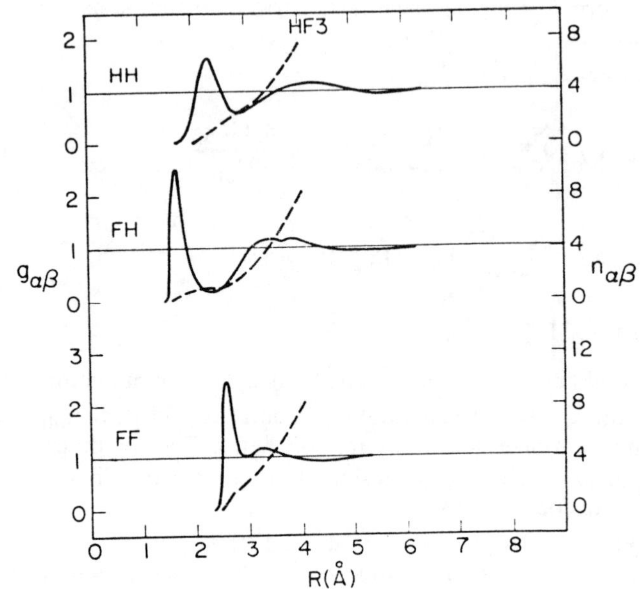

Figure 2 : Atomic pair distribution functions (full curves) and running coordination numbers (broken curves) for HF at $\rho = 1$ gr cm^{-3} ,T = 278 K (reproduced from Klein and Mc Donald (1979)).

1.4 Macroscopic static properties

The pair distribution functions introduced in the previous section allow a direct calculation of a number of macroscopic properties. If triplet and higher-order interactions between molecules may be neglected, the excess internal energy of the liquid is expressible in terms of the molecular or site-site pair distribution functions. If the inter-molecular pair potential v(1, 2) is known as a function of $\underline{R}_{12}$, $\underline{\Omega}_1$ and $\underline{\Omega}_2$, the excess energy is given by:

$$\frac{U^{ex}}{N} = \frac{1}{N} <V_N> = 2\Pi\rho \int_0^\infty <g(1,2)\, v(1,2)>_{\underline{\Omega}_1,\underline{\Omega}_2} R_{12}^2\, dR_{12} \quad (1.24)$$

while the pressure may be calculated from the virial theorem according to :

$$\frac{\beta P}{\rho} = 1 - \frac{2\Pi}{3} \beta\rho \int_0^\infty <g(1,2)\, v'(1,2)>_{\underline{\Omega}_1,\underline{\Omega}_2} R_{12}^2\, dR_{12} \quad (1.25)$$

where the prime denotes differentiation with respect to R_{12} with $\underline{\Omega}_1$ and $\underline{\Omega}_2$ held constant.

If, on the other hand, v(1, 2) is expressed as a sum of site-site potentials as in equ.(1.5), the excess internal energy takes the form :

$$\frac{U^{ex}}{N} = 2\Pi\,\rho \sum_{\alpha}\sum_{\beta} \int_o^{\infty} g_{\alpha\beta}(r)\, v_{\alpha\beta}(r)\, r^2\, dr \qquad (1.26)$$

but the pressure cannot be expressed solely in terms of the $g_{\alpha\beta}$ and $v_{\alpha\beta}$. The isothermal compressibility χ_T is related, within a standard grand canonical ensemble calculation of density fluctuations, to the long-wavelength limit of the static structure factor ; this limit is insensitive to angular correlations, so that :

$$\rho k_B\, T\chi_T = \lim_{k\to 0} S(k) = 1 + 4\Pi\rho \int_o^{\infty} [g_c(R) - 1]R^2 dR$$

$$= 1 + 4\Pi\rho \int_0^{\infty} [g_{\alpha\beta}(r) - 1]r^2 dr \qquad (1.27)$$

U and P are first order thermodynamic derivatives of the free energy. Second-order derivatives, like the specific heat, thermal expansion coefficient or bulk modulus (which is the inverse of the compressibility) are expressible as fluctuations of the Hamiltonian, the stress tensor, etc. which necessarily involve three and four-body ditribution function. In computer simulations, the statistical estimates of fluctuation quantities of the form $[\langle A^2\rangle - \langle A^2\rangle]$, which are obtained as differences of two large numbers, are generally affected by large error bars (typical relative uncertainties are 10%). In approximate calculations, three and four-body distribution functions are generally replaced by the Kirkwood superposition approximation, which may be shown to be inadequate in many applications. Only recently has an accurate approximation for the triplet correlations in atomic liquids become available, within the framework of the density functional theory of inhomogeneous liquids (Barrat et al, 1988) ; this procedure has yet to be generalized to molecular liquids.

Pair correlations also determine the static dielectric constant ε of polar liquids. A standard linear response calculation of the polarization induced in an inifinite dielectric medium by an external field leads to **Kirkwood's** formula for ε :

$$\frac{(\varepsilon - 1)(2\varepsilon + 1)}{9\varepsilon} = \frac{4\Pi}{3} \frac{\langle |\underline{M}|^2\rangle}{3Vk_BT} \qquad (1.28)$$

where $\underline{M}$ denotes the total dipole moment of the system, which is the sum of individual moments (1.6). For rigid molecules ($\alpha = 0$) of dipole moment $\mu_i = \mu\, \underline{u}_i$, $\langle|\underline{M}|^2\rangle$ is given by the Kirkwood g - factor :

$$g_K = \frac{\langle |\underline{M}|^2\rangle}{N\mu^2} = 1 + (N-1)\,\langle \underline{u}_1 \cdot \underline{u}_2\rangle = 1 + G_1 \qquad (1.29)$$

Clearly, if the expansion (1.16) in rotational ivariants is substituted in (1.29) the Kirkwood factor is directly espressible in terms of the (m n l) = (110) component, since $\underline{u}_1 \cdot \underline{u}_2$ is precisely equal to $\Phi^{110}(1, 2)$; explicitly :

$$g_K = 1 + \frac{4\Pi\rho}{3} \int_o^{\infty} h(110 ; R)\, R^2\, dR$$

$$= 1 + \frac{\rho}{3} \hat{h}(110 ; k = 0) \qquad (1.30)$$

The (m n l) = (112) component of the expansion of the pair distribution function may be shown to have the following asymptotic form (Hoye et al., 1974) :

$$\lim_{R\to\infty} h^{112}(R) = \frac{(\varepsilon - 1)^2}{\varepsilon} \frac{9k_B T}{(4\Pi)^2\mu^2\rho^2} \frac{1}{R^3} \tag{1.31}$$

which illustrates the slow decay of dipolar correlations, i.e. the absence of any screening mechanism similar to that observed in ionic liquids.

Finally, if one is interested in phase or chemical equilibria in liquids, one needs to evaluate the free energy of the liquid for a given chemical potentials of the species and phases involved. The free energies A or chemical potentials μ involve distribution functions of all orders. Many ingenious schemes have been divised to compute free energies from MC or MD simulations (Frenkel, 1986).One which is widely used is based on the following exact relation (written here for a one component liquid) (Widom, 1963 ; Shing and Gubbins, 1982) :

$$\mu^{ex} = \left(\frac{\partial A^{ex}}{\partial N}\right)_{V,T} = k_B T \,\mathrm{Log}\left[\frac{VZ_N(V, T)}{Z_{N+1}(V, T)}\right] \tag{1.32}$$

where Z_N and Z_{N+1} denote the configuration integrals for systems of N and (N + 1) molecules in a volume V at a temperature T. Now :

$$V_{N+1}(\underline{q}^{N+1}) = V_N(\underline{q}^N) + \phi \tag{1.33}$$

where ϕ is the energy of ineraction of the (N +1) th particle with all others. Substitution of (1.33) in (1.32) leads immediately to the exact relation :

$$\mu^{ex} = -k_B T \,\mathrm{Log} <\exp\{-\phi/k_B T\}> \tag{1.34}$$

so that the excess chemical potential appears as being proportional to the mean Boltzmann factor of a test particle introduced randomly into the system. This method has recently been used to calculate the ionic equilibrium constant (i.e. the pH) of water, by attempting the insertion of an extra test particle of any of the three species, H_2O, H_3O^+ and HO^- in pure water to determine the corresponding three excess chemical potentials (Guissani et al., 1988).

2. Time correlation functions and molecular dynamics in liquids

2.1 Microscopic dynamics

The first lecture was devoted to static (or equal-time) correlations in space. However, if we are interested in dynamical phenomena in liquids, like the relaxation of an initially perturbed system towards equilibrium, transport phenomena or the inelastic scattering of radiation, we are naturally led to consider time dependent correlations. The associated time (- displaced) correlation functions, which will be formally introduced in the following section, serve to characterize in a quantitative way the motions of the various degrees of freedom at the molecular level. It is customary to distinguish between "single particle" and "collective" molecular motions, although the two types are intimately coupled. Single particle (or "incoherent") dynamics include :

a) Translations of the CM ; on a short time scale, molecules vibrate in the "cage" formed by their neighbours, very much as in a solid. The characteristic vibration frequency is of the order of the Einstein frequency, which is given by (Hansen and Mc Donald, 1986) :

$$\Omega_E^2 = \frac{1}{3m\, k_B T} < |\underline{F}|^2 > \tag{2.1}$$

where $\underline{F}$ is the total force exerted on the molecule by its neighbours. For a liquid of atoms interacting by the pair potential v(r), Ω^2_E reduces to :

$$\Omega_E^2 = \frac{\rho}{3m} \int \nabla^2 v(R)\, g(R)\, d\underline{R} \tag{2.2}$$

In dense liquids near the triple point, the mean vibration period $\tau_E = \Omega_E^{-1}$ is typically of the order of 10^{-13} sec.

In a liquid, a molecule escapes from the nearest neighbour cage after a few vibrations, to diffuse through the medium, very much like a Brownian particle (cf. section 2.2). If D denotes the self-diffusion constant and $d \simeq \rho^{-1/3}$ the mean nearest neighbour distance, the characteristic time for diffusive motion is $\tau_D \simeq d^2/D \simeq 10^{-11}$ sec, i.e. about 100 τ_E.

b) Rotations of the molecular axes ; at low densities or very short times, these motions are dominated by free rotation. The characteristic temperature of rotational motion associated with a moment of inertia I is $\Theta_R = h^2/k_B I$ which for nearly all molecules (except H_2 and D_2) is negligible compared to room temperature, so that molecular rotations, just like CM translations, may be treated classically. The root mean square angular velocity, $(k_B T/I)^{1/2}$, leads to a free rotation time scale of the order of a picosecond (10^{-12} sec). In the opposite limit of strongly hindered rotations, due to strong orientational correlations, the molecular axes may undergo librational motions reminiscent of the CM vibrations ; the librational time scale is set by the root mean square torque (rather than the root mean square force), and is typically a fraction of a picosecond in the case of water. At longer times, reorientation may be considered as the result of a succession of small, uncorrelated steps; this is the Debye model of rotational Brownian motion, which is characterized by a time scale equal to the inverse of the "rotational - diffusion constant", $\tau_R = D_R^{-1} \simeq 10^{-11}$ sec for small molecules under triple point conditions.

c) As already noted earlier, intramolecular vibrations occur on a much faster time-scale. With a characteristic vibrational temperature $\Theta_v = h\omega_v/k_B$ of 10^3 - 10^4 K most vibrators are in their ground state at room temperature, with time scales $\tau_v = 1/\omega_v$ of the order of

10^{-15} sec. In many applications, vibrational degrees of freedom must be regarded as quantum degrees of freedom coupled to a "bath" of translational an rotational degrees of freedom.

Collective (or"coherent") dynamics involve the correlated motions of a large number of molecules. These are conveniently probed by scattering of electromagnetic radiation or neutrons of wavelengths λ , such that the volume λ^3 contains many molecules ($\rho\lambda^3 \gg 1$). If λ greatly exceeds the mean inter-molecular spacing, the liquid looks like a continuum and the collective dynamics may be described by the equations of macroscopic hydrodynamics ("hydrodynamic modes"). Perhaps one of the great surprises of liquid state physics is that the predictions of hydrodynamics turn out to remain quantitavely valid for spatial scales as short as ten intermolecular spacings and for time scales of little more than a picosecond. Moreover hydrodynamics is capable of describing the non-exponential decay (i.e. the non-markovian behaviour) of time correlation functions associated with individual or collective particle motion (mode coupling and long-time "tails").

Another, quite different type of motion plays an important role in hydrogen-bonded systems, namely proton transfer via quantum tunneling through the hydrogen-bonds. This mechanism, which allows protons to be exchanged between molecules, plays a fundamental role in many biological processes, and is presently the subject of intense research.

2.2 Time-displaced correlation functions

To characterize the decay of a spontaneous (thermal) or induced fluctuation of a dynamical variable A (a scalar, vectorial or tensorial function A ($\underline{q}^N(t)$, $\underline{p}^N(t)$) of the instantaneous coordinates and (or) conjugate momenta of some or all particles in the system), it is natural to investigate the time-dependence of the projection of its value at time t on its inital value, or on the initial value of some other dynamical variable B. To that purpose one defines the equilibrium time (-displaced) correlation function (TCF) of the two variables A and B as :

$$C_{AB}(t', t'') = \langle A(t')\, B^*(t'')\rangle = \langle A(t' - t'')B^*(o)\rangle \tag{2.3}$$

where the angular brackets denote an ensemble average over the initial phase, and advantage has been taken of time translation invariance (stationarity) of equilibrium averages.

Local dynamical variables are of the form :

$$A(\underline{R}, t) = \sum_{i=1}^{N} a_i(t)\, \delta(\underline{R} - \underline{R}_i(t)) \tag{2.4}$$

where a_i is any physical quantity, like the mass, the dipole moment, the linear or angular momentum, associated with molecule i. Their Fourier components are given by :

$$A_{\underline{k}}(t) = \int e^{-i\underline{k}\cdot\underline{R}}\, A(\underline{R}, t) = \sum_{i=1}^{N} a_i(t) e^{-i\underline{k}\cdot\underline{R}_i(t)} \tag{2.5}$$

A good example is the time-dependent generalization of the microscopic density (1.11), where $a_i(t) \equiv 1$. The TCF of local dynamical variables depends both on the time desplacement t = t' - t" and on the spatial displacement $\underline{R} = \underline{R}' - \underline{R}''$, if the system is homogeneous :

$$\begin{aligned} C_{AB}(\underline{R}, t) &= \langle A(\underline{R}', t')\, B^*(\underline{R}'', t'')\rangle \\ &= \langle A(\underline{R}, t)\, B^*(\underline{O}, O)\rangle \end{aligned} \tag{2.6}$$

Translational invariance in space implies that only Fourier components of oppostie wave-vectors $\underline{k}$ have non-vanishing correlation, i.e. :

$$C_{AB}(k, t) = \int e^{-i\underline{k}\cdot\underline{R}} C_{AB}(\underline{R}, t)d\underline{R}$$

$$= < A_{\underline{k}}(t)\, B_{-\underline{k}}(O) > \qquad (2.7)$$

The time evolution of any dynamical variable is governed by :

$$\dot{A} \equiv \frac{dA}{dt} = i\,L\,A \qquad (2.8)$$

where L denotes the Liouville operator :

$$L = \frac{1}{h}[H_N, \;] \qquad \text{(quantum mechanics)} \qquad (2.9a)$$

$$L = i\{H_N, \;\} \qquad \text{(classical mechanics)} \qquad (2.9b)$$

In equs. (2.9), H_N denotes the Hamiltonian of the N-particle system and the brackets denote the commutator and the Poisson bracket respectively. Equ.(2.8) has the formal solution :

$$A(t) = \exp\{i\,L\,t\}A \qquad (2.10)$$

where $A \equiv A(o)$ denotes the initial value of the dynamical variable.
One generally considers dynamical variables of zero mean. In an ergodic system any dynamical variable will be completely decorrelated from its initial value (or that of any other varaible) after a sufficiently long time interval (i.e. for times longer than all relevant relaxation times in the system) so that :

$$\lim_{t\to\infty} C_{AB}(t) = <A(t)> <B^*(o)> = <A> <B^*> = 0 \qquad (2.11)$$

This property allows us to define the spectrum of any TDF as :

$$\hat{C}_{AB}(\omega) = \frac{1}{2\Pi}\int_{-\infty}^{+\infty} e^{i\omega t}\, C_{AB}(t)\, dt \qquad (2.12)$$

For systems with "broken ergodicity" (like glasses) equ.(2.11) is not neccessarily satisfied ; the non-zero asymptotic value of CAB(t), which may be regarded as an "order parameter" (Edwards and Anderson, 1975) gives then rise to an elastic $\delta(\omega)$ - contribution to the spectrum (2.12).In theoretical calculations it is convenient to introduce the Laplace transform :

$$\tilde{C}_{AB}(z) = \int_0^{\infty} e^{izt}\, C_{AB}(t)\, dt \;\; ; \;\; \mathrm{Im}z > 0$$

$$= i\int_{-\infty}^{+\infty} d\omega\, \frac{\hat{C}_{AB}(\omega)}{z - \omega} \qquad (2.13)$$

Conversely, the spectrum (2.12) is directly related to the real part of the Laplace transform on the real axis :

$$\hat{C}_{AB}(\omega) = \lim_{\varepsilon\to o} \frac{1}{\Pi}\, \tilde{C}'_{AB}(\omega + i\varepsilon) \qquad (2.14)$$

According to (2.10) the Laplace transform may be formally expressed in terms of a propagator :

$$\tilde{C}_{AB}(z) = i <[(z + L)^{-1}A]\, B^*> \qquad (2.15)$$

This expression lends itself readily to pertubative and memory-function approximations of the TCF. The Mori-Zwanzig projection operator formalism leads to the following formally exact expression for $\tilde{C}_{AB}(z)$ (Hansen and Mc Donald, 1986) :

$$\tilde{C}_{AB}(z) = C_{AB}(t=0)\,[-iz - i\Omega_{AB} + \tilde{M}_{AB}(z)]^{-1} \tag{2.16}$$

where the frequency matrix is given by :

$$\Omega_{AB} = \langle \dot{A}\, B^* \rangle\, (\langle AB^* \rangle)^{-1} \tag{2.17}$$

and the memory function $\tilde{M}_{AB}(z)$ is the Laplace transform of the TCF of a generalized "random force". In the time variable, (2.16) takes the form of a generalized Langevin equation :

$$\dot{C}_{AB}(t) - i\Omega_{AB}\, C_{AB}(t) + \int_o^t M_{AB}(t-s)\, C_{AB}(s)ds \tag{2.18}$$

The memory function is frequently modeled in a phenomenological way (e.g. by assuming exponential decay). Among first principles theories, mode coupling theory, in which the memory function is appoximated by products of correlation functions of conserved variables (see below), has recently been successfully applied to structural slowing down in supercooled liquids near the glass transition (see e.g. Götze and Sjögren, 1989).

A particulary important class of time correlation functions are the autocorrelation functions $C_{AA}(t)$ (i.e. A = B) which are necessarily even functions of time (in classical mechanics), and which satisfy two important properties :

$$|C_{AA}(t)| \leq C_{AA}(0) \tag{2.19}$$

$$\hat{C}_{AA}(\omega) \geq 0 \tag{2.20}$$

Local variables that play an important role in the study of collective modes are the conserved variables associated with molecular variables a_i which obey microscopic conservation laws (e.g. mass, momentum and angular momentum). An example at hand is the microscopic density (1.11), which obviously statisfies the continuity equation :

$$\dot{\rho}_{\underline{k}}(t) + i\,\underline{k}\,.\,\underline{j}_{\underline{k}}(t) = 0 \tag{2.21}$$

where the microscopic current (propotional to the momentum density) is itself a conserved local variable, with $a_i(t) = \underline{v}_i(t)$ (CM velocity of molecule i). The ACF which may be constructed from these variables are the density ACF (frequently referred to as the "intermediate scattering function") :

$$F(\underline{k}, t) = \frac{1}{N} \langle \rho_{\underline{k}}(t)\, \rho_{-\underline{k}}(o) \rangle \tag{2.22}$$

and the longitudinal current ACF :

$$C_l(\underline{k}, t) = \frac{1}{N} \langle \underline{k}.\underline{j}_{\underline{k}}(t)\, \underline{k}.\underline{j}_{-\underline{k}}(o) \rangle \tag{2.23}$$

It is clear from the continuity equ.(2.21) that :

$$\lim_{k \to o} \frac{d}{dt} F(\underline{k}, t) = 0$$

More generally the ACF of a conserved variable varies more and more slowly with time as the wavelength increases ("slow variables"). Also, from (2.21), it is clear that the two ACF F and C_l are not independent, since :

$$C_l(\underline{k}, t) = \frac{1}{N} \langle \dot{\rho}_{\underline{k}}(t)\, \dot{\rho}_{-\underline{k}}(o) \rangle = -\frac{d^2}{dt^2} F(\underline{k}, t)$$

The initial values of the two ACF's are :

$$F(\underline{k}, t=0) = S(\underline{k})$$

$$C_l(\underline{k}, t=0) = \frac{k^2 k_B T}{m} \equiv \omega_o^2 \qquad (2.24)$$

In isotropic liquids, all these TCF's depend only on $k = |\underline{k}|$.

Many of the concepts introduced in this section may be illustrated on the simple example of the normalized velocity ACF :

$$Z(t) = \frac{\langle \underline{v}_i(t) \,.\, \underline{v}_i(o) \rangle}{\langle |\underline{v}_i|^2 \rangle} = \frac{m}{3k_B T} \langle \underline{v}_i(t) \,.\, \underline{v}_i(o) \rangle \qquad (2.25)$$

This even function of time admits the short time expansion :

$$Z(t) = 1 - \frac{\Omega^2}{2} E\, t^2 + O(t^4) \qquad (2.26)$$

where the Einstein frequency is given by (2.1). An elementary calculation shows that the mean square displacement of the CM of a molecule from its initial position is given by :

$$\langle |\underline{R}(t) - \underline{R}(o)|^2 \rangle = 6 \frac{k_B T}{m} \int_o^t (1-s/t) Z(s) ds \qquad (2.27)$$

so that, by identification with Einstein's law in the long-time limit, the self-diffusion coefficient is determined by the time integral of the velocity ACF :

$$D = \frac{k_B T}{m} \int_o^\infty Z(s) ds = \frac{k_B T}{m} \tilde{Z}(z=0) \qquad (2.28)$$

A part from the CM velocity ACF, one may define atomic velocity ACF's associated with the various atoms in the molecule. An example taken from the MD simulations of a model for water by Impey et al. (1982) is shown in Fig. 3. Since the motion of atom α in the molecule may be decomposed into the CM translation and a rotation around the CM (with angular velocity $\underline{\omega}$), according to :

$$\underline{v}_\alpha = \underline{v}_{CM} + \underline{\omega} \wedge \underline{r}_\alpha$$

the atomic velocity ACF is the sum of contributions from CM translations, rotations and rotation-translation coupling.

Since Z(t) is an even function, the frequency Ω occurring in the memory function equ.(2.18) vanishes. If moreover the memory function is assumed to decay infinitely rapidly, so that it reduces to a δ-function, equ.(2.18) reduces to the Langevin equ. of the theory of Brownian motion :

$$\dot{Z}(t) = -\xi Z(t) \qquad (2.29)$$

where ξ is a macroscopic friction coefficient, equal to the time integral of the memory function. Equ.(2.29) applies of course only if the basic assumptions of Brownian motion theory are satisfied, i.e. if the time-scales of the large Brownian particle and of the molecules of the solvent are sufficiently different. The resulting velocity ACF decays exponentially with a relaxation time determined by the inverse of the friction coefficient.

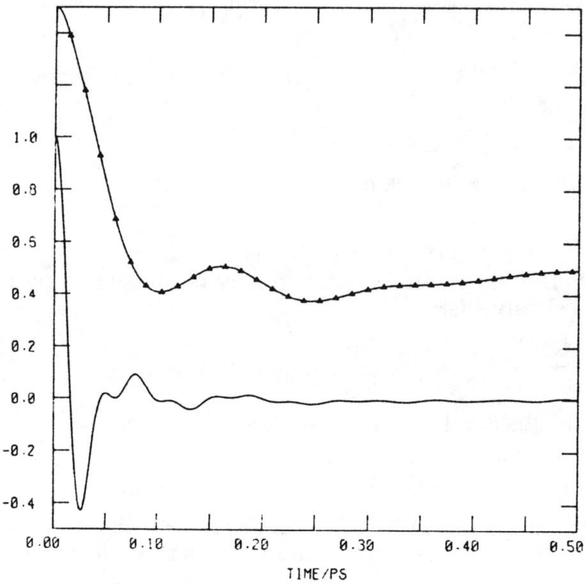

Figure 3 : Atomic velocity autocorrelation functions for H_2O at 286 K. Top : O atoms ; bottom : H atoms (from Impey et al. (1982)).

2.3 Linear response and spectroscopy

In this section we examine the response of a system to a weak external perturbation, which may be an applied force field (e.g. an a.c. electric field for the study of dielectric relaxation) or an incident radiation beam (e.g. photons or neutrons). The total hamiltonian is now of the form :

$$H_N = H_N^{(o)} + H_N^{(1)} \tag{2.30}$$

where $H_N^{(1)}$ describes the perturbation. According to equ.(2.9) the Lionville operator L splits similarily into $L^{(o)} + L^{(1)}$. If the external field F(t) couples to the dynmical variable A of the system :

$$H_N^{(1)}(t) = -AF(t) = -AF_o\, e^{i\omega t}\, e^{\varepsilon t} \tag{2.31}$$

where it is assumed that F(t) is monochromatic, and applied adiabatically from the infinite past ($\varepsilon \rightarrow 0^+$). An elementary first order perturbation calculation leads to the following expression for the non-equilibrium expectation value of a dynamical variable at time t (R.Kubo, 1966) :

$$<\Delta B(t)>_{n.e} = \int_{-\infty}^{t} \theta_{BA}(t-s)\, F(s)\, ds \tag{2.32}$$

where the after-effect function θ_{BA} is given by :

$$\theta_{BA}(t) = -\langle\{B(t), A\}\rangle = \langle B(t)\dot{A}\rangle / k_B T \; ; \; t \geq 0 \qquad (2.33)$$

the angular brackets appearing in (2.33) denote equilibrium averages. Substitution of the monochromatic form (2.31) of F(t) into (2.32) leads to the result :

$$\langle \Delta B(t) \rangle_{n.e.} = R \mathcal{X}_{BA}(\omega) F_o e^{-i\omega t} \qquad (2.33)$$

where $\mathcal{X}_{BA}$ is the complex dynamical susceptibility :

$$\mathcal{X}_{BA}(\omega) = \mathcal{X}'_{BA}(\omega) + i\mathcal{X}''_{BA}(\omega)$$

$$= \lim_{\varepsilon \to o^+} \int_o^\infty \theta_{BA}(t) e^{i(\omega + i\varepsilon)t} dt \qquad (2.34)$$

Cleary $\mathcal{X}_{BA}(-\omega) = \mathcal{X}^*_{BA}(\omega)$, so that $\mathcal{X}'$ and $\mathcal{X}''$ are, respectively, even and odd functions of frequency. Causality implies that $\mathcal{X}'$ and $\mathcal{X}''$ are related by Kramers-Kronig relations. Combination of equs. (2.33) and (2.34) leads to the "fluctuation-dissipation" theorem :

$$\mathcal{X}_{BA}(\omega) = [\langle BA \rangle + i\omega \, \tilde{C}_{BA}(\omega)] \qquad (2.35)$$

As an illustration, we calculate the mean velocity of a tagged particle subject to a homogeneous external force in the x-direction, in that case $H^{(1)} = -RX_o e^{-i\omega t} x$, and the induced velocity $\langle v_x(t) \rangle_{n.e.}$ is given by the general result (2.33) where the susceptibility is now the frequency-dependent mobility $\mu(\omega)$; since here $\dot{A} = \dot{x} = v_x$, we have immediately from (2.32) and (2.34) :

$$\mu(\omega) = \frac{1}{k_B T} \times \frac{k_B T}{m} \tilde{Z}(\omega) \qquad (2.36)$$

In particular, in view of (2.28), the static mobility $\mu = \mu(\omega{=}0)$ and the self-diffusion constant D obey the Einstein relation $\mu = D/k_B T$.

Fluctuation-dissipation expressions for other static or frequency-dependent transport coefficients, or for the frequency-dependent dielectric constant $\varepsilon(\omega)$ of a polar liquid may be derived along similar lines. More generally the inelastic cross-sections or band shapes measured by a variety ot spectroscopic techniques are directly related to well-defined TCF's (Rothschild, 1984). Thus the reorientational motion of the molecules may be probed by infrared (IR) absorption and by depolarized Rayleigh (or Raman) spectroscopy. If $\underline{u}_i(t)$ denotes the unit vector parallel to a molecular axis or to the molecular dipole moment, one defines a family of reorientational TCF's :

$$C^{(l)}(t) = \langle P_l[\underline{u}_i(t).\underline{u}_i(o)] \rangle \qquad (2.37)$$

where P_l denotes a Legendre polynominal. If one neglects collective reorientational motions (i.e. time-dependent correlations between the orientations of different molecules), which is always achieved at sufficiently high frequencies, the IR bandshape is proportional to the spectrum $\hat{C}^{(1)}(\omega)$ while the Raman bandshape is related to $\hat{C}^{(2)}(\omega)$. In the case of H_2O, for instance, the measured IR spectrum exhibits a peak around 700 cm^{-1} (i.e. 1.3 x 10^{14} sec^{-1}) while the Raman spectrum shows two peaks at 400 and 700 cm^{-1}. The interpretation of the experimental spectra is complicated by a variety of factors, including uncertainty about the contribution from vibrational relaxation and from collision-induced effects.

The frequency-dependent dielectric constant $\varepsilon(\omega)$, measured in dielectric relaxation experiments, is determined by the decay of the collective orientation of the molecular dipolar axes according to the following relation, appropriate for conducting boundary conditions (Madden and Kivelson, 1984 ; Neumann, 1986)

$$\frac{\varepsilon(\omega) - \varepsilon_\infty}{\varepsilon_o - \varepsilon_\infty} = 1 + i\omega\, \tilde{\Phi}(\omega) \tag{2.38}$$

where $\varepsilon_o = \varepsilon(\omega = 0)$ is the static dielectric constant (corresponding to the optical frequency range $\omega \geq 10^{15}$ sec^{-1}) and $\tilde{\Phi}(\omega)$ is the Laplace transform (evaluated on the real axis) of the normalized ACF of the total dipole moment $\underline{M}(t) = \sum_i \underline{m}_i(t)$ of the sample :

$$\Phi(t) = \frac{<\underline{M}(t) \,.\, \underline{M}(o)>}{<|\underline{M}|^2>} \tag{2.39}$$

In many dielectrics, including water, $\Phi(t)$ is found to decay exponentially (Debye relaxation model), except at very short times ($t < 1$ picosecond). The relation (2.38) then takes the particularly simple form :

$$\frac{\varepsilon(\omega) - \varepsilon_\infty}{\varepsilon_o - \varepsilon_\infty} = \frac{1}{1 - i\omega\, \tau_D} \tag{2.40}$$

where τ_D is the Debye relaxation time. The corresponding Cole-Cole plot of $\varepsilon''(\omega)$ versus $\varepsilon'(\omega)$ is a semi-circle. Deviations from this shape are observed on the lower left corner of the plot (corresponding to small ε' and ε''), due to short-time deviation from purely exponential decay. An example of $\Phi(t)$ computed for a rigid (non-polarizable) model of H_2O by Molecular Dynamics simulations is shown in Fig. 4 (Neumann, 1986). For non-polarizable molecules, $\varepsilon_\infty = 1$. The short-time oscillation apparent in $\Phi(t)$ is evidence of librational motion of the water molecules. Note that, as expected $\Phi(t)$ decays much faster at the higher temperature. The single-particle correlation function $C^{(1)}(t)$ decays faster than the collective TCF. The corresponding relaxation times at room temperature turn out ot be $\tau_D \simeq 6.3$ and $\tau_S \simeq 3.3$ picoseconds.

Another powerful probe of collective molecular dynamics is the quasi-elastic (or Rayleigh-Brillouin) scattering of light (Berne and Pecora, 1976), due to fluctuations of the local dielectric constant. Since the wavelength of visible light is 10^3 times larger than intermolecular distances, only collective dynamics are probed, which are independent of the details of molecular stucture, and are entirely describable by the linearized equations of hydrodynamics. This technique allows accurate measurements of macroscopic thermodynamic and transport quantities. The measured spectrum is in fact the Fourier transform of the density ACF (2.22) :

$$S(\underline{k}, \omega) = \frac{1}{2\Pi} \int_{-\infty}^{+\infty} F(k, t) e^{i\omega t} dt \tag{2.41}$$

which is called the dynamical structure factor. Note that $F(\underline{k}, t = 0)$ is identical with the CM structure factor $S(\underline{k})$, so that $S(\underline{k}, \omega)$ and $S(\underline{k})$ are related by the "elastic sum rule" :

$$S(\underline{k}) = \int_{-\infty}^{+\infty} S(\underline{k}, \omega)d\omega \tag{2.42}$$

Inelastic scattering of thermal neutrons also probes density fluctuations, but since their wavelength is comparable with the inter-molecular spacing, and since neutrons interact with the nuclei, they are sensitive to the intramolecular structure. The inelastic scattering pattern (i.e. the obseved spectrum) is a superposition of atomic dynamical structure factors $S_{\alpha\beta}(k, \omega)$, very similar to the static counterpart (1.21). Morever the observed spectrum contains a collective (or coherent) part and a single-particle (or incoherent) part. It is only in the simplest case of an atomic liquid made up of atoms which scatter neutrons coherently, that the measured spectrum is proportional to $S(\underline{k}, \omega)$ which is also seen in light-scattering experiments, although in a very different range of wavelengths and frequencies. The incoherent part of the spectrum is related to the self part of the density autocorrelation function

$$F^{(s)}_{\alpha\alpha}(\underline{k}, t) = \frac{1}{N} < \sum_i e^{i\underline{k}\cdot[\underline{r}_{i\alpha}(t) - \underline{r}_{i\alpha}(o)]} > \tag{2.43}$$

and hence probes directly the individual particle (or "self") motion. From the "self" part of equ. (2.23) it is easily shown that the spectrum $S^{(S)}{}_{\alpha\alpha}(k, \omega)$ is proportional to the spectrum $\hat{Z}_{\alpha}(\omega)$ of the atomic velocity ACF in the limit $\underline{k} \rightarrow 0$. Measurements on H_2O, thus yield direct information on the self motion of the protons, since the latter scatter neutrons predominantly in an incoherent way.

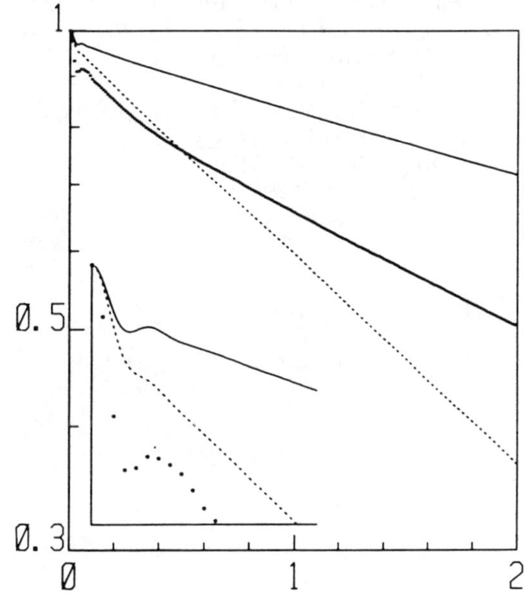

Figure 4 : Dipole moment correlation functions of H_2O. Solid and dashed lines : collective correlation function $\Phi(t)$ (equ. (2.   )) at 293 and 373 K. Dots : single-particle correlation function (equ. (2.   )). Time is in picoseconds. The short-time behaviour is shown on a 5 times enlarged scale in the inset reproduced from Neumann (1986)).

References :

Barrat, J.L., Hansen, J.P. and Pastore, G.(1988) Mol. Phys. 63, 747
Blum, L. (1972) J. Chem. Phys. 57, 1862
Blum, L. and Torruela, A.J. (1972) J. Chem. Phys. 56, 303
Carnie, S.L. and Patey, G.N. (1982) Mol. Phys. 47, 1129
Chandler, D. and Andersen, H.C. (1972) J. Chem. Phys. 57, 1930
Chandler, D. (1976) Mol. Phys. 31, 1213
Edwards, S.F. and Anderson, P.W. (1975) J.Phys. F 5, 965
Frenkel, D. (1986) in "Molecular-Dynamics Simulation of Statistical-Mechanical Systems", edited by G. Ciccotti and W.G. Hoover (North-Holland, Amsterdam)
Fries, P. and Patey, G.N. (1985) J. Chem. Phys. 82, 429
Fries, P. and Salamito, B. (1988) J. de Phys. 49, 1397
Götze W. and Sjögren, L. (1989) in "Proceedings of ILL Workshop on Dynamics of Disordered materials" (to be published)
Gray, C.G. and Gubbins, K.E. (1984) "Theory of Molecular Liquids" (Clarendon Press, Oxford)
Guissani, Y., Guillot, B. and Bratos, S. (1988) J. Chem. Phys. 88, 5850
Hansen, J.P. (1984) J. de Phys. 45, C7-97
Hansen, J.P. and Mc Donald, I.R. (1986) "Theory of Simple Liquids", 2d ed. (Academic Press, London)
Hye, J.S., Lebowitz, J.L. and Stell, G. (1974) J. Chem. Phys. 61, 3253
Impey, R.W., Madden, P.A. and Mc Donald, I.R. (1982) Mol. Phys 46, 513
Klein, M.L. and Mc Donald, I.R. (1979) J. Chem. Phys. 71, 298
Kubo, R. (1966) Rep, Prog. Phys. 29, 255
Levesque, D., Weis, J.J. and Patey, G.N. (1984) Mol. Phys. 51, 333
Madden, P. and Kivelson D. (1984) Adv. Chem. Phys. 56, 467
Montague, D.G. and Dore, J.C. (1986) Mol. Phys. 57, 1035
Neumann, M. (1986) J. Chem. Phys. 85, 1567
Rothschild, W.G. (1984) "Dynamics of Molecular Liquids" (JohnWiley, New York)
Shing, K.S. and Gubbins, K.E. (1981) Mol. Phys. 43, 717
Sprik, M. and Klein, M.L. (1988) J. Chem. 89, 7556
Wertheim, M.S. '1971) J. Chem. Phys. 55, 4291
Widom, B. (1963) J. Chem. Phys. 39, 2808

METHODS FOR CALCULATING INTERMOLECULAR POTENTIAL ENERGY SURFACES

A. J. STONE
University Chemical Laboratory,
Lensfield Road,
Cambridge CB2 1EW,
England

ABSTRACT: It is now possible to carry out accurate *ab initio* calculations on hydrogen-bonded complexes by a variety of techniques. The supermolecule approach is widely used, and is capable of high absolute accuracy, but it is subject to Basis Set Superposition Error, especially when electron correlation is taken into account, and this is a difficulty when accurate calculations of small interaction energies are required. Perturbation theory provides an alternative; it is subject to BSSE only in its description of charge-transfer effects. A disadvantage of this approach is that the perturbation expansion does not converge if the interaction is too strong, that is when the distance of approach is too short. Another disadvantage is that perturbation methods as currently implemented are 'uncoupled'; that is, the response of the electrons to the perturbation is not treated self-consistently. Nevertheless this method gives a more detailed description of the interaction than the supermolecule approach, and consequently provides more physical insight into the nature of the interaction. Both of these methods require calculations to be carried out at a wide range of dimer geometries if a full description of the potential energy surface is needed, and this is extremely time-consuming.

A useful alternative approach is to isolate the components of the perturbation expansion, namely the repulsion, the electrostatic interaction, the induction, the dispersion and the charge transfer terms, and to calculate each of them independently by the most appropriate technique. Thus the electrostatic interaction can be calculated accurately from distributed multipole descriptions of the individual molecules, while the induction and dispersion contributions may be derived from polarizabilities of the individual molecules. This approach has the advantage that the properties of the monomers have to be calculated only once, after which the interactions may be evaluated easily and efficiently at as many dimer geometries as required.

Models for the intermolecular potential emerge from this procedure almost automatically. The electrostatic, induction and dispersion energies are described in an analytic form which can be used without further approximation even in potential-intensive applications such as molecular dynamics. The repulsion is not so amenable, but it can be fitted by suitable analytic functions much more satisfactorily than the complete potential. The result is a model of the intermolecular potential that is capable of describing properties to a high level of accuracy.

J.C. Dore and J. Teixeira (eds.), Hydrogen-Bonded Liquids, 25–47.

1.1 Types of *ab initio* calculation

In order to understand the limitations of calculations on intermolecular potential energy surfaces it is necessary to review the language of *ab initio* theory. The simplest useful type of *ab initio* calculation is the *Self-Consistent Field* calculation. In the SCF ground state Ψ_0 of a closed-shell molecule certain orbitals are doubly occupied and the remainder (the *virtual orbitals*) are vacant. The resulting wavefunction has zero spin. The SCF procedure optimises the orbitals in this *Restricted Hartree-Fock* wavefunction. A more general type of wave-function is the *Unrestricted Hartree-Fock* or UHF wavefunction, in which the orbitals may be different for electrons of different spin. A full discussion can be found in the text by McWeeny & Sutcliffe, or in the revised edition by McWeeny[1]. Many standard programs are available which perform SCF calculations, and they all perform essentially the same task; although there are technical differences between them, they have to do with programming efficiency and not with the fundamental nature of the calculation.

Apart from a few specialised programs, all these calculations describe the molecular orbitals ψ_i in terms of a *basis set* of atomic functions η_r:

$$\psi_i = \sum_r c_{ri}\eta_r. \tag{1}$$

Modern calculations on molecules almost invariably use *Gaussian* basis functions of the form $x^l y^m z^n \exp(-\alpha r^2)$, where x, y and z are position coordinates relative to the nucleus, and $r^2 = x^2 + y^2 + z^2$. If α is large, the function decays very rapidly with distance from the nucleus, and so is very compact; conversely, if α is small, the function is diffuse. This function is called a Gaussian *primitive* function; atomic wavefunctions behave more like the exponential function $\exp(-\zeta r)$ (sometimes called a Slater function) than $\exp(-\alpha r^2)$, and it is necessary to use a linear combination of Gaussian primitives to describe a single Slater function. If the coefficients of the primitives are allowed to vary independently, we have an *uncontracted* basis, but it is more usual to use a *contracted* basis, in which the coefficients of the primitives in any particular basis function are fixed. This retains most of the advantages of the more flexible uncontracted basis, but greatly reduces the size of the SCF calculation.

A widely-used basis is the STO–nG type, in which each of the atomic orbitals $1s$, $2s$, $2p$, etc. is described by a contracted basis function containing n (usually 3) Gaussian primitives. This *minimal* basis is not however very flexible, because the atomic orbitals are not able to distort in response to the field of the rest of the molecule. A better basis is a *split-valence* one, such as the n–31G series of Pople and his collaborators[2], in which the core orbitals are described by a fixed contraction of n primitives, while the valence shell is described by two sets of basis functions, one containing 3 primitives and another, more diffuse, containing only one. This allows the valence orbitals to expand or contract in response to molecule formation by incorporating more or less of the diffuse basis functions. A similar type of basis is the *double-zeta* basis, in which the core basis is also split[3].

The next stage up involves the inclusion of *polarization functions.* The 6–31G* basis sets of Pople[4] include d functions on first and second row atoms, and the 6–31G** basis includes p functions on hydrogen as well. There are various DZP (double-zeta-plus-polarization) basis sets that are also used. The higher-angular-momentum functions permit angular distortion of the orbitals, which is very important when accurate energies are required.

Accurate calculation of molecular properties also calls for large basis sets. This is particularly true of response-function properties—properties such as polarizabilities which describe the response of the molecule to an external field. It has been shown that two or three sets of polarization functions are needed in order to obtain accurate dipole polarizabilities[5,6]. If we demand a comparable flexibility in the core and valence functions, we are now at the 'TZ2P' level (triple-zeta plus 2 sets of polarization functions). For a first-row atom, we now have a $[6s3p2d]$ basis, with 3 sets of $1s$, $2s$ and $2p$ functions (15 in all) and 2 sets of d functions (10 or 12 more) making 27 in all. (Each set of d functions commonly contains 6 functions, since it is simpler to use the functions $x^2e^{-\alpha r^2}$, $y^2e^{-\alpha r^2}$ and $z^2e^{-\alpha r^2}$ than to use the correct linear combinations. The sum of these three functions is spherically symmetric, and so can be regarded as an additional s function.) The largest calculations involving first-row atoms currently use basis sets such as $[7s5p3d2f1g]$, containing about 75 basis functions per atom.

The size of molecule that can be studied is limited by the number of basis functions that typical programs can handle, which is typically 200 or so. The limiting factor is usually the number of electron repulsion integrals; there are about $\frac{1}{8}N^4$ of them if the basis contains N functions, so for 200 basis functions there may be as many as 2×10^8 integrals, though usually many of them are small enough to be neglected. They are usually stored on disc or tape, and read at each iteration of the SCF calculation. Alternatively, in the so-called 'direct SCF' procedure, they are recalculated at each iteration of the SCF procedure, which is more time-consuming but does not require large amounts of slow memory.

The SCF wavefunction does not take account of 'electron correlation'; it describes the effects of electron repulsion only in an averaged way. The usual procedure for incorporating the effects of electron correlation is *configuration interaction*. We construct other wavefunctions in which some of the electrons are excited from the occupied orbitals into virtual orbitals. Most of these will not be eigenfunctions of total spin, but by taking suitable linear combinations we can construct *configuration state functions* (CSFs) Ψ_K that have definite spin. The ones that have the appropriate spin can mix with the ground state molecular orbital wavefunction, and the most general wavefunction is a linear combination of them:

$$\Psi = \sum c_K \Psi_K. \tag{2}$$

The wavefunction obtained this way is a *configuration interaction* or CI wavefunction. If all possible configuration state functions with the correct total spin are included in the sum, we have a *full CI* wave function, whose accuracy is limited only by the quality of the basis set. However it is practicable to perform full CI calculations only with very small basis sets, because the number of CSFs rapidly becomes extremely large, and the number of configuration state functions usually has to be limited in some way. The most common procedure is *singles-and-doubles CI* or SDCI, in which the only CSFs included are those that involve a single-electron or two-electron excitation from the root configuration Ψ_0. Another common approach is to use *Møller-Plesset perturbation theory*, alternatively called *many-body perturbation theory* or MBPT, in which the effects of electron correlation are taken into account by perturbation theory, commonly to second order (MP2), and sometimes to third or fourth order (MP3 or MP4).

It often happens that several CSFs have significant coefficients in the CI expansion. In the description of an ordinary single bond, for example, the configuration in which two electrons are promoted from the bonding orbital to the antibonding one plays an important rôle, especially if the bond is stretched. (In the dissociation limit, these two configurations have equal weight.) In such cases, a *MultiConfiguration Self-Consistent Field* or MCSCF wavefunction can be used[7]. This takes the same form as eq. (2), but includes only the most important configurations. The difference is that the CI coefficients c_K and the molecular orbital coefficients are optimised simultaneously. This MCSCF procedure accounts for some of the more important contributions to the electron correlation, and it can be followed by a CI calculation in which further CSFs are included but the molecular orbital coefficients are not optimised further.

The choice of a suitable calculation method and basis set is a black art that involves many compromises between accuracy and economy. It is not very difficult to calculate molecular structures with reasonable accuracy, though there are always pathological cases, but accurate determination of intermolecular interaction energies is much more troublesome.

2. The supermolecule method

A natural way to calculate the interaction energy U_{AB} between two molecules A and B is to compare the energy of the complex $A \cdots B$ with the energies of the separated molecules:

$$U_{AB} = W_{AB} - W_A - W_B. \tag{3}$$

This is the *supermolecule method.* It has a number of attractions: it is very easy to understand conceptually, and it is easy to apply, using standard *ab initio* computational techniques. Unfortunately it has some less obvious limitations. It is sometimes suggested that there are numerical disadvantages in calculating the small interaction energy U_{AB} as a difference of much larger quantities, but although there is a loss of precision it is unimportant compared with other inaccuracies. It is however true that the smallness of U_{AB} relative to W_A and W_B themselves is a source of inaccuracy, since small relative errors in the latter lead to very large percentage errors in the interaction energy. This means that the calculations must be carried out at a high level of accuracy if the results are to be of any value. However the computer time required for *ab initio* calculations increases roughly with the fourth power of the number of basis functions, so these calculations are very expensive.

Furthermore, the calculation has to be repeated at a large number of relative configurations of the two molecules if the potential surface is to be explored adequately. Six coordinates are required to specify the position and orientation of molecule B relative to molecule A, if both are non-linear. If the calculation is repeated for only four values of each coordinate (a very inadequate number) a total of $4^6 = 4096$ points will be required. There are many examples in the literature of calculations of intermolecular potential energy surfaces which use a wholly inadequate strategy for choosing points on the surface. It is common, for example, even in calculations on pairs of diatomics, where there are only three angular coordinates, to perform calculations at only a few relative orientations: often the linear, rectangular, T-shaped and crossed configurations only. Even if calculations are done for a large number of intermolecular distances, it is impossible to characterise the potential energy surface adequately from such a limited number of orientations. Fortunately it is possible to explore the surface much more efficiently[8], but a large number of points is still required.

2.1 Electron correlation and size consistency

It is also necessary to include electron correlation in any accurate calculation of intermolecular interactions, since the dispersion interaction is a correlation effect, and is usually an important component of the interaction energy. Moreover the dispersion interaction describes correlated fluctuations of the charge distributions in the two molecules, and this calls for an accurate description of the molecular polarizability, which in turn requires a large basis set containing a large number of diffuse polarization functions.

In the commonly-used singles-and-doubles CI method (abbreviated as CISD), the wavefunction is a linear combination of CSFs (eq. (2)) which includes every configuration derived by one-electron or two-electron excitation from the 'root configuration' Ψ_0. Suppose that this method is used to obtain the energies in eq. (3). The energy W_{AB} then takes into account all single and double excitations. However the energy of W_{AB} at infinite separation is the sum of W_A and W_B, and if each of these is obtained by SDCI, the wavefunction for infinitely-separated $A \cdots B$ includes configurations in which both A and B are doubly excited, so that $A \cdots B$ is quadruply excited. Thus $A \cdots B$ is described at short range by a wavefunction that contains only single and double excitations, but at long range by a wavefunction that includes higher excitations. The long-range energy is then variationally better than the energy at short range, and the strength of the interaction will be under-estimated. It might be thought that this problem could be avoided by performing a calculation on $A \cdots B$ at large separation, but this does not help, because it turns out that the correlation energy of A obtained by the CISD method is not the same for the isolated A molecule as it is for the A molecule in the infinitely-separated $A \cdots B$ complex[9].

To avoid this problem, it is necessary to ensure that the CI method is *size consistent* or *size extensive*. These two terms have slightly different meanings, but both require that the energy calculated for an assembly of molecules be the same as the sum of the energies of the individual molecules calculated separately. We have seen that the CISD method does not satisfy this requirement, but there are methods which do; they include Møller-Plesset perturbation theory and a variety of 'coupled-cluster' methods[10]. CEPA (the Coupled Electron Pair Approximation) is an approximation to the latter.

2.2 Basis set superposition error

The need for size-consistency is an illustration of a fundamental difficulty of the supermolecule method. For a single system the variation principle ensures that the better the calculation, the better the energy. When we are interested in energy *differences*, however, the variational principle does not apply. In fact, there is a very serious source of error that can be attributed to the variational principle itself.

We choose a basis for the calculation of molecule A that seems suitable for our purpose. Inevitably it is incomplete, so our wavefunction for A is not exact. The same is true of the calculation for the isolated molecule B. Now we carry out a calculation on the complex $A \cdots B$, and obtain a new energy that includes the interaction between the molecules. But in this calculation there are some new basis functions, belonging to molecule B, that were not present when we calculated the energy of the isolated A molecule, and they allow the wavefunction of molecule A to be improved variationally, so that its energy falls. This is quite separate from the true physical interaction; it is a spurious effect that occurs because the basis set that we initially chose for A was not good enough to describe the wavefunction exactly. This is *Basis Set Superposition Error*, or BSSE, and it occurs in all supermolecule calculations. The effect is to make the interaction seem more attractive than it really is.

This is an extremely troublesome problem. The standard way to deal with it is the *counterpoise correction* proposed by Boys & Bernardi[11]. The reference energy for A is calculated, not in the obvious way, but in the presence of the basis functions (but not the electrons or nuclei) of molecule B, and similarly the reference energy for B is calculated in the presence of the A basis functions. In this way, the variational improvement that arises from the presence of the 'foreign' basis functions is included in the reference calculation as well as in the $A \cdots B$ calculation. (The basis functions of the other molecule are sometimes called 'ghost' orbitals when used like this.) An obvious problem is that if we change the relative positions of the two molecules in the $A \cdots B$ complex, then we change the position of the B basis functions relative to A (and *vice versa*) so that the counterpoise-corrected calculation of the reference energies has to be repeated for every point on the surface. Clearly this increases the expense of the potential surface calculation by a factor of about 3. The effects of BSSE can be estimated and corrected for reasonably well at the SCF level, but they are bigger by an order of magnitude in correlated calculations[12]. Knowles[13] has found that for the complex $H_2 \cdots HF$, using a $[7s5p3d1f]$ basis for F and a $[5s3p1d]$ basis for H, the BSSE at the correlated level is of the order of $180\,cm^{-1}$at typical geometries, which is comparable with the well-depth of about $150\,cm^{-1}$.

It has been pointed out that in the $A \cdots B$ complex some of the orbitals of B are not available to A, because they are occupied by the electrons of B, and it has been argued that in the counterpoise calculation these orbitals should be excluded. There has been a good deal of controversy about this point, but the present view seems to be that the original prescription of Boys & Bernardi is the correct one[12]. There is no doubt that if it is necessary to use only the virtual orbitals of each molecule as ghost orbitals for the other, the counterpoise calculation is complicated greatly.

It is also evident that if a basis is used that is not capable of describing the isolated molecules correctly, then *a fortiori* it will give the wrong interaction energy, even if the effects of BSSE could be corrected precisely. It follows that the solution to the problem is not to find the ideal procedure for correcting for BSSE, but to perform the calculation in such a way that it does not arise in the first place. This requires that a sufficiently good basis set be used for the individual molecules, so that the basis functions of the other molecule cannot provide any variational improvement. Unfortunately this is not realistic even for the smallest systems at present. However it is currently possible to reduce the BSSE in SCF calculations on molecules containing first-row atoms to about $10\,cm^{-1}$ by using a very good basis, while the limit in correlated calculations is about $120\,cm^{-1}$[13].

There is a further difficulty that currently limits the accuracy of intermolecular potential calculations, even on small systems where good basis sets can be used and the BSSE correction is relatively small. If the basis set on each molecule is very good, then the basis for the complex may become nearly over-complete; that is, the basis functions are not all independent. Numerical instabilities then ensue. The difficulty seems to arise from the use of atom-centred basis functions. In order to describe the complicated distortions that may result from the presence of another molecule, it is necessary to use basis functions that are very diffuse and have high angular momentum quantum numbers, and they overlap strongly with the corresponding functions on the other molecule, especially at very short range. A better description might result from the use of basis functions centred in the region between the interacting molecules, but no way has been found so far to do this consistently.

3. Intermolecular perturbation theory

If supermolecule calculations are so troublesome, perhaps perturbation theory can provide a more satisfactory approach. The attraction of perturbation theory is that it leads directly to an expression for the interaction energy itself. Moreover it is possible to analyse the expression into a number of distinct terms that can be correlated with distinct physical effects.

3.1 Long-range perturbation theory—the perturbation

For two molecules, sufficiently far apart, the Hamiltonian for the system can be written as

$$\mathcal{H} = \mathcal{H}_A + \mathcal{H}_B + V, \tag{4}$$

where $\mathcal{H}_A$ is the Hamiltonian for the isolated molecule A, $\mathcal{H}_B$ the Hamiltonian for molecule B, and V is the interaction. V arises simply from the electrostatic interaction between the particles of A and those of B:

$$V = \sum_{i\in A, j\in B} \frac{e_i e_j}{4\pi\epsilon_0 r_{ij}}, \tag{5}$$

where e_i and e_j are the charges on particles i and j, and r_{ij} is the distance between them.

It is often useful to replace the operator (5) by its *multipole expansion*:

$$V = \sum_{tu} \widetilde{Q}_t^A T_{t\,u}^{AB} \widetilde{Q}_u^B, \tag{6}$$

where $\widetilde{Q}_t^A$ is an operator for one of the multipole moments of molecule A (charge, dipole, quadrupole, etc.) and $T_{t\,u}^{AB}$ is an interaction function. The index t is an abbreviation for the angular momentum labels lk, and the moment operator $\widetilde{Q}_{lk}^A$ is defined by

$$\widetilde{Q}_{lk}^A = \sum_{i\in A} e_i r_i^l C_{lk}(\theta_i, \phi_i), \tag{7}$$

where r_i, θ_i and ϕ_i are the spherical polar coordinates of particle i in a local coordinate system for molecule A, and C_{lk} is a modified spherical harmonic:

$$C_{lk}(\theta,\phi) = \big(4\pi/(2l+1)\big)^{\frac{1}{2}} Y_{lk}(\theta,\phi).$$

The derivation of (6) is straightforward[14,15]. It starts with an expansion of $1/r_{ij}$ in terms of spherical harmonics, via the spherical harmonic addition theorem[16]. We write the position of particle i of molecule A as $\mathbf{A}+\mathbf{r}_i$, where $\mathbf{A}$ is a local origin for molecule A, and similarly the position of particle j is $\mathbf{B}+\mathbf{r}_j$. Then

$$\frac{1}{r_{ij}} = \frac{1}{|\mathbf{R}+\mathbf{r}_j-\mathbf{r}_i|} = \sum_{k=0}^{\infty}\sum_{m=-k}^{k}(-1)^m R_{k,-m}(\mathbf{r}_j-\mathbf{r}_i)I_{km}(\mathbf{R}), \tag{8}$$

where $\mathbf{R}=\mathbf{B}-\mathbf{A}$ and $R_{km}(\mathbf{r})$ and $I_{km}(\mathbf{r})$ are the regular and irregular spherical harmonics respectively:

$$R_{km}(\mathbf{r}) = r^k C_{km}(\theta,\phi), \quad I_{km}(\mathbf{r}) = r^{-k-1}C_{km}(\theta,\phi).$$

Here C_{km} is the modified spherical harmonic defined above. Eq. (8) is valid only if $|\mathbf{r}_j-\mathbf{r}_i| < R$. Another standard addition theorem[16] states that

$$R_{k,-m}(\mathbf{r}_1+\mathbf{r}_2) = \sum_{l_1 l_2}\sum_{m_1 m_2}\delta_{l_1+l_2,k}(-1)^{k-m}\left(\frac{(2k+1)!}{(2l_1)!(2l_2)!}\right)^{\frac{1}{2}} \times R_{l_1 m_1}(\mathbf{r}_1)R_{l_2 m_2}(\mathbf{r}_2)\begin{pmatrix} l_1 & l_2 & k \\ m_1 & m_2 & m \end{pmatrix},$$

where the factor

$$\begin{pmatrix} l_1 & l_2 & k \\ m_1 & m_2 & m \end{pmatrix}$$

is a Wigner 3-j symbol[16]. Now, remembering that $R_{lm}(-\mathbf{r}) = (-1)^l R_{lm}(\mathbf{r})$, we find that

$$\begin{aligned} V &= \frac{1}{4\pi\epsilon_0}\sum_{i\in A}\sum_{j\in B}\frac{e_i e_j}{|\mathbf{R}+\mathbf{r}_j-\mathbf{r}_i|} \\ &= \frac{1}{4\pi\epsilon_0}\sum_{l_1,l_2}\sum_{m_1 m_2 m}(-1)^{l_2}\left(\frac{(2l_1+2l_2+1)!}{(2l_1)!(2l_2)!}\right)^{\frac{1}{2}} \\ &\qquad \times \sum_{i\in A} e_i R_{l_1 m_1}(\mathbf{r}_i)\sum_{j\in B} e_j R_{l_2 m_2}(\mathbf{r}_j) I_{l_1+l_2,m}(\mathbf{R})\begin{pmatrix} l_1 & l_2 & l_1+l_2 \\ m_1 & m_2 & m \end{pmatrix} \\ &= \frac{1}{4\pi\epsilon_0}\sum_{l_1,l_2}\sum_{m_1 m_2 m}(-1)^{l_2}\left(\frac{(2l_1+2l_2+1)!}{(2l_1)!(2l_2)!}\right)^{\frac{1}{2}} \\ &\qquad \times \widehat{Q}^A_{l_1 m_1}\widehat{Q}^B_{l_2 m_2} I_{l_1+l_2,m}(\mathbf{R})\begin{pmatrix} l_1 & l_2 & l_1+l_2 \\ m_1 & m_2 & m \end{pmatrix}. \end{aligned} \tag{9}$$

where in the last line we have introduced the multipole moment operators:

$$\widehat{Q}^A_{lm} = \sum_{i\in A} e_i R_{lm}(\mathbf{r}_i).$$

The multipole moment operators in eq. (9) are still referred to the global coordinate frame (as indicated by the carets). We now transform to moment operators, labelled by a tilde, that are referred to the local or molecule-fixed frame:

$$\widehat{Q}^A_{lm} = \sum_k \widetilde{Q}^A_{lk} D^l_{mk}(\Omega_A)^*,$$

where $D^l_{mk}(\Omega_A)$ is a Wigner rotation matrix whose argument Ω_A describes the orientation of molecule A. In this way we reach the following expression for the interaction operator:

$$V = \frac{1}{4\pi\epsilon_0} \sum_{l_1,l_2} \sum_{k_1 k_2} \binom{l_1+l_2}{l_1} R^{-l_1-l_2-1} \widetilde{Q}^A_{l_1 k_1} \widetilde{Q}^B_{l_2 k_2} \overline{S}^{k_1 k_2}_{l_1\, l_2\, l_1+l_2}. \tag{10}$$

In (10) $\binom{l_1+l_2}{l_1}$ is a binomial coefficient, and the S-function $\overline{S}^{k_1 k_2}_{l_1\, l_2\, j}$ is defined by

$$\begin{aligned}\overline{S}^{k_1 k_2}_{l_1\, l_2\, j} &= i^{l_1-l_2-j} \begin{pmatrix} l_1 & l_2 & j \\ 0 & 0 & 0 \end{pmatrix}^{-1} \sum_{m_1 m_2 m} \begin{pmatrix} l_1 & l_2 & j \\ m_1 & m_2 & m \end{pmatrix} \\ &\quad \times D^{l_1}_{m_1 k_1}(\Omega_A)^* D^{l_2}_{m_2 k_2}(\Omega_B)^* C_{jm}(\theta,\phi).\end{aligned} \tag{11}$$

In (11), D^l_{mk} is a Wigner rotation matrix, and the arguments Ω_A and Ω_B describe the orientations of the two molecules. The angles θ and ϕ describe the direction of the intermolecular vector $\mathbf{R}$.

It is convenient to define a new quantity $T^{A\ \ B}_{l_1 k_1, l_2 k_2}$ as follows:

$$T^{A\ \ B}_{l_1 k_1, l_2 k_2} = \frac{1}{4\pi\epsilon_0} \binom{l_1+l_2}{l_1} R^{-l_1-l_2-1} \overline{S}^{k_1 k_2}_{l_1\, l_2\, l_1+l_2}, \tag{12}$$

and in terms of this the interaction becomes

$$V = \sum_{l_1 l_2} \sum_{k_1 k_2} \widetilde{Q}^A_{l_1 k_1} T^{A\ \ B}_{l_1 k_1, l_2 k_2} \widetilde{Q}^B_{l_2 k_2}.$$

The definition (11) of the $\overline{S}$ functions suggests that they, and hence the interaction functions T^{ab}_{tu}, have to be evaluated in terms of trigonometric functions of the orientation variables. In fact this is not so; they can all be expressed in terms of direction cosines, which are obtained as scalar products of the unit vectors specifying the local coordinate system for each molecule, and no trigonometrical functions need be evaluated. Explicit formulae are given in ref. 14.

3.2 Long-range perturbation theory—first-order

If the eigenfunctions of $\mathcal{H}_A$ are ψ^A_m and those of $\mathcal{H}_B$ are ψ^B_n, then the unperturbed wavefunctions for $A \cdots B$ are $|mn\rangle = \psi^A_m \psi^B_n$. Ordinary Rayleigh-Schrödinger perturbation theory then gives the first-order energy of interaction as

$$W^{(1)} = \langle 00|V|00\rangle \tag{13}$$

$$= \int \frac{\rho^A(\mathbf{r}_1)\rho^B(\mathbf{r}_2)}{r_{12}} d^3r_1\, d^3r_2, \tag{14}$$

which is just the classical electrostatic energy of interaction between the charge distributions of A and B. It is easily evaluated if the ground state wavefunctions are known, but the calculation requires a knowledge of all the intermolecular electron repulsion integrals and is fairly time-consuming.

If instead we use the multipole expansion of the perturbation, eq. (13) becomes simply

$$U_{\text{es}}^{AB} = \sum_{tu} Q_t^A T_{t\,u}^{AB} Q_u^B, \tag{15}$$

in which Q_t^A is the expectation value of one of the multipole moments of molecule A, referred to local axes, and the interaction function $T_{t\,u}^{AB}$ takes into account the orientation dependence of the interaction as well as the distance between the molecules. It is possible, and much more useful for practical calculations, to use an alternative approach in which the indices t and u refer to real multipole moments, rather than the complex ones defined by (7). The moments are now denoted Q_{lkc} and Q_{lks}, defined, for $k > 0$, by

$$\begin{aligned} Q_{lkc} &= \sqrt{\tfrac{1}{2}}\left[(-1)^k Q_{lk} + Q_{l,-k}\right], \\ iQ_{lks} &= \sqrt{\tfrac{1}{2}}\left[(-1)^k Q_{lk} - Q_{l,-k}\right]. \end{aligned} \tag{16}$$

No transformation is needed for Q_{l0}, which is always real. The notation reflects the fact that Q_{lkc} transforms like $\cos k\phi$ and Q_{lks} like $\sin k\phi$. The factors of $\sqrt{1/2}$ ensure that a rotation of axes induces an orthogonal transformation of the moments. The first few of these moments coincide precisely with the Cartesian charge and dipole moment:

$$\begin{aligned} Q_{00} &= \sum_i e_i, \\ Q_{10} &= \sum_i e_i z_i = \mu_z, \\ Q_{11c} &= \sum_i e_i x_i = \mu_x, \\ Q_{11s} &= \sum_i e_i y_i = \mu_y, \end{aligned} \tag{17}$$

and later ones describe the quadrupole, octopole and so on. A complete list is given in ref. [14] for moments up to hexadecapole. The functions $T_{t\,u}^{AB}$ for this formulation have been tabulated for all multipole-multipole interactions up to terms in R^{-5}[14,17].

Although the expression (15) is accurate at sufficiently large separation, it converges only if the separation between the molecules is large compared with their size. This is not the case for any pair of molecules at the separations found in condensed phases or in weakly bound complexes, and it is then necessary to use a *Distributed Multipole Expansion*, which assigns charges, dipole moments, etc., to *regions* comprising single atoms or (in larger molecules) small groups of atoms. For small molecules it is helpful to treat the bonding regions separately. The electrostatic perturbation then takes the form

$$V = \sum_{ab} \sum_{tu} \widehat{Q}_t^a T_{t\,u}^{ab} \widehat{Q}_u^b, \tag{18}$$

where $\widehat{Q}_t^a$ is the operator for one of the multipole moments of region a of molecule A. The electrostatic energy becomes

$$U_{\text{es}}^{AB} = \sum_{ab} \sum_{tu} Q_t^a T_{tu}^{ab} Q_u^b. \tag{19}$$

In defining these regional moments it is necessary to specify an origin for each region, and the origin for region a is known as 'site a'. In the Distributed Multipole approach we replace the extended charge distribution of region a by a set of point multipoles at site a, rather than replacing the charge distribution of the entire molecule by a set of point multipoles at the molecular origin.

The Distributed Multipole description of a molecular charge distribution is not unique. (Even the conventional single-site multipole description is not unique, since it depends on the choice of origin[18].) There are many ways of determining these distributed multipole moments[19–24]; many authors have used distributed charges alone, but it is now widely accepted that an accurate and efficient description requires multipoles up to at least quadrupole.

3.3 Long-range perturbation theory—second-order effects

The second-order energy is, according to Rayleigh-Schrödinger perturbation theory,

$$W^{(2)} = -{\sum_{mn}}' \frac{\langle 00|V|mn\rangle\langle mn|V|00\rangle}{W_m^A + W_n^B - W_0^A - W_0^B}. \tag{20}$$

This is conventionally separated into three terms: those in which $n \neq 0$ but $m = 0$, those in which $m \neq 0$ but $n = 0$, and those in which neither m nor n is 0. These three terms are

$$U_{\text{ind}}^A = -{\sum_m}' \frac{\langle 00|V|m0\rangle\langle m0|V|00\rangle}{W_m^A - W_0^A}, \tag{21}$$

$$U_{\text{ind}}^B = -{\sum_n}' \frac{\langle 00|V|0n\rangle\langle 0n|V|00\rangle}{W_n^B - W_0^B}, \tag{22}$$

$$U_{\text{disp}} = -{\sum_m}'{\sum_n}' \frac{\langle 00|V|mn\rangle\langle mn|V|00\rangle}{W_m^A + W_n^B - W_0^A - W_0^B}. \tag{23}$$

In the first of these, we may perform the integrations over the ground-state wavefunction of B to obtain

$$U_{\text{ind}}^A = -{\sum_m}' \frac{\langle 0|V^B|m\rangle\langle m|V^B|0\rangle}{W_m^A - W_0^A}, \tag{24}$$

where $V^B = \langle \psi_0^B|V|\psi_0^B\rangle$ is the potential at A due to the unperturbed charge distribution of B. Eq. (24) then describes the response of molecule A to this potential, and is the induction energy of molecule A in the field of B. Similarly eq. (22) is the induction energy of B in the field of A.

Eq. (23) involves excitations on both molecules; it is the dispersion energy. The excited states that appear in this expression involve one-electron excitations on both molecules, and cannot appear in an SCF calculation; accordingly the dispersion energy is a manifestation of electron correlation.

3.4 Perturbation theory at short range

In long-range perturbation theory, we assume that the molecules are far enough apart that their wavefunctions do not overlap. At short range, this is no longer true, and the consequences for perturbation theory are profound. In the first place, it becomes impossible to distinguish between electrons that 'belong' to molecule A and electrons that 'belong' to B. It is therefore no longer possible to separate the Hamiltonian for the entire system in the manner of eq. (4), and there is no satisfactory way to define a perturbation operator representing the interaction between the molecules. In the second place, the wavefunctions for the unperturbed system cannot be taken to be simple products of the form $\psi_m^A \psi_n^B$; they have to be antisymmetrized with respect to all electron permutations. A more serious difficulty is that whether antisymmetrized or not, the functions $\psi_m^A \psi_n^B$ are not orthogonal to each other. Ordinary Rayleigh-Schrödinger perturbation theory assumes that the unperturbed Hamiltonian has a complete set of eigenfunctions, and that they are orthogonal.

Many versions of perturbation theory have been proposed to overcome these problems. A large number of them rely on an expansion of the perturbation equations in powers of the overlap between the functions on A and those on B. This approach appears to work when small basis sets are used, but as the basis is improved, the overlap between the functions on the two molecules becomes larger, and the expansion ceases to converge. This failure of the overlap expansion occurs with quite modest basis sets. Accordingly it is necessary to use a method that deals explicitly with the natural non-orthogonal basis functions for the problem.

If we have to use non-orthogonal wavefunctions, then the natural one-electron orbitals in which to express them are the SCF molecular orbitals of the non-interacting molecules. From these we can construct antisymmetrized (determinantal) wavefunctions in which some orbitals of each molecule are occupied. Because of the non-orthogonality of the orbitals, these determinantal wavefunctions will also be non-orthogonal. It is possible to construct a perturbation theory in which the wavefunction is expanded in terms of these determinants. Fortunately it is possible to formulate it in such a way that the separation of the Hamiltonian into an unperturbed part and a perturbation is unnecessary. The resulting Intermolecular Perturbation Theory (IMPT)[25] has been incorporated into the Cambridge Analytical Derivatives Package (CADPAC)[26].

In this theory there are some additional terms that do not arise at long range, and because of the overlap the long-range-terms are modified. The new effects are as follows:

3.4.1 Repulsion. The most important new term is the exchange-repulsion. This is a sum of two effects: an attractive exchange term which arises because the electrons can exchange between the two molecules, and a repulsive term which occurs because the electrons cannot occupy the same region of space if they have the same spin (Pauli repulsion). However it is not usually helpful to separate these two parts. The exchange-repulsion is a first-order perturbation term, so for its evaluation it is only necessary to know the unperturbed wavefunctions. Nevertheless it is time-consuming to evaluate, because all the intermolecular electron-repulsion integrals are needed, and they are different for each configuration of the complex.

3.4.2 Electrostatic Interaction. The formal expression for the electrostatic interaction is still eq. (14) at short range, but because the charge densities overlap, there are now terms that describe the effects arising from the interpenetration of the charge distributions. This is illustrated by the simple case of a He^+ ion interacting with a proton. The electrostatic interaction can be evaluated explicitly in this case, and is

$$U_{es}(He^+ \cdots H^+) = \frac{1}{R} + \left(Z + \frac{1}{R}\right) e^{-4R},$$

where R is the separation. Here we can distinguish two terms: the 'multipolar part' $1/R$ which is the classical repulsion between two unit charges at distance R, and the 'penetration term' $(Z+1/R)\exp(-4R)$ which describes the modification to the multipolar expresion that arises from the penetration of the proton within the electronic charge distribution of the He^+. We see that the latter decays exponentially with separation. It has been claimed that the multipole expansion is an *asymptotic* expansion in $1/R$, because it is impossible to find a convergent series in $1/R$ for an exponential e^{-aR}. However it is much more satisfactory to separate the multipolar part of the interaction, which converges under well-defined conditions[24,27], from the exponential penetration part, for which any attempt at an expansion in powers of $1/R$ is pointless.

3.4.3 Induction Energy. The induction energy arises in perturbation theory from single excitations from an occupied orbital of either molecule to a virtual orbital on the same molecule. Like the electrostatic interaction, the induction energy is modified by overlap of the charge distributions. At present this penetration effect has not been investigated very thoroughly, but it appears that it is not very great at typical intermolecular distances[28].

3.4.4 Charge Transfer. When two molecules overlap, it becomes possible for electron density from either molecule to flow onto the other. This effect is called charge transfer. In perturbation theory terms, it is described by excitations from the occupied orbitals of one molecule to the virtual orbitals of the other. As such, it incorporates not only the genuine physical effect of charge transfer, but the BSSE. One of the virtues of IMPT is that the effects of BSSE do not arise in most of the energy terms, and the charge-transfer interaction is the only term involving single excitations in which they do occur. It is possible to correct the calculated charge-transfer energy for these effects by a procedure similar to that used in the supermolecule method.

There is a further problem with charge transfer, however. If we could perform the calculation with a complete basis set on molecule A, then it would be possible to describe any virtual orbital of B in terms of the basis set for A. In this case, the charge transfer effects would be included completely in the induction energy for molecule A. If we were then to calculate the effects of excitations from occupied orbitals of A to virtual orbitals of B we would be counting the same effects again. In practice, the basis sets used today are too small for this to happen to any great extent, but we should be aware that in principle the charge-transfer contribution is subject to this kind of double-counting error.

3.4.5 Dispersion. At short range, perturbation theory includes contributions from several types of doubly excited state of the combined system[25]. These are (i) $AB \to AB$, (ii) $AB \to AA$ and $AB \to BB$, (iii) $AA \to AB$ and $BB \to AB$, (iv) $AA \to AA$ and

$BB \to BB$, (v) $AA \to BB$ and $BB \to AA$, where for instance $AB \to AA$ means an excitation from two occupied orbitals, one on A and one on B, to virtual orbitals that are both on molecule A. Term (i) is the dispersion interaction described at long range by eq. (23). The terms (iv) involve double excitations within a single molecule, so they describe intramolecular correlation energy. The remaining terms occur only at short range. Of these, types (iii) and (v) are dominated by BSSE effects; type (ii), which is called *charge transfer correlation*, is not subject to BSSE, but is subject to the same kind of double-counting error that arises for large basis sets in the induction energy. For moderate basis sets of the type currently used, the best perturbation-theory estimate of the dispersion energy at short range is obtained by including terms of types (i) and (ii).

3.5 Limitations of perturbation theory

It would be idle to pretend that perturbation theory is able to provide definitive intermolecular potentials. Its merits are of a different kind. Because it provides separate values for the individual contributions, it gives much more insight into the nature of the intermolecular potential than the supermolecule calculation, which merely gives a total interaction energy.

A limitation of any perturbation-theory method is that it fails if the perturbation is too strong. It is evident that as the molecules approach more and more closely, the interaction (eq. (5)) becomes stronger and stronger. There is no doubt that at some point perturbation theory will fail. In practice, it appears that second-order pertubation theory gives a good account of the interaction for distances down to σ, that is to the distance at which the repulsion becomes larger than the attraction. At distances near the potential minimum, the interaction energy to second order is very close to the supermolecule result, but the method has not been tested for the very large basis sets that are used for the best supermolecule calculations.

A further limitation is less obvious. When a molecule is placed in an external field, such as the field due to another molecule, the electrons move in response to the field. Since the SCF calculation uses a Hamiltonian (the Fock operator) which is averaged over the electron distribution, the perturbation has the effect of modifying the zeroth-order Hamiltonian. It is possible in many cases to take this modification into account, to give *coupled Hartree-Fock perturbation theory*[29]. It has not however been possible to do this so far with intermolecular SCF perturbation theory, because of the complications that arise from the non-orthogonality. Accordingly, the IMPT results are those of uncoupled perturbation theory, and are inevitably less accurate.

The third weakness of intermolecular perturbation theory is one that it shares with the supermolecule method: since it is necessary to perform calculations at a wide variety of geometries, the basis sets that can be used are inevitably inadequate. This is especially true of the induction and dispersion contributions.

4. Synthetic Potentials

For this reason, I advocate a rather different approach. We can identify a number of contributions to the interaction, but we cannot calculate them accurately by IMPT. Instead we look for better ways. The general principle is that we aim to calculate the interaction as far as possible from the properties of the individual molecules. These properties can be calculated much more accurately than the potential energy surface of the complex; remember that a calculation of the energy for a single point on the surface will require twice as many basis functions, and so roughly 16 times as much computational effort, as a calculation on the monomer. Moreover the monomer calculation has to be done only once, whereas the calculation on the complex must be repeated for several hundred points.

4.1 THE ELECTROSTATIC INTERACTION

The electrostatic interaction illustrates the method. Perturbation theory gives the expression for the electrostatic energy quoted above (eq. (14)). This can be expressed, as we have seen, in terms of the distributed multipole expansion, eq. (19), which gives an accurate and rapidly-convergent formulation of the electrostatic interaction at all configurations of the complex which do not involve any overlap of the molecular charge distributions. The moments that appear in this expression are *monomer* properties. Accordingly we can use a high-quality *ab initio* calculation on the monomer to obtain the distributed multipoles, and so get a far more accurate formulation of the electrostatic interaction than we can hope to obtain either from supermolecule calculations (where the electrostatic term is just part of the undifferentiated total energy) or from perturbation theory.

4.2 THE INDUCTION ENERGY

In the same way, the induction energy expression that we obtain from perturbation theory can be re-formulated so as to depend only on properties of the individual molecules. To do this, we start from the expression (24) for the induction energy of molecule A, and use the expression (6) for the perturbation V. We evaluate the integral over the coordinates of molecule B, and arrive at the expression

$$\begin{aligned} U^A_{\text{ind}} &= -\sum_m{}' \sum_{aa'bb'} \sum_{tt'uu'} Q^b_u T^{ba}_{ut} \frac{\langle 0|\hat{Q}^a_t|m\rangle\langle m|\hat{Q}^{a'}_{t'}|0\rangle}{W^A_m - W^A_0} T^{a'b'}_{t'u'} Q^{b'}_{u'} \\ &= -\frac{1}{2} \sum_{aa'bb'} \sum_{tt'uu'} Q^b_u T^{ba}_{ut} \alpha^{aa'}_{tt'} T^{a'b'}_{t'u'} Q^{b'}_{u'}, \end{aligned} \tag{25}$$

where

$$\alpha^{aa'}_{tt'} = \sum_m{}' \frac{\langle 0|\hat{Q}^a_t|m\rangle\langle m|\hat{Q}^{a'}_{t'}|0\rangle + \langle 0|\hat{Q}^{a'}_{t'}|m\rangle\langle m|\hat{Q}^a_t|0\rangle}{W^A_m - W^A_0}. \tag{26}$$

Here $\alpha^{aa'}_{tt'}$ is a polarizability that describes the response of the moment t at site a to a perturbation (a change in potential, field, field gradient, etc.) at site a'.

Now the expression (26) is an uncoupled formulation of the polarizability. We can replace it by a polarizability derived from coupled Hartree-Fock perturbation theory. Moreover, since it is a monomer property, we can once again perform a much better calculation than is possible on the dimer. In this way we arrive at a description of the induction energy that is far more accurate than we can obtain through either intermolecular perturbation theory, where the perturbation is treated in an uncoupled fashion, or from a supermolecule calculation, where the size of the basis is limited by the need to perform calculations at a large number of points on the potential energy surface.

There is another way in which the expression (25) might be improved upon. It gives the induction energy of molecule A in the field arising from the multipole moments of molecule B. However molecule B is also polarizable, and its moments will be modified by the presence of molecule A. If this effect is taken into account, we arrive at an expression for the induction energy that is a power series in the molecular polarizabilities[30]. In practice, the effects of molecular polarization are usually calculated in an iterative fashion; the polarized moments of each molecule are evaluated in the field due to the other molecules, and the calculation is repeated until the polarized moments are self-consistent. This however is equivalent[30] to taking some, but not all, of the terms in a perturbation series to infinite order, and moreover it is known[31] that the perturbation series for the induction energy is asymptotic, i.e. divergent. This means that the conventional iterative procedure is highly questionable, and indeed it is known to lead to singularities at short range[32]. Numerical investigation[28] suggests that the simple expression (25) is more satisfactory, provided that distributed polarizabilities are used and provided that polarizabilities up to at least quadrupole rank are included.

4.3 The dispersion energy

The expression (23) for the dispersion energy can also be replaced by a more accurate expression in terms of monomer properties. The key is the replacement of the energy denominator via the Casimir-Polder identity[33], which can be established by a simple contour integration:

$$\frac{1}{A+B} = \frac{2}{\pi}\int_0^\infty \frac{AB}{(A^2+u^2)(B^2+u^2)}du, \qquad \text{for } A>0,\ B>0. \tag{27}$$

In eq. (23) we put $A = W_m^A - W_0^A$ and $B = W_m^B - W_0^B$, and then the use of eq. (27),together with the expression (6) for the perturbation, leads to the following expression for the dispersion energy:

$$U_{\text{disp}} = -\frac{\hbar}{2\pi} T^{ab}_{tu} T^{a'b'}_{t'u'} \int_0^\infty \alpha^{aa'}_{tt'}(iu)\alpha^{bb'}_{uu'}(iu)\,du, \tag{28}$$

where the polarizability at imaginary frequency $\alpha^{aa'}_{tt'}(iu)$ is given by

$$\alpha^{aa'}_{tt'}(iu) = \sum_m{}' \frac{\Delta W_m(\langle 0|\widehat{Q}^a_t|m\rangle\langle m|\widehat{Q}^{a'}_{t'}|0\rangle + \langle 0|\widehat{Q}^{a'}_{t'}|m\rangle\langle m|\widehat{Q}^a_t|0\rangle)}{(\Delta W_m)^2 + \hbar^2 u^2}, \tag{29}$$

with $\Delta W_m = W_m - W_0$. This is a much more tractable and useful formulation than it appears to be at first sight. Remember that the multipole moment operators $\widehat{Q}^a_t$ are referred to *local* axes in the molecule, so the polarizabilities defined by (29) are also referred to local axes. Consequently no information about the relative orientation of the molecules is required to evaluate the dispersion integrals in (28), and they can be obtained once and for all for any pair of molecules, using accurate coupled Hartree-Fock calculations on the monomer to obtain the polarizabilities. The need to evaluate polarizabilities at imaginary frequency is not a problem; they can be calculated just as easily as the static polarizabilities. Moreover eq. (29) shows that $\alpha_{t\,t'}^{a\,a'}(iu)$ is very well-behaved as a function of frequency (it tends monotonically to zero as $u \to \infty$) so the dispersion integrals can be evaluated accurately by numerical quadrature using the values of $\alpha_{t\,t'}^{a\,a'}(iu)$ at a dozen or so frequencies.

The expression (28) for the dispersion energy is rather cumbersome, since it involves non-local polarizabilities $\alpha_{t\,t'}^{a\,a'}(iu)$. It can be simplified to some extent[34], but at present the connection between the theoretical expression and the site-site formulation usually used to model the dispersion interaction is incomplete.

It should eventually be possible to obtain accurate polarizabilities and dispersion integrals using correlated wavefunctions. Unfortunately there are some difficulties[35], and at present the coupled Hartree-Fock calculations seem to give the most accurate results.

4.4 Short-range effects

The modification of the interaction at short range (that is, in the region where the molecular wavefunctions overlap) has been much less fully studied, but some progress has been made. The nature of the penetration effect in the electrostatic interaction is very well understood, but accurate calculations have not been made for molecular systems. A great deal of work has been done on the short-range behaviour of the dispersion energy, and it is usual to multiply the dispersion expression by a 'damping function' to cancel the singularity that otherwise occurs as the separation tends to zero. (In this limit, the dispersion interaction becomes part of the correlation energy of the united molecule, and must remain finite.) Several authors have suggested suitable damping functions[36–38].

The repulsion energy is the most important short-range effect, but as yet there is no known way to obtain it from properties of the separate molecules. An attempt has been made to characterise the repulsive surface of a molecule by performing IMPT calculations between the molecule and a suitable test particle, such as a helium atom. Because the helium atom has only one molecular orbital and is spherically symmetrical, such calculations can be done much more easily than calculations involving two ordinary molecules. From the data for the repulsion between molecule A and the test particle, and between B and the test particle, it may be possible to construct a repulsive potential between A and B. Some limited progress has been made with this idea[39].

In practice, the short-range part of the potential is obtained from supermolecule calculations. Because of the limited size of basis that can be used for the supermolecule calculation, its accuracy is limited, but the following scheme provides a way to improve on the basic supermolecule calculation:

(i) Obtain the supermolecule energy at a range of configurations.

(ii) Evaluate the long-range part of this supermolecule energy, using the multipole moments and polarizabilities obtained for the monomers using the same basis. Apply appropriate damping functions.

(iii) Subtract this long-range contribution from the total supermolecule energy. Fit the remaining part of the energy to a suitable repulsion model.

(iv) Replace the long-range terms from the supermolecule basis with improved long-range terms obtained from a better monomer basis, again with appropriate damping functions to give the correct short-range behaviour.

The question remains of what model function to use for fitting the short-range repulsion. A very successful form is the following *anisotropic atom-atom repulsion*:

$$U_{\text{rep}} = \sum_{ab} \exp\left[\alpha(\Omega_{ab})\left(R_{ab} - \rho(\Omega_{ab})\right)\right],$$

in which the shape parameter ρ_{ab} and the hardness parameter α_{ab} depend on the relative orientation Ω_{ab} of atoms a and b. (In the general case, Ω_{ab} is a short-hand for five orientation variables.) There is now overwhelming evidence[27] of the need to introduce this kind of anisotropy into atom-atom repulsions.

5. Hydrogen-bonded complexes

Hydrogen-bonded complexes are distinguished from many other complexes by the fact that the molecules are strongly polar. Accordingly, the electrostatic interaction dominates everything else, except of course the repulsion at short range. Buckingham & Fowler[40] used an accurate distributed-multipole[19] description of the electrostatic interaction, together with a crude hard-sphere model of the repulsion, and were able to predict the relative orientations of the molecules in a very wide range of hydrogen-bonded complexes. The model does not predict distances, because the hard-sphere repulsion in effect builds the atomic separations into the model at the outset.

Subsequent detailed calculations[41,42] confirmed the importance of the electrostatic interaction, and showed that the remaining contributions to the energy vary in opposite ways—as the repulsion increases, so do the induction and charge-transfer terms, since all are associated with increased overlap of the molecular wavefunctions. The dispersion energy is usually rather insensitive to angular coordinates. As a result, in many of the cases studied, the dependence of the total energy on angular coordinates follows very closely that of the electrostatic energy.

In $H_2O\cdots HF$ and $CO_2\cdots HF$, for instance, the electrostatic energy varies very little with angle, and the total energy follows it, changing by only 1 or 2 kJ mol^{-1} over a range of 40° from the symmetrical geometry. In such cases, the position of the minimum depends on a delicate balance between the various contributions to the energy, and it is very difficult to calculate the equilibrium geometry accurately. However the fact that the potential surface is very flat emerges clearly from the calculation.

Hurst *et al.* were also able to explain the few failures of the simple Buckingham-Fowler model. In ClF/HF, for instance, there are two possible structures: the hydrogen-bonded $ClF\cdots HF$, and the 'anti-hydrogen-bonded' $HF\cdots ClF$, which is the only structure observed[43]. These both correspond to minima of the electrostatic energy, but in the simple model the hydrogen-bonded minimum is deeper. However, the use of a hard-sphere repulsion model, with standard Pauling van der Waals radii but hydrogen repulsions ignored, biases the calculation towards the hydrogen-bonded structure. The $F\cdots F$ distance is larger in the hydrogen-bonded structure than the sum of two F van der Waals radii, while the $F\cdots Cl$ distance is smaller in the other structure than the sum of the F and Cl van der Waals radii. When the electrostatic calculation is done at the correct distances, the anti-hydrogen-bonded structure is found to have a lower energy, and this result is not changed when the other contributions to the energy are included.

This is an illustration of the importance of atomic shape in determining structure. A very well-established example is Cl_2, where the crystal structure, the lattice frequencies, the liquid structure, and several thermodynamic properties of the liquid, are all well predicted by a model potential which incorporates non-spherical Cl atoms[44].

Because of the importance of the electrostatic interaction in determining the angular geometry of hydrogen-bonded complexes, a good description of the molecular charge distribution is very important. Distributed Multipole Analysis[19] provides a very economical and accurate description and has proved very useful here. The procedure has been incorporated into the CADPAC program[26] and the MCSCF-CI program of Werner & Knowles[7], and requires only a tiny fraction of the time taken for the wavefunction calculation. It is usually possible to truncate the multipole representation with very little loss of accuracy. An MCSCF-CI calculation[13] on HF, for instance, gives the molecular dipole and quadrupole to within 2% of the experimental values[45]. The calculated dipole and quadrupole are reproduced to an accuracy of 0.01% by a truncated representation comprising charges of $-0.312e$ at the F atom, $-0.136e$ at the centre of the bond, and $0.448e$ at the H atom; dipoles of $0.153ea_0$ at the F atom, $-0.19ea_0$ at the centre of the bond, and $0.0837ea_0$ at the H atom (positive values correspond to dipoles in the sense $^+HF^-$); and quadrupoles of $0.47ea_0^2$ at the F atom and $0.14ea_0^2$ at the centre of the bond. No distributed moments of higher rank are needed to give overall molecular moments that are essentially in agreement with the values from the MCSCF-CI wavefunction; the errors increase from about 1% for the octopole to about 25% at rank 10. Distributed multipoles have been calculated for much larger systems, such as cyclosporin[46], and Faerman & Price have developed transferable models that can be used to describe the electrostatic interactions in peptides[47].

Other methods have also been used[20–24]. Some authors have used point-charge models of the charge distribution, obtained by fitting to the overall multipole moments of the molecule[48,49] or to the wavefunction[50]. An alternative procedure involves fitting a point-charge model so as to reproduce the molecular electrostatic potential[51], though this procedure is inherently unsound, since the potential includes the effects of penetration which cannot be described by a multipolar model of any kind. Some authors have even used Mulliken populations as a basis for constructing point-charge models, although models derived this way do not even have the correct dipole moment, because atomic contributions to the dipole are neglected.

Distributed Multipole Analysis has the advantage that the moments are obtained in a way that optimises the convergence of the multipole expansion[19]; the advantage of simplicity of manipulation that is offered by the point-charge models is offset by the larger number of sites needed for an accurate description, and is in any case less important now that efficient techniques are available for handling model potentials containing higher-rank moments[14].

6. Many-body effects

All of the discussion so far has dealt with the pair potential. In condensed phases there are many-body effects which are sometimes very important. Probably the most venerable of these is the Axilrod-Teller triple-dipole dispersion[52], which was for a long time thought to be the only significant many-body term, at least for non-polar systems such as inert gas solids. For three molecules A, B and C this takes the form

$$U_{\rm AT} = C_{\rm AT}^{ABC} \frac{3\cos\widehat{A}\cos\widehat{B}\cos\widehat{C}+1}{a^3b^3c^3}, \tag{30}$$

where $\widehat{A}$, $\widehat{B}$ and $\widehat{C}$ are the angles and a, b and c the sides of the triangle formed by the three molecules. The coefficient $C_{\rm AT}^{ABC}$ is positive, and values are accurately known for many small molecules[53]. The sign of the geometrical coefficient depends on the shape of the triangle of molecules.

More recently, it has become clear that there are many-body effects in the short-range repulsion also[54]. The three-body contribution is negative and the four-body one is positive, at the geometries investigated, so there is some degree of cancellation.

In the case of polar molecules, it has long been known that the induction energy is strongly non-additive, and this is likely to be the most important non-additive effect in such systems. In an assembly of molecules such as a liquid, we can still partition the total induction energy between the molecules, but it takes the form[30]

$$U_{\rm ind}^A = -\tfrac{1}{2}\sum_{B\neq A}\sum_{C\neq A}\sum_{bu}\sum_{cv}\sum_{aa'tt'} Q_u^b T_{ut}^{ba} \alpha_{tt'}^{aa'} T_{t'v}^{a'c} Q_v^c. \tag{31}$$

We can write this expression alternatively in the form

$$U_{\rm ind}^A = -\tfrac{1}{2}\sum_{aa'tt'} V_t^a \alpha_{tt'}^{aa'} V_{t'}^{a'}, \tag{32}$$

where

$$V_t^a = \sum_{B\neq A}\sum_{bu} Q_u^b T_{ut}^{ba} \tag{33}$$

is one of the fields at site a due to the neighbouring molecules. Because the induction energy is bilinear in the V_t^a, it is possible for interference effects to occur; the fields due to the neighbouring molecules may cancel out or reinforce each other, and in either case the induction energy is different from the sum of the induction energies that would arise from each neighbouring molecule individually. This is a particularly significant effect in hydrogen-bonded fluids, where the inductive interaction often acts cooperatively.

7. Concluding remarks

Potential energy surfaces for hydrogen-bonded complexes are now accessible to theoretical investigation, and it is possible to calculate the features of the potential in some considerable detail, and to test them against experiment. Spectroscopy provides an important experimental testing-ground, and will allow us to refine potentials very considerably over the next few years[55]. It has often been suggested that the elaborate potentials described here are too complicated to use in molecular dynamics calculations, where speed of evaluation is paramount. It has now been shown that this view is untenable; molecular dynamics calculations with today's more elaborate potentials are not merely feasible but very efficient[44]. Indeed they are essential in any case, in view of the importance of features like anisotropic atom-atom repulsion, but in fact they are competitive with more conventional and much cruder models. We should acknowledge the important part that simple models like the Lennard-Jones potential have played in the development of our ideas about intermolecular interactions and the structure of liquids and solids, but it is now possible to move on to more realistic potentials.

References

[1] McWeeny, R. & Sutcliffe, B. T., *Methods of Molecular Quantum Mechanics* (Academic Press, London, 1969); McWeeny, R., *Methods of Molecular Quantum Mechanics* (Academic Press, London, 1989).

[2] Binkley, J. S., Pople, J. A. & Hehre, W. J. (1980) *J. Amer. Chem. Soc.* **102**, 939; Gordon, M. S., Binkley, J. S., Pople, J. A., Pietro, W. J. & Hehre, W. J. (1982) *J. Amer. Chem. Soc.* **104**, 2797.

[3] Dunning, T. H. (1970) *J. Chem. Phys.* **53**, 2823; Dunning, T. H. & Hay, P. J., in *Modern Theoretical Chemistry*, edited by H. F. Schaefer (Plenum, 1977), p. 1–27.

[4] Hariharan, P. C. & Pople, J. A. (1973) *Theor. Chim. Acta* **28**, 213; Francl, M. M., Pietro, W. J., Hehre, W. J., Binkley, J. S., Gordon, M. S., DeFrees, D. J. & Pople, J. A. (1982) *J. Chem. Phys.* **77**, 3654.

[5] Werner, H.-J. & Meyer, W. (1976) *Molec. Phys.* **31**, 855–872.

[6] van Duijneveldt-van der Rijdt, J. G. C. M. & van Duijneveldt, F. B. (1982) *Theochem.* **89**, 185.

[7] Werner, H.-J. & Knowles, P. J. (1985) *J. Chem. Phys.* **82**, 5053; Knowles, P. J. & Werner, H.-J. (1985) *Chem. Phys. Lett.* **115**, 259.

[8] Price, S. L. & Stone, A. J. (1980) *Molec. Phys.* **40**, 805.

[9] Pople, J. A., Head-Gordon, M. & Raghavachari, K. (1988) *J. Chem. Phys.* **87**, 5968.

[10] Bartlett, R. J. (1989) *J. Phys. Chem* **93**, 1697.

[11] Boys, S. F. & Bernardi, F. (1970) *Molec. Phys.* **19**, 553.

[12] Gutowski, M., van Lenthe, J. H., Verbeek, J., van Duijneveldt, F. B. & Chałasiński, G. (1985) *Chem. Phys. Lett.* **124**, 370.

[13] Knowles, P. J., private communication.

[14] Stone, A. J., in *Theoretical Models of Chemical Bonding, vol. 4*, edited by Z. B. Maksić (Springer, 1989).

[15] Stone, A. J. & Tough, R. J. A. (1984) *Chem. Phys. Lett.* **110**, 123.

[16] Brink, D. M. & Satchler, G. R., *Angular Momentum* (Clarendon Press, Oxford, 1968).

[17] Price, S. L., Stone, A. J. & Alderton, M. (1984) *Molec. Phys.* **52**, 987.

[18] Buckingham, A. D. (1967) *Adv. Chem. Phys.* **12**, 107.

[19] Stone, A. J. (1981) *Chem. Phys. Lett.* **83**, 233; Stone, A. J. & Alderton, M. (1985) *Molec. Phys.* **56**, 1047.

[20] Pullman, A. & Perahia, D. (1978) *Theor. Chim. Acta* **48**, 29.

[21] Rico, J. F., & Alvarez-Collado, J. R. & Paniagua, M. (1985) *Molec. Phys.* **56**, 1145.

[22] Cooper, D. L. & Stutchbury, N. C. J. (1985) *Chem. Phys. Lett.* **120**, 167.

[23] Sokalski, W. A. & Sawaryn, A. (1987) *J. Chem. Phys.* **87**, 526.

[24] Vigné-Maeder, F. & Claverie, P. (1988) *J. Chem. Phys.* **88**, 4934.

[25] Hayes, I. C. & Stone, A. J. (1984) *Molec. Phys.* **53**, 69; Hayes, I. C. & Stone, A. J. (1984) *Molec. Phys.* **53**, 83; Hurst, G. J. B., Hayes, I. C. & Stone, A. J. (1984) *Molec. Phys.* **53**, 107.

[26] Amos, R. D. & Rice, J. E., CADPAC: The Cambridge Analytical Derivatives Package, issue 4.0, Cambridge, 1987.

[27] Stone, A. J. & Price, S. L. (1988) *J. Phys. Chem* **92**, 3325.

[28] Stone, A. J. (1989) *Chem. Phys. Lett.* **155**, 111.

[29] Stevens, R. M., Pitzer, R. & Lipscomb, W. N. (1963) *J. Chem. Phys.* **38**, 550; Gerratt, J. & Mills, I. M. (1968) *J. Chem. Phys.* **49**, 1719.

[30] Stone, A. J. (1989) *Chem. Phys. Lett.* **155**, 102.

[31] Dalgarno, A. & Stewart, A. L. (1956) *Proc. Roy. Soc.* **A 238**, 276; Dalgarno, A. & Lynn, N. (1957) *Proc. Phys. Soc. London, A* **70**, 223.

[32] Buckingham, A. D. & Pople, J. A. (1955) *Trans. Faraday Soc.* **51**, 1173.

[33] H. B. G. Casimir and D. Polder (1948) *Phys. Rev.* **73**, 360.

[34] Stone, A. J., *Chem. Phys.*, submitted for publication.

[35] Visser, F., Wormer, P. E. S. & Stam, P. (1983) *J. Chem. Phys.* **79**, 4973; Visser, F., Wormer, P. E. S. & Jacobs, W. P. J. H. (1984) *J. Chem. Phys.* **82**, 3753; Visser, F. & Wormer, P. E. S. (1984) *Molec. Phys.* **52**, 723.

[36] Douketis, C., Scoles, G. Marchetti, S. & Thakkar, A. J. (1982) *J. Chem. Phys.* **76**, 3057.

[37] Tang, K. T. & Toennies, J. P. (1984) *Chem. Phys.* **80**, 3276.

[38] Knowles, P. J. & Meath, W. J. (1986) *Chem. Phys. Lett.* **124**, 164; *Molec. Phys.* (1986) **59**, 965; *Molec. Phys.* (1987) **60**, 1143.

[39] Stone, A. J. & Tong, C.-S., in preparation.

[40] Buckingham, A. D. & Fowler, P. W. (1983) *J. Chem. Phys.* **79**, 6426; *Canad. J. Chem.* (1985) **63**, 2018.

[41] Rendell, A. P. L., Bacskay G. B. & Hush, N. S. (1985) *Chem. Phys. Lett.* **117**, 400.

[42] Hurst, G. J. B., Fowler, P. W., Stone, A. J. & Buckingham, A. D. (1986) *Int. J. Quantum Chem.* **29**, 1223.

[43] Janda, K. C., Klemperer, W. & Novick, S. E. (1976) *J. Chem. Phys.* **64**, 2698; Novick, S. E., Janda, K. C. & Klemperer, W. (1976) *J. Chem. Phys.* **65**, 5115.

[44] Rodger, P. M., Stone, A. J. & Tildesley, D. J. (1987) *J. Chem. Soc., Faraday Transactions II* **83**, 1689; *Molec. Phys.* (1988) **63**, 173; *Chem. Phys. Lett.* (1988) **145**, 365.

[45] Gray, C. G. & Gubbins, K. E., *Theory of Molecular Fluids, vol. 1* (Clarendon Press, Oxford, 1984).

[46] Price, S. L., Harrison, R. J. & Guest, M. F. (1989) *J. Comput. Chem*, in press.

[47] Faerman, C. H. & Price, S. L., in preparation.

[48] Murthy, C. S., O'Shea, S. F. & McDonald, I. R. (1983) *Molec. Phys.* **50**, 531.

[49] Brobjer, J. T. & Murrell, J. N. (1982) *Chem. Soc. Faraday Trans. II* **78**, 1853; *Molec. Phys.* (1983) **50**, 885.

[50] Hall, G. G. (1986) *Int. Rev. Phys. Chem.* **5**, 115; Hall, G. G. & Smith, C.M. (1986) *Theor. Chim. Acta* **69**, 71; Hall, G. G. & Tsujinaga, K. (1986) *Theor. Chim. Acta* **69**, 425; Tsujinaga, K. & Hall, G. G. (1986) *Theor. Chim. Acta* **70**, 257.

[51] Momany, F. A. (1978) *J. Phys. Chem* **82**, 592; Cox, S. R. & Williams, D. E. (1981) *J. Comput. Chem.* **2**, 304; Singh, U. C. & Kollman, P. A. (1984) *J. Comput. Chem.* **5**, 129.

[52] Axilrod, P. M. & Teller, E. (1943) *J. Chem. Phys.* **11**, 299; Muto, Y. (1943) *Proc. Phys. Math. Soc. Japan* **17**, 629.

[53] Kumar, A. & Meath, W. J. (1984) *Chem. Phys.* **91**, 411.

[54] Wells, B. H. (1987) *Molec. Phys.* **61**, 1283; Wells, B. H. & Wilson, S. (1983) *Chem. Phys. Letters* **101**, 429; *Molec. Phys.* (1985) **55**, 199; (1986) **57**, 21; (1986) **57**, 421; (1989) **65**, 1363; (1989) **66**, 457.

[55] Clary, D. C., Lovejoy, C, M., ONeil, S. V. & Nesbitt, D. J. (1988) *Phys. Rev. Letters* **61**, 1576; *J. Chem. Phys.*, in press.

Quantum Effects in Hydrogen Bonded Liquids

Peter J. Rossky
Department of Chemistry
The University of Texas at Austin
Austin, Texas 78712 U. S. A.

ABSTRACT Discretized path integral simulation methods, supplemented by a wavefunction-based representation, has been applied to the determination of aspects of the structure and spectroscopy in two quantum mechanical aqueous systems. The first of these applications is the determination of the consequences of quantizing the molecular degrees of freedom of the water molecules in the pure room temperature liquid. The results provide a quantitative estimate of the significance of approximating such a system as classical and also of the size of isotope effects on the liquid structure. The latter exhibit quantitative differences from recent experiments, indicating limitations in the model used for the solvent. Second, aspects of the structure and spectroscopy of the hydrated electron are considered. Here, we treat the water classically but treat the electron quantum mechanically. The excess electron density and solvent distribution are shown to exhibit structural similarities to ionic solvation. However, it is found that the electronic state fluctuates in response to the fluid, and that these fluctuations have essential consequences for the spectroscopy of the species. In both examples, the high frequency librational motions, and the associated fluctuating fields, that are manifestations of the light hydrogen mass, polarity, and strong restoring forces characteristic of molecular displacements in hydrogen bonded liquids are found to be key ingredients in the description of the phenomena.

1. Introduction

Classical computer simulation has become a powerful and widely used tool for the elucidation of the molecular level behavior of water and aqueous solutions. Nevertheless, it is clear that quantum mechanical effects are an essential element for certain phenomena. In particular, in hydrogen bonded liquids, and water in particular, the light moments of inertia and polarity associated with the hydrogen-bonding group, combined with the relatively high effective force constants for molecular displacements, lead to relatively high frequency intermolecular motions that are also associated with relatively large fluctuations in the local fields in the liquid. The two examples in which these features play an important role that we discuss briefly here include bulk solvent properties and the solution structure of the solvated electron. In the latter case, it will be clear that the strong response of the electronic state to the fluctuating solvent field lies at the foundation of several solute properties.

For the bulk solvent, we consider isotopic substitution effects on liquid structure. In the absence of quantum effects, H_2O and D_2O would behave identically. However, such quantities as the melting point and temperature of maximum density are higher by several degrees for liquid D_2O than for H_2O.

The source of the dominant contributions to such quantum behavior can be anticipated. Liquid water, under the usual ambient conditions, is characterized by a space

J.C. Dore and J. Teixeira (eds.), Hydrogen-Bonded Liquids, 49–57.

filling network of relatively strong and directional intermolecular hydrogen bonds that exhibits a continuum of intermolecular vibrational motions. Due to the low moments of inertia of water, the highest frequency motions of this type (and hence the most quantum mechanical) are hindered rotational motions, or librations, which distort the linearity of the hydrogen bonds[2]. From either classical simulations [1] or experimental spectra [2], one can see that such motions cover a very broad range of frequencies which for H_2O is centered at around $\omega \cong 600\,cm^{-1}$ and for D_2O closer to $400\,cm^{-1}$. Hence, for a typical librational degree of freedom at 25°C, $\beta\hbar\omega \sim 2$ where β is the inverse of the product of Boltzmann's constant and the temperature and $2\pi\hbar$ is Planck's constant. Typical hindered translational motions occur at much lower frequencies and are, thus, not very sensitive to isotopic substitution.

The goal of the present pure water studies [3] was to provide a quantitative estimate of the structural ramifications of quantization of the intermolecular degrees of freedom and the corresponding zero-point amplitudes which are present. The internal vibrations of the molecules are of a sufficiently high frequency and sufficiently small amplitude that one expects that to a good approximation one can employ a rigid body description that implicitly accounts for averaging over these internal degrees of freedom [4].

While the bulk solvent properties represent a case with many moderately quantum mechanical degrees of freedom, the alternative extreme is that of a single highly quantum mechanical solute in aqueous solution. The example we consider here is that of an excess electron solvated in water. Such hydrated electrons have been the object of a great deal of experimental and theoretical analysis [5], but direct structural observation has been quite limited [6]. Hence, our primary goal in the study of this system was to characterize this structure at a level comparable to that now available for atomic ions in water [7,8], and then proceed to evaluate the physics underlying the spectroscopy of the species [5,9].

As in the case of pure water, the origin of the basic phenomenon is clear [5]. The electron is localized in the liquid by a combination of effective repulsive interactions with the solvent (resulting from the orthogonality requirements between the molecular and excess electronic wavefunctions) and attractive, primarily electrostatic, solvent-electron terms. The electronic and solvent distributions are then primarily determined by a balance between the sum of solvent-electron and solvent-solvent interaction energies, and the large electronic kinetic energy produced by the confinement of the electron to a relatively small volume.

The physics underlying the broad, featureless optical spectrum of the hydrated electron is another long standing issue [5]. The objective is to describe the excited state manifold which is manifest in the spectrum and to ascertain the relative contributions to the lineshape and intensity of such a progression of excited states compared to that of inhomogeneous broadening of individual transitions. A fully molecular calculation can reveal such detail without the bias inherent in more *ad hoc* models.

With this background on the problems to be discussed here, we outline in the next section (Section 2) some elements of the models and methods used to carry out the structural studies. We use a discretized path integral representation of the quantum degrees of freedom [10,11] to evaluate equilibrium thermal distributions and related properties via computer simulation. This approach has already been demonstrated to be a productive route for the study of important chemical systems [11]. This is supplemented by a wavefunction description of the ground and excited states [9,12] in order to evaluate spectroscopic properties. Section 3 outlines representative results for the pure water studies, while Section 4 includes results on hydrated electron structure and spectroscopy. The conclusions are presented in the final section.

We note here that very recent results from our laboratory describing aspects of electron localization dynamics [9], ultrafast transient spectroscopy [13] and the diffusive motion of the hydrated electron [14] , obtained using a ground state adiabatic dynamics algorithm, underscore the emphasis placed here on the importance of the coupling of the electronic motion to the fluctuations in the local field produced by the solvent, but we will not discuss these examples in the present article.

2. Models and Methods

There are basically three elements to the molecular-level simulation of any system. Namely, the prescription of a set of intermolecular or interparticle potential functions, the assembly of an appropriate molecular system, and the implementation of an appropriate sampling algorithm, either dynamical or essentially statistical in nature [15]. For the most part, these elements are common to both classical and quantum mechanical simulation. Typically, the models used to describe liquid water are rigid body models. They have the now standard form consisting of a spherical short-ranged Lennard-Jones interaction and a set of embedded partial electrostatic charges arranged on the nuclear framework [16-18]. For the studies of pure water, we have used the four point charge ST2 model [16], which places two partial positive charges at the proton positions and two negative charges in the locations characteristic of the positions of lone pair orbitals. The hydrated electron studies have employed the computationally more efficient SPC water model [18] to describe water-water interactions; here the total counterbalancing negative charge is localized on the single oxygen site. Both models provide reasonable descriptions of pure water structure around room temperature, although the former is somewhat too structured.

The area of electron-solvent interaction potentials is much less mature. In a practical calculation, it is necessary to treat the electrons associated with solvent molecules in the simulation implicitly and develop a so-called pseudopotential description for the electron-molecule interaction [19]. Such a potential function, well known for electron-atom interactions in solid state theory, treats the contributions from a fixed "core" of electrons in terms of an effective potential acting on the "valence" set [20]. In the present context, the pseudopotential represents the solvent electrons (and associated nuclei), and the "valence" electron is solely the single excess electron present in the solvent. In our pseudopotential, the solvent contribution is taken to be that obtained from each water molecule individually in its electronic ground state, and we do not explicitly consider any electronic relaxation contributions.

The details of the potential have been given in the literature and we do not repeat that here [19]. The potential includes three contributions. The first is a purely electrostatic term which is taken to be that produced by the charge distribution of the SPC water model. The second term is a spherically symmetric polarization term taken from electron-molecule scattering technology [21]. The third and most subtle term is an effective, repulsive, potential included to account for the requirements of orthogonality between the one-electron wavefunction describing the excess electron and those comprising the water molecular wavefunction. Our potential is analogous, on an atom by atom basis, to that used some time ago to describe the electron-helium interaction [22]. We note that several other pseudopotentials for electron-water interactions (and for the closely related ammonia system) have also been considered [23-26].

To employ these potentials in the structural studies of interest here, we use discretized path integral simulations [11]. Due to the limitations of space, we do not describe these in detail here.

The quantity of first interest is the thermal distribution of electron and/or solvent positions which follows from the Hamiltonian. It has been shown [10,11] that for this case, the distribution is equivalent to the results obtained for a completely *classical* system in which the quantum particle is replaced, in the simulation, by a cyclic chain polymer consisting of P (pseudo)particles, each connected by a pair potential to its two nearest neighbors and interacting with the other particles via the potentials described above, but reduced by a factor of P. A simulation of this classical system provides the desired spatial distributions if one interprets the polymer pseudoparticle distribution as the thermal quantum density, i.e., the spatial diagonal elements of the density matrix.

For the hydrated electron studies, we have treated the solvent *purely classically,* so that only the electron is represented by such a polymer. For pure water, each molecule in the liquid is represented by such a polymer, with each point on the polymer including both center of mass and orientational degrees of freedom [3].

All simulations are carried out for systems near room temperature. For pure water, we have considered the liquids only at fixed number density, so that the structural effects due to quantization alone will not be clouded by those resulting from the density changes that would also follow if the structures were studied at constant pressure.

To evaluate spectroscopic results for the electron discussed in this paper, we evaluate the electronic ground and excited state wavefunctions following from the set of potential surfaces sampled during the path integral simulation, as described in complete detail elsewhere [9,12]. From these states, we evaluate the excitation energies and transition dipole matrix elements.

3. Pure Water Structure

In this section we outline selected results pertaining to quantum effects on the structure of pure water at 25°C. We consider H_2O, D_2O, and classical systems. The latter may be viewed, for comparative purposes, as the large mass limit of the H_2O system. These results are discussed in detail elsewhere [3].

In the introduction we emphasized the significance of the zero point amplitude of hindered rotational degrees of freedom, so it is reasonable to expect that the degree of linearity of hydrogen bonds will increase with increasing mass. The results [3] show that the hydrogen bonds are indeed more linear in D_2O than H_2O, and, correspondingly, the classical liquid has the most linear hydrogen bonds of the three cases. The immediate ramification of such distortions is the weakening of typical pair interactions among molecular neighbors as the quantum amplitude increases, and the distribution of pair interaction energies found in the various isotopic liquids reflects this anticipated trend as well.

We next turn to intermolecular radial correlations, and focus in particular on the correlation between the molecular centers at the oxygen atoms. The relative probabilities $g_{OO}(r)$ for the three liquids show the trend observed in all the results, namely that the classical liquid is the most "structured" and H_2O is the least "structured." Although the amplitude of the changes is small (several percent), the maxima decrease, the minima are less intense, and, of most interest, the extrema occur at larger distances as the systems become more quantum mechanical. The behavior observed for $g_{OO}(r)$ is also reminiscent of the changes in water structure as the temperature is raised [16,27,28], a point we will return to shortly.

The second nearest neighbor peak (at about 4.8Å) occurs at a distance characteristic of tetrahedral molecular coordination. The shift in the relative position of the nearest neighbor peak at about 2.8Å and this second neighbor peak with increasing quantum character is then indicative of a weakening in the directionality of the intermolecular interactions.

These shifts also provide an experimentally accessible test of the results obtained in the simulation. Following the simulations just described, an elegant experiment was carried out by Egelstaff and coworkers [29] that was designed specifically to look at the isotopic scattering difference (H_2O - D_2O) for electromagnetic radiation. In this case, they used a gamma ray source, which has a number of advantages in the present context [29].

They find that the differences in the vicinity of the first (radial) peak in the transformed structure factor (predominantly $g_{OO}(r)$) are rather close to the simulated result, although the latter difference is somewhat larger. Of particular interest, however, the simulated effect on the second neighbor peak compares less well with the experiment. In particular, the experiment, at the least, indicates a much weaker effect, and the data have been interpreted [29] to imply a small *inward* shift of the peak for the more quantum mechanical liquid. This is the reverse of the result seen here for ST2 water [3]. It is interesting to note in this context that for classical liquid models of water, studies [30] have indicated that one can also observe such apparently quite different behavior for two different water models under the influence of a temperature increase. In light of the comments made above relating temperature effects to quantization, it seems likely that these subtle diffences can be manifest also in the quantum effects. Hence, a theoretical analysis corresponding to that outlined here, but for a series of models would be quite informative.

It is of substantial interest to obtain a familiar measure by which to assess the significance of quantum effects. As already indicated above a useful measure might be the temperature scale. A direct structural comparison of our results with the effects of temperature as observed experimentally via X-ray diffraction [31] leads to the conclusion that the experimental result for a 51°C temperature change and the H_2O, quantum minus classical, result are in general agreement, suggesting that the analogy between quantization effects and temperature influence is reasonable. Hence, the use of temperature differences to conceptualize quantum effects in this case is a productive route.

A related comparison has been made by Egelstaff and coworkers [29] using their gamma ray data. They find that the quantum effect (H_2O - D_2O) is apparently not as well represented by a simple temperature scaling as are the simulation data. However, they also find that a comparison of the isotopic difference in water structure to a thermodynamic difference obtained from structural data for a single isotope is useful. They find that a translation by a temperature difference of about 5^0 and by a small (0.5%) density change yield data comparable to their experiment. Hence, the concept of a thermodynamic view of the quantization effect is borne out, at least qualitatively.

4. Hydrated Excess Electron

In this section, we present some illustrative results obtained from the path integral simulation of an excess electron in bulk (*classical*) liquid water. Here, we emphasize results for the solvent orientational and radial distribution with respect to the quantum electron, and for the steady state optical spectroscopy.

4.1 SOLUTION STRUCTURE

The solvent orientational structure can be analyzed in much the same way as for ordinary atomic solutes. It is found that the solvent is bond oriented and distributed very much as it is around a negative atomic ion [7]. The radial distribution of solvent with respect to the electronic center of mass is clearly different in first appearance from ion-like behavior. For a hydrated ion, the radial correlations manifest sharply defined solvation layer structure [7,8], but for the electron both hydrogen and oxygen peaks are strongly broadened. Nevertheless, the electron is solvent coordinated in an ionic-like manner; the coordination number obtained from integration of the oxygen radial correlation function is about six. These results compare reasonably with direct structural analysis using electron spin echo measurements in an aqueous glass [6].

Considering the well defined orientational correlations found and the reasonable, and relatively small, coordination number observed, it is reasonable to attribute the apparent diffuse nature of the radial correlations to fluctuations in the shape and radius of the excess electron, rather than to a *lack* of structure. It is in this respect that the solution containing an excess electron differs most from that with a simple ion. The electron exhibits the solvation structure that would be expected of an ion which had the additional freedom to fluctuate in size and shape, while remaining compact and roughly spherical. This will be manifest further below.

4.2. STEADY-STATE OPTICAL SPECTRUM

The physical description underlying the observed broad and featureless optical absorption spectrum of the hydrated electron is a long standing issue [5] that can be directly addressed by the present theoretical studies. The calculations [9,12,31] consist of the determination of the ground electronic state and first nine excited states in the Born-Oppenheimer approximation for each of 600 configurations of the solvent sampled from the path integral simulation. The spectrum is then the envelope of the 5400 lines with intensities proportional to the corresponding electronic transition dipole matrix elements.

These results [9,31] show that the calculated spectrum, in fact, reflects the exceptional breadth and also the asymmetry evident in the experimental result [32], indicating that an analysis of the origins of these observations via the theoretical results is well founded.

The nature of the electronic states can be directly determined without calculation of the spectrum. It is found that, almost without exception, a roughly spherical s-like localized ground state is followed at higher energy by a triple of p-like states that are also bound and localized. Above the p-like states lies a band of apparently unbound delocalized states of indefinite symmetry.

The spectrum can then be understood in terms of strongly allowed excitations from the ground state to the three p states and to the higher energy band. If the three p states are ordered by energy, the spectrum can then be further decomposed into the contributions from each excitation. If the potential surface were spherically symmetric, then these three contributions would be equivalent due to the corresponding p-state degeneracy, although each excitation would still be broadened by any fluctuations in the radial potential.

The spectrum observed is dominated by the three excitations to p states, as expected. The excitation energies are, however, substantially non-degenerate, and each excitation is, further, substantially inhomogeneously broadened by the variability of the solvent surroundings. In fact, the deviation among the most probable excitation energies for the three s-p transitions contributes comparably to the breadth of each transition in determining the overall spectrum. Thus it is immediately evident that fluctuations in the solvent configuration are of critical importance to the spectral behavior.

One can then directly examine the correlation between relatively simple measures of the shape of the potential surface and the observed spectroscopic energies. (The magnitude of the transition dipole moments for the dominant s to p transitions is found to be insensitive to these considerations.) We measure the "radial" size of a solvent cavity via the radius of gyration of the ground state electronic density and the deviation from spherical symmetry via the corresponding moment of inertia tensor.

The three p states would be degenerate for a spherically symmetric, purely radial, potential, and one finds that the *average* s-p excitation energy is rather accurately predictable from the average radial extent of the ground state wavefunction. Although the width of the distribution of radii is only about 0.2 Å, or about 10% of the average value, the corresponding width in the distribution of the average excitation energy is about 0.6 eV.

The role of asymmetry in making up the remaining width lies in the energetic splitting among the p-states. When correlated against the asymmetry in the moment of inertia tensor, one again finds very good following. However, as for the radius, the typical deviation from spherical symmetry is small, corresponding to only about 8% variation in the axis lengths characterizing the electronic distribution. Nevertheless, the most probable configurations have sizeable splittings; the most probable result is 0.8 eV.

In summary, then, it is evident in this case that the lineshape for the optical spectrum critically reflects the fluctuations in the solvation structure of the quantum solute, and that these fluctuations are intimately related to the hydrogen-bonded character of the liquid.

5. Conclusions

Quantum mechanical simulation methods can now provide insight into quite complex quantum chemical systems which is presently not available via direct experimental probe. This ability is clearly expanding rapidly with emerging new techniques and new computer facilities. Hydrogen bonded liquids present an environment which is particularly rich in such quantum phenomena, both in the pure phases and in the context of solvation.

From the present examples, it is clear that although both detail and insight are emerging from such new studies, many elements of aqueous system behavior and its modelling remain to be explored. For example, the resolution of the calculated results for pure water with those now measured, by consideration of various interaction potentials, is currently being pursued in our laboratory. At the same time, the area of quantum simulation in chemical systems is still quite new, and the study of electron solvation is only a first step in a detailed understanding electronic structure in solution. Based on the already rapid growth of this field, increasingly demanding cases will no doubt be accessible to comparable analysis in the near future.

Acknowledgements

The work reported here was supported by grants from the National Institute of General Medical Sciences and the Robert A. Welch Foundation, and with the generous computational support of the University of Texas System Center for High Performance Computing. P.J.R. is the recipient of an NSF Presidential Young Investigator Award and a Dreyfus Foundation Teacher-Scholar Award.

References

1. Rahman, A.and Stillinger, F.H., *J. Chem. Phys.* **1971,** *55*, 3336.
2. Eisenberg, D. and Kauzmann, W., "The Structure and Properties of Water"; Oxford Press, London, 1969.
3. Kuharski, R. A. and Rossky,P.J., *J. Chem. Phys.* **1985,** *82*, 5164 .
4. Stillinger,F.H. *Adv. Chem. Phys.* **1975**, *31*, 1.
5. For reviews see (a) Hart, E. J.; Anbar, M. "The Hydrated Electron"; Wiley: New York, 1970; (b) "Metal-Ammonia Solutions", Lagowski, J. J.; Sienko, M. J., Eds.; Butterworths: London, 1970; (c) "Electrons in Fluids", Jortner, J.; Kestner, N. R., Eds.; Springer: Berlin, 1973; (d) "Electron-Solvent and Anion-Solvent Interactions", Kevan, L.; Webster, B. C., Eds.; Elsevier: Amsterdam, 1976; (e) Thompson, J. C. "Electrons in Liquid Ammonia", Clarendon: Oxford, 1976; (f) Schindewolf, U. *Angew. Chem. Intern. Ed.* **1978**, *17*, 887; (g) Webster, B. C. *Annu. Rep. Progr. Chem. Sec. C* **1979,** *76*, 287; (h) Feng, D.-F.; Kevan, L. *Chem. Rev.* **1980,** *80*, 1.
6. Kevan, L. *J. Phys. Chem.* **1981**, *85*, 1628.
7. Mezei, M. and Beveridge, D. L., J. Chem. Phys.**1981,** *74*, 6902; Chandresekhar, J., Spellmeyer, D. C.,and Jorgensen, W. L., *J. Am. Chem. Soc.* **1984**,*106*, 903.
8. Impey, R.W., Madden, P.A., and McDonald, I.R., *J. Phys. Chem.* **1983**,*87*, 5071
9. Rossky, P.J., and Schnitker, J.S., *J. Phys. Chem.* **1988**, *92*, 4277.
10. Feynman, R. P. "Statistical Mechanics"; Benjamin: Reading, 1972.
11. Berne, B. J.; Thirumalai, D. *Annu. Rev. Phys. Chem.* **1986**, *37*, 401.
12. Motakabbir, K.A., and Rossky, P.J., *Chem. Phys.* **1989,** *129*, 253.
13. Motakabbir, K.A., Schnitker, J., and Rossky, P.J., *J. Chem. Phys.* (in press).
14. Schnitker, J. and Rossky, P. J., *J. Phys. Chem.* (in press).
15. (a) Articles by Valleau, J. P.; Whittington, S. G.; by Valleau, J. P.; Torrie, G. M.; by Erpenbeck, J. J.; Wood, W. W.; and by Kushick, J.; Berne, B. J. in "Statistical Mechanics, Parts A and B", ed. by Berne, B. J. ; Plenum: New York, 1977; (b) Allen, M. P.; Tildesley, D. J. "Computer Simulation of Liquids", Clarendon: Oxford, 1987.
16. Stillinger, F.H.and Rahman, A., *J. Chem. Phys.* **1974,***60*, 1545 .
17. Jorgensen, W. L.; Chandrasekhar, J.; Madura, J. D.; Impey, R. W.; Klein, M. L. *J. Chem. Phys.* **1983**, *79*, 926.
18. Berendsen, H. J. C.; Postma, J. P. M.; Van Gunsteren, W. F.; J. Hermans, in "Intermolecular Forces", Pullman, B., Ed.; Reidel: Dordrecht, 1981; p. 331.
19. Schnitker, J.; Rossky, P. J. *J. Chem. Phys.* **1987**, *86*, 3462.
20. Szasz, L. "Pseudopotential Theory of Atoms and Molecules"; Wiley: New York, 1985.

21. Gianturco, F. A.; Thompson, D. G. *J. Phys. B* **1980**, *13,* 613.
22. Kestner, N. R.; Jortner, J.; Cohen, M. H.; Rice, S. A. *Phys. Rev.* **1965**, *140,* A56.
23. Jonah, C. D.; Romero, C.; Rahman, A. *Chem. Phys. Lett.* **1986**, *123*, 209.
24. Sprik, M.; Impey, R. W.; Klein, M. L. *J. Stat. Phys.* **1986**, *43*, 967.
25. Wallqvist, A.; Thirumalai, D.; Berne, B. J. *J. Chem. Phys.* **1987**, *86*, 6404.
26. Barnett, R. N.; Landman, U.; Cleveland, C. L.; Jortner, J. *J. Chem. Phys.* **1988**, *88*, 4421.
27. Mezei, M. and Beveridge, D. L., *J. Chem. Phys.* **1981**,*74,* 62 .
28. Narten, A.H., Danford, M.D. and Levy, H.A., *Discuss. Faraday Soc.* **1967,** *43*, 97.
29. Root, J. H., Egelstaff, P. A., and Hime, A., *Chem. Phys.* **1986**, *109*, 437.
30. Impey, R. W., Klein, M. L., and McDonald, I. R., J. *Chem. Phys.* **1981,** *74*, 647.
31. Bosio, L., Chen, S., and Teixeira, J., *Phys. Rev. A.* **1983**, *27*, 1468 .
31. Schnitker, J.; Motakabbir, K.; Rossky, P. J.; Friesner, R. A. *Phys. Rev. Lett.* **1988**, *60*, 456.
32. Jou, F.-Y.; Freeman, G. R. *J. Phys. Chem.* **1979**, *83*, 2383.

TRANSPORT PROCESSES, RELAXATION, AND GLASS FORMATION IN HYDROGEN-BONDED LIQUIDS

C. A. Angell
Chemistry Department
Arizona State University
Tempe, AZ 85287-1604, U.S.A.

ABSTRACT. The phenomenon of high viscosity and glass formation in supercooling hydrogen-bonded liquids is compared with that in other liquids, and peculiarities which may be associated with the presence of the specific hydrogen-bonding interaction are singled out. In the first part of the paper, these are discussed in the context of "strength" and "fragility" in liquids and linked to specific features of the topology of the potential energy hypersurfaces for such liquids. In the second part, a higher level of complexity in the behavior of certain hydrogen-bonding liquids, such as the normal alcohols and water is discussed and interpreted in terms of hydrogen bond-driven cluster formation. Large clusters of water molecules formed spontaneously in microemulsion systems are then examined spectroscopically to distinguish surface from bulk hydrogen bonds. Finally, the role of hydrogen bonding and glass formation in preservation of life in simple organisms under hostile conditions is briefly examined.

1. VISCOUS LIQUID AND GLASSY STATES OF HYDROGEN-BONDED LIQUIDS

1.1 Introduction: The Glass Transition and The Entropy Conundrum

The challenge to liquid state theory offered by the phenomenon of supercooling and glass formation has motivated much current research activity (1-7). In the extended temperature range opened up by supercooling, a variety of new phenomena are encountered. In particular, there emerges a remarkable divergent behavior in the viscosity, and in the relaxation time measured by dielectric, NMR, and light-scattering techniques. The most detailed studies of viscous liquid and glass transformation behavior have been made on liquids which are characterized by extensive hydrogen bonding, particularly glycerol and the diols. This is mainly because of their more convenient temperature ranges of liquid-like behavior, and their great resistance to crystallization.

J.C. Dore and J. Teixeira (eds.), Hydrogen-Bonded Liquids, 59–79.

An interesting aspect of supercooling in hydrogen bonded liquids is that the general rule for predicting failure to crystallize during cooling at normal rates, viz., that the liquid has a boiling point-to-melting point ratio T_b/T_m, greater than 2.0, (8,9) does not apply. Frequently, hydrogen-bonded liquids are found to be very viscous at their melting points and to supercool readily even though the melting point may be as large as 2/3 of the boiling points (i.e., T_b/T_m is only 1.5). This is presumably due to the additional resistance to the formation of critical size crystal nuclei in the body of the liquid arising from the presence of hydrogen bonds between molecules. A general discussion of crystallization kinetics, and the various devices used by physical chemists to inhibit crystalization, is given elsewhere, (9,10) and will not be discussed further in this article.

In the event that crystallization does not occur during cooling of a liquid, it is observed at some lower temperature that a sudden change in the heat capacity characteristic of the liquid state occurs at a temperature usually about 2/3 of the absolute melting point. This contrasts with the behavior in the vicinity of the melting point itself where no sign whatsoever of the passage from stable to metastable state can be seen. (Typical behavior is illustrated in Fig. 1 part a, using data for the case of ethanol). The reason for the sudden drop in heat capacity is evidently that, in the vicinity of the temperature where it occurs (called the glass transition temperature) the time scale for the rearrangement of the molecules within the liquid state is diverging. This is illustrated in Fig. 1 part (b) where we show the variation with temperature of both the viscosity and the enthalpy relaxation time, τ_H. The latter is the time scale in which a liquid recovers its equilibrium state after a very small heat pulse. It may be noted that at T_g, τ_H (which may be determined from differential scanning calorimetry experiments) is essentially the time scale for molecular rearrangments i.e., about 200 seconds. This, however, i.e., the time scale for the differential scanning calorimetry experiment when it is conducted at the common scan rate of 10°C/min. The glass transition, in other words, is nothing more than the consequence of the crossing of an internal time scale of the liquid with the time scale of the experiment being used to study it.

The latter recognition leads many to conclude that the experimentally observed glass transition is not of great interest but this conclusion is very inappropriate as further inspection of the calorimetric results quickly shows.

Fig. 1 part (c), for instance, shows that in the vicinity of the glass transition, the total entropy of the liquid determined by the appropriate integration of the liquid and crystal heat capacities (taking account of the entropy of fusion) is rapidly approaching that of the crystal. In other words, if the decreasing heat capacity observed in the normal calorimetric experiment (due to the finite experimental time scale) were not to occur at T_g, it would have to occur at a temperature not far below T_g in order to avoid a thermodynamic conundrum viz., the disordered phase entropy becoming lower than that of the (ordered) crystal phase (or even becoming negative at a somewhat lower temperature). This is a description of the famous Kauzmann Paradox (12) which has been much discussed in the glass science literature (13-17). There is as yet no generally accepted theory which accounts for the tendency of the liquid entropy to vanish at temperatures far above 0°K. This is one of the active areas of theoretical investigation at this moment.

What is needed at this time is at least an accurate description of the behavior of the liquid relaxation times with decreasing temperature.

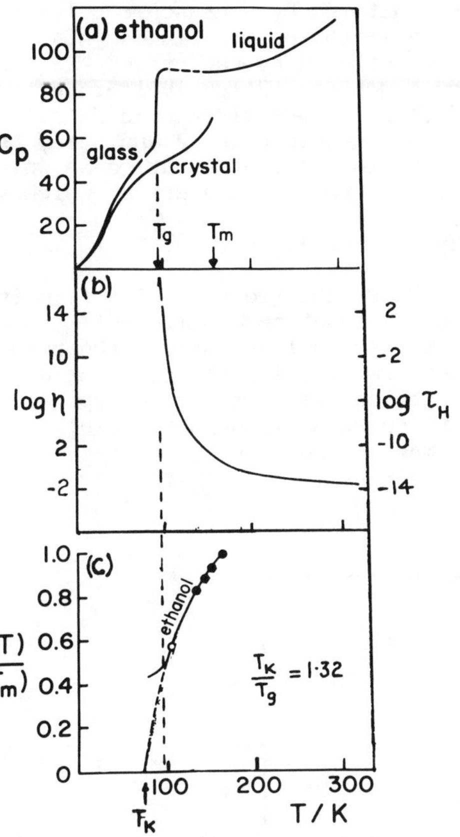

Figure 1 (a) Calorimetric signature of the glass transition, using the case of ethanol (data from ref. 3) to show increase of heat capacity over value of crystal, occurring at T_g.

(b) Variation of log viscosity and enthalpy relaxation time associated with approach to the glass transition.

(c) Tendency of excess entropy of liquid over crystal to disappear at T_K on cooling below the normal glass transition without change of heat capacity.

1.2 Relaxation Times and Scaled Viscosities

The relaxation behavior of viscous liquids is conveniently monitored by the simply measured shear viscosity η_s which is proportional to the average shear stress relaxation time τ_s. The enormous variation with temperature, of the characteristic relaxation time which may be observed in glassforming liquids is illustrated for the much-studied case of glycerol in Fig. 2. It is particularly notable that the relaxation time for mechanical stresses (which is monitored by ultrasonic relaxation experiments) is essentially identical with the more fundamental relaxation time for the enthalpy, (monitored by the recent AC calorimetry technique of Nagel and coworkers (2)). Their results are shown in Fig. 3, along with the dielectric and ultrasonic data.
The failure of the Arrhenius equation seen in Figs. 2 and 3 may be simply rectified using a modified version of that equation in which a finite temperature, T_0, replaces 0 K as the temperature at which the relaxation time goes to infinity,

$$\eta \text{ (or } \tau) = A \exp (B/(T-T_o)), \qquad (1)$$

where A and B are also constants. This modified Arrhenius form is the well known Vogel-Tammann-Fulcher (VTF) equation and its ability to fit the data for glycerol is shown by the solid line through the points in Fig. 2. (The dotted line is a power law fit, see ref. 2). As a second example, the linearization of relaxation time data for a typical polyalcohol butanediol for which there exist light scattering relaxation data over 10 orders of magnitude, may be quoted.

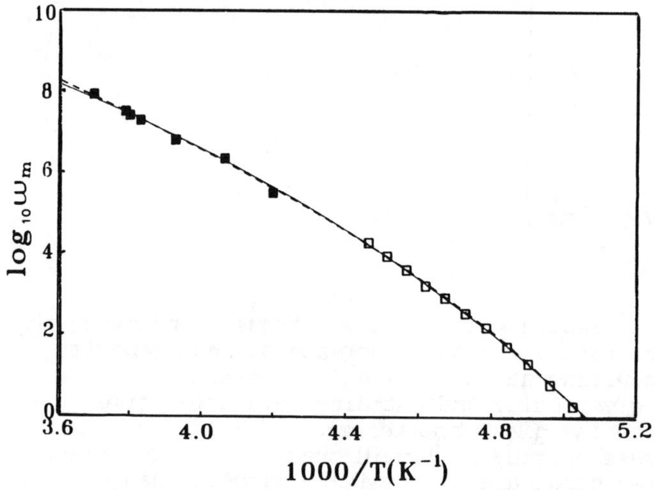

Figure 2. Variation of mechanical and electrical relaxation times with temperature for glycerol (data from ref. 2).

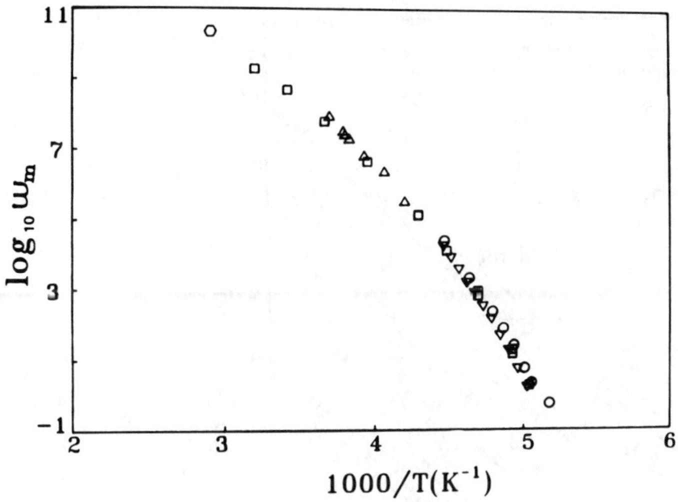

Figure 3. Comparison of dielectric (□) and mechanical (△) relaxation times, with the more fundamental enthalpy relaxation time (▽) for glycerol (data from ref. 2 (b,c).

We prefer to write equation 1 in the normalized form,

$$\eta = A \exp DT_0/(T-T_0). \qquad (2)$$

In this form, which can be justified theoretically (see below) the difference between all liquids obeying the VTF equation, or even approximating it, is caused by the parameter D. This is seen most readily when experimental data are plotted in the normalized form using the scaling temperature T^* as the temperature for some fixed viscosity, e.g. $\eta = 10^{13}$ poise, or fixed relaxation time, e.g. $\tau = 10^2$ s. Fig. 4 shows viscosity data for a wide variety of liquids that are plotted in the reduced form, and establishes the "strong and fragile" pattern for liquid relaxation processes which has been discussed in several articles (19-22). The insert to Fig. 4 shows how this pattern may be reproduced by Eq. (2) with D values falling between 3 and 50. Note that the alcohols we have discussed fall in the middle of the diagram. Fig. 4 requires that, for the case D = 12 which is characteristic of this group, T^*/T_0 should have the value 1.33. It is striking that this is the average value found for the Kauzmann temperature of the series of polyalcohols $[CH_2OH]_n$ where n = 1-6, studied by the author and D.L. Smith (23) (using purely thermodynamic measurements as in Fig. 1). Furthermore, we note that this is also the ratio T_g/T_K obtained for ethanol in Fig. 1. This is all consistent with the original suggestion of Gibbs (4) that the relaxation time τ_H (and also τ_s when shear relaxation is fully coupled to structure) is governed by the state of liquid disorder represented by the configurational

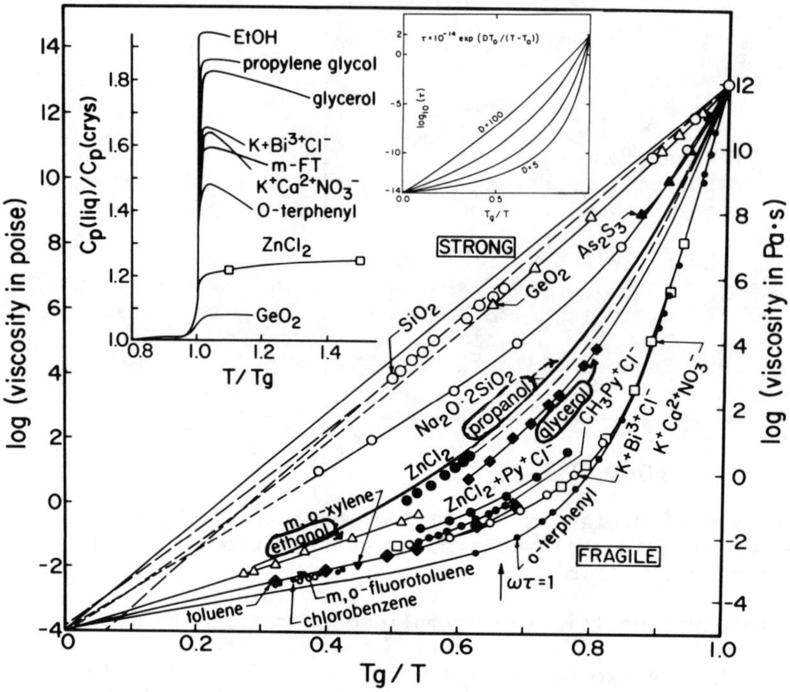

Figure 4. T^*-scaled Arrhenius plot of viscosity data for a wide variety of liquids, displaying the range of "strong" to "fragile" behavior observable in the liquid state (adapted from ref. 19). Here T^* is the temperature at which the viscosity reaches 10^{13} poise. Inserts show variation of heat capacity through the glass transition for the same liquids, and plots of equation (2) for D values between 5 and 100.

entropy S_c (which may be obtained from purely thermodynamic data). Indeed, the Adam-Gibbs theory (24) coupled with some observations on the temperature dependence of the liquid excess heat capacity, ΔC_p to be discussed below, leads directly to the normalized form of the VTF equation, Eq. (2) above.

Adam and Gibbs examined a modification of transition state theory, invoking cooperatively rearranging groups of molecules, to derive the expression

$$\tau = \tau_0 \exp C/TS_c \tag{3}$$

where the constant C contains a quantity $\Delta\mu$ representing the height of the energy barrier over which the cooperatively rearranging group must pass. The failure of the Arrhenius equation seen in Fig. 2 is, according to this picture, a consequence of the temperature dependence of the liquid configuration entropy. This suggests that the quantity D in Eq. (2) should be related to the excess of liquid over crystal heat capacity manifested at the calorimetric T_g since

$$S_c = \int_{T_K}^{T} \Delta C_p/T \, dT \qquad (4)$$

The data seen in the insert to Fig. 4 for inorganic and molecular liquids are broadly consistent with this expectation. <u>However the alcohols we have discussed (which are highlighted in Fig. 4) are clearly exceptional</u>. The particularly large ΔC_p seen for the alcohols would suggest exceptional fragility with D ~ 2, while in fact the D value describing their behavior in Fig. 4 is intermediate. We can interpret this seeming discrepancy in a natural way in terms of characteristic features of the 3N + 1 dimensional potential energy hypersurface (25,26) for the case of hydrogen-bonded liquids but must first review the Adam-Gibbs theory and its connection to Eq. 2 in more detail.

1.3 Adam-Gibbs Theory and the Potential Energy Hypersurface

Adam and Gibbs (24) argued that in viscous liquids the packing of molecules was too tight for individual molecules to undergo changes of orientation without the cooperation of their immediate neighbors. The fundamental process by which perturbing stresses are relaxed in this case must involve the <u>cooperative</u> movement of groups of molecules. It was supposed, in their treatment, that the size of the group necessary to execute a change from one local configuration to another, (which must have a minimum configurational degeneracy of 2), should increase with decreasing temperature as the packing density increased. They showed how this increasing size of the cooperative group could be quantitatively related to the total degeneracy, measured by the configurational part of the entropy, S_c, and how this in turn, would cause departures from the Arrhenius equation, via Eq. (4) above.

To evaluate S_c according to Eq. 4, we must have a functional form for ΔC_p as a function of temperature. Although any attempt to be quite general in this respect will inevitably fail, there does seem to be a functional form of exceptionally simple nature, $\Delta C_p = k/T$, which describes the behavior of a great many molecular liquids(27,28). Data for a group of aromatic and saturated hydrocarbons (28) are shown in Fig. 5, the inset to which shows that the proportionality constant k is a linear function of the number of carbons in the molecule in question. If we introduce this functional form into equation 4, we obtain immediately the relation

$$S_c = \int_{T_K}^{T} k/T^2 \, dT = k(T-T_K)/T_K \qquad (5)$$

Introduction of Eq. (5) into Eq. (3) then leads to the VTF equation Eq. (1) in the modified form which we introduced above, Eq. (2). The important insight to be gained from this derivation is that the parameter D, which distinguishes the reduced viscosity behavior of liquids according to the strong and fragile pattern of Fig. 4 contains <u>two</u> quantities characteristic of the individual liquid, k and $\Delta\mu$, one of which (k) is thermodynamic in nature, and the other, ($\Delta\mu$) kinetic. In

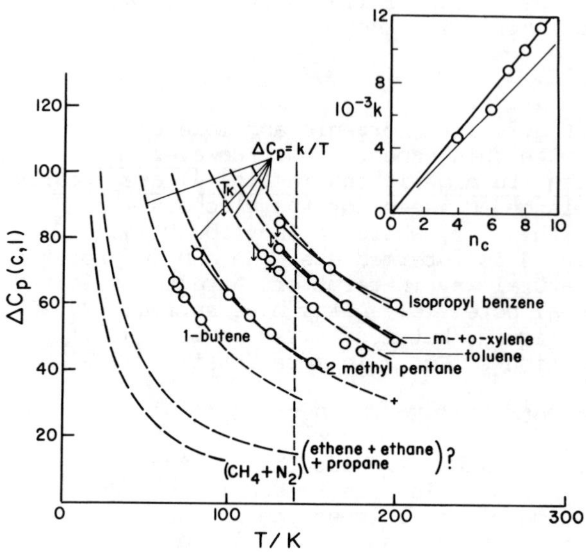

Figure 5. Variation of the excess of liquid over crystal heat capacities of hydrocarbon liquids as T_g is approached from above (from ref. 28).

view of their large heat capacity changes at T_g, it would therefore seem that the alcohols find themselves located in the middle of the diagram, Fig. 4, as a consequence of unusually large values of the free energy barrier $\Delta\mu$ to rearrangement of the critical size cooperative group. This seems reasonable, since a characteristic feature of cooperative rearrangements in H-bonded liquids must be the necessity to break a number of hydrogen bonds during passage over the energy barrier. In other words, more than just a local density fluctuation is required to permit the rearrangement.

We have previously attempted to distinguish strong from fragile liquids in terms of the density of minima in the potential energy hypersurface in the 3N + 1 dimensional configuration space of the N particle molecular systems (21). In view of the above discussion, we should now distinguish the hydrogen bonded liquids from other liquids with high densities of configurational states by the relatively large kinetic barriers between minima of different depths on the 3N + 1 dimensional hypersurface descriptive of the potential energy of the N molecules in configuration space to which we referred to above. The distinction between strong liquids and fragile liquids on the one hand, and fragile liquids with large free energy barriers between minima and the more common fragile liquids on the other, is illustrated by the simplified free energy surfaces represented in Fig. 6. Fig. 6 part (a) gives a two dimensional illustration of the topology of the hypersurface

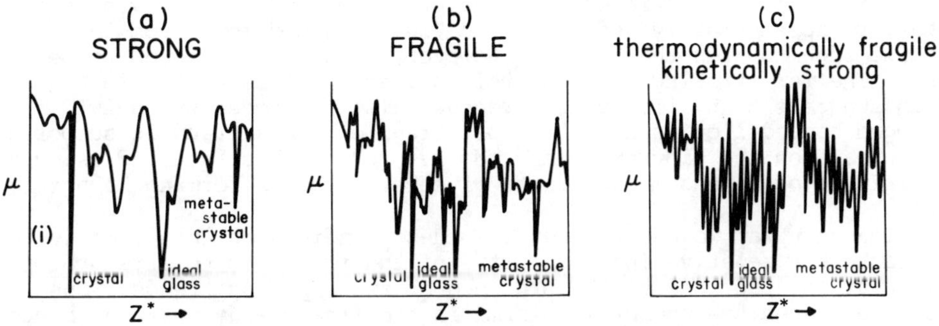

Figure 6. Two dimensional representations of the potential energy hypersurface for (a) strong and (b) fragile liquids and (c) fragile liquids which appear strong in Figure 4 due to the existence of high energy barriers between adjacent minima on a fragile (high density of configuron states) surface (from ref. 29).

characteristic of strong liquids where the requirement of tetrahedral coordination of the network centers places strong limitations on the arrangements of the other particles and thus restricts the number of distinct minima on the potential energy surface. Fig. 6 part (b) shows the situation where the coordination number restriction is lifted or where the presence of an asymmetric molecular shape leads to a proliferation of alternative distinct arrangements of about the same energy.

As discussed elsewhere (21,28) the higher density of configurational states in these cases is consistent with the larger configurational entropy at a given kinetic energy kT above the energy characteristic of the deepest minimum on the surface (in which the system would reside at T_K, if time were allowed for it to find it). Of course, it is the larger degeneracy of the surface which drives the system to a high entropy in a short temperature interval, hence which gives the large excess liquid heat capacity characteristic of the fragile liquid. Fig. 6 part (c) (29) shows the case where the surface is characterized by the same high density of minima as in part (b) but where the minima are separated by large free energy barriers, as in our picture of the hydrogen bonded liquids.

We must now ask whether all hydrogen bonded liquids, irrespective of structure, will be intermediate in character because of the large barrier crossing term $\Delta\mu$ discussed above.

To probe this question, Alba and the author (29) undertook the study of a series of disubstituted benzenes in which the CH_3 groups of the fragile liquid m-xylene could be replaced with increasing strongly interacting groups. For instance, changes were made from $-NH_2$ opposed by fluorine (m-fluoroaniline), through $-NH_2$ substituting for the methyl group (m-toluidine), to -OH opposed by fluorine (m-fluorophenol) to -OH unopposed by fluorine (m-cresol). It was anticipated that with increasing strength of intermolecular hydrogen bonding, the viscosity behavior (appropriately reduced as in Fig. 4) would move from the fragile edge towards the center. However, this was not found to be the case. The few existing data for this series are illustrated in Fig. 7 which suggests that, if any differences exist, the most strongly hydrogen bonded liquids are, the most fragile. This trend is supported by new data on the dihyroxy compound, resorcinol (best-studied using a 1:1 mixture of o- and m-isomers to give better glass-forming behavior) (30).

The only way we can rationalize these findings at this time is to attribute the relative insensitivity (of the fragility to strength of molecular interaction) to the presence of two opposing trends. One of these increases the number of minima in the free energy surface, hence causes a shift to greater fragility with increasing hydrogen bonding strengths, and the second (the hydrogen bond), increases the barrier crossing height, hence shifts the behavior in the opposite direction.

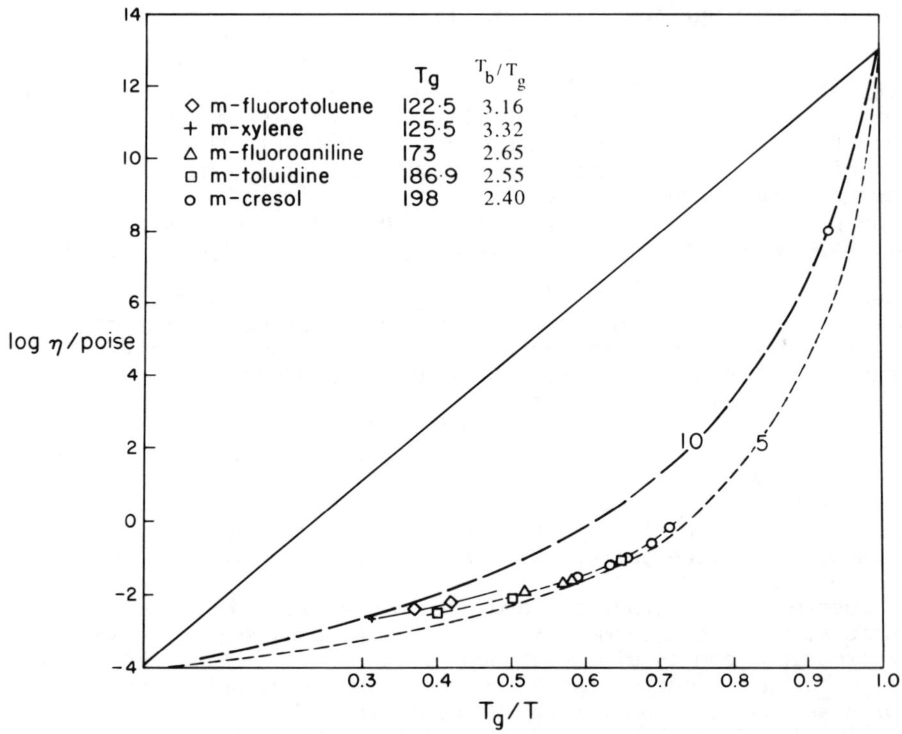

Figure 7. Variation with temperature of the viscosity of various disubstituted benzene liquids of a T_g -scaled Arrhenius plot. Here T_g is the calorimetric value defined by Figure 1 (a). Legend gives T_b/T_g values for each case. (From ref. 29).

Evidence that this may be the case is found in the contrast of Fig. 7 with Fig. 8 which shows data for hydrocarbon molecules only. Important correlating information is included in the legend to Fig. 8. There we give the ratio T_b/T_g which is a measure of the reduced liquid range of each molecular substance. We note that for the hydrocarbons, there is an excellent correlation between the liquid range and the strength of the liquid. The higher the boiling point to glass transition temperature ratio, the "stronger" the liquid appears to be. Turning to the legend of Fig. 7 we see that again there is a wide variation in liquid range T_b/T_g. Furthermore, the value for dihydroxy

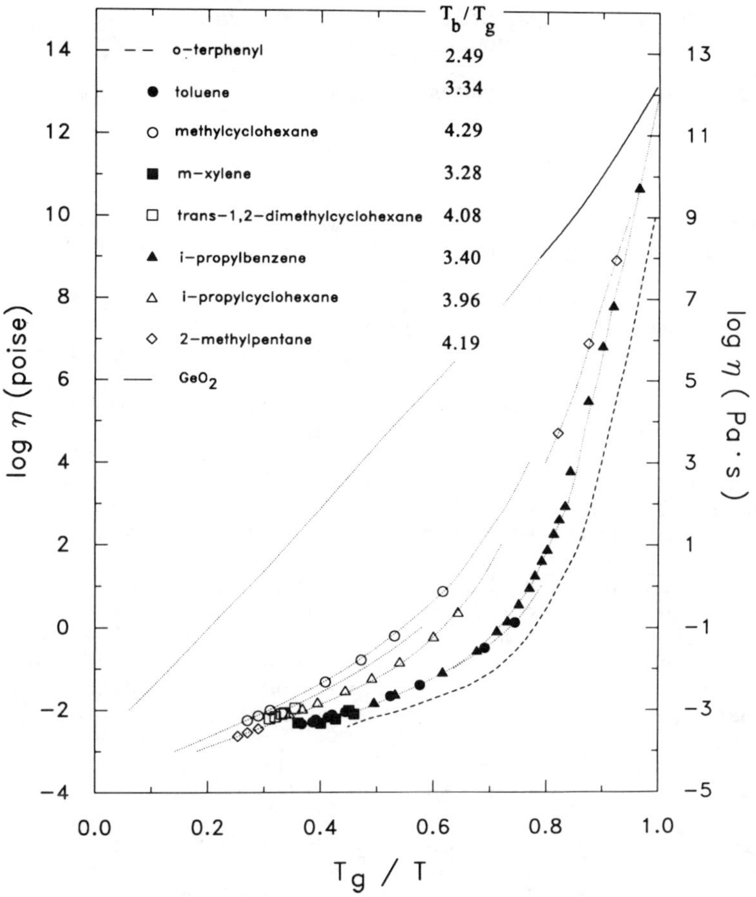

Figure 8. Variation of viscosity of with temperature of various hydrocarbon liquids, whose heat capacity behavior is seen in Fig. 5, using same scaling as in Fig. 7. Legend gives T_b/T_g values for each case. (Adapted from ref. 28).

benzene (resorcinol) is smaller than for any other liquid including the most fragile case, o-terphenyl (30). In the absence of an abnormally large barrier-crossing energy term in Eq. 4, this liquid would presumably be the most fragile on record. The effect of hydrogen bonding in the disubstituted benzenes, then, is consistent with that for the normal alcohols and diols; the difference in fragility is mainly due to the fact that the aromatic molecules on which the hydrogen bonding group resides in the benzene derivatives are intrinsically fragile.

The above observations and rationalizations are necessarily very qualitative at this time and much more data must be acquired before a clear picture can be expected. It is anticipated that further laboratory studies guided by the "picture" we have outlined here for hydrogen-bonded liquids, will be helpful in improving overall understanding of viscous liquid phenomenology.

2. CLUSTERING PHENOMENA AND OTHER SPECIAL FEATURES IN HYDROGEN BONDED LIQUIDS

2.1 Introduction

In part 1, we described some general thermodynamic and dynamic characteristics of glassforming molecular hydrogen bonded liquids which, while differing considerably one from the other, behaved in a more or less normal manner. In the present part we will pick out some cases where we believe there are some special effects or functions, due directly to the hydrogen bonding phenomenon, which warrant treating the liquids or systems concerned separately. Firstly, in section 2.2, we will discuss the cases of normal alcohols in which there seem to be very significant differences in relaxation times for different types of perturbations imposed on the system, due to the presence of clustering phenomena. This section will include a discussion of such phenomena in the most important of all hydrogen bonded liquids, water. In section 2.3 we will look at some experiments on systems with variable-sized clusters which are controlled by the experimentalist through the preparation of microemulsions. In the microemulsions we will consider the hydrogen bonded liquid under consideration will be water. Finally, in section 2.4, we will consider some cases of hydrogen-bonded liquids in which the position in temperature of the glass transition proves to be of considerable biological significance, being the factor which determines the organism's ability to survive under hostile climatic conditions.

2.2 Clustering Phenomena In Hydrogen Bonded Liquids

A common finding for relaxation of viscous liquids is that, after suffering some perturbation from the equilibrium state, the average relaxation time is non-Arrhenius in character, as described in part 1 of this paper. A second common finding, not discussed in part 1, is that the relaxation process cannot be described using a single relaxation time. Rather, a function implying either an intrinsic nonexponentiality of the relaxation process, or the presence of a distribution of a simultaneous relaxation processes, must be incorporated in the description.

Rather striking exceptions to this general rule are found in the case of normal alcohols such as propanol and pentanol which are easily supercooled into the glassy state. An accompanying phenomenon, which suggests an explanation for this abnormal behavior, is that the relaxation of the liquids under mechanical stresses is much faster than that under electrical stresses, i.e., the shear or volume relaxation times are considerably shorter than the dielectric relaxation times (31). Factors of up to 150 seem to describe the relative relaxation times of the much-investigated case of propanol. This distinction and its origin has been discussed in some detail in a recent paper by Floriano and the author (32). It may also be seen in the comparison of light scattering-based mechanical relaxation times (33) with dielectric relaxation times from very high frequency studies of n-pentanol described in the paper of Salvetti in this volume (34). Appropriate to the higher temperatures of the latter studies, the ratio of relaxation times observed is smaller than that observed for propanol at temperatures where the relaxation times are in the microsecond regime, but the difference is still almost two orders of magnitude. For instance, at 25°C, Salvetti shows (in his Fig. 6) that τ_D = 800 ps, while Fig. 5 of ref. 33 shows that the mechanical relaxation time, is 8.8 ps. These results are combined with lower temperature data for propanol (35,36) in Fig. 9.

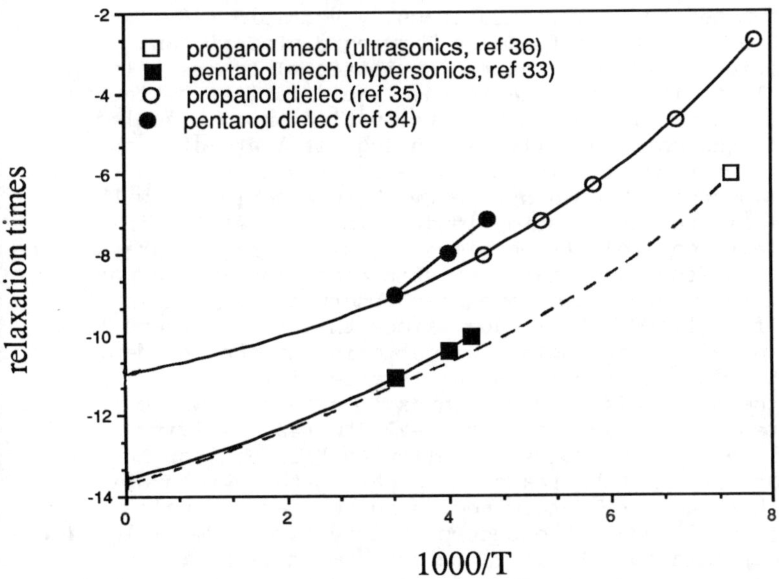

Figure 9. Comparison of mechanical and dielectric relaxation times for two normal alcohols, propanol and pentanol, showing ~2 order of magnitude gap attributed to clustering phenomenon. Compare with data in Fig. 2 for case of glycerol.

Although it cannot be told from the last-mentioned high frequency mechanical relaxation studies, the lower frequency ultrasonic measurements (36) show clearly that the mechanical relaxation is highly nonexponential, while the dielectric relaxation is always exponential. The explanation which has been given to this striking difference between the two relaxation processes (32) is that the dielectric relaxation reflects the behavior of hydrogen bonded clusters of molecules which dominate the dielectric strength while the mechanical relaxation is occurring principally through molecules on the outside of the clusters which respond much more rapidly to the applied stress. In this respect, the clusters are behaving like the massive dipoles (large ion-paired species) studied in nonpolar solvents by Williams and colleagues (37). Exactly the same distinctions, viz., slower relaxation and exponential relaxation, were observed for the large species. This is understood to be a result of the large species being obliged to relax by a Brownian diffusion process, which carries no memory of sequential events. An explanation of the alcohol anomaly in terms of clusters was supported, in the work of Floriano and the author (32), by demonstrating that small additions of electrolytes (e.g. $LiClO_4$) which break up the clusters by successfully competing for the alcohol groups as coordinating ligands, reduce the dielectric relaxation time and also lead it to become nonexponential in nature i.e., lead to normal behavior. A similar conclusion can be drawn from the incisive NMR studies of ^2D relaxation for alcohol, methylene, and methyl deuterons by Versmold (38).

The observation that the difference between dielectric and mechanical relaxation times (and their distributions) can be related to the presence of clustering phenomena in the liquid is interesting in connection with a previously unresolved observation on supercooled water (39). Among the many anomalies which suggest that water should undergo a phase transition of some sort at -45°C (39) (due to a virtual completion of the hydrogen bonded open network in association with the divergence of the cluster correlation length) is a difference in mechanical and electrical relaxation times. If dielectric relaxation and volume relaxation had the same characteristic time constant, a maximum in the Brillouin line width would have been observed at -21°C in the careful studies of Migliardo et al (40), since this is the temperature at which phonons generated by inelastic scattering of the incident light beam would have the same frequency as the inverse dielectric relaxation time. In fact, the Brillouin line continues to broaden down to the lowest temperatures of observation, viz., -27.2°C (40) implying a much faster mechanical relaxation process. Following Floriano and Angell (32), we note that care must be taken in comparing these relaxation times to ensure that they are determined using the same formalism, and that the formalism utilized is the susceptibility formalism employed in normal dielectric relaxation studies. In other words, the relaxation time for the compressibility, not that for the bulk modulus, should be compared with the dielectric relaxation time.

It appears in this case that the two relaxation times must differ by about an order of magnitude. Again, the natural explanation would seem to be that the dielectric strength is carried by the dipole-correlated clusters of water making up the hydrogen bonded network "patches" (41).

In this case, the difference in relaxation times should increase as the temperature decreases and the "patches" become exponentially larger. Certainly a difference in temperature dependences of dielectric and viscous flow processes is observable, though it will be necessary to obtain more detailed information on both processes before the full validity of our proposed picture can be established.

2.3 Water in Nanoscopic Systems and Near Surfaces--Microemulsion Studies

It is possible to study some unusual aspects of hydrogen bonded liquids by containing them as the dispersed phase in a "water-in-oil" type microemulsion. When a suitable surfactant is chosen it is possible to have a microemulsion in which the only hydrogen bonds in the whole system are those internal to the dispersed phase. An example is the three component water-in-oil microemulsions obtained using the surfactant dimethyl didodecyl ammonium bromide (DDA^+Br^-). Such microemulsions have been described in some detail by Evans and coworkers (42), and have been the subject of an overtone infrared spectroscopy study by the author and coworkers (43). We discuss them briefly here in order to demonstrate the range of hydrogen bond strengths which can be observed when water molecules are forced into conditions of weak interaction of hydrocarbon surroundings. We can also demonstrate that the same type of weak hydrogen bonds can be observed in certain unemulsified aqueous solutions.

The region of composition in the ternary system, water-octane-DDA^+Br^-, in which clear microemulsion phases are found, is shown in Fig. 10. According to electrical conductivity studies, the major part of this microemulsion region consists of bicontinuous microtubular water dispersions. However, at the extreme of high water content, where the liquids become insulating, it appears that an isolated droplet phase exists.

Information on the type of H_2O-neighbor interactions which can be found in this system is given by the overtone infrared spectra shown in Fig. 11 which shows data for one of a series of compositions along the line AB in the diagram. The dashed line in Fig. 10 shows the spectrum of bulk water in the same wave length range scaled onto the microemulsion spectrum at the wavelength 1.5μm. Note that above $\lambda = 1.5\mu m$ the spectrum of the microemulsion is indistinguishable from that of pure water. The series of peaks obtained by subtracting the bulk water spectrum (shown as a dotted line in Fig. 11) from the microemulsion spectrum apparently represents the water at the surface of the microemulsion droplet since this spectrum is increasing in intensity as the total water content decreases.

It is not clear from the spectra of Fig. 12 whether or not this series of peaks represent three different and distinct populations of weakly hydrogen bonded water molecules. However, the presence of the same three peaks in a different intensity distribution in the spectra of compositions along the line CD of Fig. 11, suggests that they, in fact, do. It will take more systematic work with matrix phases of different compositions to determine what specific interactions give rise to each component of the spectrum.

It is interesting to observe that (44) a spectrum with many of the same features as Fig. 12 can be obtained from solutions in the binary system, water + polypropylene glycol, as well as from -OH terminated PPG

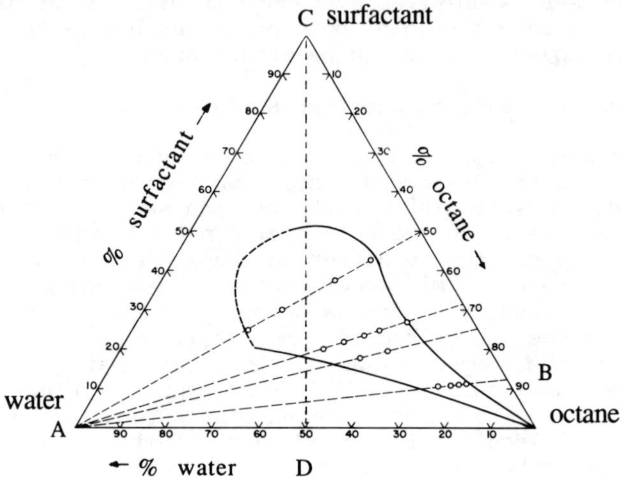

Figure 10. Phase diagram for water-in-oil microemulsion system in which only the water contains H-bonds (from refs. 42 and 43).

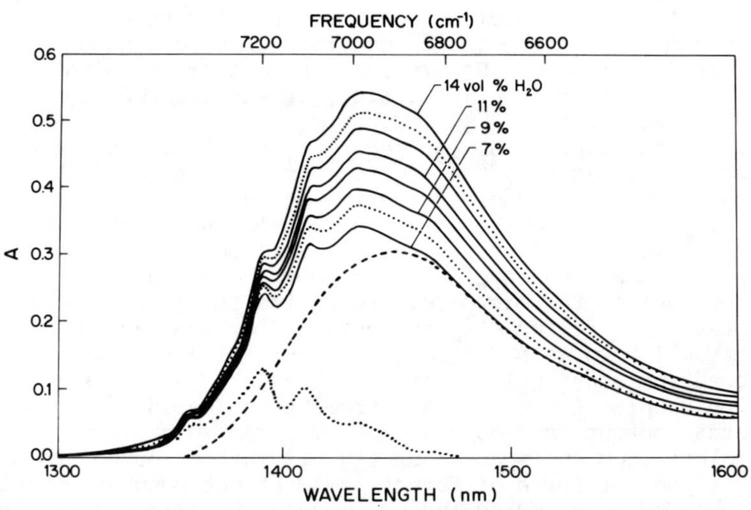

Figure 11. Overtone IR spectra of water-in-oil microemulsion containing 40 vol % water (on line AB of Fig. 10) compared to bulk water scaled onto spectrum at 1.5 micron (dashed line). Spectrum for surface water molecules obtained by differencing, is shown as dotted line. (From ref. 43).

itself. In this case we study the uncoupled -OH oscillators by using H_2O/D_2O solutions. Fig. 12 shows the overtone IR spectrum of uncoupled -OH oscillators in key compositions in this system. Clearly, PPG and its solutions in water contain a range of high frequency oscillators not present in water itself. Although the peak frequencies of this spectrum (which presumably is due to water molecules forced into contact with the hydrocarbon components of the PPG molecule) do not correspond exactly with those in the microemulsion, a common origin is strongly suggested. Since the binary solution is a simpler system than the microemulsion, it is easier to assign the frequencies observed to a particular structural origin. We tentatively suggest that the highest frequencies are due to $-OH \cdots CH_2-$ interactions as illustrated in Fig. 12 inset.

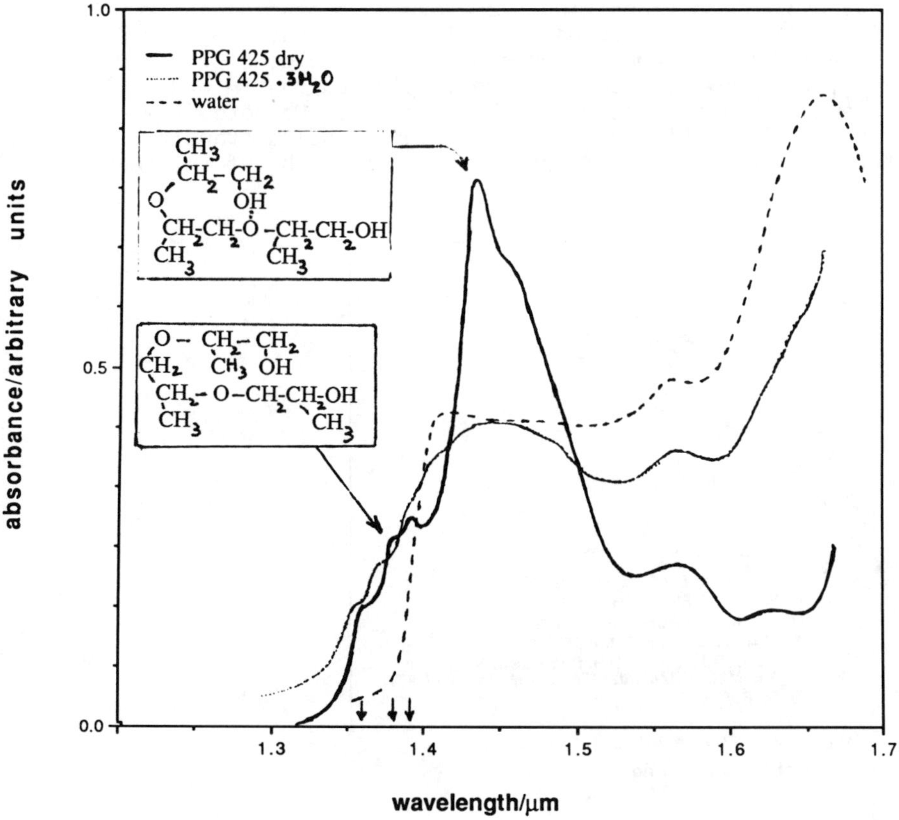

Figure 12. Overtone IR spectra of polypropylene glycol + H_2O/D_2O solutions, showing the presence of non-bonding -OH groups similar to those in the microemulsions of Fig. 11. Insert shows probable interactions responsible.

2.4 Hydrogen Bonded Liquids and Glasses in a Life-Preserving Role in Nature

Appropriate to this acount of special effects of H-bonding observable in pure liquids, is a brief discussion of the role of hydrogen bonding and glass formation in the preservation of organisms under harsh conditions in nature. It has been observed that certain desert insects can survive for long periods in drought conditions and then recover full viability after drought-breaking rains. Investigation of the biology of these insects has shown that in dry times they can dehydrate non-destructively and reversibly in association with the in-cell synthesis of the disaccharide sugar, trehalose. Since trehalose is just one of a number of possible sugars of formula $(CH_2O)_{12}$ we consider it of interest to determine why nature chooses this sugar as the one to synthesize. In Fig. 13, we show glass transition temperatures measured (45) for the series of $(CH_2O)_{12}$ sugars, trehalose maltose, and sucrose, and compare them with data for the sugar glucose and the polyalcohol glycerol. The latter is synthesized by insects in very cold regions in order to protect themselves from destruction by freezing. We note from Fig. 13 that it is trehalose which exhibits the highest glass transition temperature in the presence of water. We suggest, therefore, that it is

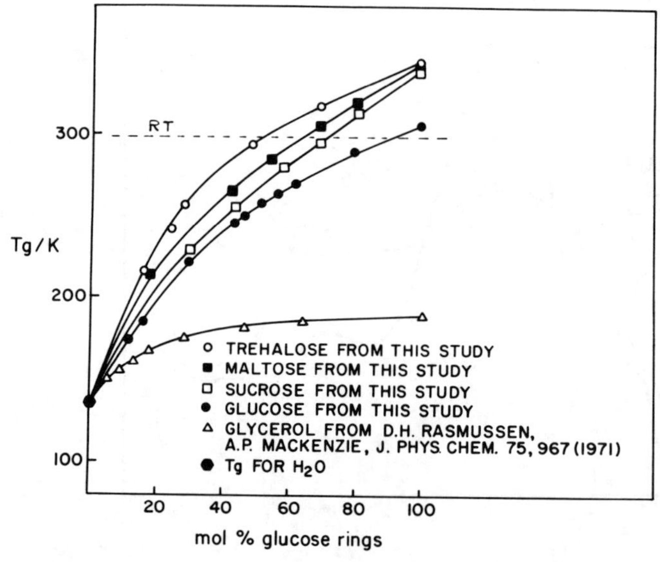

Figure 13. Glass transition temperatures for aqueous solutions of various sugars and glycerol, showing unusually high T_g values for **trehalose** solutions, the sugar chosen by nature for intracellular synthesis for drought survival capability. (From ref. 45).

the high viscosity, and the attendant suppression of crystallization of solid phases, which is the important element of the protection offered by trehalose to dehydrating insects. We see from Fig. 13 that, when during dehydration, the water content has decreased to one molecule per glucose ring in the trehalose solutions, the glass transition temperature has risen to above ambient temperature. Thus it would appear that the partially dehydrated insects actually obtain their state of suspended animation by passing into the vitreous state.

2.5 Concluding Remarks

In this section, we have given several examples of unusual aspects of structure and relaxation in liquids which arise specifically from the existence of the hydrogen bonding interaction. There are, of course, many others which could have been chosen, particularly in the field of molecular biology, but these are dealt with rather completely in other contributions to this volume. Thus we have chosen to discuss those most familiar to us. We have dealt with them here very briefly; additional detail may be found in the original literature quoted in the text.

ACKNOWLEDGEMENTS: We are indebted to the National Science Foundation under Grants Nos. CHE-8318416 and DMR 8805597 for support of the diverse research efforts described in this paper. Details of the work of C. Alba on disubstituted benzenes, D. Fields and J. L. Green on microemulsions, and J. L. Green on polypropylene oxide water solutions, will be given in papers currently in preparation.

REFERENCES:

1. See, for instance, "Dynamic Aspects of Structural Change in Liquids and Glasses," Ed. C.A. Angell and M. Goldstein, Ann, NY, Acad. Sci., Vol. 484, NY Acad. Sci. NY (1986).
2. a) Birge, N. and Nagel S., Phys. Rev. Lett. **54**, 2674 (1985).
   b) Birge, N., Phys. Rev. B **34**, 1641 (1986).
   c) Jeong Y.H., Nagel S., and Bhattacharya S., Phys. Rev. A **34**, 602 (1986).
3. Suga, H., and Seki, S., J. Non-Cryst. Solids, **16**, 171, (1974).
4. Goetz, A. Sjolander, J. Phys. Solid State Phys. **52**, 3038, (1984).
5. Wolynes, P.G., "Aperiodic Crystals: Biology, Chemistry and Physics in a Fugue with Stretto," for H. Frauenfelder, honor volume (preprint).
6. Magazu, S., Majolino, D., Mallamace, F., Migliardo, P., Alliotta, F., Vasi, C., d'Aprano, A., and Donato, D.J., Mol. Phys. **66**, (1989).
7. Yan, Y-X, Cheng, L-T., and Nelson, K.A., J. Chem. Phys. **88**, 6477, (1988).
8. Turnbull, D., Cohen, M. H., J. Chem. Phys. **52**, 3038, (1970).

9. Angell, C.A., Busse, L.E., Cooper, E.E., Kadiyala, R.K., Ghelfenstein, M., Szwarc, H., and Vallal, A., J. de Chim. Phys. 82, 267 (1985).
10. Angell, C.A., Choi, Y., J. Microscop. 141. **251**, (1986) (text of Plenary Lecture).
11. Moynihan, C. T., Macedo, P.B., Montrose, C.J., Gupta, P.K., DeBolt, M., Dill, J.J., Dom, B.E., Drake, P.W., Easteal, A.J., Elterman, P.B., Moeller, R.P., Sasabe, H.A., and Wilder, A., Ann. NY, Acad. Sci. **279**, 15 (1977), New York Acad. of Sci, NY, eds. M. Goldstein and R. Simha.
12. Kauzmann, A.W., Chem. Rev. **43**, 219 (1948).
13. Angell, C.A., J. Chem. Ed. **47**, 583 (1970).
14. Angell, C.A., Sichina, W., Ann. NY Acad. Sci., **279**, 53 (1976).
15. Anderson, P.W., in "Ill-Condensed Matter," Eds, Balian, R., Maynard, R., and Toulouse, G., (Les Houches), North Holland, 171, 1979.
16. Gibbs, J. H., in "Modern Aspects of the Vitreous State," Butterworth Sci. Publ. Ltd. London (1960), Chapter 7.
17. Stillinger, F.H. J. Chem. Phys. **88**, 7818, (1988).
18. Drake, P.W., Dill, J.F., Montrose, C.J., Meister, R.J., Chem. Phys **67**, 1969 (1977).
19. Angell, C.A., <u>Relaxations in Complex Systems</u>, ed. K. Ngai and G.B. Wright, National Technical Information Service, U.S. Dept. of Commerce, Springfield, VA 22161.
20. Angell, C.A., J. Non-Cryst. Sol., **102**, 205 (1988)
21. Angell, C.A., J. Phys. Chem. Sol. **49**, (8), 863 (1988).
22. Murthy, N.N., J. Phys. Chem. **89**, 93, 3347.
23. Angell, C.A., Smith, D.L., J. Phys. Chem. **86**, 3845, (1982).
24. Adam, G., Gibbs, J.H., J. Chem. Phys. **43**, 139 (1965).
25. Goldstein, M., J. Chem. Phys. **51**, 3728, (1969).
26. Stillinger, F.H., J. Chem. Phys. **80**, 2742, (1984).
27. Privalko, V.P., J. Phys. Chem. **84**, 3307, 1980.
28. Alba, C., Busse, L.E., and Angell, C.A., J. Chem. Phys. (in press).
29. Alba, C. and Angell, C.A.,
30. Senapati, H. and Angell, C.A. to be published.
31. Litovitz, T.A., McDuffie Jr. G.E., J. Chem. Phys. **39**, (3), 729 (1963).
32. Floriano, M.A., and Angell, C.A., J. Chem. Phys. **91**, 2537, (1989).
33. Magazu, S., Majolino, D., Mallamace, F., Migliardo, P., Aliotta, F., Vasi, C., d'Aprano, A., and Donato, D.J., Mol Phys. **66**, (1989).
34. Salvetti, G. This volume, p.
35. Davidson, D.W., Cole, R.H. J. Chem Phys. **19**, 1484, (1951).
36. Litovitz, T.A., and Sette, D., J. Chem Phys. **21**, 17 (1953).
37. Davis, M., Hains, P.J., and Williams, G., J. Chem. Soc. Faraday Trans **69**, 1785 (1973).
38. Versmold, A., in "Molecular Liquids," Eds. A.J. Barnes et al (Reidel, Dordrecht) (NATO ASI Series), Plenum, NY, 1987.

39. Angell, C.A., "Supercooled Water," Ann. Res. Phys. Chem. **34**, 593-630 (1983).

40. (a) Maisano, G., Migliardo, P., Aliotta, F., Vasi, C., and Wanderlingh, F., Phys. Rev. Lett. **52**, 1025 (1984).
(b) Magazu,S., Maisano, G., Majolino, D., Mallamace, F., Migliardo, P., Alliota, F., and Vasi, C., J. Phys. Chem. **93**, (1989).

41. Stanley, H.E., Teixiera, J., J. Chem. Phys. **73**, 3034, (1980).

42. (a) Chen, S.J., Evans, D.F., and Ninham, B.W., J. Phys. Chem. **88**, 1631, (1984).
(b) Blum, F.D. Pickup, S., Ninham, B.W., and Chen, S.J., J. Phys. Chem. **89**, 711, (1985).

43. Fields, D.L., Green, J.L., and Angell, C.A., J. Phys. Chem. (to be submitted).

44. Green. J.L. and Angell, C.A., (to be published).

45. Green, J.L. and Angell, C.A., J. Phys. Chem. **93**, 2880, (1989).

THERMODYNAMIC PROPERTIES OF SOME H-BONDED LIQUIDS IN THEIR UNDERCOOLED AND / OR OVERCOMPRESSED STATES

C.Alba-Simionesco
Laboratoire de Chimie-Physique
11 rue Pierre et Marie Curie
75231Paris cedex 05, France

ABSTRACT. Some thermodynamic properties of hydrogen bonded liquids are presented using the piezothermal method, which allows the direct measurement of the expansivity as a function of temperature and pressure. In their undercooled and overcompressed states, new features are revealed and provide a lower temperature limit for the existence of the phase. The latter corresponds to the Kauzmann temperature, calculated from Cp data, and the temperature divergence of the viscosity from V-T-F equation. Further developments of the metrology defining a glass transition pressure and an equivalent Kauzmann pressure are proposed.

1. INTRODUCTION

Our description of liquids is based on the observation of thermodynamic quantities such as the expansivity, the heat capacities at constant pressure and constant volume and the isothermal compressibility as a function of pressure and temperature. In previous papers [1,2], we have reported a set of empirical equations derived from such molecule as n-butane, CO2, argon, n-hexane, which exhibit simple forms:

- as a function of pressure, the expansivity follows a square root law

$$\alpha = A / (P - P_\lambda)^{1/2} \tag{1}$$

where A and P_λ are pressure-independent coefficients. In spite of its generality, this law has not yet received theoretical support. When extended to take into account the temperature dependence, equation 1 may be written in the following phenomenological form

$$\alpha = \alpha_0 (P_0 - P_\lambda)^{1/2} / (P - P_\lambda)^{1/2} \tag{2}$$

where $P_\lambda = f(T)$ specifies the pseudocritical line of the divergence of α and where the pair (α_0, P_0) defines the coordinates of a single intersection point for all the isotherms. This feature was already observed by Bridgman[3] at the beginning of the century. Its interest is related to the behavior of the heat capacity as a function of pressure, according to the following equation :

$$(\partial Cp / \partial P)_T = - V T (\alpha^2 + (\partial \alpha / \partial T)_P) \tag{3}$$

the temperature derivative $(\partial \alpha / \partial T)_P$ is generally positive at low pressures and becomes negative at higher pressures; it controls the decrease, the existence of a minimum, and the possible increase of Cp as a function of P.

J.C. Dore and J. Teixeira (eds.), Hydrogen-Bonded Liquids, 81–88.

While all liquids investigated - pure or homogenous mixtures- conform to equation 1,their T-dependence may differ from that predicted by equation 2,and consequently the Cp-behavior along an isotherm; in this respect additional aspects were observed with the study of Toluene, because of its relatively low melting point Tm with regard to its parent molecule, the benzene. Therefore our thermodynamic description of the liquid state was extended to a second temperature regime,called in figure1(b) the"low or intermediate"range approaching Tm, which shows a multiplicity of crossing points of the isotherms decreasing from 1kbar to atmospheric pressure when temperature decreases, while equation 2 is now considered as describing the "hot" region of the liquid.

Hydrogen bonded liquids ,wich are frequently reluctant to crystallise and good glass-formers , allow us to distinguish a third T-domain in the undercooled state between Tm and the glass transition temperature Tg .Below further attention will be given to this domain . The other interesting aspect of the undercooled and overcompressed liquid thermodynamics has a more theorical character ; the particular behavior of the derivative coefficients in these areas leads to the notion of an "Ideal" Glass Transition, described as a structural phase transition as the Kauzmann Paradox[4], which provides at atmospheric pressure a lower temperature limit for the existence of a supercooled liquid; it will be referred here to as a "limit" temperature* and a "limit" pressure, the actual observation of which is prevented by the vitrification.

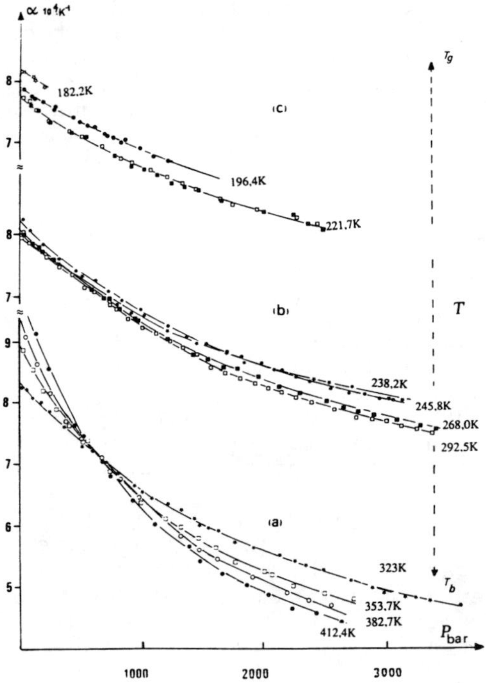

Figure 1. General pattern of several isotherms of the expansivity α as a function of pressure in the case of liquid m-FluoroAniline :(a)-"hot" range common to all liquids, with a single crossing point of all the isotherms. (b)-"intermediate" range when T approaches the low melting region.(c)-"undercooled" regime approaching Tg, where α increases when T decreases.

2. METHOD AND APPARATUS: PIEZOTHERMAL ANALYSIS AND ISOTHERMAL CALORIMETERS

Most expansivity determinations involve volume measurements under pressure[5,6]. It is also possible to use thermodynamic means[7]: considering a temperature rise due to an adiabatic pressure increase and measuring separately the heat capacity per unit volume, the thermal expansion coefficient may be computed. The metrology used here allows the first direct measurement of α as a function of T and P. Developed by L.Ter Minassian[8,9] the piezothermal method is based on the determination of the heat of compression δQ[10] under isothermal conditions.Using the Maxwell relations , the quantity of heat liberated by one mole of substance after a variation ΔP of the pressure on a sample may be written as :

$$\delta q = TdS = -\alpha VTdP \tag{4}$$

where δq, S and V are the molar quantities for heat, entropy and volume. In an effective calibrated volume Ve with n=Ve/V moles one defines directly the expansivity α as a function of T and P from the experiment :

$$\delta q' = n\,\delta q = -\alpha\, VeT\ dP \tag{5}$$

The final result is obtained by integrating over a not too large pressure interval ΔP in order to neglect the variation of α within this interval, and may be written as:

$$(\alpha - \alpha_r) = -\Delta Q / Ve\ T\ \Delta P \tag{6}$$

α_r is a corrective term arising from the contribution of the vessel to the overall thermal effect.

Consequently the other thermodynamic derivatives are obtained from the following relations :

the molar volume

$$V(T,P) = V(Tr,P)\exp\int_{Tr}^{T}\alpha\ dT \tag{7}$$

the heat capacity Cp at constant P, see equation 3

$$\Delta Cp = Cp\ (Pr,T) - Cp(P,T)$$

the isothermal compressibility

$$\Delta K_T = \int_{Tr}^{T} -(\delta\alpha/\delta P)_T\ dT \tag{8}$$

where Pr and Tr are the reference pressure and temperature.

In order to obtain these quantities with useful accuracy in

the temperature range of 90-350K, this requires a high performance isothermal calorimeter of a new type, based on a pneumatic principle and conceived by Ter Minassian and Milliou[11]. It permits a precision of 2°/°° in the ΔQ measurement. A H.P. continous shunted tube constitutes the calorimetric cell , with a cône-to-cône closure at one end while the other is connected to a H.P. system , strain gauges and compressor; this cell was build for the study of viscous liquids and ensures an hydrostatic pressure transmission on the sample. The effective working volume Ve of the cell is calibrated using the expansion or compression of perfect gas in the limit of 1-10 bar , where $\alpha T=1$; then $Ve=Qe/\Delta P$.For higher temperatures, a conventional flux calorimeter was used.

The experimental procedure consists of introducing the liquid under study into the whole high pressure system,- calorimetric cell, tube and H.P. pump-, and waiting for thermal equilibrium to be reached at the working temperature. The pressure is then varied by steps of 50-200 bar, the quantity ΔQ being measured at the end of each step. The experiments are performed along isotherms up to a pressure of 4 kbar, within the temperature range 180-420 K .

3. SUPERCOOLED AND OVERCOMPRESSED m-FLUOROANILINE (m-FA) BY PIEZOTHERMAL ANALYSIS

We have recently extended the piezothermal method to the study of undercooled and overcompressed liquids near their glass transition. We will present some data for m-FA which show common behavior with other molecular liquids on which reports may be found in the litterature[12].

m-FA, a simple disubstitued benzene ring compound, is a good glassformer which never cristallises and is well suited for the observation of liquid properties in the vicinity of the glass transition [13]. In this temperature range new phenomena are encountered :

-i- a systematic increase of the expansivity, irrespective of the pressure, when the temperature is lowered as it is shown in figure 1(c). The consequence of this is a sharper decrease of the molar volume as Tg(173K) is approached following a non-linear T-dependence, see figure 2(a). The tendency of the liquid volume, as well as its entropy, is to fall much more rapidly than that of the crystal; analysed with reference to the classical approache of Free Volume theories[14], we may define an exces volume as

$$v_{free}=V_{liq}(T,P_{atm})-V_o(T_k,P_{atm}) \qquad \text{and}$$

the proportion v_{free}/V_o in figure2(b)

Because no crystal data exist for m-FA, we have choosen to

analyse our data with reference to the Kauzmann temperature Tk estimated from similar molecular liquids giving Tg/Tk=1.20 in reference(13), at which these quantities tend to zero.

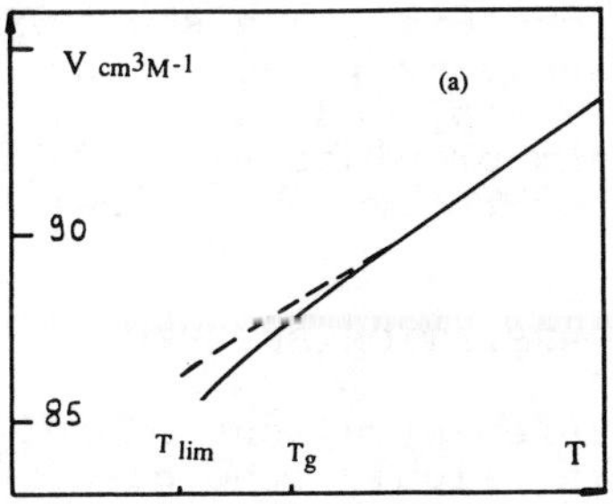

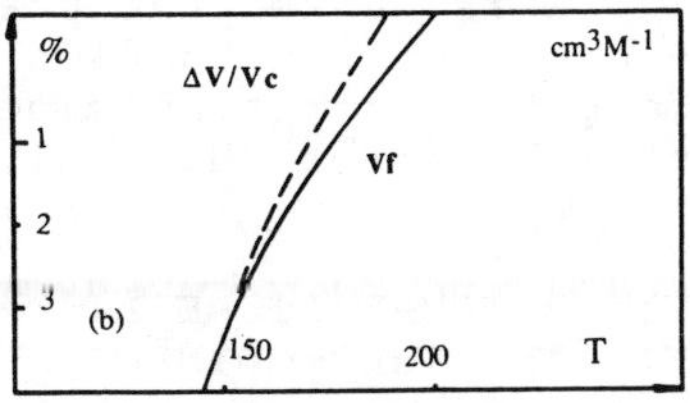

Figure 2. Consequence on the molar volume of the behavior of α with T at Patm:(a) V varies rapidly, (b)the exces volume and proportion tend to zero at a finite temperature equal to Tk.

-ii- an increase of the specific heat Cp with the pressure at constant temperature, starting from the atmospheric pressure as shown in figure 3

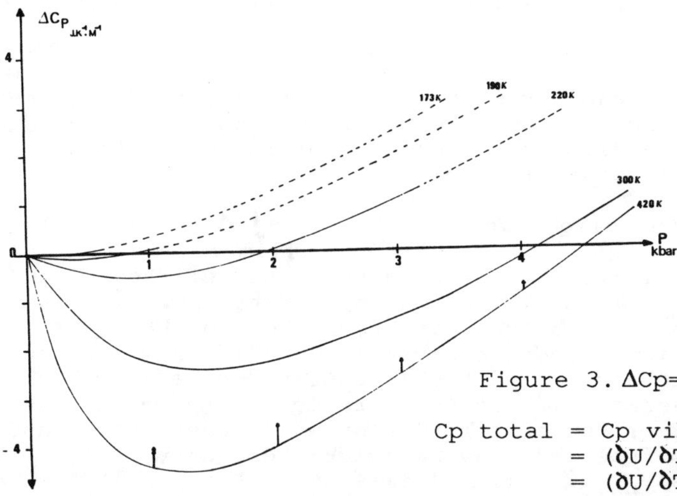

Figure 3. $\Delta Cp = Cp(T,P) - Cp(T,Patm)$

$$
\begin{aligned}
Cp\ total &= Cp\ vib - Cp\ struct \\
&= (\partial U/\partial T)_P + P(\partial V/\partial T)_P \\
&= (\partial U/\partial T)_P - T(\partial S/\partial P) \qquad (9)
\end{aligned}
$$

If we decompose the heat capacity into two contributions, vibrational and structural, this may be explained either as an ordering effect associated with an entropy decrease together-or-with a reduction of the interaction energy[15,16]. The distinction between an entropy and an energy description of the transition needs more experiments in a wide range of P and T. The behavior of α as a function of temperature , as well as the behavior of Cp as a function of pressure, reflects structural changes. They are correlated to the abrupt increase of the viscosity vs T implying a highly cooperative viscous flow process. In this sense the thermal expansivity , describing the co-variance of the fluctuations of the entropy and the volume [17], is a good macroscopic detector and provides a thermodynamic key for the analysis of cooperative phenomena.

4-" LIMIT " TEMPERATURE AT Patm AND CORRELATION BETWEEN KINETIC AND THERMODYNAMIC PROPERTIES

m-FA is not a special case : a literature survey on one of the most exhaustively studied glassforming liquid, the glycerol shows the same behavior of the expansivity and the density in the highly viscous range. With very accurate volumic measurements of A.K.Schulz [18] in the glassy , supercooled liquid and crystal states at Patm, we may define now an exces proportion of the supercooled liquid over the stable crystal as the following ratio:

$$r_{ex} = V_{liq}(T,P) \;/\; V_{crys}\,(T,P) \qquad \text{for } T<Tm$$

$$rex = \exp \int_{T_{lim}}^{T} (\,\alpha_{liq} - \alpha_{crys}\,)\, dT \qquad (10)$$

this quantity varies rapidly with temperature and tends to one at a finite temperature T_{lim} denoted by the crossover of liquid and crystal volumes; a "limit" temperature of 137 K is then computed with the following principle:

$$\text{Ln}\,(rex)_{Tm} = \text{Ln}\,(Vliq \;/\; Vcrys)_{fusion} = \int_{Tlim}^{Tm} \Delta\alpha \;dT \qquad (11)$$

where $(rex)_{Tm} = (Vliq/Vcrys)_{fusion}$ is the ratio of the molar volume of respectively the liquid and the crystal at the melting point Tm and $\Delta\alpha$ is the difference between the expansivity of the liquid and the crystal up to Tm.From data on glycerol and m-FA, it appears that the quantity $\Delta\alpha$ varies in a simple hyperbolic fashion with temperature $\Delta\alpha$=D/T shown in figure 4. In this case, the temperature at which the crystal and liquid volumes become the same may be calculated analyticaly

$$(Vliq\;/Vcrys)_{fusion} = (Tm\;/Tlim)^{D} \qquad (12)$$

Interestingly enough, this is close to the point To of 135 K obtained by best fitting of the V-T-F equation for the enthalpy relaxation time [19] as a function of T , and to Kauzmann temperature calculated by vanishing of configurational entropy from Cp data Tk of 137K [20]. In this sense, both theories -entropy [21] and free volume- are in agreement.

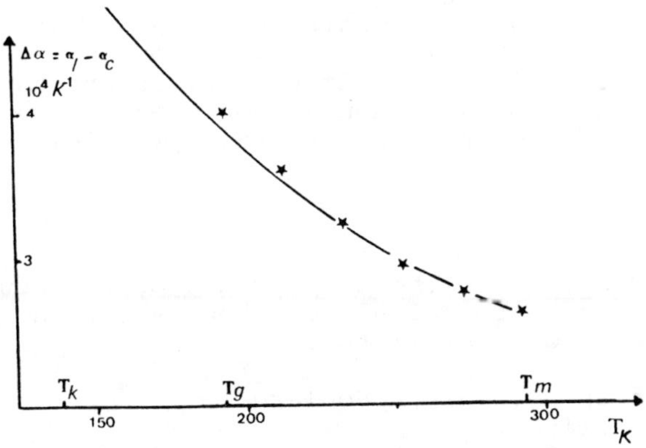

Figure 4. Hyperbolic T-dependence of $\Delta\alpha$ at Patm for glycerol from (18)

Consequently a parallelism can be established between kinetic and thermodynamic coefficients, during the cooling process of the liquid ; it is interpreted as the result of a structural change in the phase involving low potential energy arrangements of molecules as well as an increase of short and intermediate order. It is difficult indeed to specify the type of order which appears however this latter essentially depends on the molecular shape and on the intermolecular forces [12b].

5. CONCLUSION: THE GLASS TRANSITION PRESSURE Pg AND THE LIMIT PRESSURE Plim

Most of the recent theoretical activity has been focussed on the kinetic aspects of the supercooled state and proposes a mode-coupling[22] approach to the behavior of viscous liquid around its glass transition. Many experiments using mainly neutron scattering techniques[23] are in a good agreement with the mode-coupling theory and predicts a critical dynamical instability at some temperature, which lies distinctly well above the real glass transition Tg.

However, keeping in mind the underlying thermodynamic factors implied by the existense of such temperatures as Tk and Tlim, only high accuracy experimental data can build up the links between these two appraoches. Only few measurements have been made of viscoelastic behavior of liquids under high hydrostatic pressure because of the experimental difficulties involved, but the growing interest in the effect of pressure on the thermodynamic properties encourage the development of new technics in the pressure range of 4 or 7 kbar. Thus an extension of our measurements would allow us to observe the thermodynamic manifestation of a glass transition Pg along an isothermal path and consequently the first definition of a "limit" pressure Plim. If such point as T limit, T k, To exists at constant pressure, a"Plimit" -or Kauzmann - should exist at constant temperature, and be estimated with the same principles.

REFERENCES

* The terms of T_{lim} and P_{lim} were choosen because of the new character of the set of data used to calculated them and then to compare them to the other "IGT" temperature

(1)-C.Alba ,L. Ter Minassian, A.Denis,A.Soulard J.Chem.Phys.**82**,(1),384,(1985)

(2)-L.Ter Minassian,K.Bouzar,C.Alba,J.Phys.Chem.,1988, **92**,487

(3)-P.W. Bridgman,Proc. Am. Acad. Arts Sci.**49** , 3,(1913)

(4)-W.Kauzmann,Chem.Rev.43,219,(1948)

(5)-P.W.Bridgman, Collected experimental papers,volIV, Harvard University Press,1964

(6)-A.Quach,R.Simha,J.Appl.Phys.,**42**,4592,(1971)

(7)-O.Sandberg,P.Andersson,G.Bäckström,J.Phys.E,**10**,474,(1977)

(8)-L. Ter Minassian and co J.Chim.Phys.**67**,265 (1970)

(9)-L. Ter Minassian ,P.Pruzan, J.Chem.Phys.**75**,3064,(1981)

(10)-H.B.Callen "Thermodynamics" Wiley Ed.,1960

(11)-L.Ter Minassian,F.Milliou J. Phys. E **16** , 450 (1983)

(12)-a-C.Alba-Simionesco,L.Busse;D.List;C.A.Angell,J.Chem.Phys in press;-b-C.Alba-Simionesco,C.A.Angell(to be submitted)

(13)-C.Alba-Simionesco,to be published

(14)-M.H.Cohen and D.Turnbull,J.Chem.Phys.**31**,1164,(1959); M.H.Cohen and G.S.Grest, Phys.Rev. B**20**,1077,(1979).

(15)-J.Jäckle Phil.Mag.B 1987,**56**,113-127

(16)-M.Goldstein, J.Chem.Phys,**51**, 3728, (1969)

(17)-L.Landau,E.Lifschitz,"Statistical Physics",(1967),ed Moscou

(18)-A.K.Schulz,J.Chim.Phys. **6**, 324,(1954)

(19)-N.O.Birge,J.R.Nagel Phys.Rev.Lett.**54**,2674,(1985)

(20)-C.A.Angell,D.L.SmithJ.Phys.Chem.,**86**,3845,(1982); C.A.Angell Phys.Chem.Solids 49,863-871,(1988)

(21)-G.Adams ,J.H. Gibbs J.Chem.Phys.**43**,139,(1965)

(22)-W.Götze,L.SjörenZ.Phys.B**65**,415,(1987);U.Bengtzelius, W.Götze,A.Sjolander J.Phys.Chem.**17**,5915,(1984)

(23)-"Dynamics of disordered materials",Springer Proceedings in Physics vol 37,1989

X-RAY DIFFRACTION STUDIES OF LIQUIDS

L. BOSIO
E.S.P.C.I.
10, rue Vauquelin
75231, Paris Cedex 05, France

ABSTRACT. After a survey concerning the basic features of the extended X-ray absorption fine structure and X-ray diffraction, a brief description of the experimental methods is given. The steps to convert the raw data into a reduced intensity function are also described. The structural description of molecular liquids as given by Fourier transformation is then analysed with a special reference to hydrogen-bonded liquids.

1. INTRODUCTION.

In studies of hydrogen-bonded liquids, it may seem paradoxical to refer to X-ray studies when it is known that hydrogen atoms are difficult to locate by X-ray methods. However, we shall see that X-ray experiments can indicate clearly the presence of hydrogen-bonded groups and also give complementary information to that obtained by other methods. The structure of a molecular liquid is only expressible in terms of distribution functions [1] depending on the distance between molecular centres and depending also on the orientation of the molecules. But the angular correlation function cannot be measured experimentally and all we can obtain is a superposition of weighted pair correlation functions related to all atomic pairs involved in intra- and intermolecular contributions. To identify these pairs, a large number of experimental data from the measurements where the atoms have different scattering lengths are required: in other papers, neutron scattering [2] and electron diffraction studies [3] have been considered; here we are concerned with methods using X-rays and we shall begin by recalling some fundamental properties of X-rays.

X-ray radiation is a part of the electromagnetic spectrum in the approximate wavelength range 0.01 to 500 Å, sandwiched between γ-radiation at shorter wavelengths and ultraviolet radiation on the high wavelength side. X-ray tubes in which a stream of high-energy electrons impinges on an anode are the most common sources of X-rays but high flux X-ray beams are now provided by synchrotron radiation sources: synchrotron radiation is based on the emission of electromagnetic waves by accelerated charged particles at relativistist speeds.

For diffraction purposes, the X-ray wavelength λ is restricted to the range 0.5 - 2.5 Å; since the energy E associated with the radiation is given by $E = hc / \lambda$ (where h is the Planck constant, c the light velocity) the energy range of incident photons is 5 to 25 keV. This energy is much higher than the thermal energy (kT = 25 meV at 300 K); so, it is very

J.C. Dore and J. Teixeira (eds.), Hydrogen-Bonded Liquids, 89–100.

difficult to measure the change in energy of scattered photons and this is the reason why X-rays cannot be used as a tool to study dynamic processes.

X-rays have the velocity of light so that, when they pass through a liquid, they see a frozen landscape: the description of the structure therefore will arise from the superposition of several snap-shot pictures averaged over time and space. When an X-ray beam passes through a material medium, its intensity is reduced by a variety of effects which may be grouped under two general headings: absorption and scattering.

2. ABSORPTION.

The absorption relationship giving the intensity I that is transmitted through a substance of thickness t versus the incident intensity I_o has the familiar form:

$$I = I_0 \exp(- \mu t) \qquad (1)$$

and this equation defines the total linear absorption coefficient μ . The incident X-ray beam is attenuated by several processes such as coherent and Compton scattering, considered below but, essentially, the absorption is due to the photoelectric effect: we shall see that an electron can interact with a photon whose electromagnetic wave field imparts to that electron a certain acceleration. Suppose that an incident photon has a sufficient energy $h\nu$ to eject an inner electron from an atom (Fig. 1a): this may happen when $h\nu > E_b$, if E_b is the energy required to knock an electron out of a given shell. The absorbed radiation is used to eject the electron, now called "photoelectron" which carries away any excess energy in the form of kinetic energy. Therefore the X-ray photon disappears completely and the absorption coefficient exhibits an abrupt increase at the so-called "absorption edge" (Fig. 1b). According to quantum-mechanical considerations, the probability of ejection is largest for quanta just sufficient for the process and it is small for a photon of energy very much in excess of that of the absorption edge: this is the reason why, between two absorption edges, μ decreases when the energy of the photon increases.

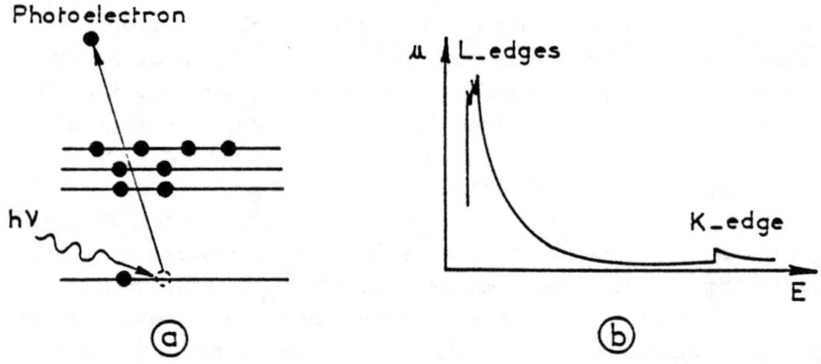

Figure 1. a) Absorption of a photon in the inner shell of an atom. b) Variation of the absorption coefficient as a function of the incident energy.

The photoelectron that escapes from the excited atom can probe the local structure: the energy state of the escaping electron is represented as an outward-moving spherical wave centered around the absorbing atom (Fig. 2a). If the atom is surrounded by other atoms, the electron wave will be back-scattered by those atoms. It is the addition of such waves that gives rise to the fine structure above an absorption edge, called EXAFS (Extended X-ray Absorption Fine Structure); such modulations can extend up to about 1 keV beyong the edge and have a magnitude of typically less than 10% of the edge jump (Fig. 2b). The frequency of the EXAFS oscillations will depend on the distance between the absorber and its near neighbors, while the amplitude of the oscillations will depend on the number and type of neighbors. Because of the small effective electron escape depth, EXAFS cannot provide long-range information and therefore is, in principle, suitable to study the liquid structure: one can probe the local environment around a particular element by employing X-ray energies near an absorption edge of the element of interest.

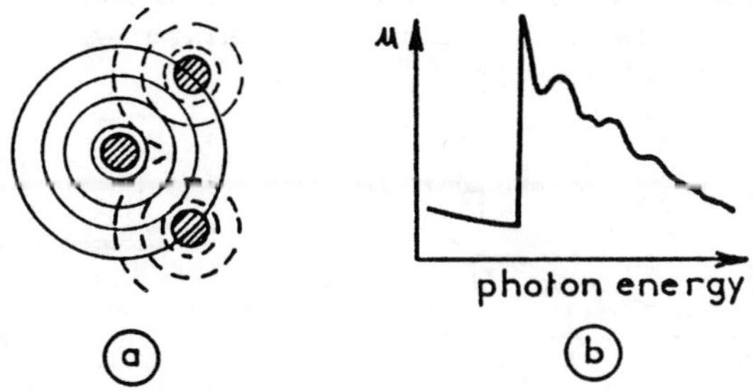

Figure 2. a) Schematic of the radial portion of the photoelectron wave (solid line: outgoing wave; dashed line: back-scattered wave). b) An X-ray absorption spectrum

An experimental set-up for measuring EXAFS modulations is very simple: after passing through a monochromator which selects a given incident energy, the beam intensity I_o is monitored in an ion chamber; the beam then passes through the sample and is observed in another ion chamber which measures the transmitted intensity I versus the incident energy (the EXAFS modulations can also be detected by measuring the intensity of the fluorescence yield).

Unfortunately, the presence of an atomic scattering phase shift term in the argument of the sine function describing the EXAFS oscillations [4] gives rise to a complication in deducing the local structure. Moreover, for the subject in which we are interested, the atoms able to form hydrogen bonds, namely fluorine, oxygen and nitrogen, have K-edges located at very low energies: 686, 532 and 402 eV respectively so that the measurements require a vacuum around the sample, a condition not consistent with the liquid state of the compound of interest. However, it is worthy of note that X-ray Raman scattering by hard X-rays [5] is a very promising method for the determination of the local structure around light elements.

3. SCATTERING.

3.1. Coherent scattering. Scattering by electrons.

Consider an X-ray beam whose wave has the form:

$$\underline{E} = \underline{E}_o \exp (- iwt) \qquad (2)$$

(w is the angular frequency) encountering a particle of charge e and mass m (Fig. 3a). Let $\underline{k}_o$ be the wave vector, $|k_o| = 2\pi/\lambda$. If the particle is moving slowly compared with the velocity c of light, it will receive and acceleration γ from the electric field so that

$$\underline{\gamma} = e / m \ \underline{E} \qquad (3)$$

and the accelerated particle must, in turn, radiate an electromagnetic wave in all directions; the factor 1/m in equ. 3 shows why, if the particle is an atom, electrons are the only effective scatterers (the lightest nucleus, the proton, has the same magnitude of charge as the electron but 1837 times the mass). If the incident beam is unpolarized, one can readily show that the wave scattered by an electron is polarized: the intensity I_e

scattered by one electron at a distance r in a direction 2 θ to the direction of the incident beam is given by the familiar Thomson equation:

$$I_e = I_o r_e^2 / r^2 \ (1 + \cos^2 2\theta) / 2 \qquad (4)$$

where I_o is the intensity of the incident beam, $r_e = 2.82\ 10^{-15}$ m is the classical radius of one electron; $P(\theta) = (1 + \cos^2 2\theta) / 2$ is called the polarization factor.

Note that in X-ray measurements, the intensity (that is the energy per unit area per unit time) is usually given in units of the classical scattering power of an electron which is abbreviated as electron units (e.u).

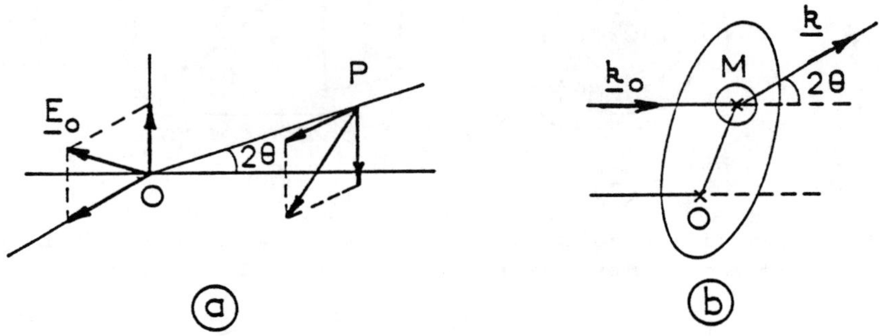

Figure 3. a) Scattering of an unpolarized beam by a single free electron. b) Scattering by an elemental volume.

Consider now a collection of electrons confined in a volume V and illuminated by the X-ray beam (Fig. 3b). If $\rho(\underline{r})$ represents the charge density in electron units per volume, each volume element dV around the point M, located at a distance r from the origin O, scatters at point P in the direction 2 θ a wave whose the wave vector is $\underline{k}$. By defining the scattering vector

$$\underline{q} = \underline{k}_o - \underline{k} \quad \text{or} \quad |q| = 4\pi \sin\theta / \lambda \qquad (5)$$

it is easy to show that the wave at P has the form:

$$\rho(\underline{r}) \exp[-i(wt + \underline{q}\,\underline{r})]\, dV \qquad (6)$$

The amplitude scattered by the volume element dV is then $\rho(\underline{r}) \exp(-i\,\underline{q}\,\underline{r})\, dV$ and the resultant amplitude in the direction 2 θ is proportional to

$$a = \int \rho(\underline{r}) \exp(-i\,\underline{q}\,\underline{r})\, dV \qquad (7)$$

In other words, the scattered amplitude is proportional to the Fourier transform of the electron-density. In the case of a spherically symmetrical distribution, equ. 7 can be written:

$$a = \int \rho(r)\, r^2 (\sin q r)/(q r)\, dr \qquad (8)$$

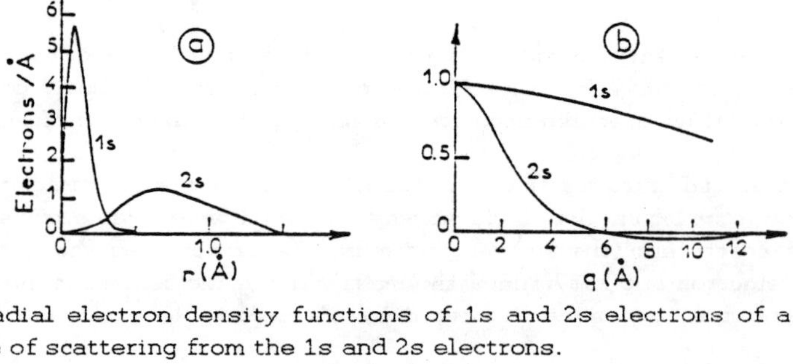

Figure 4. The radial electron density functions of 1s and 2s electrons of a carbon atom. b) The amplitude of scattering from the 1s and 2s electrons.

As an application, consider the electronic cloud around an atom, a carbon atom for example. If the radial electron-density function of 1s and 2s electrons have the shape reproduced in figure 4a, then the amplitudes of scattering from 1s and 2s electrons are given by the Fourier transforms shown in figure 4b and by adding the contribution of all electrons we obtain the so-called atomic scattering factor f_o for the carbon atom.

In the forward direction (q = 0) the scattered waves from electrons are in phase, and f_o is equal to the atomic number Z (Fig. 5); as q increases, f_o falls off quite rapidly because the dimensions of the electronic cloud are comparable with the wavelength of X-rays. Figure 5 compares the curves f_o / Z for atoms frequently encountered in hydrogen-bonded liquids: the more tightly the electrons are bound to the nucleus, the greater will be the concentration of electron-density towards the centre of the atom: this is the reason why the hydrogen atom - in addition to the fact that it possesses only one electron - is so bad a scatterer (note that this situation does not occur in neutron scattering because the scatterer is the nucleus which can be considered as a dimensionless point).

This dependence of the atomic scattering factor on the kind of atom will cause a difficulty for compounds since it is not possible to define an electronic scattering factor f_e such that f_o would be equal to Z f_e.

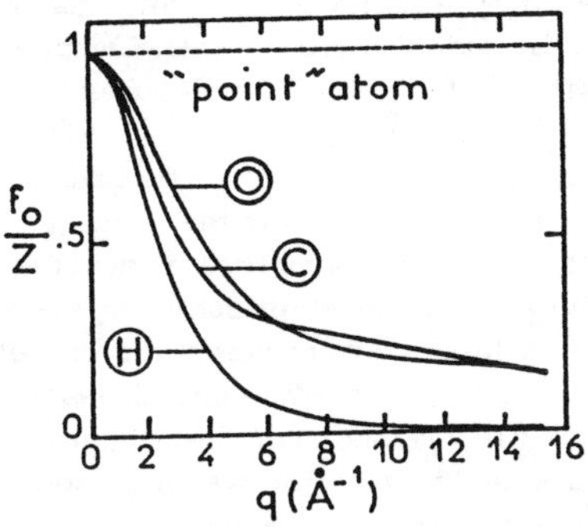

Figure 5. Atomic scattering factor f_o divided by the atomic number Z for H, C and O (solid lines) and for a point atom (dashed line).

3.2. Anomalous scattering.

In the discussion so far, it has been assumed that the electrons are free and scatter X-rays according to the Thomson treatment. The scattering power of a bound electron may be greater or less than that of a free electron and the phase of the scattered wave may be different. These effects may be taken into account by representing the atomic scattering factor f as a complex number:

$$f = f_o + \Delta f' + i \Delta f'' \quad (9)$$

where $\Delta f'$ and $\Delta f''$ are called the real and imaginary terms of anomalous scattering; their values together with f_o are tabulated [6]. The dispersion corrections depend on the X-ray energy but are appreciable only at an absorption edge where the angular frequency of a certain kind of electrons is equal to that of the incident wave. An important consequence of anomalous scattering is that it is possible to change the value of one of the atomic scattering factors of a polyatomic system by choosing the incident energy on both sides of an absorption edge: thus one can determine differential distribution functions [7], rather like the isotopic substitution used in neutron scattering [8].

3.3. Incoherent scattering (Compton effect).

A collision between an incident photon and an electron is not purely elastic and there is an exchange of energy and momentum; after collision, the photon has a slightly smaller energy, the wavelength λ' is larger so that

$$\Delta\lambda = \lambda' - \lambda = h / mc \quad (1 - \cos\theta) \qquad (10)$$

The modified X-rays carry no information about the structure and must be correctly removed; the modified scattering per atom, I_{mod}, is given in Tables [9].

Finally, taking into account the recoil electron velocity (Breit correction) and a quantum mechanical consideration (Dirac correction), it is necessary to multiply I_{mod} by the so-called Breit-Dirac recoil factor: $R = (\nu' / \nu)^n$, where ν and ν' are the frequencies of the incident and incoherently scattered X-rays respectively, and $n = 2$ or 3.

4. SCATTERING BY A LIQUID.

4.1. Molecular pair correlations.

Consider a volume V containing N molecules. If we are concerned with non-spherical particles, the amplitude scattered by any molecule depends, through $\rho(\underline{r})$ and $\underline{q}$ (equ. 7), on its orientation given by the Eulerian angle Ω : the amplitude scattered must now be written $a(\underline{q}, \Omega)$ and the intensity scattered by N molecules is:

$$I = \Sigma \langle a(\underline{q}, \Omega_j)\, a^*(\underline{q}, \Omega_k) \exp(-i\, \underline{q}\, \underline{r}_{jk}) \rangle \qquad (11)$$

if $\underline{r}_{jk} = \underline{r}_j - \underline{r}_k$; the brackets denote an average over the grand canonical ensemble. Equ. 11 gives a complete description of the system; however, $a(\underline{q}, \Omega)$ must be known but is not accessible from experiments. To calculate the molecular scattering factors, approximations are required [10,11]. But the use of these approaches is severely limited and it is very difficult to obtain direct information about the angular correlations in a liquid by this formalism. The problem is only simplified in two special cases [12]: i) when the particles have an effective sphericity, that is, a spherical distribution of electron-density. Such a situation occurs in X-ray measurements on liquid water: orientational correlation between neighboring molecules must be expected to play an important role but it is not directly detectable with X-rays because the water molecule has an electron-density very nearly spherical, ii) when the particles have a dynamical sphericity; here, the orientation of pairs of molecules is independent of their separation; this situation may occur in systems of low density which is not usually the case in the liquid state.

These two conditions are satisfied when one considers the atoms as scattering particles and that is the reason why correlations are studied by using the formalism based on the partial pair correlation functions dependent on the spatial relationship of individual atoms as described in subsequent sections. However, a large number of different and independent experimental data are required to identify the atomic pairs.

4.2. Atom pair correlations.

If the scattering particles are atoms (with electron-density distributions assumed spherical even inside a molecule), equ. 11 becomes:

$$I = \Sigma \langle f_j(q)\, f_k^*(q) \exp(-i\, \underline{q}\, \underline{r}_{jk}) \rangle \qquad (12)$$

where f_j and f_k are the scattering factors of the atom pairs. The total liquid structure factor $S_M(q)$ may be defined as the scattered intensity related to one scattering molecule, including intra- and intermolecular correlations, so that:

$$S_M(q) = 1 / N \langle \Sigma f_j f_k^* \exp(-i\, \underline{q}\, \underline{r}_{jk}) \rangle / (\Sigma f_j)^2 \quad (13)$$

The term (Σf_j) squared at the denominator of equ. 13 is a sharpening term introduced to take into account the fact that the atomic scattering factors fall off when q increases. This is an approximate method which defines an average scattering factor per electron [13]. Since i) a liquid is a macroscopically isotropic material and ii) the pair correlation function is by far the most important among all the distribution functions, the conversion of the structure factor to a real space representation can be made by calculating the inverse Fourier sine transform (cf. equ. 8) which can be written after some arrangements:

$$r\,[g(r) - 1] = 1 / (2\pi^2 \rho_o) \int q\,[S_M(q) - 1] \sin qr \, dq \quad (14)$$

where g(r) is the pair correlation function which represents the probability of finding a pair of atoms with a separation r ; ρ_o is the molecular density.

5. EXPERIMENTAL TECHNIQUES.

In any diffraction experiment, a sample is exposed to collimated radiation, and the diffracted intensity is recorded at various q values either by varying θ or λ since $q = 4\pi \sin \theta/\lambda$.

5.1. Monochromatic incident beam.

This is by far the most widely used mode; the white radiation of a synchrotron beam or the contaminated characteristic emission from an X-ray tube is monochromated by a crystal monochromator. In the transmission mode (Fig. 6a), a convenient method is to use in the primary beam a bent crystal with an asymmetric cut to focus the emergent radiation onto the slit set in front of the detector. In this transmission geometry, the liquid must be contained in a cell equipped with transparent windows. In the reflection mode (Fig. 6b) one can obtain diffraction patterns from the free surface of a liquid by the use of a θ - θ diffractometer: the X-ray tube and the detector rotate at the same angular rate and in opposite sense about a horizontal axis lying in the surface of the liquid.

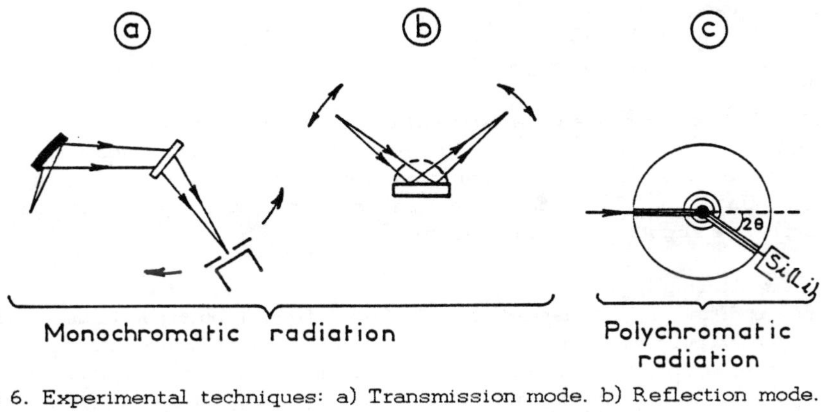

Figure 6. Experimental techniques: a) Transmission mode. b) Reflection mode. c) Energy dispersive diffraction method.

In both techniques the scattered photons are counted by a scintillation or a proportional counter in conjunction with a pulse height analyser. Usually, the counter is moved around the sample and the scattered X-rays are collected at equidistant steps with increments Δq typically equal to 0.05 Å. One can save time by using a curved position sensitive detector since the counting processes are performed simultaneously at different angular positions; in this case, a well-collimated beam illuminates the liquid contained in a Lindemann-glass capillary. To avoid air scattering at low angles, the sample must be placed in an evacuated tank or in a helium gas environment.

5.2. Polychromatic incident beam.

In some special cases - for example, in studies of the effect of pressure on the structure of liquid water [14] - it is advantageous to use an X-ray energy analyser set at a fixed angle θ (Fig. 6c). Since the incident beam is white radiation, the spectrometer measures photon energy and q varies as a function of λ . The limitation of this technique arises in the data reduction which is more complicated when compared with a scanning method.

6. CORRECTIONS AND DETERMINATION OF ABSOLUTE INTENSITY.

For determining the radial distribution function of a liquid, the raw scattered intensity I_{meas} must be corrected and put on an absolute scale. Many papers have been devoted to methods of carrying out such a treatment [15]; we restrict ourselves to the main corrections corresponding to the sections developed previously and to the case of an incident monochromatic beam.

6.1. Background B(θ).

If the experiments have been performed in an evacuated sample environment, with a sample holder equipped with suitable windows, the scattered background is greatly reduced. A scan with an empty cell must be performed: the measured scattering of the empty cell, reduced by the transmittance of the liquid, gives the background correction $B(\theta)$.

6.2. Polarization P(θ).

For an unpolarized primary beam (i.e. in absence of a crystal monochromator) the polarization factor $P(\theta)$ is that appearing in the Thomson relation (equ. 4). If a monochromator is used, then

$$P(\theta) = (1 + \cos^2 2\theta_o \cos^2 2\theta) / (1 + \cos^2 2\theta_o) \qquad (15)$$

where θ_o is the Bragg reflection angle of the crystal used as monochromator.

6.3. Absorption A(θ) correction and variation of effective irradiated volume G(θ).

These factors are dependent on the diffractometer and sample geometry. In the transmission mode, the corrections are relatively simple either for a cylindrical sample [16] or flat slab [15]; in the reflection mode, these terms are more complicated when the sample is a slightly absorbing material [17].

6.4. Incoherent scattering I_{inc}.

The Compton modified I_{mod} scattering, calculated from tabulated values, must be multiplied by the proper recoil factor R. If the monochromator is located in the diffracted beam - an arrangement often used to eliminate the fluorescence radiation when the energy of the incident beam is slightly higher than that corresponding to the absorption edge of one of the components - the incoherent intensity distribution is modified and the transfer response of the monochromator must be determined.

6.5. Multiple scattering.

The parasitic scattering (essentially the secondary scattered intensity) depends on the instrumental parameters and several kinds of correction have been proposed [18]. In fact, by choosing samples of relatively small thickness, such a correction can be neglected in many cases.

To summarize, the corrected I_{cor} and the normalized intensity I are given by:

$$I_{cor} = [I_{meas} - B(\theta)]\, P(\theta)^{-1} A(\theta)^{-1} G(\theta)^{-1} \quad (16)$$

$$I = \alpha I_{cor} - I_{inc} \quad (17)$$

where α is a factor transforming the corrected intensity from arbitrary units to electron units.

6.6. Normalization factor.

The normalization factor α may be determined by two methods: i) at large values of q, the interferences are greatly damped and the observed intensity must approach that of independent atoms, that is $\Sigma f_j^2 + I_{inc}$, then $\alpha = (\Sigma f_j^2 + I_{inc}) / I_{cor}$; ii) the second method [19] compares the integrals of the intensity functions taken over the whole accessible space, range 0 to q_m, then

$$\alpha = [\int q^2 (\Sigma f_j^2 + I_{inc})\, dq - 2\pi^2 \rho_0] / [\int q^2 I_{cor}\, dq].$$

A refinement of α may be obtained by minimizing the oscillations in the low r range of the Fourier transform within a region where it is known that g(r) is featureless.

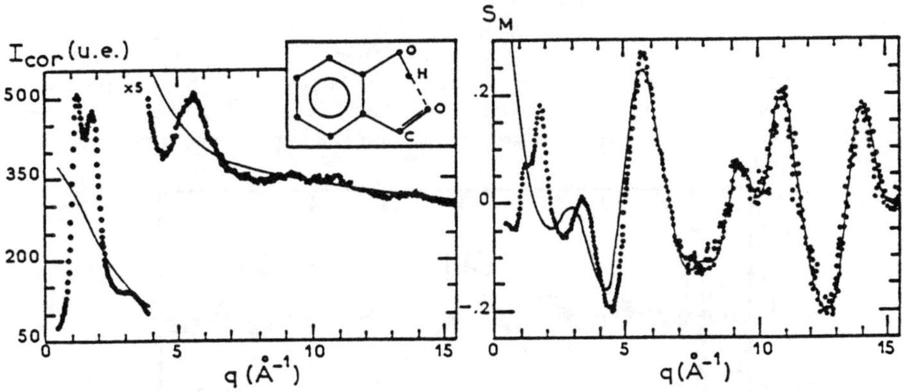

Figure 7. Salicylic aldehyde: a) Scattering intensity from X-ray diffraction (points), scaled to the self scattering from independent atoms in one molecule (solid line). b) q-weighted structure function (points) and molecular structure function q F_1 (solid line).

As an example, figure 7a shows the observed corrected intensity $I_{cor}(q)$ relative to salicylic aldehyde at room temperature; this liquid has been chosen as an illustration of a chelated molecule: a hydrogen bond connects the oxygen atom of the aldehyde group to a hydrogen atom located in the ortho position (inset Fig. 7a), the O...O distance being equal to 2.65 A [20]. The experiments were carried out on a slab sample (4 mm thick, λ = 0.7 Å)) in the transmission geometry.

7. INTERPRETATION OF RESULTS IN TERMS OF RADIAL DISTRIBUTION FUNCTIONS.

Whenever we are concerned with molecular liquids whose molecules contain M atoms, it is convenient to separate the total liquid factor $S_M(q)$, equ. 13, into the term arising from correlations in the same molecule, $F_1(q)$, and that from correlations in different molecules, $D_M(q)$, so that

$$S_M(q) = F_1(q) + D_M(q) \qquad (18)$$

Usually, the configuration of the molecule is known and $F_1(q)$ may be written:

$$F_1(q) = 1/(\Sigma f_j)^2 \; \Sigma f_j f_k J_o(q.r_{jk}) \exp(-\gamma_{jk} q^2) \qquad (19)$$

with γ_{jk}, the r.m.s. variation in the mean distance r_{jk}, $J_o(x) = \sin x / x$ and summation over all atoms in the molecule (we have neglected the terms related to any anharmonicity).

The q-weighted molecular structure function F_1 relative to salicylic aldehyde as calculated from the parameters determined by X-ray measurements on the crystalline form [20] is shown in figure 7b by a solid line: despite the complexity of the molecule and the fact that no parameter has been adjusted, for $q > 6$ Å^{-1}, we see that $S_M(q)$ and $F_1(q)$ are similar (here, all the γ_{jk} values were set equal to zero and this apparent rigidity of the molecule is perhaps the signature of the hydrogen bond inside the molecule).

The distinct structure factor $D_M(q)$, calculated according to equ. 18, decays to zero very rapidly. When, between molecules, hydrogen bonds exist, significant changes appear in D_M: figure 8 shows the q-weighted distinct functions for butyl alcohol $C(CH_3)_3OH$ and for neopentane $C(CH_3)_4$ near their respective melting points, as determined by Narten and Sandler [21]: for $q > 8$ A^{-1}, the amplitude of the function D_M for neopentane is smaller than the uncertainty of an average data point; in contrast, for the tertiary butyl alcohol, D_M continues to oscillate in the high q-range. This can also be seen in real space by computing the Fourier transform $g_L(r)$ defined by:

$$r\,[\,g_L(r) - 1\,] = 1 / (2\pi^2 \rho_o) \int q\, D_M \sin qr\, dq \qquad (20)$$

The intermolecular radial distribution function related to butyl alcohol exhibits a peak at 2.74 Å – not present in neopentane – which is attributed to the hydrogen-bonded O...O between molecules. Similar observations have been reported by other authors and concerning studies on formic, acetic or propionic acids [22] and liquid formamide [23].

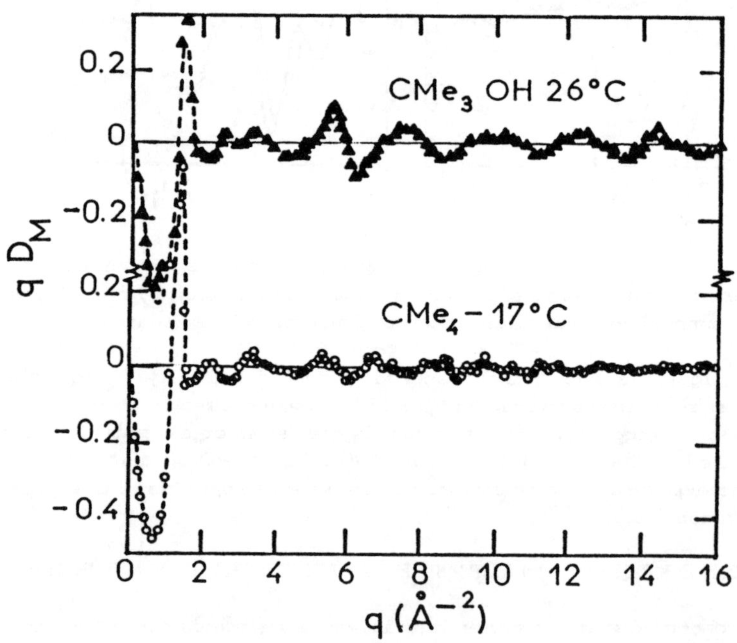

Figure 8. The q-weighted distinct structure function for CMe_3OH (top) and CMe_4 (bottom). After ref. 21.

Actually, $g_L(r)$ is a composite function which contains contributions from all the partial correlation functions weighted according to the scattering length-values and the relative concentration. By combining electron, X-ray and neutron diffraction studies (with isotopic substitutions in the last case), it is possible to change the contrast between the scattering species and then to obtain partial correlation functions related to given atomic pairs. An illustration of such a procedure, concerning liquid water, is given in this book [8].

8. ISOCHORIC TEMPERATURE DIFFERENTIAL OF THE X-RAY STRUCTURE FACTOR IN WATER.

In water, the structural variation versus temperature is expected to be small, and if the diffraction data are treated in the usual way (that is, after all corrections) the weak differences will be blurred and smoothed after Fourier transformation. For a precise comparison it is better to use the difference techniques as pointed out by Dore [2]. But to avoid corrections for density variation with temperature, one can take advantage of the fact that water has an anomalous behaviour with a density maximum so that it is possible to perform experiments at two different temperatures with the same density [24]. This isochoric temperature differential offers an accurate technique for studying local oxygen-atom ordering by X-ray diffraction when the temperature is decreased.

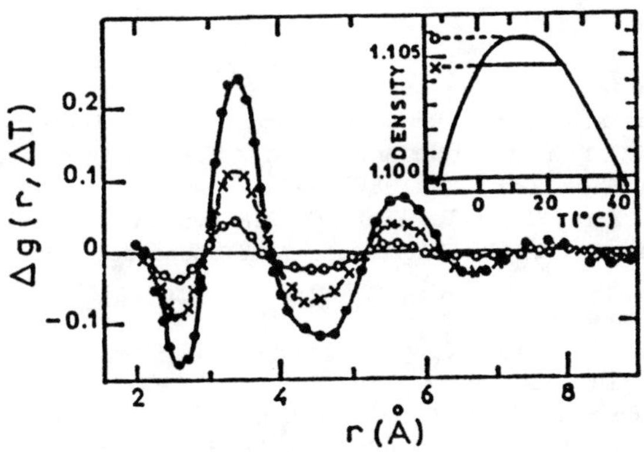

Figure 9. $\Delta g(r, \Delta T)$ for heavy water as calculated from X-ray measurements performed at three pairs of temperatures. (The inset shows the density curve and the positions of the measured temperatures).

X-ray structure factors for heavy water have been measured at 40.0, 23.5, 15.5, 11.2, 7.1, 0.2 and -11.0°C. From these data, isochoric temperature differentials were constructed for pairs of temperatures ΔT = 51.0, 23.3 and 8.4 °C around the maximum density point at 11.2 C. Figure 9 shows $\Delta g(r, \Delta T)$ for the three measured pairs of temperature; a striking feature is the proportionality between the structural variations and the temperature differences; therefore, it can be stated that the thermodynamic anomalies of water are not related to a structural short-range peculiarity. From these measurements, another conclusion can be drawn and consists of a substantial rearrangement of the molecular centres at the level of the second-neighbor shell. As stated above, X-rays cannot give any direct information about the molecular orientational correlations; nevertheless, it can be concluded from these data that the second nearest neighbors have a tendency to move away when the temperature is lowered. This corresponds to an increase of O-O-O angle towards the tetrahedral coordination value and such a result supports the inference made from neutron scattering measurements in highly supercooled water [8].

9. CONCLUSION.

The application of X-rays to probe the structure of liquids, and specially of hydrogen-bonded liquids, has proved to be a valuable technique but its ability to furnish useful information depends upon obtaining precise data for coherent intensity to large values of q and upon attention to possible errors in data treatment. For water, the determination of structural information is greatly facilited by the fact that the distibution of electron density in a molecule is almost spherical: in such conditions, the molecule can be treated as a homoatomic system, the pair correlation function thus yielding information about the molecular centres (which are in fact not very different from oxygen atoms). For molecular liquids, it it possible to give a reasonable account of the intramolecular structure, but the interpretation of intermolecular structure remains ambiguous because one has to determine many pair correlation functions. This difficulty can partly be overcome by using several other diffraction techniques and also EXAFS data.

REFERENCES.

[1] J.P. Hansen, this volume.
[2] J. Dore, this volume.
[3] G. Palinkas, this volume.
[4] See for example: "X-ray Absorption. Principles, Applications, Techniques of EXAFS, SEXAFS and XANES" ed. by D.C. Koningsberger and R. Prins, Wiley-Interscience 1988.
[5] K. Tohji and Y. Udagawa, Phys. Rev.B 36 (1987) 9410.
[6] International Tables for X-ray Crystallography IV. The Kynoch Press 1974.
[7] P.H. Fuoss, P. Eisenberger, W.K. Warburton and A. Bienenstock, Phys. Rev. Lett. 46 (1981) 1537.
[8] M-C. Bellissent-Funel, this volume.
[9] International Tables for X-ray Crystallography III. The Kynoch Press 1968. H.H.M.Balyuzi, Acta Cryst. A31 (1975) 600.
[10] M.D. Zeidler "Diffraction Studies on Non-Crystalline Substances" ed. by I. Hargittai and W.J. Orville-Thomas, Elsevier Scientific Publishing Cy 1981, p. 271.
[11] K.E. Gubbins and C.G. Gray, Molec. Phys. 23 (1972) 187.
[12] A.H. Narten and H. A. Levy, J. Chem. Phys. 55 (1971) 2263.
[13] B.E. Warren "X-ray Diffraction" Addison-Wesley Publishing Cy, 1969.
[14] Yu.E. Gorbaty and Yu.N. Demianets, Molec. Phys. 55 (1985) 571.
[15] See for example, F. Hajdu and G. Palinkas, J. Appl. Cryst. 5 (1972) 395.
[16] A.P. Kendig and C.J. Pings, J. Appl. Phys. 5 (1965) 1692.
[17] M.E. Milberg, J. Appl. Phys. 1 (1967) 64.
[18] G.D. Wignall, J.A.J. Jarvis, W.E. Munsil and C.J. Pings, J. Appl. Cryst. 7 (1974) 366.
[19] J.B. Krogh-Moe, Acta Cryst. 9 (1956) 951. N. Norman, Acta Cryst. 10 (1957) 370.
[20] P. Bourre-Maladiere, Compt. Rend. 237 (1953) 825.
[21] A.H. Narten and S.I. Sandler J. Chem. Phys. 71 (1979) 2069.
[22] T.V. Gorbunova, V.V. Shilov and G.I. Batalin, Zhurn. Strukt. Khim. 14 (1973) 424.
[23] E. Kalman, I. Serke, G. Palinkas, M.D. Zeidler, F. Wiesmann, H. Bertagnolli and P. Chieux, Z. Naturforsch. 38a (1983) 231.
[24] P.A. Egelstaff, J. A. Polo, J.H. Hahn and S.H. Chen, Phys; Rev. Lett. 47 (1981) 1733. L. Bosio, S.H. Chen and J. Teixeira, Phys. Rev. A27 (1983) 1468.

NEUTRON DIFFRACTION TECHNIQUES

P. Chieux
(Institut Laue Langevin, Grenoble, France)

and

J.C. Dore
(Physics Laboratory, University of Kent, Canterbury, England)

Abstract

Structural properties of hydrogen-bonded liquids can be investigated by thermal neutron diffraction. The basic formalism linking the measured intensity distribution to the liquid structure factor $S_M(Q)$ and the partial pair correlation functions $g_{\alpha\beta}(r)$ is described. Current experimental techniques using steady-state (reactor) and pulsed (accelerator) neutron sources are presented and the analysis procedures discussed in relation to the precision of the final data. Special methods using temperature variation and isotopic substitution are considered and some actual examples are briefly discussed. Small angle scattering studies can be used to give structural characteristics over a wider spatial domain and a short description of recent work on liquid mixtures is included. The review ends with a brief consideration of likely future developments.

1 Introduction

The use of X-ray diffraction for the study of crystallographic structures is a well-established and widely-known technique. In more recent times the use of thermal neutrons, which have comparable wavelengths in the region of 1Å, has played an increasingly important role in structural studies of both the ordered and disordered states of condensed matter. This type of work began after the construction of research reactors in the 1950's and has undergone continual development such that neutron diffraction techniques are often preferred for the

J.C. Dore and J. Teixeira (eds.), Hydrogen-Bonded Liquids, 101–116.

study of hydrogen-bonded liquids. The last decade has also seen alternative methods of neutron production introduced through the use of electron and proton accelerators. The purpose of this short review, is to outline these developments and to describe briefly the instrumental features and data-analysis procedures which give the required structural information.

2 Diffraction formalism

Thermal neutrons are scattered isotropically (s-wave scattering) and the outgoing spherical wave for a heavy nucleus has an amplitude $b\frac{e^{i\mathbf{k}.\mathbf{r}}}{r}$ where **k** is the wave vector and b is the bound-atom coherent scattering amplitude. The scattered wave comprises coherent and incoherent components depending on the nuclear properties (spin, isotope etc) but the diffraction pattern will be dependent only on the interference of coherently-scattered waves. The phase difference $\Delta\phi_{ij}$ between two waves scattered by two nuclei separated by a position vector $\mathbf{r}_{ij}$ is given by

$$\Delta\phi_{ij} = \mathbf{Q} \cdot \mathbf{r}_{ij} \qquad (2.1a)$$

where Q is the elastic scattering vector

$$\mathbf{Q} = \mathbf{k}' - \mathbf{k} = \frac{4\pi}{\lambda} \sin\frac{\theta}{2} \qquad (2.1b)$$

and λ is the incident wavelength (see Fig.1).

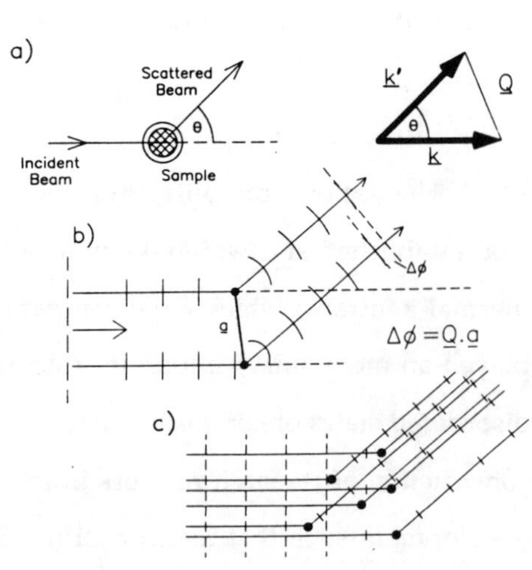

Figure 1: Scattering of coherent radiation by a disordered assembly of point scatterers.

a) geometrical features;
b) phase relation;
c) summation of waves.

For an assembly of scattering centres, the full diffraction pattern is given by the summation over all scattered waves so that the differential scattering cross-section for an assembly of N identical nuclei can be written as

$$\frac{d\sigma}{d\Omega}(\theta,\lambda) = \frac{1}{N} < \sum b_i\, b_j\, e^{\mathbf{Q}.\mathbf{r}_{ij}} > \tag{2.2}$$

where the summation extends over all pairs of nuclei, i,j, and the brackets represent and ensemble average. In the static approximation it is convenient to define a structure factor by

$$\frac{d\sigma}{d\Omega}(Q) = b^2 S(Q) \tag{2.3}$$

for an assembly of identical scatterers.

In general the sample will contain different atomic components and the structure-factor must be expressed as a sum over different species. For molecular liquids [1,2] it is also convenient to use the molecule as the basic unit and to define the structure factor as a sum of intra-molecular and inter-molecular terms, i.e.

$$S_M(Q) = f_1(Q) + D_M(Q) \tag{2.4}$$

where $f_1(Q)$ is a molecular form-factor

$$f_1(Q) = \frac{\Sigma b_i b_j \sin Qr_{ij}\, e^{-\gamma_{ij}Q^2}}{(\Sigma\, b_i)^2} \tag{2.5}$$

The structure of the liquid, as distinct from that of the molecule, is related to the real space distribution by the Fourier-Bessel transform

$$d_{L(r)} = 4\,\pi r \rho_M\, [\,\bar{g}(r) - 1] = \frac{2}{\pi} \int_0^\infty Q\, D_M(Q) \sin Qr\, dQ) \tag{2.6}$$

The function $\bar{g}(r)$ is a composite function given by

$$\bar{g}(r) = \Sigma\, c_\alpha{}^2 b_\alpha{}^2\, g_{\alpha\alpha}(r) + 2\Sigma\, c_\alpha\, c_\beta\, b_\alpha\, b_\beta\, g_{\alpha\beta}(r) \tag{2.7}$$

where α and β represent different atomic species, c_α is the fractional concentration ($\Sigma\, c_\alpha = 1$) and the summation is taken over all distinct species. The $g_{\alpha\beta}(r)$ functions are the partial pair correlation functions which may be used to define the detailed structure of the liquid. In the case of molecular liquids, and particularly hydrogen-bonded liquids, there will be some form of orientational correlation and the liquid structure can be defined in terms of a different function $G(r_c, \omega_1, \omega_2)$ where ω_1 and ω_2 represent the orientation of molecular axes for a separation of molecular centres, r_c. This function can give more information about the structure and is readily computed from simulation studies but the most direct way of experimental investigation is by diffraction studies so that the $g_{\alpha\beta}(r)$ formalism is most suitable.

3 Neutron diffraction measurements

a) Neutron Facilities

The neutron has a half-life of approximately 14 mins so that neutron beams for effective experimental usage must be produced directly from nuclear reactions. The spectrum of thermal neutrons produced in a nuclear reactor operating at a little over room-temperature has a Maxwell-Boltzmann with a peak in the 1Å region that is ideally suited to diffraction studies. Research reactors have now been constructed in many countries but the High-Flux Reactor [3] at the Institut Laue Langevin, Grenoble, France, which became operational in 1972 represents the most advanced facility and provides a wide range of different instruments for neutron studies. The D4 diffractometer used mainly for structural studies of disordered materials is on a hot source beam line providing wavelengths of 0.5 and 0.7Å. More recently there has been a development of pulsed neutron sources, which originally used primary electron beams (~ 50-100 MeV) to produce high energy neutrons by photofission processes and which are subsequently thermalised in a polyethylene moderator. Examples of these electron linac facilities are based at AERE Harwell, England, (Helios) [5] and at Tohoku, Japan [6]. It was also realised that higher neutron fluxes could be obtained from primary proton beams (800MeV) by the spallation process and work [7] at the Argonne Lab, USA with the Intense Pulsed Neutron Source (IPNS) provided much of the initial basis for the design of further facilities such as ISIS [8] at the Rutherford Appleton Laboratory, England and KENS [9] at KEK, Japan.

The neutron spectrum from these sources contains an epithermal contribution at short wavelengths and the overall distribution can be modified by the moderator. A typical spectrum is shown in Fig 2. Although access to these central facilities is, in some senses, dependent on the available instrumentation and demand for beam-time, there is now a wide international user community involved with experiments that range from fundamental physics to complex biological systems.

Figure 2: Typical Neutron Spectrum for a pulsed source.

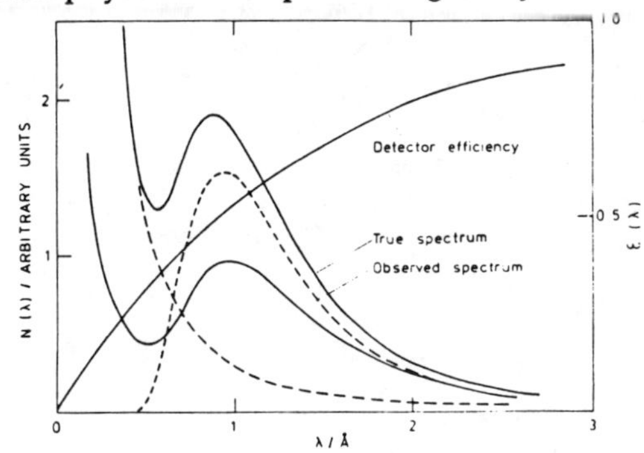

b) Diffraction techniques with a steady-state neutron source

The use of a fixed wavelength from reactor sources enables the diffraction pattern to be measured from the angular distribution of scattered neutrons. The single detectors used in early experiments were later replaced by multidetectors and eventually by 'banana-type' detectors which cover the whole angular range. The layout is schematically shown in Fig.3 and the 7C2 instrument based on the Orphée reactors (CEN, Saclay) is shown as an example in Fig.4. The angular distribution is recorded (usually on a dedicated computer) for a pre-set number of monitor counts and measurements are made for the sample, the empty container, background and a standard vanadium bar. Vanadium is an almost completely incoherent scatterer and is used to calibrate the detector efficiencies and/or to provide an absolute normalisation of the data.

c) Diffraction studies with a pulsed neutron beam

The time-of-flight method uses a pulsed "white" beam of neutrons and the instruments are based on fixed angle detectors as shown schematically in Fig. 5; one of the early instruments (TSS Mark II) used on the Harwell Electron Linac is shown

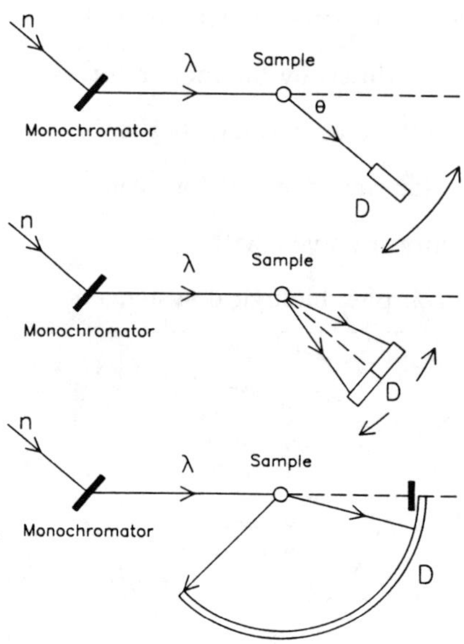

Figure 3: Basic layout for a diffraction measurement a) single counter; b) multi-detector; c) banana-type detector.

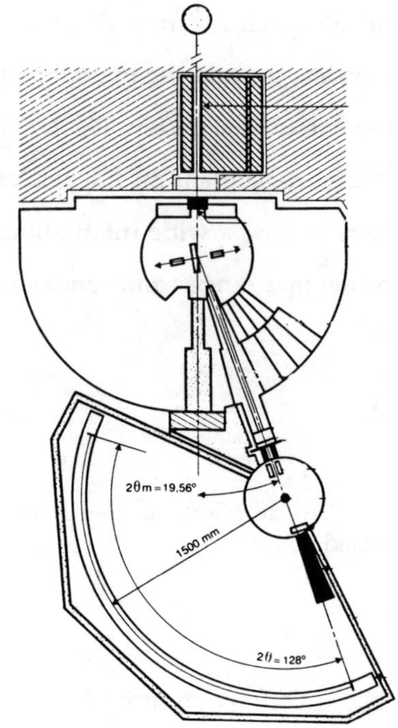

Figure 4: The 7C2 diffractometer on the Orphée reactor,

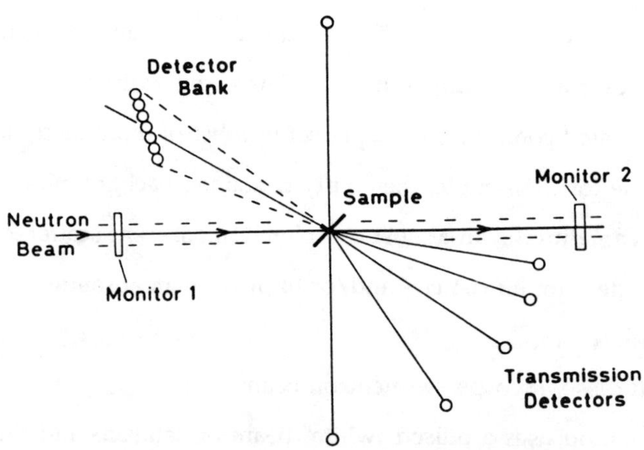

Figure 5: The basic layout for a time-of-flight diffracton measurement.

in Fig.6. The spectrum of scattered neutrons is recorded for a preset monitor count using the same samples as for the reactor case. Since the incident spectrum is "under-moderated" (Fig.2b) the measurements extend to higher Q-values. The scattering from the vanadium standard now serves as a measurement of the incident neutron spectrum. A technical comparison of the steady-state and pulsed neutron methods is beyond the scope of this review and the reader is referred to the earlier paper of Dore and Clarke [10] for presentation of principles and more recent work by Bermejo et al [11] for discussion of current issues.

d) Small angle neutron scattering

Measurements at low Q-values (≤ 0.5 Å^{-1}) are normally associated with the study of heterophase systems where the interference features are characteristic of large-range behaviour [12] but there are a few cases where the SANS intensity profile for hydrogen-bonded liquids or liquid mixtures is of interest. The schematic layout of the D17 instrument at ILL is shown in Fig.7. The neutrons with a fixed wave-length in the 6-15 Å range are detected in an area multidetector and since the scatter-ing is radially symmetric, the counts are re-grouped to give the SANS intensity, after appropriate correction for transmission factors through the sample. Corresponding time-of-flight instruments have been built on pulsed sources but the useful beam intensities are generally much lower by an order of magnitude or more; an example is the LOQ instrument [13] on the SNS.

4 Special Methods

i) Temperature and pressure studies

Many hydrogen-bonded liquids show significant structural change in relation to external parameters such as temperature and pressure. It is often convenient to study these effects directly by using first order difference methods. If the structure factor is measured at a reference temperature T_o(or pressure, P_o) are subsequently at a range of different temperatures, T (or pressures, P), the difference function can be written as

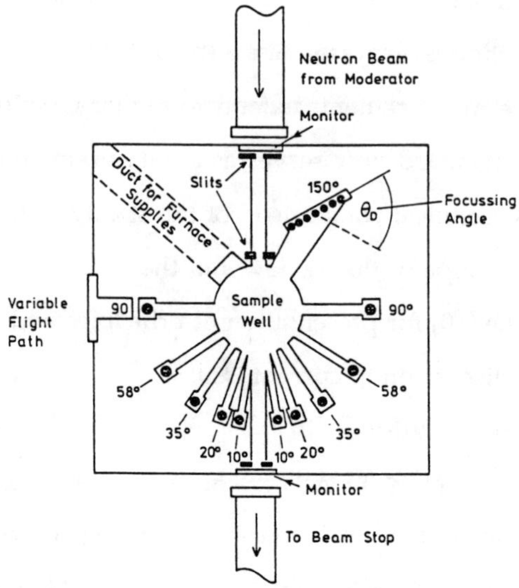

Figure 6: The total scattering spectrometer (TSS) installed on the Harwell Electron Linac.

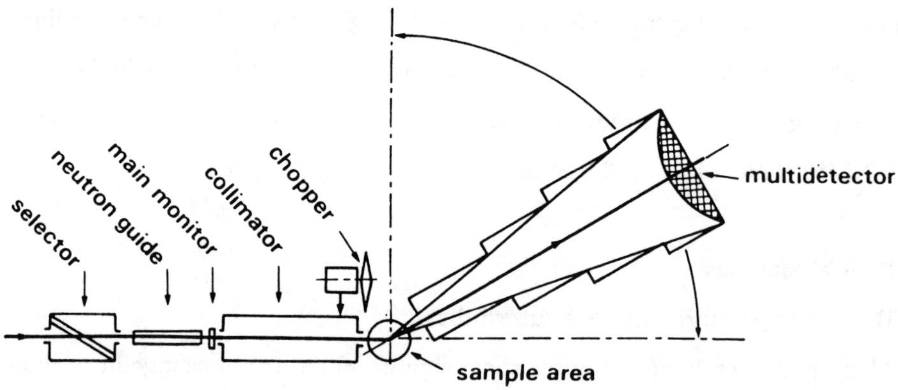

Figure 7: The D17 small angle scattering instrument at ILL.

$$\Delta S_M(Q,T) = S_M(Q,T) - S_M(Q,T_o) = \Delta D_M(Q,T)$$

If the molecular form-factor, $f_1(Q)$ is independent of temperature (pressure) the resulting function is representative of change in the intermolecular correlations. The Fourier-Bessel transform therefore becomes:-

$$\Delta d_L(r,T) = 4\pi r\,[\rho \Delta \bar{g}\,(r,T) + \Delta\rho\,[\bar{g}(r) - 1] = \frac{2}{\pi}\int_o^{\infty} Q\,\Delta D_M(Q,T)\sin Qr\,dQ$$

which gives the changes arising from the structural rearrangement $\Delta g(r,T)$ and the density $\Delta\rho$. Due to the density maximum in water it is also possible to choose pairs of temperatures in which the density is unchanged and in this special case, the isochoric temperature derivative [14] can be evaluated.

ii) Partial pair correlation functions.

In a multi-component system of n independent scatterers there are $\frac{1}{2}$ n(n+1) partial pair correlation functions, $g_{\alpha\beta}(r)$, and it is possible to evaluate the separate functions if the scattering lengths can be varied in a systematic way by the use of isotopic substitution. A list of b-values [15] for several elements is given in Table 1 and shows that there are several elements which are well suited to this approach (e.g. Cl and N) but others are more difficult (e.g. C) or impossible (O); hydrogen is a special case as the large change in b-value (D $\rightarrow$ H) is also associated with a huge increase in incoherent scattering. For hydrogen-bonded liquids the H/D substitution method is obviously of particular significance and will be discussed in other papers dealing with various liquids including water as a special case. Other applications occur using different isotopes in liquid mixtures and aqueous solutions [16] where specific features of hydration are investigated.

Another possibility of isolating individual $g_{\alpha\beta}(r)$ functions is to combine data from different radiation probes, particularly X-ray diffraction [17]. Although some useful work has been done along these lines, there are often difficulties associated with instrumentation differences, experimental corrections and systematic uncertainties in the combination of the different datasets. In present circumstances, the neutron isotopic substitution is the most reliable method but future developments in the use of synchrotron X-rays could change this situation [18].

Table 1: A compilation of coherent neutron scattering lengths for various isotopes relevant to studies of hydrogen-bonded liquid.

Z	Symbol	Natural Abundance(%)	b_{coh} (fm)	σ_{coh} (barn)	σ_{incoh} (barn)
1	H	99.98	-3.74	1.8	~80
	D	0.02	6.67	5.6	2.0
	T		4.7	2.8	?
6	C	(100)	6.65	5.6	0
	^{13}C	1.1	6.0	5.5	1.0
7	N	(100)	9.40	11.1	0.3
	^{15}N	0.4	6.5	5.3	?
8	O	(100)	5.83	4.2	0
	^{17}O	0.04	5.78	4.2	?
	^{18}O	0.2	5.84	4.5	0
16	S	(100)	2.85	1.02	0.2
	^{33}S	0.76	4.7	3.00	?
	^{34}S	4.22	3.48	1.54	0
17	Cl	(100)	9.58	11.5	5.9
	^{35}Cl	75.5	11.8	17.5	?
	^{37}Cl	24.5	2.6	0.8	?

5 Data Analysis

i) Diffraction experiments

The recorded datasets for 'sample + container', 'empty container', 'vanadium standard' and 'background' are combined to give the normalised differential scattering cross-section $d\sigma/d\Omega(\theta,\lambda)$ for the liquid sample under investigation. The procedure requires careful treatment for sample attenuation and multiple-scattering effects which can influence the accuracy of the final results if they are inappropriately evaluated. In the case of D (and H)-containing liquids there are additional corrections due to inelasticity effects (Placzek corrections) but a detailed discussion of the varied procedures for calculating these terms goes beyond the scope of this review. However, it is useful to give some indication of the way the diffraction pattern is affected since the shape of the S(Q) curve corresponds to what would be observed if the scattering was purely elastic.

A full evaluation of the corrections would require a knowledge of the dynamic structure factor, $S(Q,\omega)$ [19] and if this was known to the required accuracy there would be no need to measure S(Q)! It is therefore necessary to use approximations to model the molecular motions and these are based on the assumption of normal mode vibrations of the molecular and free recoil in the scattering process. Corrections to the self-terms are Q-dependent and are different for steady-state and pulsed neutron experiments. For reactor studies there is a characteristic fall-off in intensity as the scattering angle is increased. This effect is smaller for short wavelength neutrons but is particularly severe for light atoms such as H and D. In comparison, self term corrections for pulsed neutron are also dependent of geometric factors related to flight path ratios and are generally lower in the high Q-value region (for fixed θ) due to the use of short wavelengths [20]. The inelasticity effects also apply for corrections to the intra-molecular interference terms which are affected by molecular recoil, causing a small shift in the oscillatory pattern. These corrections are important for water but can sometimes be neglected for heavier molecules. In general, the corrections are reduced in magnitude for lower θ and higher energy (shorter λ). The optimisation of count-rate against these factors is a complex matter which is still an active field of study.

After the analytic corrections have been applied, and incoherent scattering subtracted, the cross-section data can be converted to the structure factor, $S_M(Q)$. The next step is to fit the $f_1(Q)$ form-factor to the high-Q data using eqn.2.5 with r_{ij} and γ_{ij} values as fitting parameters. The precision with which these values can be extracted is dependent on the statistical accuracy of the data, the maximum value of the Q-range and the complexity of the molecular structure. The $f_1(Q)$ curve can then be extrapolated to lower Q-values and subtracted to determine the inter-molecular function $D_M(Q)$ that characterises the liquid structure. Although $D_M(Q)$ can be compared directly to computer predictions it is usual to convert it to a real-space relation using eqn.2.6 since the data have a finite Q-range. Care is sometimes needed to avoid truncation errors [21] in evaluating the transform which can produce spurious oscillations in the resulting composite $d_L(r)$ function. The resolution in real space is also inversely proportional to the maximum Q-value which is why short wavelength beams are preferred. This conventional inversion method has been used for many years but new procedures based on 'maximum entropy' [24] and 'reverse Monte Carlo' techniques [25] are emerging as alternatives.

6 Examples of recent studies

In order to illustrate some of the basic principles, it is convenient to consider the present status of current investigations and limitations in the accuracy of the final data.

a) Water

Liquid water remains one of the most challenging problems for structural studies although substantial advances have been made in recent years. Temperature variation measurements have now been made into the deeply super-cooled regime for liquid D_2O and the results have been of great importance in understanding the evolution of hydrogen-bonded networks; a review of this work is given by Bellissent-Funel [24]. The separation of the three partial functions $g_{OO}(r)$, $g_{OH}(r)$ and $g_{HH}(r)$ has proved controversial [25] and raised important questions about both the treatment of inelasticity corrections and the non-equivalence of hydrogen and deuterium in these systems. These details are beyond the scope of this review and concern a full evaluation of error propagation and consistency checks for both reactor and accelerator-based techniques.

b) Methanol

Methanol (CH_3OH) has four non-equivalent atomic sites and therefore ten independent partial functions. It is impractical to determine all these functions but the main interest can be focussed on the hydroxyl group such that a selected substitution of (OH/OD) samples can yield useful information over determination of the datasets enables various consistency checks to be made as discussed by Montague et al., [26].

c) Glycerol

Glycerol is of particular interest as an organic glass-former and has been used in many systems for dynamic studies as discussed by Angell [27]. From the structural point of view, it poses many problems due to the large number (7) of non-equivalent atoms and torsional flexibility of the submolecular backbone. Temperature variation studies of D-glycerol have shown that the structure does not change greatly over a temperature regime where dynamic and transport properties vary in magnitude by many decades. In this case there are many intra-molecular parameters defining the $f_1(Q)$ pattern and results are needed for a wide Q-range. Published data [28] using reactor techniques having a Q_{max} of ~ $20Å^{-1}$ do not give sufficient spatial resolution and sensitivity so that pulsed neutron techniques are particularly useful in this case. Some preliminary data [29] from the LAD diffractometer are shown in Fig.8 illustrating the improved spatial resolution that can be obtained for the intra-molecular peaks ($\lesssim$ 3 Å) with the extended Q-range resulting from the use of a larger scattering angle in pulsed neutron measurements.

d) Water-alcohol mixtures

Much interest in hydrogen-bonded liquids is devoted to the properties of liquid mixtures as the thermodynamic properties can show large deviations from regular solution behaviour. Diffraction studies are complicated by the added complexity of additional terms and variable concentration although some attempts have been made in recent years for low-concentration studies of water-alcohol mixtures [30]. Another aspect relates to the behaviour at very low Q-values and the use of SANS techniques. On approach to a phase transition the concentration fluctuations contribute to loss of uniform scattering throughout the volume of the sample and a significant temperature-dependent scattering results. The first published studies by Teixeira and D'Arigo [31] showed that alcohols of different chain-length has substantially different properties. The particular case of t-butanol/water has recently been investigated in more detail [32] and a plot of the preliminary data is given in Fig.9. There has been some suggestion that partial micellization could occur in some systems due to clustering of the apolar regions of the alcohol molecule but no convincing evidence has yet been found for this effect. As the additional studies of water/alcohol mixtures develops this type of experiment will provide essential information for the full interpretation of the results. Another special application of SANS concerns studies of super-cooled water [25] where the observed density fluctuations may be interpreted on the basis of a H-bond percolation model [24].

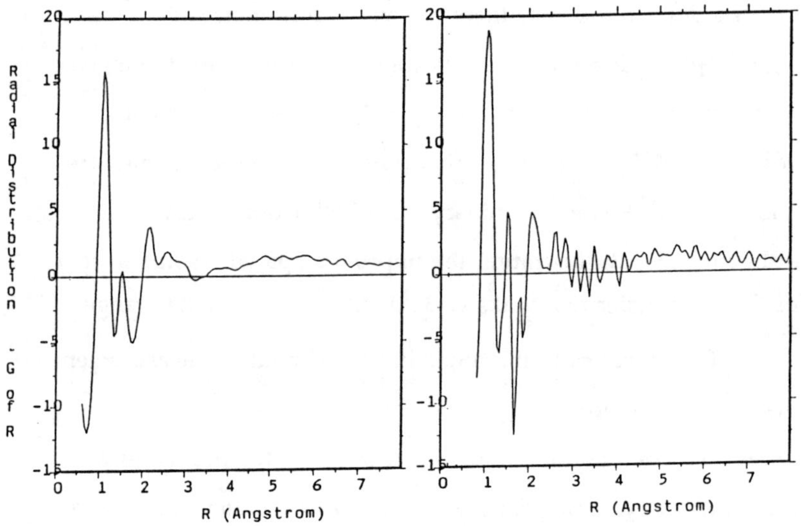

Figure 8: Preliminary g(r) data for D-glycerol from the LAD diffractometer; a) 25°; $Q_m \sim 20$ Å^{-1}; b) 35°; $Q_m \sim 40$ Å^{-1}.

Figure 9: SANS intensity for t-butanol/water (D_2O) [10 mole%] at A) 308K; B) 293K; c) 283K.

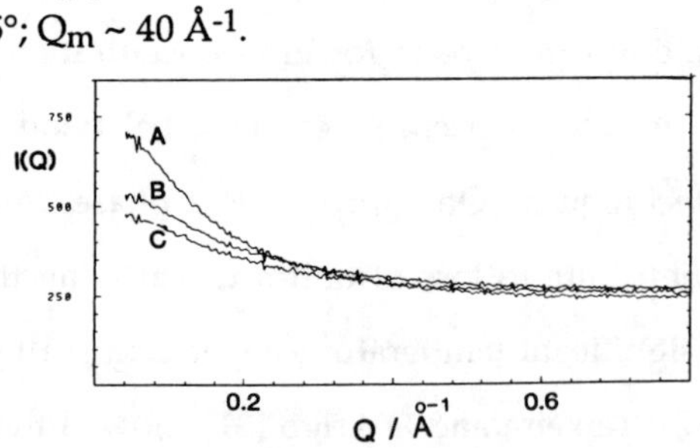

7 Future Developments

The main instrumentation developments will probably concern higher statistical accuracy (particularly for short wavelengths and experiments using isotopic substitution) and improved methods for handling inelasticity corrections. Plans have already been made for a re-designed hot-source diffractometer D4C at ILL for use at short wavelengths [27]. The specific advantages in epithermal pulsed neutron beams are being incorporated into small angle diffractometers such as SANDALS [28] at the Rutherford Appleton Laboratory and GLAD at the Argonne Laboratory. Both of these instruments should be ideal for studies of H or D containing materials by reducing the inelasticity effects; they should be operational in 1990.

From the scientific viewpoint there are various directions that can be taken and emphasis will probably be divided between the study of simple systems under extreme conditions (P, T etc.), more complex molecules, liquids in confined geometry and liquid mixtures.

References

1. J.G.Powles, Adv.Phys., **22**, 1 (73).
2. J.C.Dore, p.383, in 'Molecular Liquids' Vol.135, NATO ASI Series C, Reidel (Pub), 1984.
3. 'Neutron Research Facilities at the ILL High Flux Reactor' B.Maier (ed) 1986.
4. Various moderator characteristics are described in the series of international meetings on Advanced Neutron Sources: ICANS Proceedings.
5. 'Helios' is a 50 MeV electron linac facility at the Atomic Energy Research Establishment (AERE), Harwell, England.
6. An electron linac similar to that in ref.5 at the Tohoku University, Japan was also used for condensed matter research.
7. The Intense Pulsed Neutron Source (IPNS) is based on a 500 MeV proton synchrotron at the Argonne National Laboratory, Illinois, USA.

8. 'ISIS' is a Spallation Neutron Source based on a 800 MeV proton synchrotron at the Rutherford Appleton Laboratory, Chilton, England.
9. KENS is a proton synchrotron used for high energy particle studies and spallation neutron measurements based at KEK, Japan.
10. J.C.Dore and J.H.Clarke, Nuc.Inst. & Meth. **136,** 79 (76).
11. F.J.Bermejo, J.C.Dore, W.S.Howells, P.Chieux and E.Enciso, Physica **B156,** 154, (89).
12. SANS studies are typically for Q $\lesssim$ 0.3 $Å^{-1}$ and relate to spatial features with a range $\gtrsim$20Å.
13. R.K.Heenan, 'Low-Q' instrument, LOQ on 'ISIS'; RAL Report.
14. P.A.Egelstaff, et al. Phys.Rev.Lett., **47,** 1733 (81).
15. V.F.Sears 'Thermal neutron scattering cross-sections for condensed matter research' Chalk River Report AECL-8490, 1984.
16. G.W.Neilson in 'Water and Aqueous Solutions', Colston Papers No.37. Adam Hilger (pub), 1986.

    G.W.Neilson and J.E.Enderby, in 'Aqueous Solutions' NATO ASI series C, Reidel (pub), 1987.
17. J.C.Dore, p.1 in 'Neutron and X-ray Scattering: Complementary Techniques', IoP Conference Series No.101, Adam Hilger (pub).
18. The Synchrotron Radiation Source is a 2 GeV electron accelerator based at the Daresbury Lab, Warrington, England.
19. H.S.Chen; this volume.
20. J.G.Powles, Mol.Phys., **37,** 623 (79) and earlier papers.
    W.S.Howells, J.de Phys., Suppl.C7, p.81, 1984.
21. J.Waser and V.Schomacher, Rev.Mod.Phys., **25,** 671 (53).
    F.J.Bermejo et al., Physica Scripta, **38,** 78, 1988.
22. A.K.Soper, p.189 'Static and Dynamic Properties of Liquids', Springer Proceedings in Physics, **40,** 1989.
23. R.C.McGreevy, p.41 in 'Neutron and X-ray Scattering' see ref [17].
24. M-C. Bellissent-Funel, this volume.
25. J.C.Dore p.3 in 'Water Science Reviews' Vol.I, Cambridge (pub), 1985.
26. D.G.Montague, this volume.
27. A.Angell, this volume.
28. M.Garawi, J.C.Dore and D.C.Champeney, Mol.Phys., **62,** 475 (87).
29. W.S.Howells, private communication.
30. D.G.Montague, J.C.Dore and P.Chieux, in preparation.
31. G.D'Arrigo and J.Teixeira, J.Chem.Soc.Faraday Tran3., **86,** 1503 (90).
32. D.Huxley, J.C.Dore and P.Chieux, unpublished data.

RECENT STRUCTURAL STUDIES OF LIQUID D_2O BY NEUTRON DIFFRACTION

M.-C. Bellissent-Funel
Laboratoire Léon Brillouin (CEA-CNRS)
CEN-Saclay
91191 Gif-sur-Yvette cedex, France

Abstract. We present our recent structural studies of liquid D_2O performed by neutron diffraction. The results cover a large temperature range extending from 95°C down to -31.5°C i.e. close to the temperature of homogeneous nucleation and particularly in the low temperature range where liquid water exhibits an abnormal behaviour. The structural changes versus temperature are discussed using a difference function technique. We show that the isochoric temperature derivative function $\Delta d_L(r,\Delta T)/\Delta T$ where $d_L(r,T)$ is the pair correlation function at T, has a universal shape for distances larger than 2 Å, indicating a systematic change in spatial correlations over the density maximum. The structural investigation of deeply supercooled water has been achieved by using emulsions with water droplets of some μm diameter. At -31.5°C the pair distribution function of water shows an increase of the spatial correlations which confirms the fact that the structure of deeply supercooled water tends towards that of low density amorphous ice.

Finally, the method of H/D substitution is described and it is demonstrated how this method can improve the determination of the pair partial functions, when applied to D_2O/H_2O mixtures containing relatively small amounts of H_2O (α_H = 0.0, 0.03, 0.05, 0.10, 0.20). We make a comparison of the results obtained at room temperature with earlier experimental results and with computer simulation studies.

1. INTRODUCTION

The detailed structural arrangement of molecules in liquid water has been of fundamental interest for a considerable time. A comprehensive survey is presented in many reviews(1-5), showing how the structural study of liquid water by using various spectroscopic methods has evolved over the years.

Water molecules interact through hydrogen bonds for which the characteristic energy and the directionality are at the origin of the unusual anomalous properties (6-8) of liquid water such as density, self-diffusion, isothermal compressibility and relaxation times. It is known that hydrogen bonding plays a major role in the structure of water ; thus it was important to extend the study of the structure of water in the deeply supercooled state where it is expected to have an increase of the number of intact hydrogen bonds as suggested by the description of the molecular arrangement in water by a continuous random network of tetrahedrally coordinated H-bond(9). On the other hand, the key to understanding the properties of hydrogen bonding can be found in the comparison with amorphous ice. Amorphous ice is the

J.C. Dore and J. Teixeira (eds.), Hydrogen-Bonded Liquids, 117–128.

disordered solid and thus differs from liquid water primarily in terms of the molecular motions. Therefore, comparison between the structure of deeply supercooled water and that of amorphous ice[10,11] is interesting.

The purpose of this lecture is to give the more recent account about the temperature variation of the structure of liquid water, in a wide range of temperature, and down to -31.5°C i.e. close to the temperature of homogeneous nucleation. In addition, some new determination of the pair partial functions at room temperature, will be presented.

2. THEORY

The basic theory for studies of neutron scattering from molecular systems has already been reviewed in the lecture by Chieux and Dore[12] and only the essentials will be repeated here.

The liquid structure factor $S_M(Q)$ can be written in terms of a molecular form factor $f_1(Q)$ and a difference function $D_M(Q)$ which contains all the intermolecular contributions, i.e.

$$S_M(Q) = f_1(Q) + D_M(Q) \tag{1}$$

For D_2O molecules, $f_1(Q)$ is given by :

$$f_1(Q) = \Big[b_O^2 + 2b_D^2 + 4b_O b_D\ j_o(Qr_{OD})\ \exp(-\gamma_{OD}\ Q^2)$$

$$+ 2b_D^2\ j_o(Qr_{DD})\ \exp(-\gamma_{DD}Q^2)\ \Big] \Big/ (b_o + 2b_D)^2 \tag{2}$$

where r_{OD} is the intramolecular oxygen-deuterium distance and $r_{DD} = 2r_{OD} \sin\frac{\beta}{2}$ for an intramolecular bond angle β, $j_o(x) = (\sin x/x)$ and $\gamma_{ij} = \frac{1}{2}\left\langle u_{ij}^2\right\rangle$ where $\left\langle u_{ij}^2\right\rangle$ is the mean-square amplitude of displacement from equilibrium position due to the normal mode vibration of the molecule ; b_O and b_D are the coherent scattering lengths of oxygen and deuterium atoms ($b_O = 0.5805 \times 10^{-12}$ cm, $b_D = 0.6674 \times 10^{-12}$ cm).

The total pair correlation function g(r) is related to the Fourier transform of $S_M(Q)$ by :

$$d(r) = 4\pi r\rho_M\ [g(r) - 1] = \frac{2}{\pi}\int_0^\infty Q\ [S_M(Q) - S_M(\infty)]\ \sin Qr\ dQ \tag{3}$$

where $S_M(\infty) = \left(b_O^2 + 2b_D^2\right) / (b_O + 2b_D)^2$ is the asymptotic value of $f_1(Q)$ at large Q and ρ_M is the molecular number density.

The function d(r) is a combination of the separate partial correlation functions and includes sharp peaks due to the intramolecular distances. It is more convenient to remove the intramolecular terms by subtracting the molecular form factor from $S_M(Q)$ to give $D_M(Q)$, which may be transformed by :

$$d_L(r) = 4\pi r\rho_M \ [g_L(r) - 1] = \frac{2}{\pi}\int_0^\infty Q \ D_M(Q) \ \sin Qr \ dQ \qquad (4)$$

to give the pair correlation functions $d_L(r)$ and $g_L(r)$ for the intermolecular terms only.

The composite function $g_L(r)$ (or $d_L(r)$) consists of a weighted sum of partial terms which for neutron diffraction by D_2O is expressed by :

$$g_L(r) = 0.489 \ g_{DD}(r) + 0.421 \ g_{OD}(r) + 0.090 \ g_{OO}(r) \qquad (5)$$

where the dominant contributions concern the DD and OD spatial correlations.

In contrast with X-ray scattering which is mainly sensitive to the oxygen positions[1], neutron diffraction can provide the pair correlation between the relative positions of deuterium and oxygen and thus gives information on the orientational correlations of water molecules.

For measurements at different temperatures, the function $\Delta D_M(Q,\Delta T)$ is defined as :

$$\Delta D_M(Q,\Delta T) = S_M(Q, T_1) - S_M(Q, T_2) \qquad (6)$$

with $\Delta T = T_1 - T_2$ and where it is assumed that the molecular conformation remains unchanged.

The difference function $\Delta D_M(Q,\Delta T)$ may be transformed to give the change in real space by :

$$\Delta d_L(r,\Delta T) = \frac{2}{\pi}\int_0^{Q_M} Q \ \Delta D_M(Q, \Delta T) \ M(Q,Q_M) \ \sin Qr \ dQ \qquad (7)$$

where $M(Q,Q_M)$ is a modification function suitably chosen for the range of data extending to Q_M.

We shall define the isochoric temperature derivative (ITD) functions as : $\Delta D_M(Q,\Delta T)/\Delta T$ (relation 8) and $\Delta d_L(r,\Delta T)/\Delta T$ (relation 9) for a pair of temperatures T_1 and T_2 such that the density of water is the same.

3. EXPERIMENTAL PROCEDURE AND DATA ANALYSIS

The experiments have been performed at the reactor Orphee at Saclay, on the 7C2 spectrometer[13] which is equipped with a BF_3 position sensitive detector with 640 cells ; the angular step between two adjacent cells is equal to 0.2 degrees which leads to a maximum diffraction angle 2θ of 128 degrees. With the two used wavelengths, the following scattering vector ($Q=4\pi \sin\theta/\lambda$) ranges are covered :

$$0.4 \ Å^{-1} < Q < 16 \ Å^{-1} \quad \text{for } \lambda = 0.7 \ Å$$
$$0.25 \ Å^{-1} < Q < 10 \ Å^{-1} \quad \text{for } \lambda = 1.1 \ Å$$

the wavelength $\lambda = 1.1$ Å used in order to extend the measurements in the wavelength $\lambda = 1.1$ Å used in order to extend the measurements in the low-Q region with a good resolution.

According to the temperature range of the experiment, several kinds of samples have been used : bulk D_2O (Ø = 8 mm)[14], D_2O in capillaries (Ø = 0.8 mm)[15] and D_2O in emulsions (Ø = 8 mm)[16][17] in order to increase the degree of supercooling. The emulsion was constituted with a mixture of deuteroheptane and CCl_4 with sorbitan as a surfactant ; the average size of the water droplets was around some µm.

After the usual absorption[18] and multiple scattering[19] corrections, and the normalisation procedure using a vanadium rod, we obtain the total scattering cross-section $\frac{d\sigma}{d\Omega}$ (Q) which contains both $S_M(Q)$ and the incoherent scattering contribution together with the inelasticity effects. Except for the extraction of the pair partial functions where the inelasticity corrections were done using the formalism of Powles[20], we used an empirical method based on a Q-polynomial expansion and already applied with success to the data analysis of high density amorphous ice[11] and recently to the determination of a single $S_M(Q)$ of liquid water from experiments at three different wavelengths[21].

4. TEMPERATURE VARIATION STUDIES

4.1. Structure factors $S_M(Q)$[14][17]

In Figure 1, the derived structure factors $S_M(Q)$ are plotted versus Q, at 95°C, 75°C, 18°C, 5°C, -10.5°C, -18.5°C and -31.5°C together with that of low density amorphous ice[11] for comparison. A striking feature of the neutron structure factor is its extremely strong temperature dependence, characterized by a displacement of the main diffraction peak position Q_o towards smaller Q values with decreasing temperature, as already observed in previous experiments mentionned in the references 16 and 22. Moreover, we observed a sharpening of this main diffraction peak with decreasing temperature. The evolution of Q_o as a function of temperature is given in Figure 2. It appears that Q_o tends towards the value 1.70 $Å^{-1}$ which is characteristic of low density amorphous ice. Moreover, the slope of the $Q_o(T)$ function increases with decreasing temperature. The slope of $Q_o(T)$ in the lower temperature regions correlates with the decrease in bulk density and emphasizes the link between the microscopic and macroscopic features.

Let us now focus our attention in the low-Q part of the structure factor $S_M(Q)$ at 5°C and 18°C[14] which is given in Figure 3 : the inset shows more in detail the low Q region and gives the calculated values of $S_M(Q)$. One observes clearly the increase of $S_M(Q)$ with decreasing Q, the anomaly being more pronounced at 5°C with a minimum occuring clearly around 0.5 $Å^{-1}$. The absolute scale enables the determination of a small correlation length of the order of 6 Å within the same hypothesis used before for the treatment of the "normal" contribution to the compressibility[22]. This anomalous

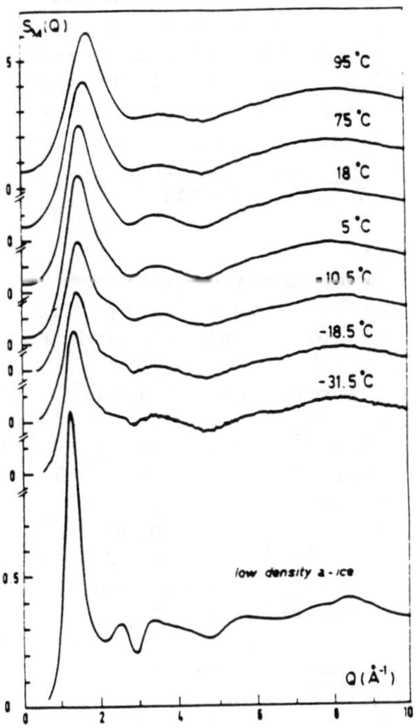

Figure 1. *Structure factors $S_M(Q)$ of D_2O obtained respectively at 95°C, 75°C, 18°C, 5°C, -10.5°C, -18.5°C, -31.5°C. The structure factor $S_M(Q)$ of low density amorphous ice is given for comparison (ref. 10).*

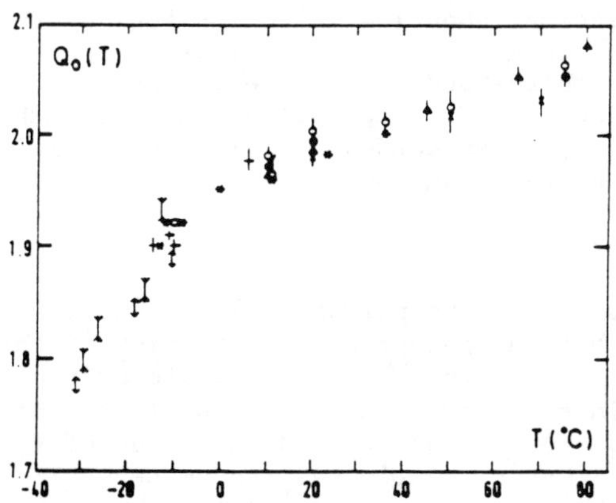

Figure 2. *Evolution of the main diffraction peak $Q_o(T)$ of $S_M(Q)$ versus the temperature.*
↕ *-10.5°C, -18.5°C, -31.5°C (ref.17)*
For the other symbols, see ref.16.

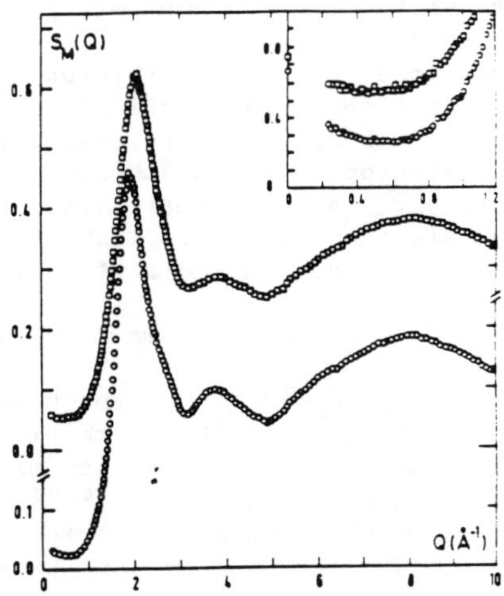

Figure 3. *Structure factors $S_M(Q)$ of D_2O obtained respectively at 5°C (dots) and 75°C (squares). Inset : structure factor of water at small wavevector transfer, at 5°C and 75°C. Note the decrease of $S_M(Q)$ with increasing Q at low temperature.*

behaviour already obtained by X-ray scattering[22] has just been confirmed by neutron scattering[14] and indicates the presence of density fluctuations with a correlation length ξ that increases up to about 8 Å as T is decreased to 253 K. This result supports the theoretical description[9] and the simulation of hydrogen bond network of water[23]. Tiny "patches" of this network, perhaps those in which all bonds are intact, are of lower density than the rest of the network and hence give rise to anomalous density fluctuations.

4.2. Pair correlation functions $d_L(r)$[16][17][21]

In Figure 4, the pair correlation function of low density amorphous ice is plotted together with that of water at 27°C and that of supercooled water at -10.5°C and -31.5°C.

In the small-r range, when the temperature is lowered, some features characteristic of amorphous ice show up : for example, two peaks appear at 1.75 Å and 2.30 Å the first one corresponding to the hydrogen-bond distance and the second one to the intermolecular D-D distance. In the large-r range one observes additional small oscillations, which are not present at -10.5°C and an out-of-phase behaviour for the broad oscillations ($r > 10$ Å). These features are similar to those observed in low density amorphous ice. Neutron diffraction is sensitive to both position and angular correlations and $d_L(r)$ is dominated by the partial functions $g_{OD}(r)$ and $g_{DD}(r)$ (see equation (5)). Because X-ray data show that oxygen-oxygen distances are not strongly temperature dependent[24], the observed strong temperature evolution of g(r) is mainly due to the temperature dependence of the g_{OD} and g_{DD} pair correlation functions. The increased effect of hydrogen-bonding therefore produces enhanced orientational correlations between neighbouring molecules and, as the hydrogen-bonded clusters increase in size, there is a greater spatial correlation between second- and third-bonded molecules. However, the assembly still possesses liquid characteristics so that the correlations of the transient hydrogen-bonded assemblies are less developed than in the static structure of amorphous ice.

4.3. Comparison with amorphous ice

In order to interpret the behaviour of the pair correlation function of water at -31.5°C, we recall that an appropriate description of the structure of amorphous ice has been given in terms of a continuous random network model (CRN)[10]. In this model, the basic building unit is a water pentamer. This is equivalent to a description of the molecular arrangement in water by a continuous random network of tetrahedrally-coordinated H-bonds[9].

Figure 5 gives the comparison between the experimental $d_L(r)$[10] and the theoretical one[25] for amorphous ice. A good agreement is observed, at least at small and medium r.

Further support for this viewpoint is provided from some recent neutron diffraction experiments on hyperquenched solid water[26] where it is shown that rapid cooling of liquid droplets in a supersonic jet produces a form of vitreous ice which has a structure very similar to that of vapour-deposited amorphous ice. The fully hydrogen-bonded network is therefore a natural end product to the structural evolution of the system with decreasing temperature if crystallisation can be suppressed.

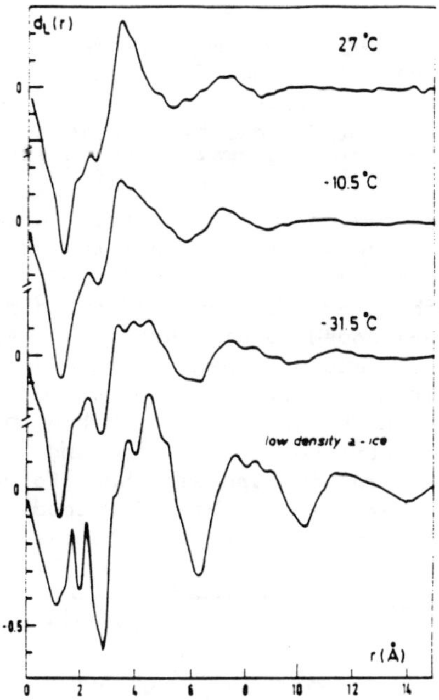

Figure 4. *Pair correlation functions* $d_L(r)$ *of* D_2O *obtained respectively at 27°C, -10.5°C, -31.5°C.*
The function $d_L(r)$ *of low density amorphous ice is given for comparison (ref. 10).*

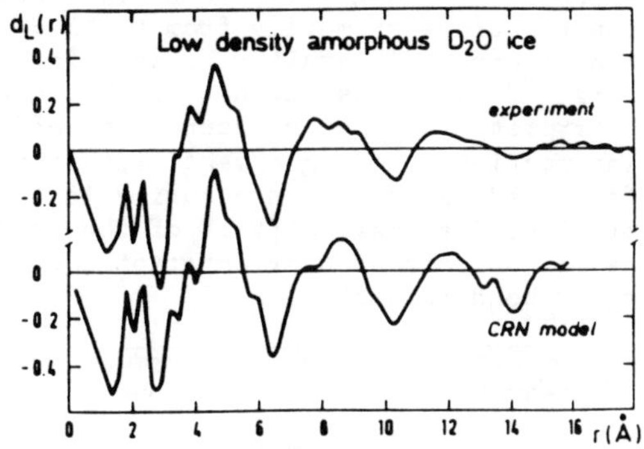

Figure 5. *Comparison between the experimental* $d_L(r)$ *of low density amorphous ice and that obtained from the continuous random network model (ref. 25).*

5. THE ISOCHORIC TEMPERATURE DERIVATIVE FUNCTION

Liquid water is a convenient system for measuring the isochoric temperature derivative (ITD) of any property because it has the rare property of showing a maximum in the density[6-8] versus temperature curve (Inset, Fig. 6). (ρ_{max} = 1.1060 g/cm^3 at 11.2°C for D_2O). This feature allows one to perform experiments at two different temperatures but at the same density. Many measurements have been conducted by X-ray[4][27] and neutron scattering[3][5][15] in the supercooled regime. We give here the functions $\Delta D_M(Q, \Delta T)/\Delta T$ (relation 8) and $\Delta d_L(r, \Delta T)/\Delta T$ (relation 9) for two pairs of temperatures (-9°C, 36°C) and (0°C, 23.5°C) where the density conditions were satisfied. The two $\Delta D_M(Q, \Delta T)/\Delta T$ functions for these pairs are shown in Figure 6. The displacement of the main diffraction peak is responsible for the sharp peaks in the region of 2 $Å^{-1}$ but there is some residual structure in the curves extending up to 10 $Å^{-1}$. At higher Q-values there is small variation beyond the statistical uncertainties of the measurements.

Fourier transformation of $\Delta D_M(Q, \Delta T)/\Delta T$ gives the $\Delta d_L(r, \Delta t)/\Delta T$ curves shown in Figures 7.a and 7.b. There is significant structural variation over a range extending up to 10 Å. In Figure 7.c, the two curves (7.a, 7.b)$\Delta d_L(r, \Delta T)/\Delta T$ can be superimposed except for the first peak near 2 Å. This indicates that there is a systematic linear variation in the pair correlation functions (at constant density) over this temperature region suggesting that there is a universal ITD curve which effectively defines the structural changes over the 2-10 Å region. The peak at 2 Å in $\Delta d_L(r, \Delta T)/\Delta T$ arises from changes in the nearest neighbour correlations and contains contributions primarily from the hydrogen-bonded OD distance (~ 1.8 Å) and there is a smaller change in this composite peak for the data (Fig. 7.b) corresponding to the higher ΔT value. It is possible to interpret this as a saturation effect in the local hydrogen bonding interactions, in agreement with the spectroscopic work[28] and the four fold coordination occuring in the crystalline and amorphous forms of ice. The main effect of the other changes can be understood in terms of increased orientational correlation over larger distances in the H-bond network.

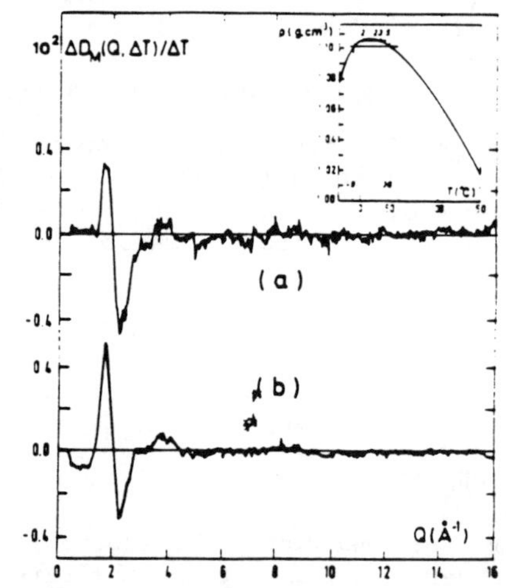

Figure 6. *Isochoric temperature derivative (ITD) functions $\Delta D_M(Q,\Delta T)/\Delta T$.*
a) between 2°C and 23.5°C
b) between -9°C and 36°C.
Inset : density of D_2O as a function of temperature. Taken from G.S. Kell, J. Chem. Eng. Data 20, 97 (1975).

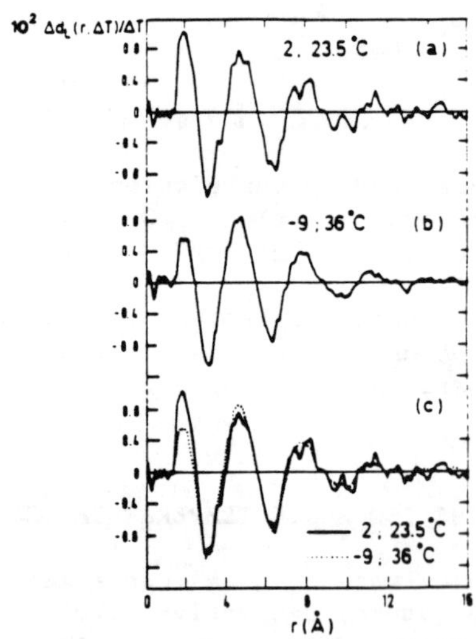

Figure 7. *Real-Space distributi[on] functions $\Delta d_L(r,\Delta T)/\Delta T$ for isochoric temperature derivati[ve] (ITD) functions.*
a) between 2°C and 23.5°C
b) between -9°C and 36°C.
The overlapping data are sho[wn] below (c).

6. THE PARTIAL PAIR FUNCTIONS

Much work[3-5] has been performed in order to determine the partial pair functions using two different ways. In principle, three correlation functions must be measured by observing three diffraction patterns with varying scattering amplitudes for the O and H (or D) atoms ; this has been done by using separate experiments with X-rays, neutrons and electrons[29]. Due to the different evaluation procedures and instrumental resolution associated with these three experiments, the partial functions finally obtained are subject to systematic uncertainties.

The second way of obtaining these partial pair functions was to exploit the difference in the neutron scattering amplitudes for H and D together with the assumption that H_2O and D_2O are identical. Therefore the isotopic substitution method was carried out with at least three samples covering pure H_2O, D_2O and a H_2O/D_2O mixture. But the amounts of H_2O used in these experiments were high so that the uncertainties in data on H_2O were large mainly because of the large incoherent scattering. On the other hand, it is now more widely accepted that quantum effects[30] do play a role in the structure of water and that H and D therefore behave differently. The preceding remarks explain why the quality of the results obtained by either the three radiations or the isotopic substitution method are expected to be similar. A way to improve the quality of the result when using the isotopic substitution method is to reduce the H-component in order to minimise the effects of H/D non equivalence, inelasticity corrections and incoherent scattering. We have recently adopted this approach[31]. Several measurements have been made, at room temperature, for the following combinations : pure D_2O, and the mixtures H_2O/D_2O such that the concentration α_H of H_2O in the H_2O/D_2O mixtures has the values 0.03, 0.05, 0.10 and 0.20. It is clear that the differences in the weight factors are smaller than in the previous studies and therefore a careful treatment of the data is needed. Instead of combining three composite pair correlation functions, it is possible to combine two pair correlation functions g(r) by adopting an oxygen-oxygen partial correlation function obtained from X-ray measurements on D_2O[27]. By using this partial the two equations can be solved simultaneously and the O-D and D-D partials can be extracted. Moreover by combining each two partials extracted for each combination and weighting them according to the statistical accuracy of the measurements, we can deduce an average partial function $g_{OD}(r)$ and $g_{DD}(r)$. The average $g_{OD}(r)$ and $g_{DD}(r)$ are shown in Figure 8 together with the $g_{OO}(r)$. Further work on the treatment of these datasets is in progress.

DISCUSSION

A checklist comparison of experimental data for $g_{OH}(r)$ and $g_{HH}(r)$, including some results of simulation is given in the reference 5. There are now new measurements due to Soper and Phillips[32] and those presented here . In each case the problems due to multiple scattering have been successfully treated and the results are found to be in good overall agreement for r-values up to 8 Å as shown in figures 9a and

9b. The most obvious difference is in the hydrogen bond position and the first peak in the D-D partial which is at a smaller r-value in our results compared to those of Soper and Phillips.

Although the oxygen-oxygen peak in Soper and Phillips data is larger than that of the X-ray function adopted in this work (Figure 9c), the two $g_{OD}(r)$ partials show good agreement in the 2.5 - 3.5 Å region where the O-O peak is subtracted. The other discrepancies arising between the two partials are due to the uncertainties involving the O-O correlation function.

The O-D and D-D partial functions exhibit a negligible structure for r-values > 5 Å in agreement with the data of Soper and Phillips ; but they disagree with the data of Palinkas et al[29] and the data of Reed and Dore[33]. Moreover the high hydrogen bond peak in the $g_{OD}(r)$ data of Thiessen and Narten[34] is not observed.

A detailed comparison of all the datasets for water still remains to be carried out.

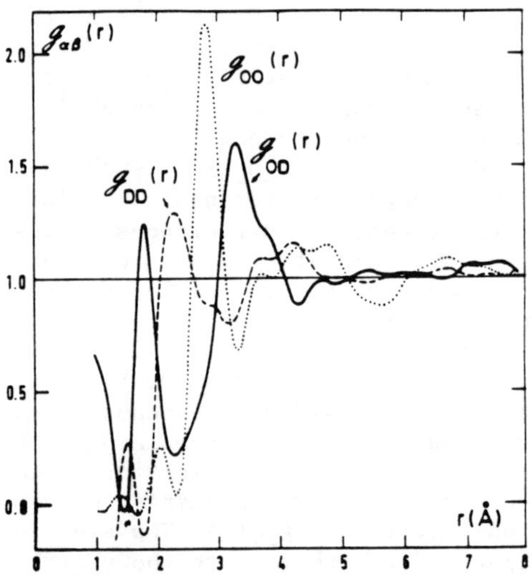

Figure 8. *The average partial pair correlation functions (ref.31). $g_{OD}(r)$ solid line, $g_{DD}(r)$ dashed line, $g_{OO}(r)$ dotted line (ref. 27).*

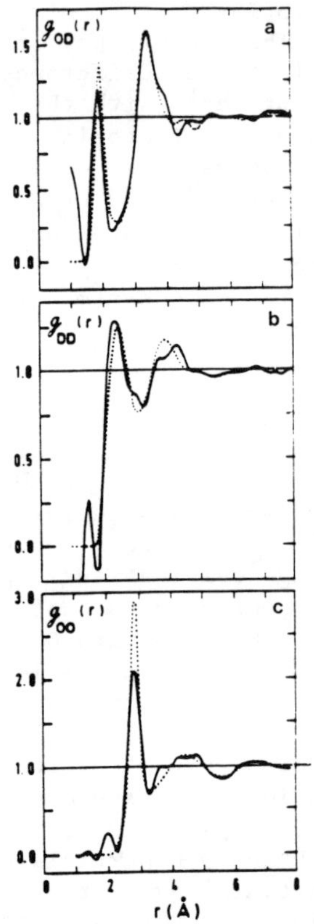

Figure 9. *a) Comparison between $g_{OD}(r)$ curve (solid line) and th of Soper and Phillips (dotted li (ref. 32).*
b) Comparison between the g_{DD} curve (solid line) and that Soper and Phillips (dotted li (ref. 32).
c) Comparison between the $g_{OO}(r)$ curve (ref.27) (solid line) and of Soper and Phillips (dotted li (ref.32).

7. CONCLUSION

In this lecture, we gave a review of the recent structural work performed on liquid D_2O as a function of temperature and using neutron diffraction. In particular, we show the existence of a universal ITD (isochoric temperature derivative) curve with which it is possible to describe the structural changes of liquid water over the 2-10 Å range.

The measurements, at low temperature, confirm the increasing spatial correlations in deeply supercooled liquid water as the temperature is decreased and the tendency to evolve towards the structure of low-density amorphous ice. These data should provide an important basis for a comparison with predictions based on different forms of the interaction potential between water molecules.

Finally, the new determination of the pair partial functions using the isotopic substitution method with samples containing small amounts of H_2O in the H_2O/D_2O mixtures has been successfull and will be extended to other temperatures, and particularly for the supercooled state.

References

1. A.H. Narten and H.A. Levy, in Water : a comprehensive treatise, Vol. 1, edited by F. Franks (Plenum Press, New York, NY) (1972)
2. D.I. Page, in Water : a comprehensive treatise, Vol. 1, edited by F. Franks (Plenum Press, New York, NY) (1972)
3. P.A. Egelstaff, Adv. Chem. Phys. **53**, 1 (1983)
4. S.H. Chen and J. Teixeira, Adv. Chem. Phys. **64**, 1 (1985)
5. J.C. Dore, in Water Science Reviews, 1, edited by F. Franks (Cambridge University Press) (1985)
6. C.A. Angell, in Water : a comprehensive treatise, Vol. **7**, edited by F. Franks (Plenum Press, New York, NY) (1982)
7. E.W. Lang and H.D. Ludeman, Angew. Chem. Int. Ed. Engl., **21**, 315 (1982)
8. C.A. Angell, Ann. Rev. Phys. Chem. **34**, 593 (1983)
9. A. Geiger, F.H. Stillinger and A. Rahman, J. Chem. Phys. **70**, 4185 (1979) ;
   H.E. Stanley and J. Teixeira, J. Chem. Phys. **73**, 3404 (1980)
10. M.R. Chowdhury, J.C. Dore and J.T. Wenzel, J. Non Cryst. Solids, **53**, 247 (1982)
11. M.-C. Bellissent-Funel, J. Teixeira and L. Bosio, J. Chem. Phys. **87**, 2231 (1987)
12. P. Chieux and J.C. Dore, this book
13. J.P. Ambroise, M.-C. Bellissent-Funel and R. Bellissent, Phys. Rev. Appl. **19**, 731 (1984)
14. L. Bosio, J. Teixeira and M.-C. Bellissent-Funel, accepted for publication to Phys. Rev. A. Rapid Communications.
15. M.A.M. Sufi, Ph. D. Thesis, University of Kent, 1986 ;
    M.A.M. Sufi, J.C. Dore, J. Teixeira, M.-C. Bellissent-Funel, L. Bosio and P. Chieux, to be published.

16. M.-C. Bellissent-Funel, L. Bosio, J.C. Dore J. Teixeira and P. Chieux, Europhys. Lett. **2**, 241 (1986)
17. M.-C. Bellissent-Funel, J. Teixeira, L. Bosio and J.C. Dore, accepted for publication to J. Phys. C : Condensed Matter
18. H.H. Paalman and C.J. Pings, J. Appl. Phys. **39**, 2635 (1962)
19. A. Menelle, Thesis, Université Paris VI (1987)
20. J.G. Powles, Mol. Phys. **36**, 1181 (1978)
21. M.-C. Bellissent-Funel, J. Teixeira, L. Bosio and J.C. Dore, to be published.
22. L. Bosio, J. Teixeira and H.E. Stanley, Phys. Rev. Lett. **46**, 597 (1981)
23. H.E. Stanley, J. Teixeira, A. Geiger and R.L. Blumberg, Physica, **106A**, 260 (1981)
24. L. Bosio, J. Teixeira, J.C. Dore, J. Steytler and P. Chieux, Mol. Phys. **50**, 733 (1983)
25. P. Boutron and R. Alben, J. Chem. Phys. **62**, 4898 (1975) ; M.G. Sceats, M. Stavola and S.A. Rice, J. Chem. Phys. **70**, 3927 (1979)
26. A. Halbrucker, E. Mayer, L. O''Mard, J.C. Dore and P. Chieux, in preparation
27. L. Bosio, S.H. Chen and J. Teixeira, Phys. Rev. **A27**, 1468 (1983)
28. G.E. Walrafen in Water : a comprehensive treatise, Vol. **1**, edited by F. Franks (Plenum Press, New-York, NY) 1972
29. G. Palinkas, E. Kalman and P. Kovacs, Mol. Phys. **34**, 525 (1977)
30. J.H. Root, P.A. Egelstaff and A. Hime, Chem. Phys. **109**, 437 (1986) ;
R.A. Kuharski and P.J. Rossky, Chem. Phys. Lett. **103**, 357 (1984)
31. M. Garawi, Ph.D. Thesis, University of Kent, 1987
32. A.K. Soper and M.G. Phillips, Chem. Phys. **107**, 47 (1986)
33. J. Reed, Ph.D. Thesis, University of Kent, 1981 ;
J. Reed and J.C. Dore, unpublished data.
34. W.E. Thiessen and A.H. Narten, J. Chem. Phys. **77**, 2656 (1982) ; A.H. Narten, W.E. Thiessen and L. Blum, Science, **217**, 1033 (1982)

H/D SUBSTITUTION IN NEUTRON DIFFRACTION

D. G. Montague*
Willamette University
Salem, OR 97301, USA, and
IPNS Division, Argonne National Laboratory,
Argonne, IL 60439, USA.

ABSTRACT. This paper presents the mathematical descriptions of isotopic substitution in terms of neutron scattering lengths and pair correlation functions illustrated by some experimental results. Simple arguments lead to an algorithm for H/D substitution techniques in multi-component systems. The accuracy of the measurements of the coherent part of the scattering and the validity of the corrections are the crucial elements in obtaining useful information. The dominating problem with the corrections is our inability to adequately correct for inelasticity effects because of the relatively large incoherent scattering cross-section for hydrogen. Some examples are given of numerical subtraction techniques with implications for more accurate measurements on two new neutron difractometers being built for measurements on liquids.

1. INTRODUCTION

Isotopic substitution is a well-established technique used to separate the total pair correlation into the component parts. Although n(n+1)/2 measurements are needed to completely specify an n-component system, it is often possible to do a smaller number of selected experiments to study those parts of the material of interest. Most often the limits of instrument time and the availability of suitable isotopes are more serious limitations than the number of independent measurements required to completely specify the system. In some cases interaction substitution can be of great benefit; for example, X-ray scattering which is dominated by the atoms with larger Z can be used as the equivalent of a zero scatering length neutron diffraction measurement for hydrogen substitution.

2. MATHEMATICAL DESCRIPTION

Neglecting the questions of corrections and accuracy for the moment and following the notation of [1], the scattering cross-section per molecule for a multi-component system can be written

*Work supported by U. S. Department of Energy under the Faculty Research Program administered by the Division of Educational Programs at the Argonne National Laboratory.

J.C. Dore and J. Teixeira (eds.), Hydrogen-Bonded Liquids, 129–137.

$$\frac{d\sigma}{d\Omega} = \sum_{ij} c_i c_j \bar{b}_i \bar{b}_j S_{ij}(Q) + \sum_i \{c_i \overline{b_i^2} - (c_i \bar{b}_i)^2\} \tag{1}$$

$$= \sum_{ij} c_i c_j \bar{b}_i \bar{b}_j \{S_{ij}(Q) - \delta_{ij}\} + \sum_i c_i \overline{b_i^2} \tag{2}$$

where,

$$\sigma^{coh} = 4\pi(\sum_i c_i \bar{b}_i)^2, \quad \sigma^{coh} + \sigma^{inc} = 4\pi \sum_i c_i \overline{b_i^2}, \tag{3}$$

$$\langle b\rangle = \sum_i c_i \bar{b}_i, \quad \text{and} \quad \langle b^2\rangle = \sum_i c_i \overline{b_i^2}. \tag{4}$$

so that we may write the coherent part of the scattering as,

$$\frac{d\sigma^{coh}}{d\Omega} = \frac{d\sigma}{d\Omega} - [\langle b^2\rangle - \langle b\rangle^2] = \sum_{ij} c_i c_j \bar{b}_i \bar{b}_j S_{ij}(Q) \tag{5}$$

which is defined in terms of the total correlation function, S(Q),

$$\langle b\rangle^2 S(Q) = \sum_{ij} c_i c_j \bar{b}_i \bar{b}_j S_{ij}(Q) \tag{6}$$

so that the distinct part or interference part of the scattering can be written,

$$\frac{d\sigma^{int}}{d\Omega} = \frac{d\sigma}{d\Omega} - \langle b^2\rangle = \langle b\rangle^2\{S(Q)-1\} = \sum_{ij} c_i c_j \bar{b}_i \bar{b}_j \{S_{ij}(Q) - \delta_{ij}\} \tag{7}$$

An idealized representation of the three parts of the cross-section is shown in figure 1. The information that we want is the conformation of the atoms in r-space and this is contained only in the oscillatory part of the scattering. Thus it is possible to obtain some structural information from the oscillatory features resulting from experiments even when the accuracy of the normalisation is not particularly good.

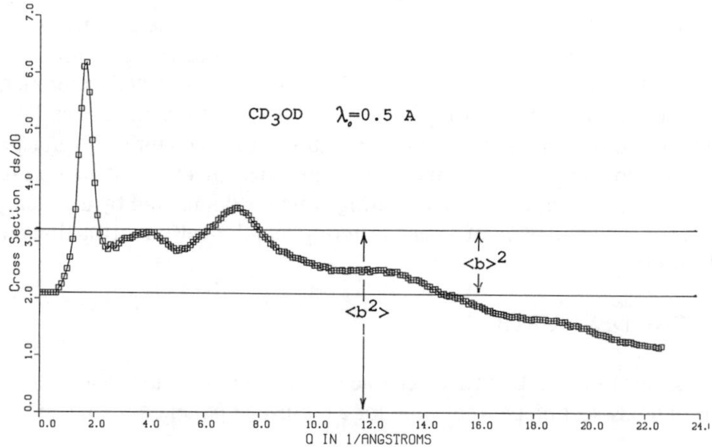

Figure 1. A representation of the results of a neutron diffraction experiment showing the coherent and incoherent cross-sections.

We write the interference part of the scattering for a generalised molecule, as ROH, ROD, and ROX; where the X stands for a mixture of ROH and ROD. We assume that the mixing is randomly uniform and does not affect the conformation of the molecule. Then,

$$\frac{d\sigma}{d\Omega}_H^{coh} = c_H^2\bar{b}_H^2\{S_{HH}(Q)-1\} + 2c_H\bar{b}_H \sum_{i\neq H} c_i\bar{b}_i S_{Hi}(Q) + \sum_{i,j\neq H} c_i c_j \bar{b}_i\bar{b}_j\{S_{ij}(Q)-\delta_{ij}\} \quad (8)$$

$$\frac{d\sigma}{d\Omega}_D^{coh} = c_H^2\bar{b}_D^2\{S_{HH}(Q)-1\} + 2c_H\bar{b}_D \sum_{i\neq H} c_i\bar{b}_i S_{Hi}(Q) + \sum_{i,j\neq H} c_i c_j \bar{b}_i\bar{b}_j\{S_{ij}(Q)-\delta_{ij}\} \quad (9)$$

$$\frac{d\sigma}{d\Omega}_X^{coh} = c_H^2\bar{b}_X^2\{S_{HH}(Q)-1\} + 2c_H\bar{b}_X \sum_{i\neq H} c_i\bar{b}_i S_{Hi}(Q) + \sum_{i,j\neq H} c_i c_j \bar{b}_i\bar{b}_j\{S_{ij}(Q)-\delta_{ij}\} \quad (10)$$

Solving for $S_{HH}(Q)$-1, yields

$$S_{HH}(Q)-1 = \{1/(\bar{b}_D-\bar{b}_X)\left[\frac{d\sigma}{d\Omega}_D - \frac{d\sigma}{d\Omega}_X\right] - 1/(\bar{b}_D-\bar{b}_H)\left[\frac{d\sigma}{d\Omega}_D - \frac{d\sigma}{d\Omega}_H\right]\}/\{c_H^2(\bar{b}_X-\bar{b}_H)\} \quad (11)$$

where this correlation function contains only intermolecular terms as long as there is no change in the RO- part of the molecules in the processes of isotropic change and in mixing the two isotopic species. Application of this technique as illustrated in figure 2 for CD_3OD, CD_3OH, and mixtures of CD_3OD with CD_3OH. [2]

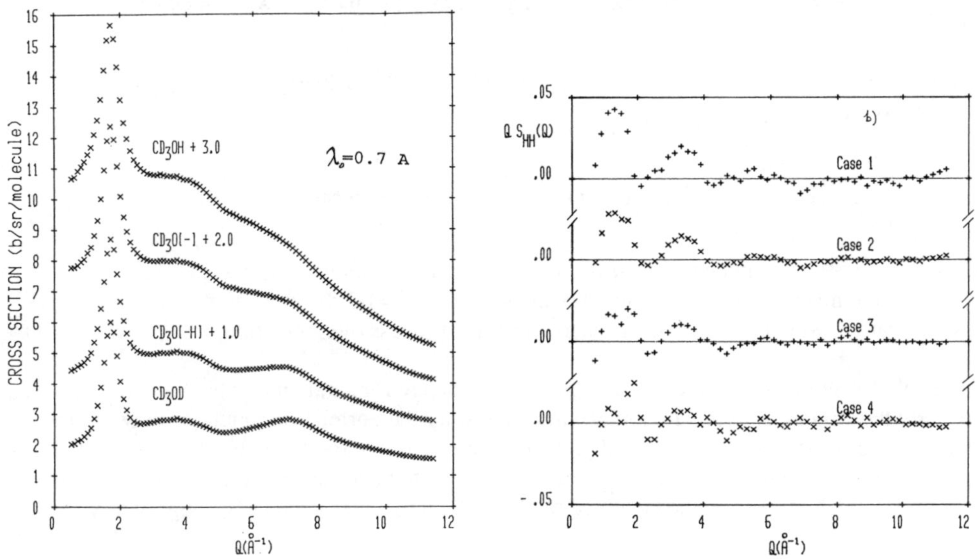

Figure 2. a) Scattering cross-sections for CD_3OD, CD_3OH, and mixtures of CD_3OD and CD_3OH; b) The resulting HH distribution fuction. Reference [2].

Next let us consider an aqueous mixture and assume as before that random, non-disruptive mixing occurs: y ROH + (1-y) H_2O. Use the symbol a to denote the intramolecular contributions, x for intermolecular, R for the molecule, W for water, and y for the fraction of ROH molecules to the total of water plus ROH molecules. Then the constant c_i^2 becomes $y(c_i^R)^2$ for the molecules and $(1-y)(c_i^W)^2$ for water.

$$\frac{d\sigma^{int}}{d\Omega} = y\bar{b}_H^2\{S_{HH}(Q)-1\} + 2y\bar{b}_H \sum c_i^R\bar{b}_i\{S_{Hi}^{aR}(Q)+S_{Hi}^{xR}(Q)\}$$
$$+ y\sum_i c_i^R c_j^R \bar{b}_i\bar{b}_j\{S_{ij}^{aR}(Q)\delta_{ij}+S_{ij}^{xR}(Q)\} + (1-y)\bar{b}_O^2\{S_{OO}^{xW}(Q)-1\}$$
$$+ (1-y)\bar{b}_H^2\{S_{HH}^{aW}(Q)+2S_{HH}^{xW}(Q)-1\} + (1-y)2\bar{b}_H\bar{b}_O\{S_{HO}^{aW}(Q)+2S_{HO}^{xW}(Q)\}$$
$$+ \sqrt{y(1-y)}\ \{\bar{b}_H^2 S^x_{H^R H^W}(Q)+2\bar{b}_H\bar{b}_O S^x_{H^R O^W}(Q)+2\bar{b}_H\sum_{i\neq H} c_i^R\bar{b}_i S^x_{i^R H^W}(Q)\}$$
$$+ \sqrt{y(1-y)}\ 2\bar{b}_O\sum_{i\neq H} c_i^R\bar{b}_i S^x_{i^R O^W}(Q) \qquad (12)$$

If we consider the cross-section to be a continuous function of the variable y, then we can observe the behavior by taking partial derivatives with respect to y;

$$\frac{\partial}{\partial y}\frac{d\sigma^{coh}}{d\Omega} = \{\text{R-terms}\}-\{\text{W-terms}\}+(1-2y)/2/\sqrt{y(1-y)}\{\text{cross-terms}\} \qquad (13)$$

where the R-terms are those terms in equation (2) multiplied by y, W-

terms by (1-y), and cross-terms by $\sqrt{y(1-y)}$.

$$\frac{\partial^2}{\partial y^2}\frac{d\sigma^{coh}}{d\Omega} = -1/4[y(1-y)]^{-3/2}\{\text{cross-terms}\} \qquad (14)$$

Thus the cross-terms can be obtained from the first order differences by making the experimental measurements for different values of y. At y=1/2 the cross-terms vanish in the first derivative, which provides a means to check the process of subtraction.

For alcohol-water mixtures there is a large density change for small amounts of alcohol in water. To study the structures in this region, we must consider the correlation functions between the various atoms of the liquid. For small y, the contributions due to the intermolecular correlations between ROH molecules; that is, S^R_{ij}, must be small. In the absence of clustering the terms S^R_{ij} must vanish as the volume occupied by the ROH molecules decreases; roughly as y^3. Then,

$$\frac{\partial^2}{\partial y^2}\frac{d\sigma^{coh}}{d\Omega} = 6y\{\text{R-terms}\} - 1/4[y(1-y)]^{-3/2}\{\text{cross-terms}\} \qquad (15)$$

where the first terms is much smaller than than the second and varies more rapidly with y. Thus the ROH molecule-water interactions dominate the external correlations for the ROH molecule. Figure 3 shows subtraction results for methanol and water.[3]

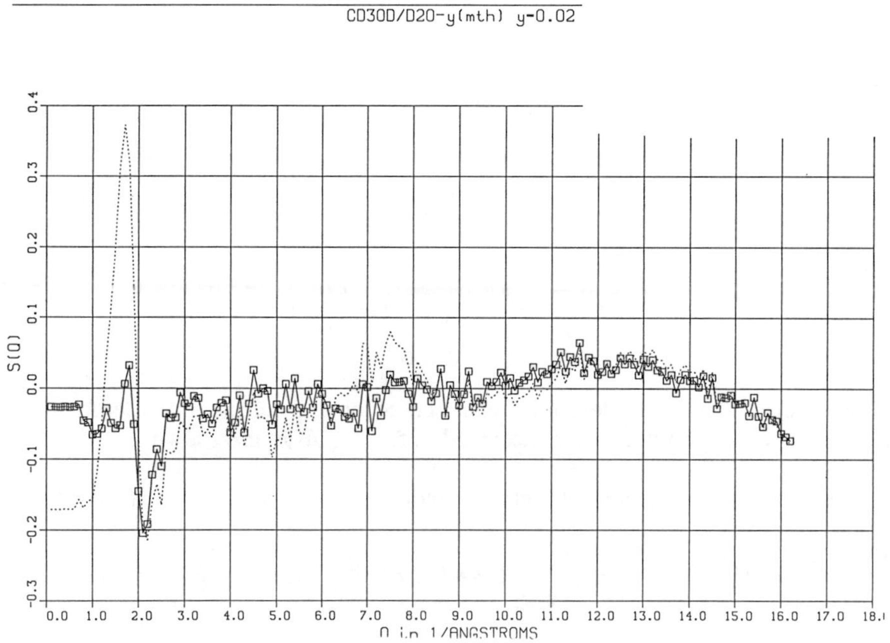

Figure 3. The HH correlation function from measurements on methanol and water. From reference [4].

3. INELASTICITY CORRECTIONS

It is inelasticity effects which cause the most difficulty in the subtrtactions for light isotopes. Several calculations and approximation schemes are used.

Method 1) The first and most commonly used is due to Placzek[5]; it does not work satisfactorily for water, as is shown in figure 4 as a function of angle, for neutrons of wavelength 0.5 A. The situation is even worse for lower energy neutrons. The Placzek approximation works well for atoms with large mass and for incoherent cross-sections equal to or less than the coherent cross-sections. In general calculations for hydrogen containing molecules are not reliable. The application of this correction term can be expressed as

$$\frac{d\sigma}{d\Omega} = \sum_{ij} c_i c_j \bar{b}_i \bar{b}_j \{S_{ij}(Q) - \delta_{ij}\} + P(\Theta,\lambda) \sum_i c_i \overline{b_i^2} \tag{16}$$

where coherent inelasticity effects have been ignored. The term P(Θ,λ) is the Placzek correction which depends on the neutron flus distribution and diffractometer geometry as well as angle sample atom masses and neutron scattering lengths.

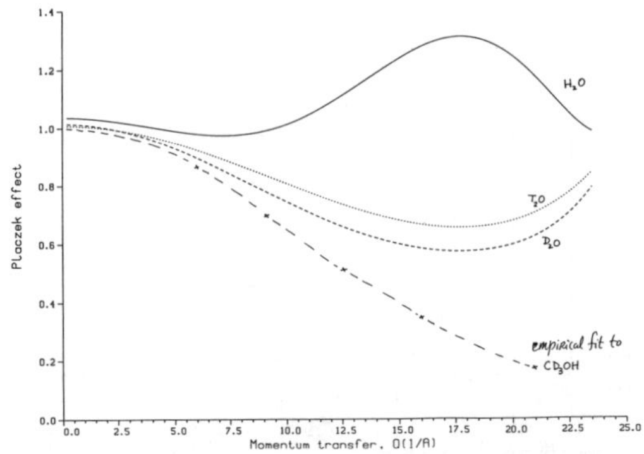

Figure 4. Placzek inelasticity corrections for H_2O as a function of momentum transfer for a reactor neutron source. The result of a parameterized fit fo CD_3OH is also shown.

<u>Method 2)</u> Direct subtraction of time-of-flight differential cross-section at a fixed angle.[6];

$$x \left.\frac{d\sigma}{d\Omega}\right]_{H_2O} + (1-x)\left.\frac{d\sigma}{d\Omega}\right]_{D_2O} - \left.\frac{d\sigma}{d\Omega}\right]_{X_2O} = 4x(1-x)(\bar{b}_H-\bar{b}_D)\, c_H^2 \left.\frac{d\sigma}{d\Omega}\right]_{HH} + \Delta \qquad (17)$$

where Δ is the term due to non-cancellation of inelasticity effects in the self terms and in the O-H and O-D distinct terms. The largest contribution comes from the hydrogen self scattering, is slowly varying, and has a small value of less than 0.025 barns/molecule. The results when Δ is not subtracted are shown in figure 5a for water [6] and figure 5b for ethylene glycol[7]. The limitation of this method is that only the partial correlation function for the substituted species is accessible.

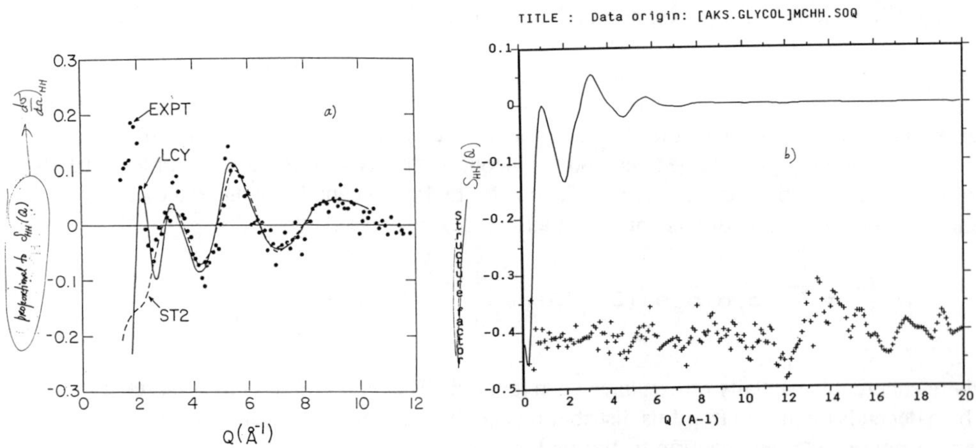

Figure 5. a) The correlation function for water. From reference [6] b) the HH correlation function for ethylene glycol. From reference [7]; the Maximum Entropy results are shown in the solid line; the residual as + symbols.

Method 3) For neutrons with fixed wavelength, the correction $P(\Theta,\lambda)$ is parameterized as $1-AQ^2+BQ^2$; that is, as an expansion of the first and second order terms of the Placzek approximation[2]. Inherent in this method is the need to fit the correlation function to

$$\sum_i c_i c_j b_i b_j \sin(QR_{ij})/Q/R_{ij}\; e^{-\gamma_{ij}Q^2} \tag{18}$$

for each atom-atom pair. Results obtained using this method suggest some differences in the OH bond length; this may be an indication of a coherent inelasticity effect. Calculations using the Powles approximation indicate the dominant effect in coherent scattering is in the term involving OH. Figure 4 illustrates the use of the $1-AQ^2+BQ^4$ expansion to approximate the $P(\Theta,\lambda)$ term and is shown as the dashed line.

Method 4) If sufficient data as a function of Θ and λ are obtained, such as from a diffractometer on a time-of-flight neutron source, the effective $S(Q,\omega)$ can be reconstructed[9]. Two neutron diffractometers, SANDALS[10], figure 6, at the Rutherford Appleton Laboratory, Chilton, England, and GLAD[11], figure 7, at the Argonne, USA, are under construction in order to make low angle and short wavelength neutron measurements. These two new diffractometers have been designed for liquids and amorphous materials with detectors at low angles and the capability for measuring neutrons in the 0.2 A to 4.0 A range. The Placzek inelasticity calculation for H_2O is shown in figure 8 for such instruments as a function of detector angle; each line is shifted upwared by 0.1 from the preceding line and covers the same wavelength range; i.e., $0.2\ A < \lambda < 4.0\ A$. Note that the region for which $P(\Theta,\lambda)$ is within 1% of unity extends from about $0.5\ A^{-1}$ to $26\ A^{-1}$ over the range of angles from 1° to 23.5^o.

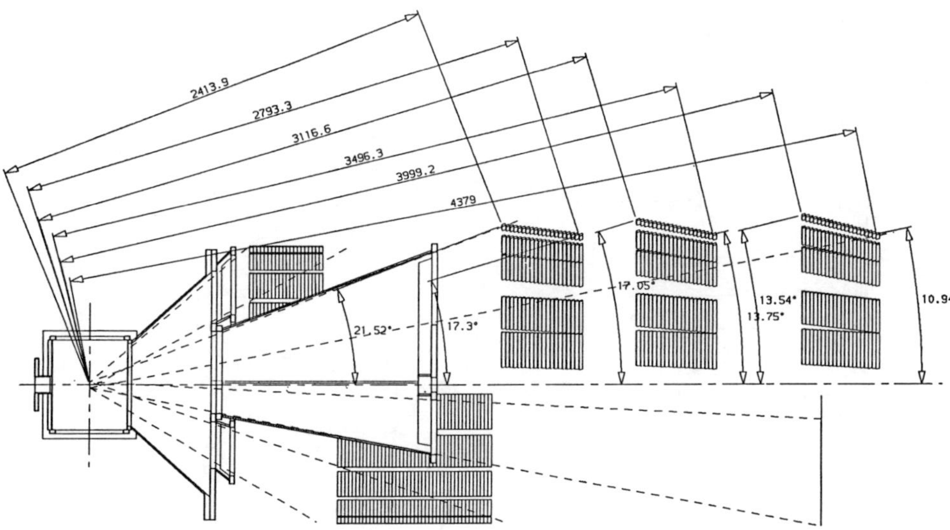

Figure 6. The liquid and amorphous materials diffractometer, SANDALS, at the Rutherford Appleton Laboratory[9].

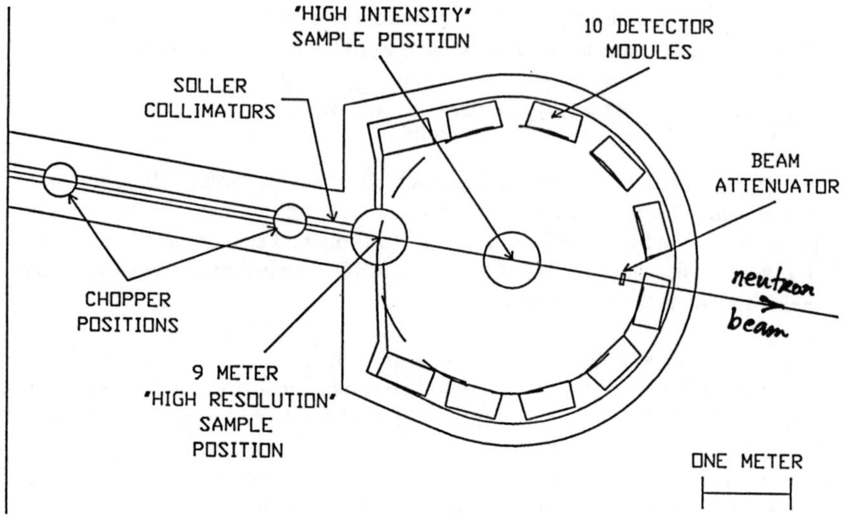

Figure 7. The liquid and amorphous materials diffractometer, GLAD, at the Argonne National Laboratory[10].

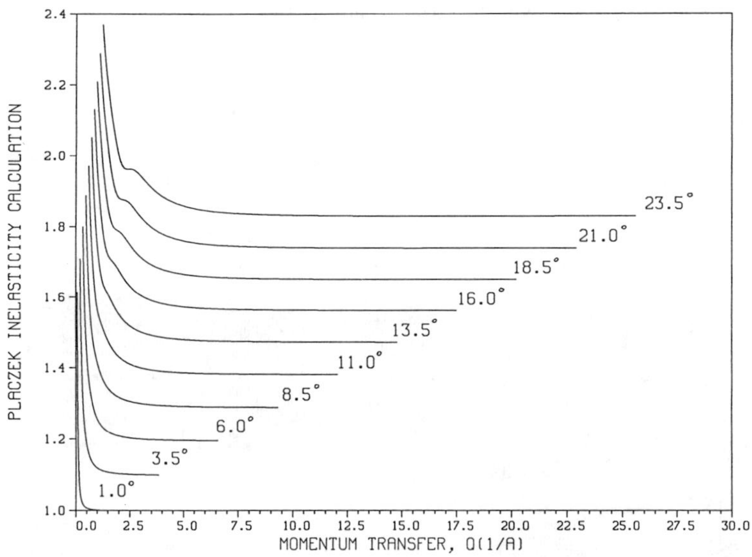

Figure 8. Placzek inelasticity calculations for H_2O as a function of momentum transfer and scattering angle for a time-of-flight neutron source. The neutron wavelength varies from 0.2 A to 2.0 A for each angle.

4. CONCLUSIONS

Techniques for isotopic substitution on hydrogen bonded systems enable us to obtain useful information on molecules containing hydrogen. Strong intermolecular hydrogen-hydrogen correlations are observed in water, methyl alcohol, and mixtures with neutron diffractometers on existing time-of-flight and reactor sources. New instruments will provide much higher statistical accuracy and greater precision in the measurements of separated pair correlation functions, as well as other atom-atom correlations in molecular liquids.

REFERENCES

[1] Powles, J. G. (1973) 'The Structure of molecular liquids by neutron scattering',Advances in Physics 22, 1-56
Price, D.L. (1981) 'Analysis of time-of-flight neutron diffraction data from isotropic samples', IPNS Note 19, Argonne National Laboratory
Chieux, P. (1978) 'Liquid structure investigation by neutron scattering', in Dachs, H. (ed.) 'Neutron Diffraction Topics in Current Physics', Springer-Verlag, pp. 271-302
[2] Montague, D.G., Dore, J.C., and Cummings, S. 'Structural studies of liquid alcohols by neutron diffraction III. CD_3OH, CD_3OD, and CD_3OH/D mixtures', Molecular Physics, 53 1049-1066
[3] Dore, J.C., Chieux P., and Montague, D.G.
[4] Sears, V.F. (1986) 'Neutron scattering lengths and cross-sections', in Skold, K. and Price, D.L. (eds.), Methods of Experimental Physics: Neutron Scattering, Vol. 23A, Academic Press, pp. 521-550
[5] Placzek, G. (1952) Physical Review, 86, 377-382
[6] Soper, A.K. and Silver, R.N. (1982) Physical Review Letters, 49,471-475
[7] Soper, A.K. and Montague, D.G.
[8] Bermejo, F.J., Mompean, F.J., Santoro, J., Steytler, D.C., and Dore, J.C. (1981) 'Molecular structure of liquid CD_3OD. A pulsed neutron study', Journal of Molecular Liquids, 33, 188-202
[9] Johnson, M.W. (1986) 'The determination of the complete neutron scattering law $S(k,\omega)$ from total scattering experiments using computed tomography', RAL-86-041, Rutherford Appleton Laboratory
[10] Soper, A.K. (1988) 'Future perspectives for liquids and amorphous materials diffraction at ISIS', Proceedings of the International Conference on Advanced Neutron Sources, X, October 1988
[11] Crawford, R.K., Price, D.L., Haumann, J.R., Kleb, R., Montague, D.G., Carpenter, J.M., Susman, S., and Dejus, R.J. (1988) 'GLAD: the IPNS Glass, Liquid, and Amorphous Materials Diffractometer', Proceedings of the international Conference on Advanced Neutron Sources, X, October 1988

STRUCTURAL ANALYSIS OF LIQUID FORMIC ACID USING NEUTRON DIFFRACTION

G. I. SWAN
Physics Laboratory, University of Kent at Canterbury
Canterbury, Kent, CT2 7NR, United Kingdom

Abstract. The structure of liquid formic acid is investigated using neutron diffraction. H/D isotopic substitution techniques are employed to obtain partial structural information. The bond lengths obtained from the intramolecular analysis are consistent with those obtained by other methods. The intermolecular structure consists of molecules linked extensively through hydrogen bonds of bond length 1.8Å to form long chains.

1 Introduction

Formic (or methanoic) acid ($HCOOH$) is the lightest member and the strongest acid of the carboxylic acid group ($C_nH_{2n+1}COOH$). Hydrogen bonding has long been thought to play a major influence in the overall intermolecular structure.

In the gas phase, formic acid exists as an equilibrium between ring dimers and monomers [1,2,3]. The ring dimer consists of two monomers joined together through two symmetric hydrogen bonds (see figure 1) and is predominant at room temperature.

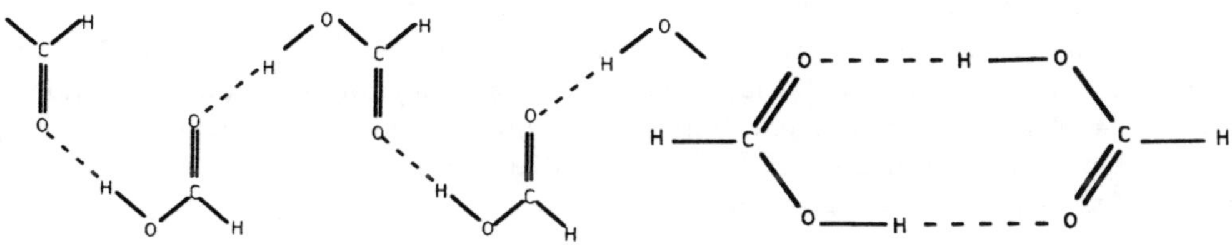

Figure 1: Schematic Formic Acid Structures. Left; Solid phase chains (as obtained from X-ray diffraction [4]). Right; Gas phase ring dimers.

X-ray diffraction studies of crystalline formic [4] and acetic acids indicate that the molecules are linked through hydrogen bonds into long chains (see figure 1) to form a large polymer (or chain) structure. Similar X-ray studies on heavier carboxylic acids find that the molecules occur as ring dimers.

J.C. Dore and J. Teixeira (eds.), Hydrogen-Bonded Liquids, 139–146.

An X-ray diffraction study of liquid formic, acetic, and propionic acids has shown them all to be largely hydrogen bonded with each molecule forming two hydrogen bonds [5]. It was concluded that the structure of the liquid corresponded to a dynamic equilibrium between chains and ring dimers, with the equilibrium displaced towards an increasing proportion of ring dimers on passing from formic to propionic acid. Other studies suggest that the displacement of the equilibrium is more marked. Both dielectric and Raman measurements show a ring dimer structure for acetic acid [6,7], but not for formic acid [8,9].

Infrared [10] and NMR [11] measurements show that pure liquid formic acid is polymeric, and that on dilution with carbon tetrachloride, splits to form dimeric units which on further dilution split into monomers. The self diffusion coefficient and the molecular reorientation time for the pure formic acid at $25°C$ [11] are found to be $1.04 \times 10^{-5} cm^2 s^{-1}$ (compared with $2.30 \times 10^{-5} cm^2 s^{-1}$ for water) and $\approx 5ps$ respectively indicating that the chains themselves are dynamic structures with a reasonably fast exchange of molecules between them.

Recently, a neutron diffraction study of liquid formic acid [12,13] obtained some surprising results; notably that the OH group is rotated about 50° out of the molecular plane. However, the results do not adequately describe the data obtained, and in particular, the hydrogen bonding component was not included in the fitting procedure or the analysis. It was thus considered useful to do further neutron diffraction measurements.

2 Experimental and Data Reduction

The neutron diffraction data was obtained in France using the $7C_2$ spectrometer [14], on the Orphée nuclear reactor at CEN Saclay. The diffraction patterns of the fully deuterated (DCOOD), hydroxyl deuterated (HCOOD), and a mixed ((H/D)COOD) formic acid samples were all measured at $18°C$, $34°C$, and $49°C$, although this article will only be concerned with some of the results obtained at $18°C$. The mixture was produced to obtain an average coherent scattering at the formic site of zero ($\bar{b}_{H/D} = 0$ for a deuteron mole fraction of 0.359) to give explicit partial structural information. Unfortunately, the initial samples supplied contained 5% water to stabilise the formic acid, and there was insufficient $DCOOH$ left after the water extraction process for isotopic substitution at the hydroxyl site.

The data was transferred to the University of Kent for reduction and analysis. The data reduction procedure for each of the formic acid data sets mainly consists of the subtraction of the sample cell and background contribution, absorption and multiple scattering corrections, and an approximate normalisation to the differential scattering cross section $\frac{d\sigma}{d\Omega}(Q)$. The details of these processes are given in [15].

3 Data Fitting Procedure

The differential scattering cross section (per molecule) $\frac{d\sigma}{d\Omega}(Q)$ can be conveniently expressed as

$$\frac{d\sigma}{d\Omega}(Q) = \frac{d\sigma}{d\Omega}(Q)_{self} + \frac{d\sigma}{d\Omega}(Q)_{intram} + \frac{d\sigma}{d\Omega}(Q)_{interm}. \tag{1}$$

The self contribution is given by

$$\frac{d\sigma}{d\Omega}(Q)_{self} = \left(\frac{\sigma_{incoh}}{4\pi} + \sum_{i=1}^{5} \overline{b}_i^2 \right) \times P(Q) \tag{2}$$

where the recoil (or inelasticity) correction is found to be given satisfactorily by the relation

$$P(Q) = 1 - AQ^2 + BQ^4 - CQ^6. \tag{3}$$

with the parameters A, B, and C determined empirically.

The intramolecular interference contribution is given by

$$\frac{d\sigma}{d\Omega}(Q)_{intram} = \sum_{i,j=1\ j>i}^{5} 2\overline{b}_i\overline{b}_j \frac{\sin(QR_{ij})}{QR_{ij}} exp^{-\frac{1}{2}<u_{ij}^2>Q^2} \tag{4}$$

where R_{ij} is the separation distance and $< u_{ij}^2 >$ is the mean square amplitude of thermal vibration between the ith and j^{th} atomic nuclei within the formic acid molecule.

In Q-space, the intramolecular and self contributions are obtained by varying the molecular conformation (bond lengths and angles) and $P(Q)$ to fit the experimental data in the high-Q region where the intermolecular contribution $\frac{d\sigma}{d\Omega}(Q)_{interm}$ is negligible. However, it is found that the fit is substantially improved by including a contribution due to hydrogen bonding $\frac{d\sigma}{d\Omega}(Q)_{HB}$, such that

$$\frac{d\sigma}{d\Omega}(Q)_{HB} = f_{HB} 2\overline{b}_O\overline{b}_D \frac{\sin(QR_{HB})}{QR_{HB}} exp^{-\frac{1}{2}<u_{HB}^2>Q^2} \tag{5}$$

where the parameter f_{HB} refers to the average number of hydrogen bonds per molecule in the liquid (where a value of unity corresponds to each molecule involved in two hydrogen bonds) and is a variable in the fit. The fitted data sets at $18°C$ are shown in figure 2. The data sets at $34°C$, and $49°C$ were fitted simultaneously with those at $18°C$ with identical intramolecular parameters.

A novel approach was employed to increase the sensitivity and consistency of the fit by simultaneously fitting (at high-Q) the difference of the interference contributions between the isotopic mixtures. This reduces the number of intramolecular terms in equation 4 from ten to four (ie. only those terms which involve the site of the H/D substitution), as well as eliminating the hydrogen bond contribution (which involves the hydroxyl rather then the formic hydrogen). The difference fit shown in figure 3 demonstrates the high precision of the experiment. Full details of this and other parts of the fitting procedure are given in [15].

The bond lengths obtained from this study are shown along with those obtained from other studies in the solid, liquid and gas phases in table 1. In particular, the bond lengths obtained by X-ray diffraction [5] are all between 2% and 4% greater than those found in this study. This systematic difference is probably due to the effects of the inelastic nature of the neutron scattering experiment on the interference terms. If so, the two studies are in good agreement considering that both studies also found that each molecule is on average involved in two hydrogen bonds.

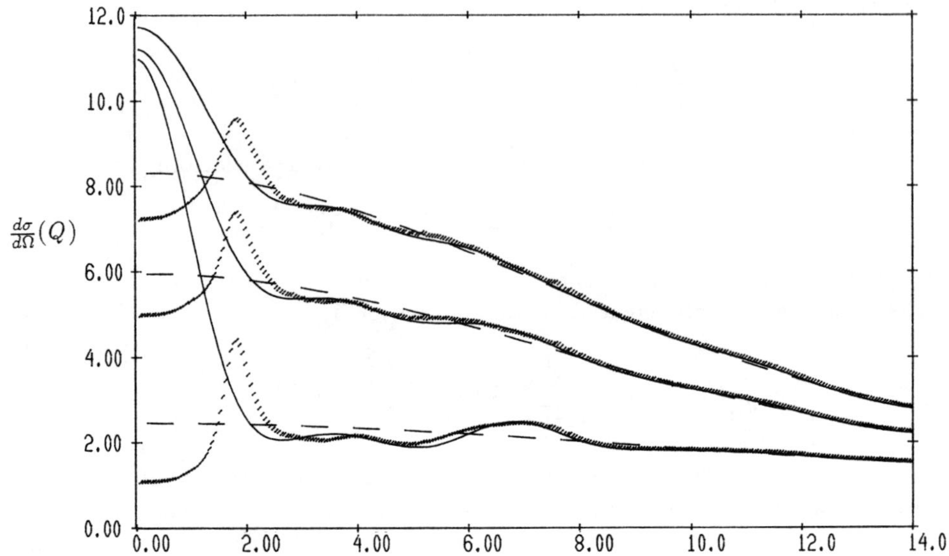

Figure 2: Form Factor Fit (—), Self contribution (- - -), and Experimental data (.....) at 18°C. From bottom to top: DCOOD, MCOOD, and HCOOD

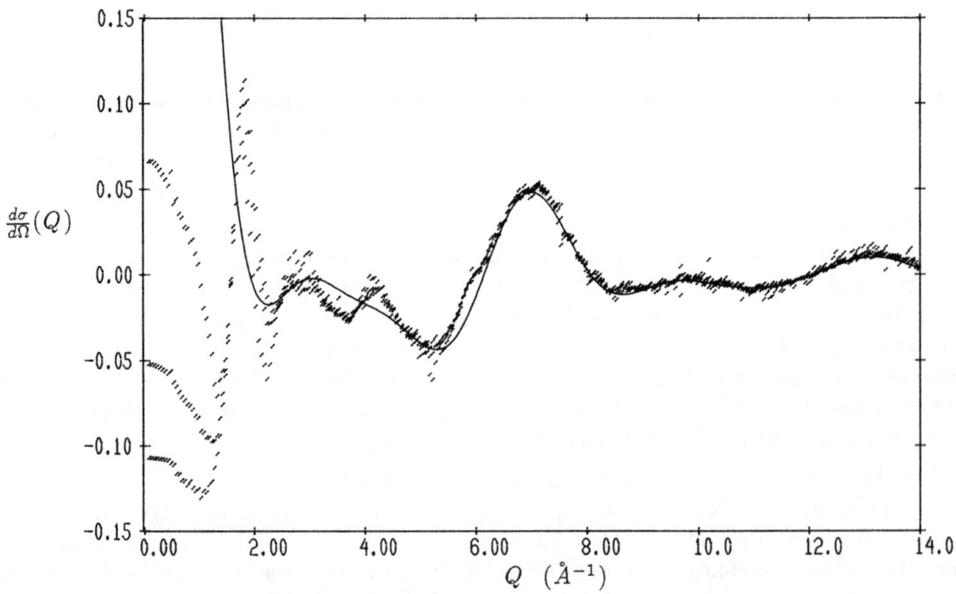

Figure 3: Difference Form Factor Fit (—) and Difference Experimental Data Sets (.....) at 18°C.

$R_{ij}(\AA)$	Solid X-ray [4]	Liquid Phase n-diff This Work	n-diff [12]	X-ray [5]	Gas e-diff [2]
C=O	1.221	1.189	1.23	1.217	1.220
C–H_{For}	0.92	1.078	1.10	1.106	1.082
C–0	1.311	1.314	1.38	1.361	1.323
O–H_{Ac}	0.88	0.944	1.03	0.984	1.036
H_{Ac} – –O	1.75	1.820	—	—	—
O–H_{Ac} – –O	2.625	—	—	2.703	2.703

Table 1: Comparison of formic acid bond lengths obtained by various authors.

4 Analysis of the Intermolecular Distribution Functions

The intermolecular contribution $\frac{d\sigma}{d\Omega}(Q)_{interm}$, can now be extracted using equation 1 and Fourier transformed to give the intermolecular distribution function

$$d_L(r) = 4\pi r\rho(g(r)-1) \tag{6}$$

$$= \frac{2}{\pi\sum_1 \bar{b}_i \sum_2 \bar{b}_j}\int_0^{10\AA^{-1}} Q\frac{d\sigma}{d\Omega}(Q)_{interm}\sin(Qr)\mathrm{sinc}\left(\frac{\pi Q}{10.0}\right)dQ \tag{7}$$

where g(r) is the pair correlation function and ρ is the molecular number density. The summation is over all the atoms which contribute to the intermolecular interference term from two different molecules labeled 1 and 2 and the *sinc* term is a modification function (Lorch [16]) to minimise truncation effects (mostly low frequency oscillations) in r-space.

The intermolecular distribution functions ($d_L(r)$) for the three isotopic mixtures at $18°C$ are shown in figure 4, and are almost identical to those at $34°C$ and $49°C$. The fitting procedure has already indicated that on average each formic acid molecule forms approximately two hydrogen bonds (corresponding to $f_{HB} \approx 1$) which implies that the formic acid molecules exist as *dimers*, *chains*, or a mixture of the two. The shaded parts (of figure 4) centred on $1.8\AA$ indicate the hydrogen bond contributions, and the damping of the oscillations as r increases correspond to increasing disorder in the liquid.

It is important to note that the hydrogen bonding represented in the $d_L(r)$ function by the peak at $1.8\AA$ is not a product of the data fitting and analysis procedure. The $d_L(r)$ functions obtained where no hydrogen bonding is included (corresponding to $f_{HB} = 0$ in the fitting procedure) are shown by dotted lines in figure 4 which clearly demonstrate that hydrogen bonding in formic acid is extensive.

5 Analysis of the Partial Distribution Functions

The measurements on three isotopic mixtures enable the calculation of the experimental partial distributions $d_L(r)_{RR}$, $d_L(r)_{RF}$, and $d_L(r)_{FF}$ which give the maximum explicit information possible from this experiment. The carboxylic acid head group ($COOD$) is identical for $DCOOD$, $MCOOD$, and $HCOOD$, and is denoted by the symbol R. The formic

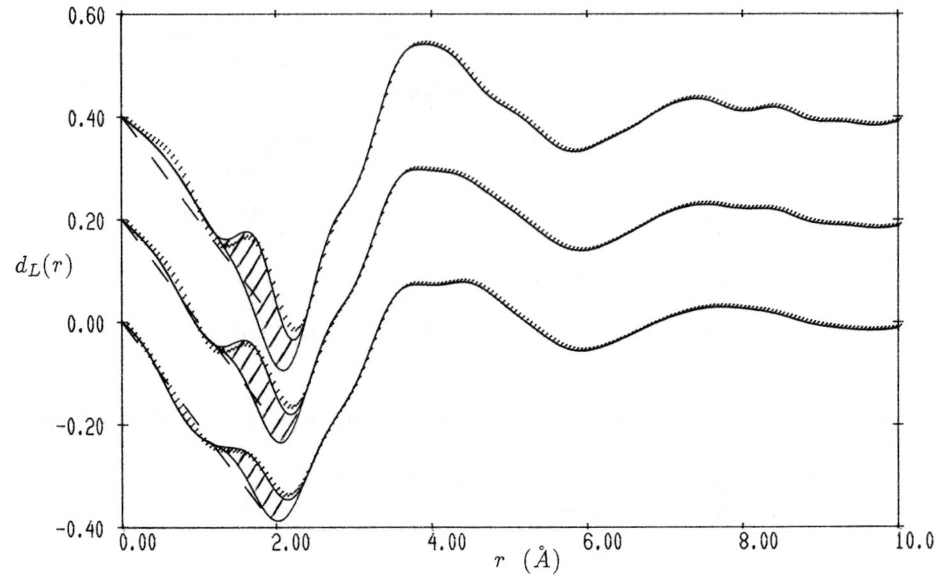

Figure 4: $d_L(r)$ and density line (- - -) for formic acid at 18°C. From bottom to top: $DCOOD$, $MCOOD + 0.2$, and $HCOOD + 0.4$. The shaded section is the hydrogen bond contribution. The dotted line arises from a fit which assumes no hydrogen bonding.

hydrogen site (being the site of the H/D isotopic substitution) is denoted by the symbol F. Figure 5 shows the $d_L(r)_{RR}$, $d_L(r)_{RF}$, and $d_L(r)_{FF}$ partials at 18°C.

The peak positions of 3.80Å for $d_L(r)_{RR}$, 4.65Å for $d_L(r)_{RF}$, and 4.05Å for $d_L(r)_{FF}$ give an insight into the relative orientations of the hydrogen bonded molecules. A comparison of the peak positions of the RR and RF partials shows that for two adjacent molecules the R part of one molecule is (on average) closer to the R rather than the F part of the other molecule by almost 1Å. This indicates (as expected) that the proton donor site of the hydrogen bond is occupied by the hydroxyl hydrogen rather than the formic hydrogen.

The size and shape of the peak at 4Å in the $d_L(r)_{FF}$ function indicates strong and sharp correlations between formic hydrogens at this distance. The formic hydrogens within closed dimers (as seen in the gas phase) are just over 6Å apart. As this correlation is not seen in the data, the dimer species can not exist in significant proportions. However, the chain structures of formic acid (as seen in the solid phase) have formic-formic correlations in the region of 4Å, and are not inconsistent with any of the data produced so far. This suggests that liquid formic acid exists predominantly in molecular chain structures.

6 Conclusions

The results shown in this paper demonstrate the extensive hydrogen bonding and polymeric chain structure of the liquid. Further information on the nature of the chain structure from this data could probably be gained from reverse Monte Carlo simulation techniques [17].

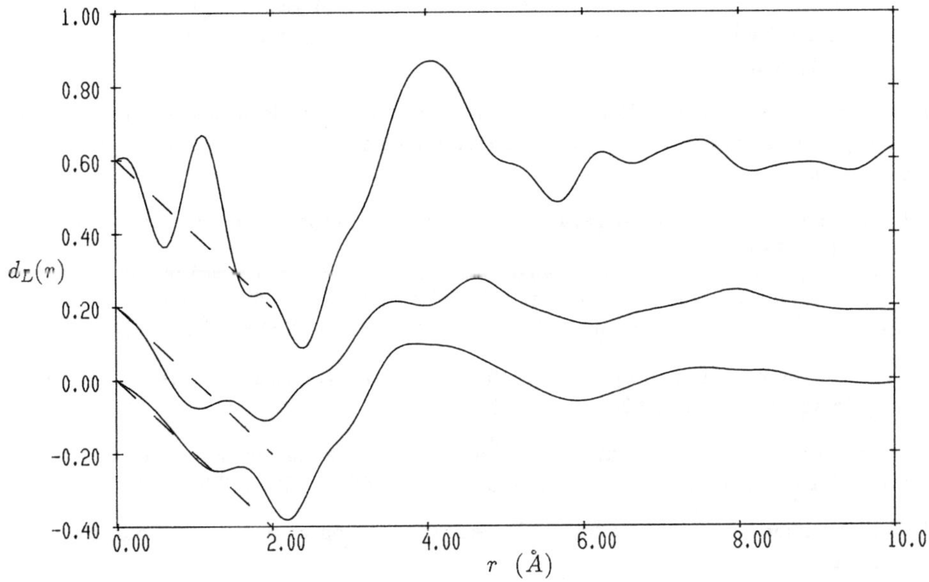

Figure 5: $d_L(r)$ for the three experimental partials of formic acid at $18°C$. From bottom to top: $d_L(r)_{RR}$, $d_L(r)_{RF} + 0.2$, and $d_L(r)_{FF} + 0.6$.

Neutron diffraction measurements using the H/D isotopic method at the hydroxyl hydrogen site will give more specific information on the hydrogen bond environment and hence on the chain structure. Hopefully, both approaches will be pursued in the near future.

A full discussion of liquid formic acid can be found in [15] where both these results and other results from the analysis are contained. In particular, strong evidence indicates that the molecule is predominately planar and in the trans conformation (as in figure 1), and the temperature difference analysis shows a relatively small reduction in hydrogen bonding as the temperature is increased.

7 Acknowledgements

I would like to thank John Dore and Marie-Claire Bellissent for useful discussions concerning this data, Dave Huxley for the continual use of his graphics programs, and NATO for the financial support given to attend this conference.

References

[1] L. Pauling and L. O. Brockway. The structure of the carboxylic group. I. the investigation of formic acid by the diffraction of electrons. *Proc. Nat. Acad. Sci*, 20:336, 1934.

[2] A. Almenningen, O. Bastiansen, and Tove Motzfeldt. A reinvestigation of the structure of monomer and dimer formic acid by gas electron diffraction technique. *Acta Chem. Scand.*, 23:2848, 1969.

[3] L. G. Bonner and R. Hofstadter. Vibration spectra and molecular stucture IV. infra-red absorption spectra of the double and single molecules of formic acid. *J. Chem. Phys*, 6:531, 1938.

[4] I. Nahringbauer. A reinvestigation of the structure of formic acid (at 98K). *Acta Cryst*, B34:316, 1978.

[5] T. V. Gorbunova, V. V. Shilov, and G. I. Batalin. X-ray diffraction study of formic, acetic, and propionic acids in the liquid state. *Russ. J. Struct. Chem.*, 14:388, 1973.

[6] C. P. Smyth and H. E. Rogers. The dielectric polarization of liquids. VIII. acetic and butyric acids. *J. Amer. Chem. Soc*, 52:1824, 1930.

[7] O. Faurskov and P. -A. Lund. Intermolecular raman active vibrations of hydrogen bonded acetic acid dimers in the liquid state. *J. Chem. Phys*, 78:652, 1983.

[8] J. F. Johnson and R. H. Cole. Dielectric polarization of liquid and solid formic acid. *J. Amer. Chem. Soc*, 73:4536, 1951.

[9] S. M. Blumenfeld and H. Fast. Low frequency spectra of solid and liquid formic acid. *Spectrochim. Acta*, 24A:1449, 1968.

[10] D. Chapman. The infrared spectra of liquid and solid formic acid. *J. Chem. Soc.*, 225, 1956.

[11] A. Kratochwill and H. G. Hertz. NMR relaxation study and molecular pair distribution functions of the system formic acid/chloroform. *J. Chimie Physique*, 74:814, 1977.

[12] H. Bertagnolli, P. Chieux, and H. G. Hertz. The structure of the formic acid molecule in the liquid state from neutron diffraction measurements involving five isotopically different species. *Ber. Bunsenges. Phys. Chem.*, 88:977, 1984.

[13] H. Bertagnolli and H. G. Hertz. Pair configuration of maximum occurrence probability in liquid formic acid as determined from neutron scattering and proton magnetic relaxation results. *Ber. Bunsenges. Phys. Chem*, 89:500, 1985.

[14] J. P. Ambroise, M. C. Bellissent-Funel, and R. Bellissent. Neutron scattering determination of local order in amorphous and liquid systems using a position sensitive detector. *Revue Phys. Appl.*, 19:731, 1984.

[15] G. I. Swan. *Structural Studies of Liquid Formic Acid and Water in Porous Silica by Neutron Diffraction.* PhD thesis, Univesity of Kent at Canterbury, 1989.

[16] E. Lorch. Neutron diffraction by germania, silica and radiation silica glasses. *J. Phys. C (Solid St. Phys.)*, 2:229, 1969.

[17] R. L. Mcgreevy and L. Pusztai. Reverse monte carlo simulations: a new technique for the determination of disordered structures. *Molecular Simulation*, 1:359, 1988.

STRUCTURAL STUDIES OF WATER NEAR AN INTERFACE

A.K.SOPER
Neutron Science Division
Rutherford Appleton Laboratory
Chilton, Didcot,
Oxon, OX11 0QX
England

ABSTRACT. Neutron diffraction, using the isotope substitution technique, is used to examine the structure of water confined within pores of sizes ranging between 5Å and 25Å. The smectite clay, montmorillonite, and vermiculite are used in this study, because they have a layered structure and absorb water, causing the well known swelling characteristic of clay minerals immersed in water. The clay layers retain their orientational correlation as water intercalates and this allows the water structure to be probed in different directions, normal and parallel to the surface. Neutron diffraction experiments on these materials and the associated data analysis techniques are described in some detail. For vermiculite there is evidence for a water layer very close to the surface, with protons readily penetrating the surface. For montmorillonite although the interfacial water has a bulk-like local structure it shows a pronounced oscillation in average density over several molecular diameters normal to the surface.

1. INTRODUCTION

The behaviour of water near a surface has been the subject of several computer simulation [1-3] and many experimental studies [4-7]. The computer studies generally use a rectangular box with periodic boundary conditions, with two opposite walls of the box forming the solid surfaces. Layered materials such as the smectite clays [8] or vermiculite [9] therefore are a convenient experimental substrate to test theories and simulations because the surfaces are smooth and parallel at the atomic level, and the thickness of the water layer can be controlled over a wide range from 3Å to as much 100Å or more, depending on the precise details of charge and ion distribution in the clay layer [10]. The underlying properties of clay minerals have been known for many years [10] but only fairly recently have the full implications of using these materials to study the structure and dynamics of interfacial water been realized. Ross and Hall [11] and

J.C. Dore and J. Teixeira (eds.), Hydrogen-Bonded Liquids, 147–158.

Sposito and Prost [6] have reviewed the use of various techniques to characterise the water structure in clays. The present article will present some new neutron diffraction data which exploit isotope substitution to extract details of the interfacial water structure.

Discussion of water or any liquid between solid interfaces is complicated by the fact that the water molecules are able to diffuse, at least in the space between the surfaces, so the system is neither a pure crystal nor a pure liquid. Although this statement may seem trivial it presents a major difficulty for the analysis of diffraction data since neither the techniques of crystallography nor those used to analyse liquid structure data are applicable. In particular an ensemble of water distributions may be compatible with the diffraction data and this ensemble has to be evaluated as part of the analysis. In the present work a Monte Carlo simulation is used to generate such an ensemble of distributions.

2. THEORY

Clays exhibit at least three stages of intercalation as water enters the lattice. In the initial crystalline phase with basal plane spacings up to about 20Å the layers generally form a regular lattice [9,10]. There is evidence for mixed spacings at some water contents, but generally the spacings are given approximately by the expression $d \simeq (9.6 + 2.8j)$ Å, where j adopts the values 0,1,2,3 and sometimes 4. Note that the increment in d-spacing corresponds to the diameter of a water molecule. There is no evidence for spacings between these values. In this case a series of discrete (001) Bragg reflections are seen in the diffraction pattern.

As further water enters the Bragg peaks are replaced by a regular but diffuse diffraction pattern analogous to a 1-dimensional fluid [10]. To a large extent the individual clay layers become spatially uncorrelated in a direction normal to the surface but retain a considerable degree of orientational correlation out to large spacings.

Finally as yet more water is added the layers disperse to form a suspension or gel depending on the ionic concentration, and with little or no orientational correlations.

The structure factor obtained by diffraction from a simple linear monatomic chain has been discussed by Mohan [12]. This theory can be applied to the clay problem by treating each clay layer as an "atom" in the simple chain. However the simple theory must be generalized to take account of the internal structure of the clay layer itself as well as the structure in the water molecules which are associated with the clay. For clarity we will only deal with the case of an infinite chain: this approximation is not serious because the peak broadening due to disorder in the clay layering is normally much larger than finite size effects on the peak shape.

Mohan shows that the structure factor (or diffraction intensity) for a linear chain of N atoms is

$$I(Q) = \frac{1}{N}\sum_{k,k'=1}^{N}\exp\{iQ(k-k')d\}\ \exp\{-\frac{Q^2<(uk-uk')^2>}{2}\}. \quad (1)$$

where d is the equilibrium separation of the atoms, and u_k is the relative displacement of the k'th atom from its equilibrium position. The angular brackets indicate an ensemble average over the states of the system. Expression (1) can be simplified by noting that for this model, which assumes a harmonic disordering mechanism,

$$<(u_k-u_{k'})^2> = \sigma^2 |k-k'|, \quad (2)$$

where σ is a width associated with the disorder. For the case of clay in water the disorder is caused by the variable number of water molecules between successive surfaces.

With (2) the structure factor evaluates to

$$I(Q) = \frac{1 - \exp(-Q^2\sigma^2)}{1 + \exp(-Q^2\sigma^2) - 2\mathrm{Cos}(Qd)\exp(-Q^2\sigma^2/2)} \quad (3)$$

The diffraction pattern from the first (crystalline) stage of hydration is obtained when $\sigma \rightarrow 0$ and corresponds to infinitely sharp Bragg peaks, while the second stage (long range disorder) corresponds to non-zero σ. As can be seen in figure 1 even fairly modest disordering (σ=0.1d) leads to a very rapid damping of the peaks in S(Q), in accordance with the measurements of Norrish [10].

The more general case of clay surfaces which have internal structure and which have water molecules adsorbed on them can be derived from these results by defining a local density profile, $l_k(Z)$, as a function of distance Z from the centre of the k'th clay platelet. It leads to an immediate generalization of equation (1):

$$I(Q) = \frac{1}{N}\sum_{k,k'=1}^{N}\exp\{iQ(k-k')d\}\ \exp\{-\frac{Q^2\sigma^2|k-k'|}{2}\}\ L_k(Q)L_{k'}(-Q) \quad (4)$$

where

$$L_k(Q) = \int_{-\infty}^{\infty} l_k(z)\ \exp(iQz)\ dz. \quad (5)$$

For the crystalline material (σ=0 in the model), and in the absence of diffuse scattering between Bragg peaks, the local form factors, $L_k(Q)$, are the same for all k, and so can be taken outside the summation in (4). Hence the integrated Bragg intensities for the simple linear lattice are modified by a form factor:

$$I(Q) = N*L(Q)*L(-Q). \tag{6}$$

For the disordered case, it can be seen from figure 1 that for $Q\sigma \geq 4$, only the k=k′ term is significant, i.e. provided Q is large enough the diffraction pattern is insensitive to correlations between atoms on different clay layers, only those between atoms on the same layer. In that case the diffraction intensity is now continuous in Q, but has a form very similar to (6):

$$I(Q) = \langle L_k(Q)*L_k(-Q)\rangle_k, \tag{7}$$

where the angle brackets indicate an ensemble average over k. The main difference between (6) and (7) is that in (6) all the layers are to be treated as the same, while in (7) variations can occur from one layer to the next. Examples of both types of diffraction are given below.

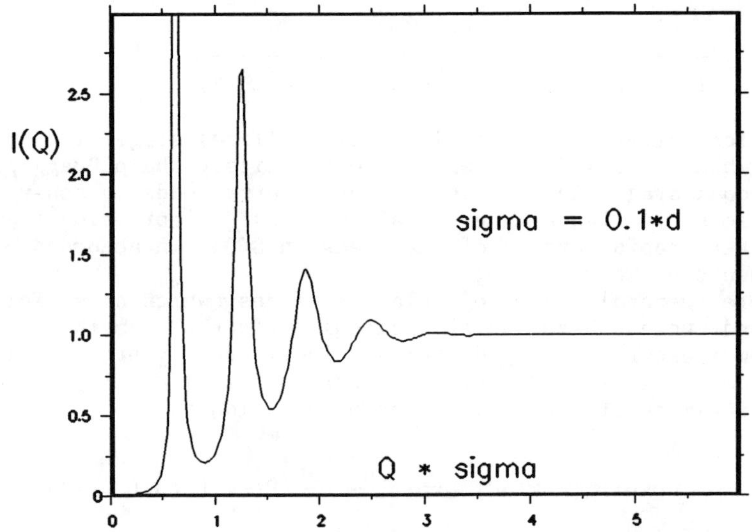

Figure 1. Structure factor, equation (3), for a harmonically disordered linear chain of atoms with σ, the standard deviation of the disorder, = 0.1 x the average atomic spacing.

A further, straightforward, generalization of these equations is needed for multicomponent systems, such as the case of water in clay, [13], but the above results serve to demonstrate the principle features of diffraction from these systems.

3. MONTE CARLO SIMULATION

Due to the absence of phase information there is no direct method of reducing the measured diffraction patterns to density profile. In addition and particularly for the disordered case there may well be a large number of different local distributions which contribute to the measured structure factor, as described above. Therefore in order to obtain a representative sample of particle distributions which are compatible with the data a one-dimensional Monte Carlo simulation is performed on the diffraction data, in which, instead of the energy being used as the criterion for deciding whether to accept or reject each configuration, the fit to the data, as measured by the R-factor, is tested instead. The square of the R-factor for the j'th configuration is defined as

$$R_f^2 = \sum_Q (D(Q) - M_j(Q))^2 \; / \; \sum D^2(Q) \tag{8}$$

where D(Q) is the measured intensity dataset, and $M_j(Q)$ the corresponding model intensity calculated using equations (4) - (7). The probability for acceptance or rejection of each configuration is based on a Boltzmann-like factor in the usual way:

$$\begin{aligned} P_{j\to j+1} &= \exp\{-\lambda \Delta R_f^2) \qquad \text{if} \quad \Delta R_f^2 > 0. \\ &= 1 \qquad\qquad\qquad\; \text{if} \quad \Delta R_f^2 < 0. \end{aligned} \tag{9}$$

where ΔR_f^2 is the change in R-factor from the j'th to j+1'th configuration. The parameter λ is a multiplier which determines how close the model distribution is to fit the data - it plays the same rôle as temperature in a conventional simulation.

In practice the model intensities must be modified from their ideal values, equation (4), by a form factor, A(Q), which arises from the angular mosaic of the clay particles and the finite range over which the atoms in each clay layer are correlated [13]. This form factor causes the diffraction intensities to fall as $\sim 1/Q^2$ and so tends to smear out the features in real space. This broadening in the calculated density profiles is caused by the sample and normally has little to do with the resolution of the diffractometer used to make the diffraction measurements.

4. EXPERIMENTAL

4.1. Hydrated vermiculite

A single crystal of vermiculite from Lano, Texas (surface charge density approximately 0.93 electrons per 5.6X4.6Å unit cell [14]), with approximate dimensions 10mm x 10mm x 1.5mm was initially saturated with Ni^{++} ions. Subsequently it was soaked in heavy water for about 3 weeks prior to the diffraction experiments. During the diffraction experiment the humidity was held at 85% which corresponds to about 4.5 water molecules per clay unit cell, and a layer spacing of 14.4Å. After the data had been accumulated it was then soaked in light water for a similar period and the measurements repeated. Thus two datasets were taken one with the water deuterated, the other with the water protonated. Diffraction data were taken on the LAD diffractometer at the ISIS spallation neutron source, U.K., with the diffraction vector, Q, normal to the (001) diffraction planes. Up to 30 orders of reflection could be measured in this way. After normalization to the incident neutron spectrum the data were corrected for detector efficiency and attenuation in the sample. Finally integrated Bragg intensities were calculated for each sample. These are shown in figure 2 and the diffraction contrast in going from H to D is seen clearly in this case. A more detailed description of the experiment is in preparation [15].

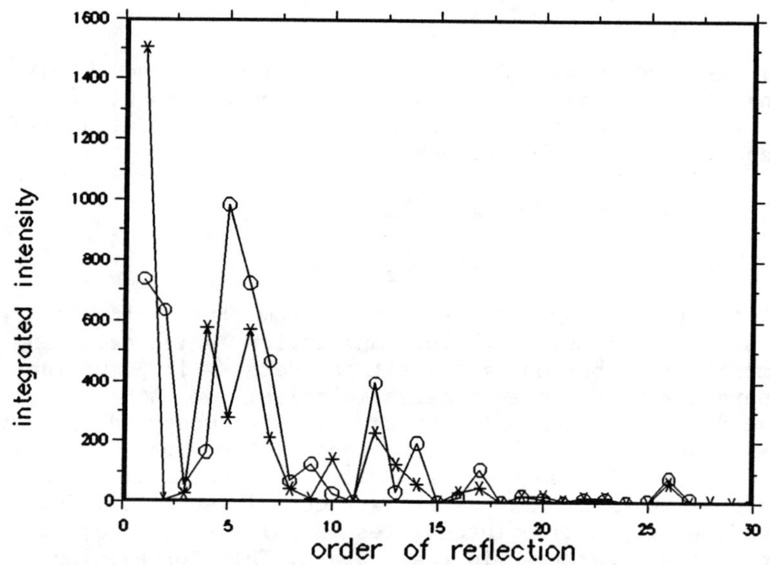

Figure 2. Integrated (001) diffraction intensities for protonated (stars) and deuterated (circles) nickel vermiculite at 85% relative humidity.

The diffraction intensities were analysed into two separate density profiles, one for water oxygens plus clay atoms, the other for water hydrogens, using the Monte Carlo technique described in section 3. The results are shown in figure 3. The R-factor for the simulation was 3%.

In the simulations the underlying clay structure was assumed to be known from crystallographic data [9]. The clay layer positions were allowed to refine over the range ±0.2Å about their crystal positions as the calculation proceeded, but in practice the final positions were

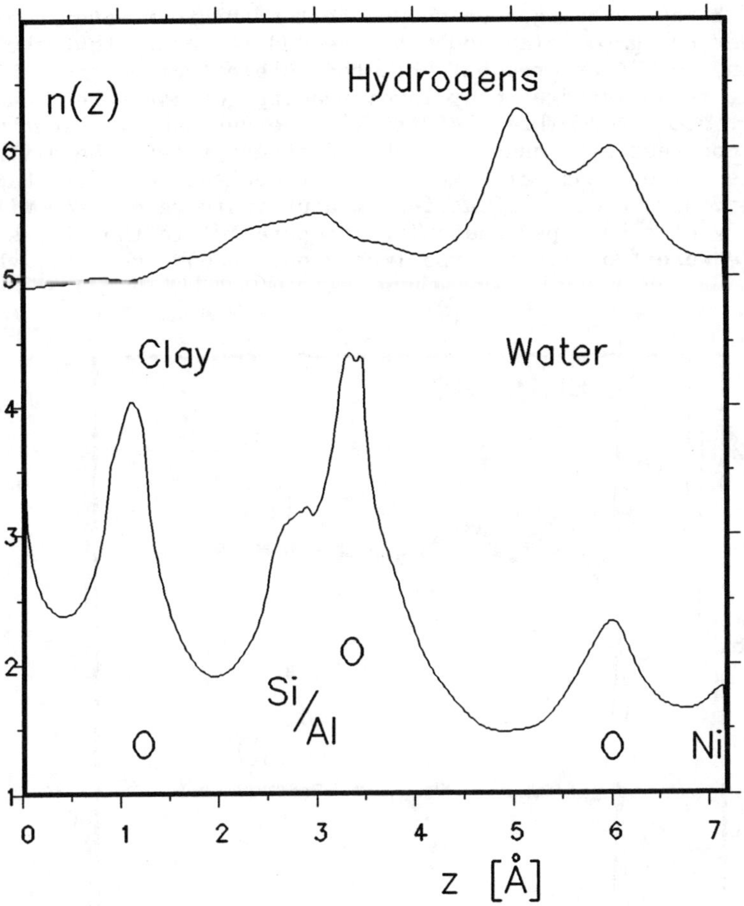

Figure 3. Calculated density profiles for water in vermiculte. The hydrogen distribution (top) was calculated separately from the water oxygen plus clay profile.

close to their accepted values. However no assumptions were made about the water positions which initially started out as uniformly distributed.

4.2. Montmorillonite

For these experiments a sample of Wyoming bentonite (surface charge density ~0.4 electrons per unit cell [14]), after appropriate purification and selection of particles of diameter <2μm, was saturated with Na^+ ions, and formed into a partially oriented aggregate by gradual precipitation from suspension over a period of several weeks.

An initial X-ray investigation showed that a minimum water content of 1gm of water per gm of clay would be needed to ensure that the sample would be in the disordered but oriented state (see section 2). This corresponds to an average clay layer spacing of 34Å. The same clay sample was used for three diffraction experiments, each with a different contrast of H and D in the adsorbed water. The diffraction measurements were made at the McMaster Nuclear Reactor, Hamilton, Canada. In this case diffraction measurements were made with the diffraction vector both perpendicular and parallel to the clay surface, and so there were two sets of experiments one in each geometry. Because

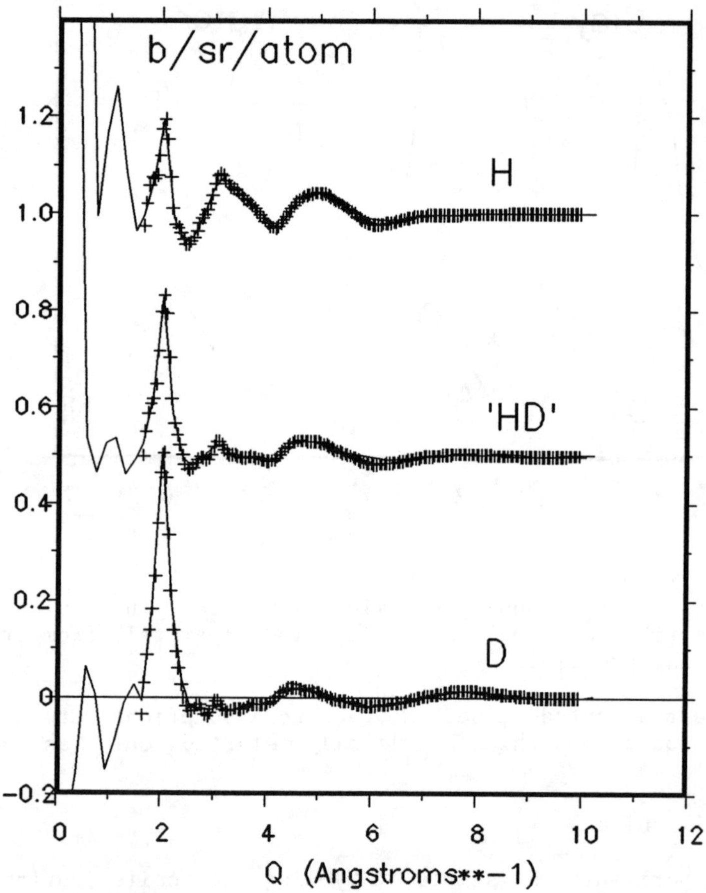

Figure 4. Interference total differential scattering cross sections, normalized per atom, for light water in Na-montmorillonite (top), heavy water in the same (bottom) and a 50:50 mixture of heavy and light water (middle). The single atom (incoherent) scattering has been subtracted.

these data were measured on a constant wavelength diffractometer it was necessary to employ half-angling to keep the Q vector either normal or parallel to the clay sheets. The half angling technique however limited the data to Q values in the range $1.5Å^{-1} < Q < 10Å^{-1}$. Figure 4 shows the measured interference differential cross sections for Q normal to the clay surface as a function of H/D content, after correcting for attenuation and multiple scattering, and after removal of the single atom (incoherent) scattering.

It is seen in figure 4 that the measured differential cross sections appear somewhat different to those seen in bulk water [16]: the major effect when Q is normal to the surface is a sharp first peak at all hydrogen contents near $Q=2Å^{-1}$ which is not seen in bulk light water, and is much broader for deuterated bulk water samples. This is also in contrast to the case when Q is parallel to the surface (data not shown here) where the curves look very similar to their bulk counterparts. Therefore these results already suggest a structuring effect on the water normal to the clay surface.

As for the vermiculite the diffraction intensities were analysed to two density profiles, one for H and one for "O". At the water content of this experiment (1gm of water per gm of clay) montmorillonite is in the spatially uncorrelated phase, so equation (12) can be used to analyse the diffraction data over the measured Q range. The calculated density profiles, figure 5, were averaged over 2000 configurations to ensure that a representative ensemble had been taken, and were checked by repeating the simulation several times: the differences between successive simulations were insignificant. Note that because of this averaging technique much of the local order around each atom has been removed in the curves shown in figure 5, and the residual structure must therefore arise from density oscillations near the surface. As before the underlying clay structure was assumed at the outset but allowed to refine within ±0.2Å of the accepted positions. The water oxygens and hydrogens were allowed to approach the clay surface to within 3.48Å of the centre of the clay layer, which was the outer limit allowed for the clay oxygens. This constraint on the water positions, which was not imposed in the vermiculite simulation, was used to compensate for the lack small Q information in the diffraction data. In any case the results were not very sensitive to the choice of constraints. The R-factor for these simulations was held at 7%.

5. DISCUSSION OF RESULTS

The interpretation of the vermiculite density profiles is fairly straightforward: the well known clay peaks at 0Å (Mg,Al), 1.07Å (O) ,2.73Å (Si and Al) and 3.3Å (O) are clearly visible, plus a peak near 6Å and fifth peak halfway across the unit cell at 7.2Å. It is generally accepted that the cations in vermiculite lie midway between the clay surfaces, in which case there would be a Ni hydration sphere with 6 water oxygens about 1.2Å before and after the Ni peak. The areas of the two peaks are fully consistent with this interpretation, which

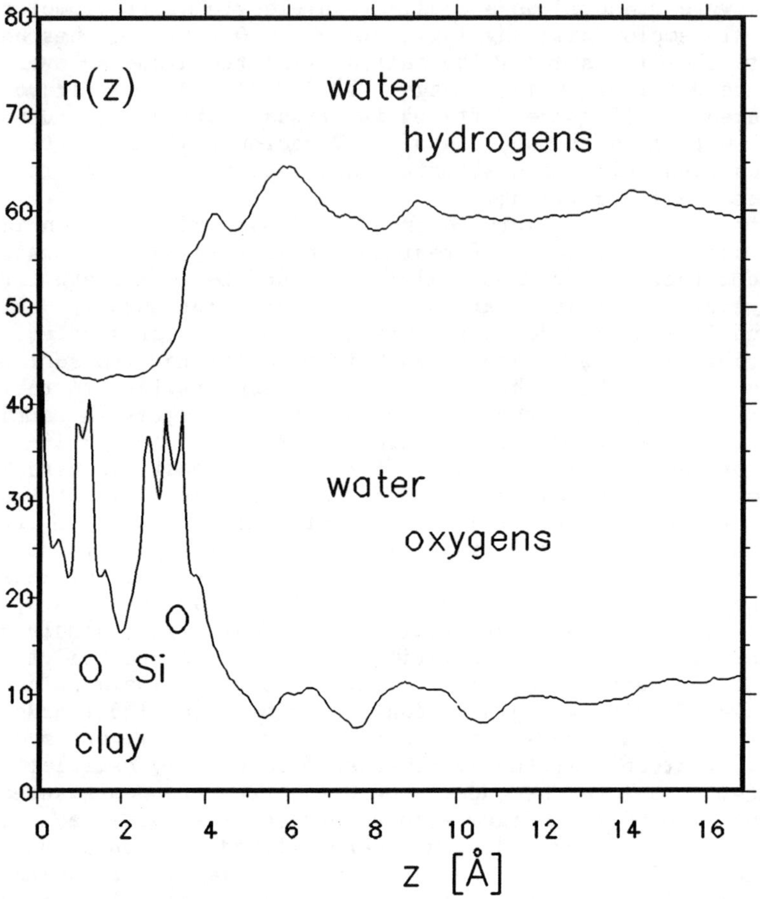

Figure 5. Calculated density profiles for water in Na-montmorillonite as described in the text.

is further supported by the existence of two hydrogen peaks near the oxygen peak. (No assumption has been made in the simulation about the geometry of the water molecules: the hydrogen distribution is adjusted independently from the oxygen distribution.) Clearly therefore the water associated with the cation is well oriented with respect to the surface.

An unexpected feature of these data is that the clay peak at 3.3Å appears to be a composite of a sharp clay peak plus a broader peak at nearly the same distance. The area of this additional peak is between 0.5 and 1 water oxygens per unit cell greater than if it consisted of clay oxygens alone. In addition there is an increase in the hydrogen density in the same region. Therefore the diffraction data imply that a layer of water molecules is embedded in the clay surface. This is believed to be the first time that such a layer has been observed, and apparently accounts for the additional absorbed water not coordinated to the central cation. From the broadness of the hydrogen distribution it appears that this embedded layer of water molecules is not strongly oriented by the surface.

For the montmorillonite data, which were taken for a much thicker water sheet than for vermiculite, the water is clearly less correlated with the surface. Even so there is still a residual density oscillation of ~±25% of bulk density which extends several molecular diameters into the liquid: broad peaks can be seen at 6.2Å, 9Å and ~12Å. Previous computer simulations [1,2] have indicated that the oscillations have a smaller amplitude than this. However these simulations do not include the fact that the clay surface is negatively charged and so a detailed comparison is not possible.

Note that there is apparently an increase in hydrogen density near z=4Å from the centre of the clay layer, and the "O" distribution also appears to increase in density in this region. However the integrated peak areas in the clay region indicate an additional oxygen density of less than 0.5 oxygens per unit cell above what would be expected for the clay alone, so within the limitations of the data, particularly with the lack of small Q data, the evidence in support of an embedded layer of water in montmorillonite is weak. A more likely explanation for the increase in hydrogen density is that some molecules near the surface are quite strongly oriented by the surface. Overall however there is significantly less orientational correlation of water molecules with the montmorillonite surface than in vermiculite.

6. CONCLUSION

Two experiments on water intercalated in vermiculite and montmorillonite have been described. It is quite clear from the results shown here that water is modified structurally in a direction normal to the clay surface. The extent and size of this modification depends on the detailed stoichiometry of the clay layer. The more highly charged vermiculite with a narrow pore spacing of ~5Å has a strong structuring effect with most of the water forming hydration spheres around the central cations. There is however evidence for a layer of water molecules embedded in the clay surface. For the weaker charged montmorillonite, which can expand indefinitely, water retains its liquid like characteristic, but a residual density osccillation is seen proceeding at least 3 molecular diameters into the liquid. The question of whether montmorillonite would show the same structuring effect on water as vermiculite for the same pore dimension can only be resolved by further experimentation.

7. ACKNOWLEDGEMENT

The vermiculite experiments were performed in collaboration with

N.T.Skipper and J.D.C. McConnell (Oxford). The work on montmorillonite was performed with the assistance of J.M. Gromek and H.E. Poll (Guelph).

8. REFERENCES

[1] Lee,C.Y., McCammon,J.A., and Rossky,P.J., (1984)'The structure of liquid water at an extended hydrophobic interface', J Chem Phys 80, 4448-4455.
[2] Christou,N.I., Whitehouse,J.S., Nicholson,D., and Parsonage,N.G.,(1985)'Studies of high density water films by computer simulation', Mol. Phys.,55,397-410.
[3] Aloisi,G., Guidelli, R.,Jackson,R.A.,Clark,S.M., and Barnes,P.,(1986)'The structure of water at an interface', J Electroanal. Chem., 206,131-137.
[4] Hawkins,R.A., and Egelstaff,P.A.,(1980)'Interfacial water structure in montmorillonite from neutron diffraction experiments', Clays and Clay Minerals, 28,19-28.
[5] Adams,J.M.,and Riekel,C.,(1980),'One dimensional neutron diffraction study of vermiculite',Clays and Clay Minerals,28,444-445.
[6] Sposito,G. and Prost,R.(1982)'Structure of water adsorbed on smectites',Chemical Reviews, 82,553-573.
[7] Israelachvili,J.N. and Pashley,R.M. (1983),'Molecular layering of water at surfaces and origin of repulsive hydration forces', Nature,306,249-250.
[8] Pinnavaia,T.J., (1983) 'Intercalated clay catalysts', Science,220,365-371.
[9] Brindley,G.W. and Brown,G. (eds.), (1980) Crystal structures of clay minerals and their X-ray identification, Mineralogical Society Monograph No.5, Mineralogical Society, London
[10] Norrish,K., (1954) 'Swelling of montmorillonite', Disc. Farad. Soc.,18,120-134
[11] Ross,D.K. and Hall,P.L. (1980) 'Neutron scattering methods of investigating clay systems', in J.W.Stucki and W.L.Banwart (eds.), Advanced Chemical Methods for Soil and Clay Minerals Research, Reidel, Boston, chap. 2.
[12] Mohan,M.M. (1986) 'Structure factor of a single chain of finite size', Phys. Rev. B34,8915-8921.
[13] Soper,A.K.,Gromek,J.M. and Poll, H.E. (1989), in preparation.
[14] Newman, A.C.D. (ed.) (1987), Chemistry of clays and clay minerals, Mineralogical Society Monograph No. 6, Longman, London.
[15] Skipper,N.T., Soper,A.K., and McConnell,J.D.C. (1989) 'The structure of interlayer water in Ni-vermiculite', in preparation.
[16] Soper,A.K. and Phillips,M.G. (1986) 'A new determination of the structure of water at 25°C', Chem. Phys. 107,47-60.

SANS AND QENS STUDIES OF VYCOR CONTAINING D_2O/H_2O MIXTURES

J.J. TUCK, D.K. ROSS and J-C. LI
School of Physics and Space Research, The University of Birmingham, Edgbaston, Birmingham, B15 2TT, UK

B. SARKISSIAN
Institut Laue-Langevin, 156X, Centre de tri, 38042, Grenoble, France.

ABSTRACT. The diffusion of water within a given pore and through adjacent pores has been studied in the porous silica, Vycor, using small angle neutron scattering and small-angle quasi-elastic neutron scattering. The spatial range measured - between 10 and 1000 Å - is below that observed by light scattering or spin-echo NMR but covers the physically interesting transition region between intra-pore and macroscopic (inter-pore) diffusion. Using porous Vycor glass fully saturated with a mixture of D_2O/H_2O water, which has been matched to the scattering density of the matrix to suppress the coherent small-angle scattering, we can measure $<r^2(t)>$, i.e. the time-dependent mean square displacement of the average proton from its position at t = 0. The measurements were made on IN11 - the Spin Echo Neutron Spectrometer (SENS) at ILL - which is designed to measure scattering at small angles and extremely high energy resolution, ie. exactly the present requirement. In this paper we discuss the use of both the Small Angle Neutron Scattering (SANS) and the SENS techniques for studying the behaviour of liquids in porous materials and present our preliminary results.

1. Introduction

1.1 NEUTRON SCATTERING STUDIES OF POROUS SOLIDS

The properties of porous Vycor silica glass have been extensively studied both because of their intrinsic interest and because Vycor provides a convenient model system for other porous silicas and for porous solids in general. Thus, there have been a number of investigations into its structural [1-3] and other properties [4-8]. Aspects of a porous solid that are currently of particular interest are the description of individual pore geometry, the relative position of neighbouring pores and the representation of the interconnections between adjacent pores. The properties of the pore walls before and after dehydration and of the chemisorbed water are also of interest. Neutron elastic, quasi-elastic and inelastic scattering can play important roles in such studies. When a liquid such as water (a suitable mixture of light and heavy water) is introduced, one can use Small-Angle Neutron Scattering (SANS) in combination with contrast matching techniques to study aspects of the pore geometry. Thus, the neutron scattering density for the water can, for example, be matched to that of either the silica matrix or a vacuum. Matching to the silica gives a measure of the relative scattering density of the silica; matching to a vacuum allows one to examine changes

J.C. Dore and J. Teixeira (eds.), Hydrogen-Bonded Liquids, 159–170.

in the surface geometry, in particular, in the surface roughness, as the water is removed [3]. In addition, the behaviour of liquids confined within small pores is itself an interesting problem. Thus, how the density, freezing point and equilibrium vapour pressure vary with pore size, how the liquid interacts with the surface of the solid and how the condensation mechanism during absorption differs from the evaporation process during desorption, can all be investigated with SANS.

Small–Angle Quasi–Elastic Neutron Scattering (SAQENS) offers a further unique technique for studying porous solids in that it provides otherwise unobtainable information about the diffusion of the liquid through the pores on a microscopic scale. This is currently a topic of considerable interest, partly due to the practical importance of low rates of liquid flow through porous rocks under pressure gradients in the oil and water industries and partly because of the development of fractal and similar descriptions of the processes involved. When quasi–elastic measurements are extended towards low Q by means of the spin echo technique, one can measure diffusion over a local network of pores and, hence, can investigate the interconnectivity of these pores through a determination of $\langle r^2(t)\rangle$ over the appropriate length range. Such results can be used to test appropriate theories of either the traditional kind involving tortuosity factors or those based on more recent speculations about the relationship between the fractal description of a system and the form of $\langle r^2(t)\rangle$ within the range of r values where a fractal description is appropriate [9–11]. As oil shales [12] and other rocks [13] have been shown to exhibit fractal behaviour, i.e. the small angle scattering intensity varies as Q^{-n} with $3<n<4$ for different shales, these interesting ideas may have considerable practical importance.

In recent papers, Benham et al [3] and Li et al [14] have described their SANS measurements on porous Vycor. In the present paper, we intend to focus on the SANS matching technique and the neutron spin echo (SENS) technique and to present some initial results using the latter method.

1.2 MATCHING THE SCATTERING LENGTH DENSITY OF THE VYCOR MATRIX USING H_2O/D_2O MIXTURES

Since SANS measures the scattering density fluctuations across many atoms or molecules, very small density variations between the Vycor matrix and the pores can be detected and any tendency for condensation to take place in pores of a particular size group can also be seen. In order to highlight the features to be studied, the choice of the matching point is extremely critical. Contrast matching can be achieved by varying the H_2O/D_2O ratio. Because we know the scattering length densities $\rho_{H_2O} = -0.57\mathrm{x}10^{-10}$ cm^{-2} and $\rho_{D_2O} = 6.36\mathrm{x}10^{-10}$ cm^{-2} (based on the bulk densities of light and heavy water), we can choose to make the scattering density of water, ρ_w, satisfy either of the following conditions:

	Match point 1	$\rho_w = \rho_{vycor}$	(64% D_2O)
or	Match point 2	$\rho_w = \rho_{air} = 0$	(8% D_2O)

When the scattering length density of water is matched to that of porous Vycor (Match Point 1), the pores will be entirely invisible to neutrons and there will be no coherent small angle scattering. Consequently, only those pores from which water has been removed during an absorption/desorption process, or during two–phase fluid flow through the pores under pressure, will produce coherent scattering. On the other hand, as mentioned above, Match Point 2 is particularly useful for looking at the changes in the structure of the matrix as hydration proceeds [3].

Optimizing the H_2O/D_2O ratio to minimise the SANS is not difficult in principle. Porous Vycor has a density of 1.5 and absorbs a weight of water equivalent to 28% by volume if we assume that the water has its normal density. However, the density of water in small pores is not necessarily the same as that for the bulk state and it is, of course, temperature dependent. In practice, therefore, the matching has to be done empirically. Because the intensity at a particular Q varies as the square of the scattering density difference, we can in principle get the match point by plotting the square root of the intensity against the H/D ratio – choosing the signs so as to obtain a straight line – the match point being the intercept with the x axis. In practice, of course, this simple picture only applies if the two media have exactly constant densities and so we have progressed through three stages in order to achieve the best match. First, on the D17 spectrometer at the ILL we used the value calculated from the bulk densities, namely 57% D_2O. These measurements showed that this was not the best matching ratio although it gave a more sensitive measurement of density change on freezing than would a perfectly matched system [3]. We then carried out a coarse matching test using LOQ at ISIS (Rutherford–Appleton Laboratory), which showed that the ideal match lay between 60% and 70% D_2O. This range was then further refined, again using LOQ. Fig. 1 shows the best matched sample, containing 64±1% D_2O water at room temperature, in comparison with a sample containing 8% D_2O, the zero scattering density mixture. On the assumption that the discrepancy is due to a reduction in the density of the absorbed water, we obtain an average water density reduction of 10%.

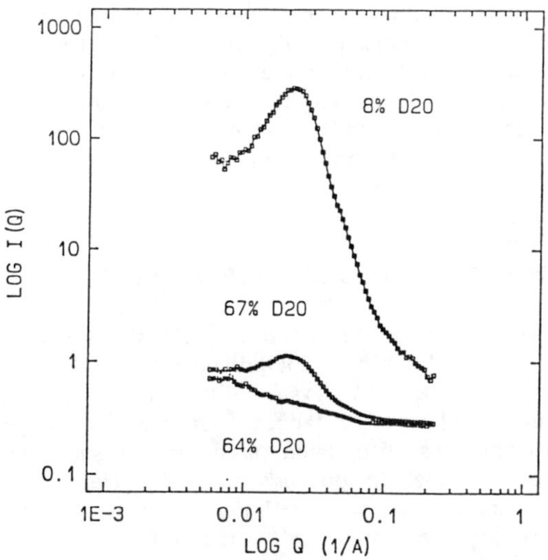

Fig. 1. SANS from Vycor saturated with 64% D_2O water (matched to the Vycor) compared with data for the same sample saturated with 67% and 8% D_2O water (zero scattering length density).

The typical SANS pattern from dry porous Vycor, like the 8% sample, shows a peak at $Q = 0.025$ $Å^{-1}$ which characterises the spatial distribution of pores in the sample. This distribution was originally determined by the spinodal decomposition process that occurred in the original borosilicate glass prior to etching out the B_2O_3 component. As can be seen, the curve for the matched sample is very flat at the spinodal peak position with a coherent scattering intensity that is certainly lower than the incoherent background. Compared with the 8% sample, the amplitude of the peak is reduced by

3 orders of magnitude. This dramatic reduction in intensity implies that the scattering length density of the Vycor matrix has been matched to within 1%. It is important to note from this result that, not only has the peak been matched out, but also that there is a complete absense of scattering at higher Q – in contrast to the results of Wiltzius et al [15] who used a mixture of C_6H_{12} and C_6D_{12} in a similar experiment. This observation indicates that, with matched water, the boundary layer has exactly the same scattering density as the bulk liquid in the centre of the pores and as the bulk silica, implying that water gets incorporated uniformly into the surface layer, exactly compensating for the loss of silica. Other evidence clearly shows that the structure of this surface layer must be complex. The appearance of surface fractal scattering when the air-dried material is baked [3] would suggest that the "chemisorbed" water that is incorporated in the surface layer forms a gel (i.e. individual water molecules have replaced the B_2O_3 units that had previously been etched away) and that in this gel, "molecules" of both silica and water occupy the same volume as in their respective bulk phases.

In addition to the strongly absorbed water incorporated in the surface, it is commonly supposed that multilayers of some kind of structured water are formed on the surface of the silica. This process corresponds to the region of the isotherm that follows the BET behaviour very precisely with a temperature dependence that scales closely with relative humidity, implying that the intermolecular interaction is similar to that in bulk water [16]. In this surface region, we know that many of the physical properties of water must nevertheless be different from those of bulk water or ice, in particular the density [17] and the vibrational behaviour [18]. It is therefore quite surprising that all these changes can take place in the surface layer without affecting the scattering length density. Recent IINS measurements on TFXA at ISIS [19] for Vycor equilibrated with a small number of layers of water molecules at 20K indicate that the boundary layer must contain many broken H–bonds between water molecules (water-like structure). This might suggest that the boundary layers do not change density on freezing and hence that the SANS from a sample matched in the liquid phase would show structure in the Porod region below the eventual freezing point. This point has, however, still to be investigated.

1.3 SENS MEASUREMENTS OF THE TRACER DIFFUSION COEFFICIENT

For many years, light scattering has been used to study the dynamics and structures of colloidal suspensions, where both the particle size and the spatial correlation lengths are of the order of the wavelength of light (4000–7000 Å). Tracer diffusion coefficients measured by this technique are obviously measured over distances that are very large compared to the fractal range. Light scattering has also been used to measure the tracer diffusion coefficient of optically active molecules in porous solids [9]. Here, one is also interested in the chemical diffusion coefficient of the liquid as this can be related to the flow rate of the liquid through the solid under the influence of a chemical potential via the compressibility. For a fractal system with a lower characteristic dimension of around the pore dimension (20–30Å) and an upper limit of some hundreds of Angstroms, small angle X-ray and neutron scattering are the appropriate techniques to use. Neutron scattering is particularly useful because of the contrast variation technique and the possibility of using quasi-elastic scattering. The spin echo spectrometer, IN11, is specifically designed for this type of experiment with very high resolution in energy transfer (10^{-7} eV) at small angles of scatter (low Q). The principle of this technique is similar to the NMR technique of the same name. The very high energy resolution is achieved by making two time measurements on the same neutron as opposed to the more usual time–of–flight technique of defining the time origin by pulsing the beam. In the latter technique, the use of a short enough pulse and a long enough flight path to achieve the above resolution would involve sacrificing so much intensity as to make the measurement impossible. Full details of the operation of the instrument can be found elsewhere [20,21] but the basic principles are as follows. A roughly monochromatic beam is polarised with its direction of polarisation parallel to the beam direction. This now passes along the incident path in a fixed vertical magnetic field, H_Z, which causes the neutron spin to precess with the appropriate Larmour frequency. At the sample, the spin direction is reversed and the scattered neutron passes through an equal path length in the same field, H_Z, so that the precession direction is reversed and the precession "unwinds" itself by the time the neutron reaches the spin analyser and the detector. The essential point here is that, although all the different neutron energies in the incident beam will precess to a different extent, after an elastic scatter they will all exactly unwind. Now, very small changes in either of the magnetic fields will cause the detected intensity to vary as a cosine function; thus if the minimum intensity is zero, the polarisarion is complete, ie the polarisation for elastic scattering $P_0(Q,t)$ is defined by the equation

$$P_0(Q,t) = [I_{max}(Q,t) - I_{min}(Q,t)] / [I_{max}(Q,t) + I_{min}(Q,t)] \quad (1)$$

Here, t is called the fourier time and is given by $\pi N/E_0$ where N is the number of turns and E_0 is the mean incident energy. Thus, if both magnetic fields (precession rates) are increased together in steps and the polarisation is remeasured at each value, the polarisation will only decrease slowly, due to field inhomogeneities, etc.. If now, we introduce a quasi–elastic scatterer, the velocities after scattering will be spread out relative to the incoming value and the unwinding process will no longer be precise so that the polarisation will be reduced, written $P_m(Q,t)$. Clearly, the higher the precession rate, the greater the loss of polarisation. The properties of the quasielastic scatterer are therefore given by

$$P(Q,t) = P_m(Q,t)/P_0(Q,t) \quad (2)$$

It turns out that P(Q,t) is proportional to S(Q,t), the fourier transform of the usual scattering function, $S(\underline{Q},\omega)$ and for many purposes the former is actually a more useful function for comparing with models.

2. Theory

A quasi–elasic neutron scattering experiment measures the intensity of neutrons scattered into unit solid angle and unit energy interval around E. This intensity may be expressed as

$$\begin{aligned} d^2\sigma/d\Omega d\omega = k/k_0[\{(1-x)\sigma^{inc}{}_H/4\pi + x\sigma^{inc}{}_D/4\pi \\ + x(1-x)(b_H-b_D)^2\}S^{inc}(Q,\omega) \\ + \{(1-x)\langle b_H\rangle + x\langle b_D\rangle + 0.5b_O\}^2S^{coh}(Q,\omega) \\ + (\rho_{water} - \rho_{Vycor})^2S(Q)] \end{aligned} \quad (3)$$

where $Q = k - k_0$ and $h\omega = E - E_o$

In writing these expressions, we have separated the isotope and the spin contributions to the incoherent cross sections, where <> implies averaging over the spin states. This has been done so that we can introduce x, the deuterium fraction, explicitly and because isotope and spin incoherence have to be dealt with separately in spin analysis experiments. $S_{inc}(Q,\omega)$ and $S_{coh}(Q,\omega)$ are the dynamic incoherent and coherent structure factors of the system. In the latter term, we have omitted the coherent elastic structure factor, which is included in S(Q), the overall elastic structure factor. $\sigma^{inc}{}_H$ and $\sigma^{inc}{}_D$ are the spin incoherent cross sections and $<b_D>$ and $<b_H>$ are the spin averaged scattering lengths for deuterium and hydrogen respectively while b_O is the scattering length for oxygen, which has been introduced into the coherent scattering term on the assumption that $2\pi/Q \gg$ the size of the molecule. ρ_{water} and ρ_{Vycor} are the scattering densities of water and Vycor respectively – ie the number densities of formula units in water and silica respectively times the average scattering lengths of the respective elements per formula unit. In the present case, we know that the energy transfer is very small, i.e. $|\mathbf{k}_0| = |\mathbf{k}|$, and hence we may take $Q = (4\pi/\lambda)\sin(\theta/2)$ and $k = k_o$. In the SENS experiment, the polarisation in the scattered beam can be measured at different angle θ for fixed λ. In order to relate this to the scattering cross sections in (3) above, we must first note that spin incoherence is associated with a 2/3 chance of spin flip, which, of course reverses the sign of the contribution to the polarisation. Secondly, we note that we measure S(Q,t) which is the fourier transform of $S(Q,\omega)$. The x dependence of the various factors contributing to the magnitude of P(Q,t) are illustrated in Fig. 2.

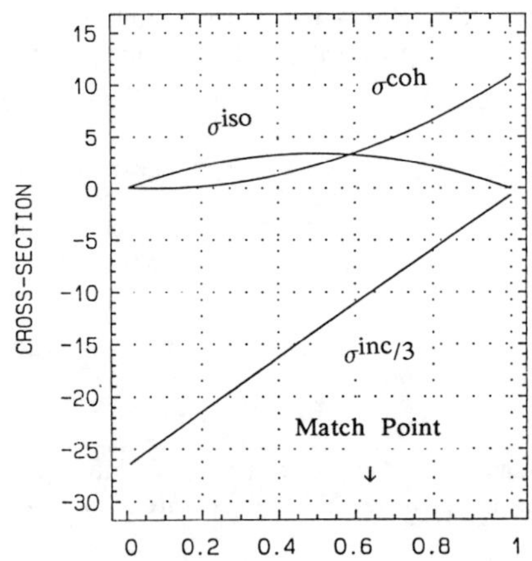

Fig 2. The x dependence of the spin incoherent and isotope incoherent contributions to the polarisation as defined in equation (2). The variation of the coherent cross section and the position of the SANS matching point in silica are also shown.

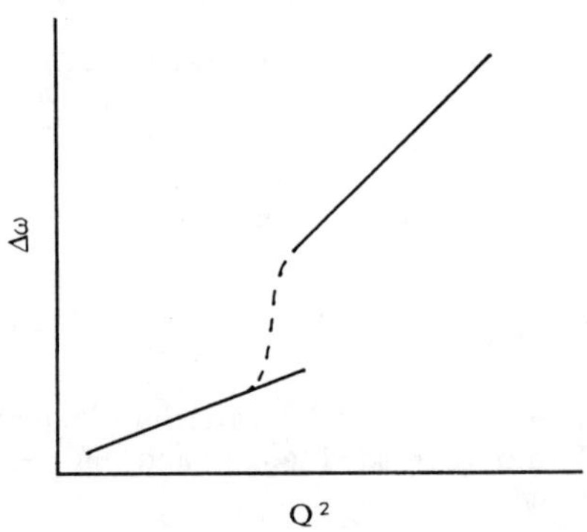

Fig 3. The general form of the variation of the FWHM of $S^{inc}(Q,\omega)$ with Q for a liquid in a porous material

The coherent and incoherent neutron scattering functions, $S_{coh}(Q,\omega)$ and $S_{inc}(Q,\omega)$, are properties only of the scattering system. Measurements of these functions can yield valuable information on the microscopic and macroscopic dynamic properties of a fluid in a porous medium. In the limit of small Q, (ie $1/Q \gg$ upper linear dimension for fractal behaviour), the coherent scattering function, $S_{coh}(Q,\omega)$, is related to the macroscopic chemical diffusion coefficient, D_c [14], by

$$S_{coh}(Q,\omega) = \pi^{-1} D_c Q^2/((D_c Q^2)^2 - \omega^2) \tag{4}$$

and the incoherent scattering function $S_{inc}(Q,\omega)$ is similarily associated with the macroscopic tracer diffusion coefficient, D_t [22], by

$$S_{inc}(Q,\omega) = \pi^{-1} D_t Q^2/((D_t Q^2)^2 - \omega^2) \tag{5}$$

In the incoherent QENS experiment, the coherent QENS could have been eliminated by choosing the appropriate value of x, i.e. we could choose $\langle b_{water} \rangle = 0$, but this has the disadvantage that S(Q), the small angle elastic contribution, is large at this value of x and would swamp the incoherent term below some Q value that is, by definition, of interest to us. We have therefore chosen to work with $\rho_{water} \simeq \rho_{Vycor}$. This means that there is a finite coherent QENS contribution. We have investigated this contribution and it would seem to be small and too fast to be seen in the present experiment [23] so that the time dependent contribution will be determined by the incoherent scattering. This, in turn, is determined by the self correlation function, $G_s(r,t)$, the probability distribution in space r at time t of a molecule that was at the origin at time t=0. From this, S(Q,t) is obtained by fourier transformation in space

$$S(Q,t) = \int G_s(\underline{r},t) \exp(i\underline{Q}.\underline{r}) d\underline{r} \tag{6}$$

Now, it is reasonable to describe the behaviour of $G_s(r,t)$ in terms of the time dependence of the mean square deviation of a gaussian which may be written $\langle r^2(t) \rangle$. This will, after a short time, become $6D_\ell t$ where D_ℓ is the local diffusion coefficient within the pore (D_ℓ is presumably comparable to the value of D_t in the bulk liquid). At later times, the diffusion process will slow up as the molecule has to find its way out of the pore, until at long time, $\langle r^2(t) \rangle$ becomes $6D_m t$, the macroscopic tracer diffusion coefficient measured over distances that are large compared with the upper limit for fractal behaviour. If we represent the full width at half maximum of $S^{inc}(Q,\omega)$ as $\Delta\omega(Q)$, then $\Delta\omega(Q)$ must vary as shown in Fig. 3. At small Q, it varies as $D_m Q^2$ ie with a small gradient and at larger Q, it varies as $D_\ell Q^2$, ie with a much higher gradient. In the intervening range, it passes through a transition region so as to join up with the asymptotic forms. In this region, the peak shape is no longer lorentzian.

Neutron spin–echo permits direct time–domain measurement of the scattering function S(Q,t). For our system, t is typically ns, and the polarisation measured for a purely incoherent scatterer in a symmetric neutron spin–echo scan is given by

$$P(Q,t) = \frac{\int S_{inc}(Q,\omega) \cos(t(\omega-\omega_o)) d\omega}{\int S_{inc}(Q,\omega) d\omega} \tag{7}$$

Now, if $S(Q,\omega)$ is as in equation (5) above,

$$P(Q,t) = \exp(-D_t Q^2 t) \quad (8)$$

The coefficient of the exponent is thus just the line width measured in the normal energy transfer experiment.

3. Sample Preparation and Experimental Details

The well-characteried system, Porous Vycor, here used in its powdered form, is marketed by Corning. It consists of a borosilicate glass which has been allowed to undergo partial spinodal decomposition into its components. The B_2O_3 component is then removed by chemical etching, leaving about 28% by volume of interconnected pore space with a mean pore size of about 30 Å and an internal surface area of about 250 m^2/g. A mixture of 57% D_2O:43%H_2O was prepared. Three samples were dried by heating in a vacuum and were then equilibrated at 4%, 16% and 96% relative humidity, respectively. Using a glove bag, they were loaded into aluminium sachets and then sealed into cylindrical aluminium containers. Measurements were then performed at two temperatures (273 K and 300 K).

Small angle scattering data were obtained in two ways, first from the total count-rate in the detector with the analyser switched off and second by adding the spin-up and spin down count-rates. The measurements were made at angles between 1 and 16 degrees and the incident wavelength of 4.8 Å was used. This gives Q values from 0.045 to 0.36 $Å^{-1}$. The spin-up and spin-down components are respectively given by

$$I_+(Q) = I_{coh}(Q) + I_{inc}{}^{isotope} + (1/3) I_{inc}{}^{spin}(Q) \quad (9)$$

$$I_-(Q) = (2/3) I_{inc}{}^{spin}(Q) \quad (10)$$

where the I(Q) terms are the measured intensities and are proportional to the corresponding cross sections defined in equation (3). Note that experimental complications such as imperfect polarisation have been omitted [21]. Because $I_{coh}(Q)$ includes the small angle scattering which would swamp the incoherent scattering in the relevant Q range, it is necessary to approximately match scattering densities by use of the above D_2O/H_2O ratio.

4. Results and Discussion

The total intensities I(Q) (polariser on, analyser off) were measured as a function of wavevector Q for the three samples with different water content at a temperature of 273 K. Fig. 4 shows these data on a log-log plot. As one can see, the intensity decreases with increasing water content because of the partial matching of the filled regions. At high Q, the curves tend to flatten out due to the dominant contribution from incoherent scattering. Fig. 5 shows the non spin-flip and spin-flip intensities for a spin-up incident beam, $I_+(Q)$ and $I_-(Q)$, as a function of Q for sample 3 at 273 K. As expected, $I_+(Q)$ more or less follows the shape of the SANS component shown in Fig. 1 and $I_-(Q)$ is much flatter. This implies that the spin-down intensity is dominated by incoherent scattering from the hydrogen, the rise at low Q being due to imperfect polarisation analysis multiple scattering etc. Moreover, by taking $I_{coh}(Q)=0$ at the high Q end, we can estimate $I_{inc}(Q)$ from $I_-(Q)$ by equation 8 and this increases from sample 1 to 3 in ratio 1:1.75:2.18 which is rather less strongly that the estimated water contents. This suggests that there may be a significant isotropic background contribution included in this intensity. From Fig. 5 it will be noted that the spin-down does indeed exceed the spin-up at higher Q's.

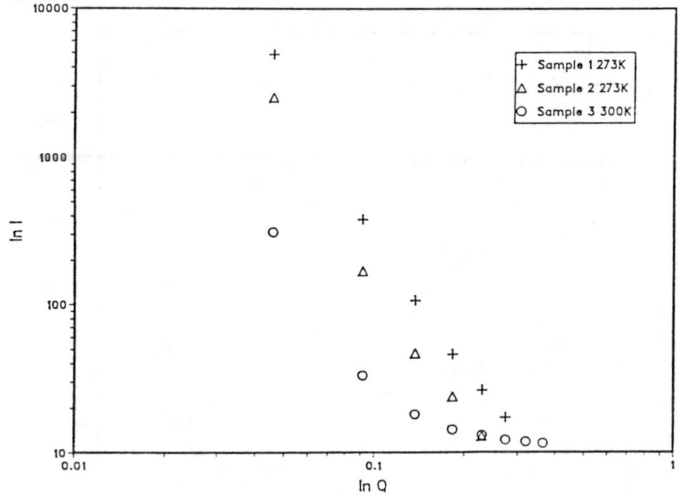

Fig. 4. The total spin up intensity plotted as a function of Q on a log-log scale for the three samples equilibrated with 57% D_2O:43% H_2O at progressively increasing humidities

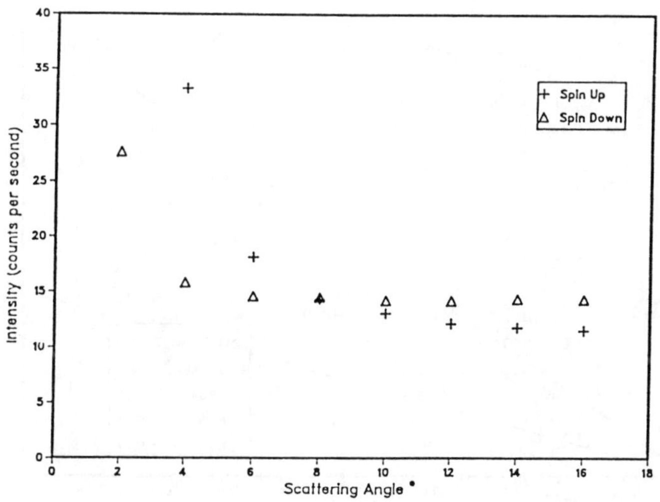

Fig. 5. Non-spin-flip and spin-flip components of the scattering from sample 3, (97% RH, 57% D_2O) at zero guide field (t = 0), demonstrating that the spin flip component is dominated by the isotropic incoherent scattering.

The main object of the present experiment was to measure the decay of the incoherent echo with fourier time. At reasonably high Q (> 0.1 Å^{-1}), where the non spin flip scattering is only just greater than the spin flip scattering, and at temperature near to 273 K or above, the echo is observed to increase with increasing time (field) as shown in Fig. 6. Statistical errors of about 0.05 counts/sec have been omitted for

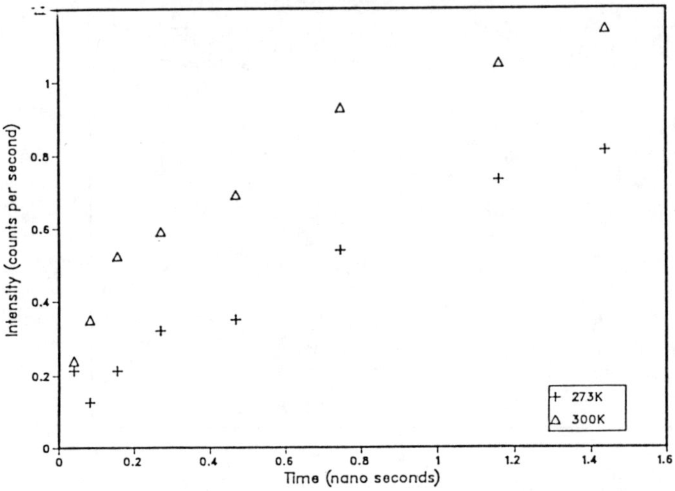

Fig. 6. Spin echo versus fourier time at 273K and 300K for sample 3 (RH 97%, 57% D_2O).

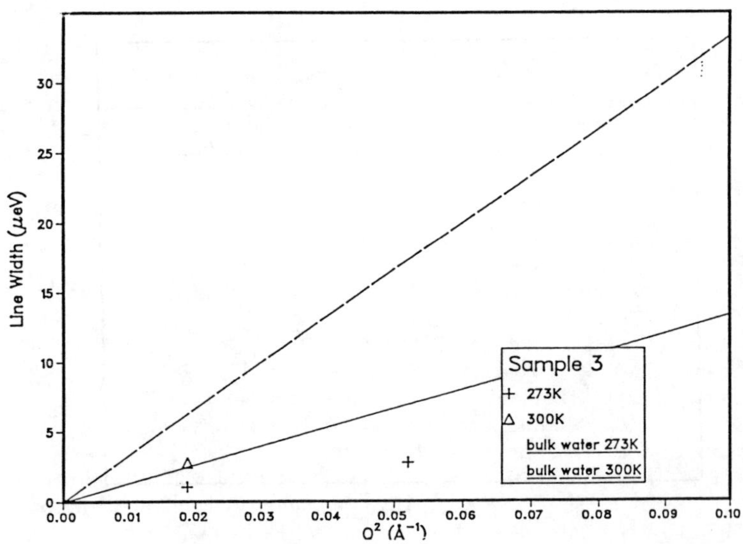

Fig. 7. Plot of line width in μeV against Q^2 at 273K and 300 K. The full and dashed lines indicate the expected broadenings for bulk water.

clarity. The total echo increases with t because the spin–up component, being elastic, is independent of t while the spin–down component, being due to scattering from the diffusing molecules, gives rise to a reducing negative component of the echo.

From such time–domain measurements, exponential decay times can be estimated. These are plotted in Fig. 7 as a function of Q^2 so that values of the D_t can be estimated for the two measuring temperatures. The present Q range corresponds to local diffusion and it can be seen that the apparent diffusion coefficients are about half those for bulk water.

5. Conclusion

The results presented above are only from preliminary data obtained in a first experiment. The accuracy of the data analysis was hampered by the residual coherent SANS which made the initial and final values of the echo difficult to determine. Moreover, measuring the diffusion over a cluster of pores will imply running at the longest available wavelength on IN11 and will involve doing preliminary experiments on a SANS instrument to further reduce the extent of the elastic SANS contribution. We believe, however, that this experiment demonstrate the utility of neutron spin–echo time domain measurements in the study of diffusion in a porous system. More detailed SANS studies of the system as a function of contrast and relative humidity and of inelastic neutron scattering from H_2O absorbed in the surface layer of silica along with measurements of the isotherms for absorption and desorption of water obtained with a computer–controlled micro–balance, are currently in progress.

6. References

[1] D.W. Schaefer and K.D. Keefer, Phys. Rev. Lett. 53 (1984) 1383.

[2] J–C. Li, D.K. Ross, P.L. Hall and R.K. Heenan, Physica B156&157 (1989)

[3] M.J. Benham, J.C. Cook, J–C. Li, D.K. Ross, P.L. Hall and B. Sarkissian, Phys. Rev. B39 (1989) 633.

[4] M.T. Bishop, K.H. Langley and F.E. Karasz, Phys. Rev. Lett. 57 (1986) 1741.

[5] P. Levitz and J.M. Drake, Phys. Rev. Lett. 58 (1987) 686.

[6] J.R. Beamish, A. Hikata and C. Elbaum, Phys. Rev. B27 (1983) 5848.

[7] J. Warnock, D.D. Awschalom and M.W. Shafer, Phys. Rev. Lett. 57 (1986) 1753.

[8] U. Even, R. Rademan, J. Jortner, N. Manor and R. Reisfeld, Phys. Rev. Lett. 59 (1987) 284.

[9] W.D. Dozier, J.M. Drake and J. Klafter, Phys. Rev. Lett. 56 (1986) 197

[10] S. Alexander and R. Orbach, J. Physique Lett. 43 (1982) L625.

[11] R. Rammal and G. Toulouse, J. Physique Lett. 44 (1983) L13.

[12] P.L. Hall, D.F.R. Mildner and R.L. Borst, Appl. Phys. Lett. 43 (1983) 252.

[13] A.J. Katz and A.H. Thompson, Phys. Rev. Lett., 54 (1986) 197.

[14] J–C. Li, M.J. Benham, L.D. Howe and D.K. Ross, IOPP Short Meeting Series on "Neutron and X–ray scattering: Complementary techniques" in press.

[15] P. Wiltzius, F.S. Bate, S.B. Dierker and G.D. Wignall, Phys. Rev. A36 (1987) 2991.

[16] M.J. Benham, private communication.

[17] F.M. Etzler, J. Colloid Interface Sci, 92 (1983) 43.

[18] P.G. Hall, A. Pidduck and C.J. Wright, J. Colloid Interface Sci, 79 (1981) 339.

[19] J–C. Li, M.J. Benham, S.M. Bennington, D.K. Ross and J. Tompkinson, IOPP Short Meeting Series on "Neutron and X–ray scattering: Complementary techniques" in press.
[20] F. Mezei, Z Physik 255 (1972) 146.
[21] F. Mezei and others in "Neutron spin echo" Lecture Notes in Physics, 128, Springer–Verlag, Berlin, 1980, Ed. F.Mezei.
[22] G.H. Vineyard, Phys. Rev. 110 (1958) 999.
[23] D.K. Ross and J–C. Li, to be published.

Molecular Dynamics Simulation Studies of the Hydrogen Bond Network in Water

ALFONS GEIGER AND PETER MAUSBACH

Fachbereich Chemie der Universität Dortmund;
Otto-Hahn-Str., D-4600 Dortmund, West-Germany
and
Institut für Physikalische Chemie der RWTH Aachen;
Templergraben 59,D-4600 Dortmund, West-Germany

1 Introduction

It is generally accepted that the unusual properties of water are largely due to the structure and dynamics of its hydrogen-bond network [1,2,3]. There has been much speculation about its connectivity pattern, but unfortunately there is no possibility to observe the hydrogen-bond network directly by experiments. This dilemma can be overcome by using realistic computer simulations [4]. Such calculations reveal the picture of an infinite random network, which is subject to continuous and rapid restructuring [5]. Fig. 1 gives a characteristic grafical representation of the hydrogen-bond network, obtained from a molecular dynamics simulation.

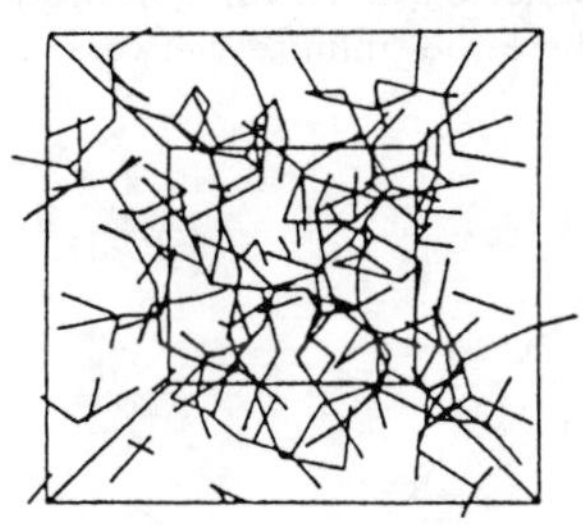

Figure 1: Illustration of the hydrogen-bond network. Configuration of 216 water molecules in a periodic box.

It is possible to describe the connectivity of this random network quantitatively by using lattice percolation theory [6,7,14]. By studying the changes that are produced by varying temperature and density, one can contribute to the understanding of various effects observed in metastable water and aqueous solutions.

J.C. Dore and J. Teixeira (eds.), Hydrogen-Bonded Liquids, 171–183.

2 Molecular dynamics simulations outline

In the following, we discuss the results of two series of molecular dynamics simulations. In one series, the density is kept fixed at $\varrho = 1.0 g/cm^3$ and the temperature is varied in eight steps from 287 to 235 K. In the other, at an approximately constant temperature of T = 273 K, the density is varied stepwise from $\varrho = 1.0 g/cm^3$ to $0.7 g/cm^3$.

We compare the results of these 'supercooling' and 'stretching' series with a simulation of 'normal' water at $\varrho = 1.0 g/cm^3$ and T = 284 K. In all cases we performed constant energy and constant density (N,V,E) simulations [8] for systems of 216 water molecules which interact via an ST2 pair potential [9]. The direct molecular interactions up to the cutoff distance $r_c = 7.8 Å$ were combined with a reaction field approximation for more distant parts of the system [10]. Before the actual simulations were started, equilibration runs were applied, which extended up to 300 ps as the systems became more viscous with decreasing temperature or density. For more details see refs. [11] and [12].

3 Hydrogen-bond definition

The interaction of water molecules is described by a continuous potential surface, which does not allow us to distinguish naturally between 'broken' and 'intact' hydrogen bonds. It is more appropriate to think of a smooth unimodal distribution of strong, weak and non-existing hydrogen bonds. To treat such systems with traditional bond percolation theory [2,13], where an unambiguous definition of an intact bond is necessary, the following approach has proven to be very successful.

To start, a reasonable and practicable definition is imposed: two molecules i and j are considered to be hydrogen-bonded whenever their interaction energy V_{ij} is below a negative threshold value V_{HB} and their mutual oxygen-oxygen separation is less than 3.5 Å. The arbitrary value V_{HB} is then varied gradually from strongly negative values, close to the absolute minimum of the interaction potential, up to values close to zero [5,7]. In this way, we can observe the connectivity pattern produced by the strongest bonds and the growth of the network, as weaker and weaker bonds are included.

First, only few bonds will be found, the average number of hydrogen-bonds per water molecule n_{HB} will be close to zero, and only very small aggregates of connected molecules will be observed. With increasing (less negative) V_{HB}, n_{HB} will increase, as will the number of the hydrogen-bonded aggregates and their average size S. When V_{HB} approaches zero, n_{HB} approaches values close to four; the bonds will then form a totally connected, spacefilling network. At some intermediate value of V_{HB} a bond percolation threshold, where the 'infinite' network appears for the first time, will have been passed.

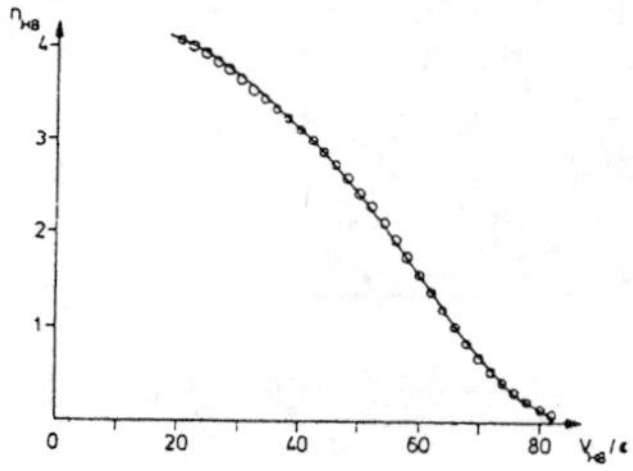

Figure 2: Average number of hydrogen-bonds n_{HB} per water molecule as function of energy threshold V_{HB}.

A quantitative description of the behaviour outlined above is given in the following figures for the system at T = 284 K and $\varrho = 1.0 g/cm^3$. Fig. 2 shows n_{HB} as a function of the threshold value V_{HB} in units of the ST2-parameter $\epsilon = 0.317 J/mole$.

As one can see, for very strict definitions only few (strong) hydrogen-bonds are found, whereas for V_{HB} values close to zero n_{HB} even exceeds the value of four which one expects from crystalline ice. This indicates the presence of 'bifurcated' H-bonds, at least according to the above definition and if weak bonds are also considered.

4 Hydrogen-bonds in normal water

Assuming independent forming and breaking of H-bonds on a lattice of four-bonded water molecules, one expects a mole fraction of

$$f_j = \binom{4}{j} p_j (1-p)^{4-j} \tag{1}$$

water molecules, which have exactly j bonds. p is the bond forming probability.

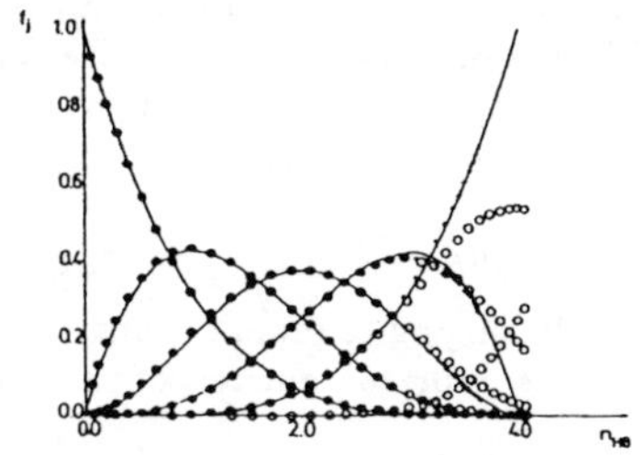

Figure 3: Fraction f_j of water molecules with j intact H-bonds (j = 0 ... 5). Circles: MD results, full line: binomial distribution.

In Fig. 3 the circles show values of f_j, obtained from the MD simulation for various choices of V_{HB}, plotted against the average number of H-bonds n_{HB} (V_{HB}). The full lines represent the binomial distribution, eq. (1), with $p = n_{HB}/4$. Up to $n_{HB} \approx 3$, one observes very good agreement, whereas for larger n_{HB}, when weak H-bonds are also permitted, increasing deviations are observed. This is due to the appearance of five-bonded water molecules.

In a next step, we study larger molecular aggregates. We distinguish between bond-*networks* of water molecules and aggregations of four-bonded molecules, which we call *clusters*. As the assumption of independent breaking of bonds on an ice lattice works well for f_j, the occurance probability of small to medium-sized networks and clusters can be calculated by combination theory [7]. As an example, the probability of finding clusters of *three* connected four-bonded molecules (weight fraction of water molecules, belonging to such clusters) is given by

$$W_3^c = 18p^{10}(1-p^3)^8 \tag{2}$$

and the corresponding values from the simulation are compared in Fig. 4.

The plot is equivalent to Fig. 3, with $p = n_{HB}(V_{HB})/4$. Again, a very good agreement is observed, although a small but systematic deviation is present, which will be discussed later. The bell shape of the curve, which one observes for all W_n, can be understood as follows: Starting with a very strict definition and gradually relaxing it, we find at first few aggregates of a given size n, which multiply with increasing $n_{HB}(V_{HB})$. Finally, there will be many of these networks and clusters, which start to merge, thus decreasing again W_n.

When approaching the percolation threshold, the size of the aggregates increases strongly. For large aggregates the probability W_n can no more be given analytically (at least, when an underlieing

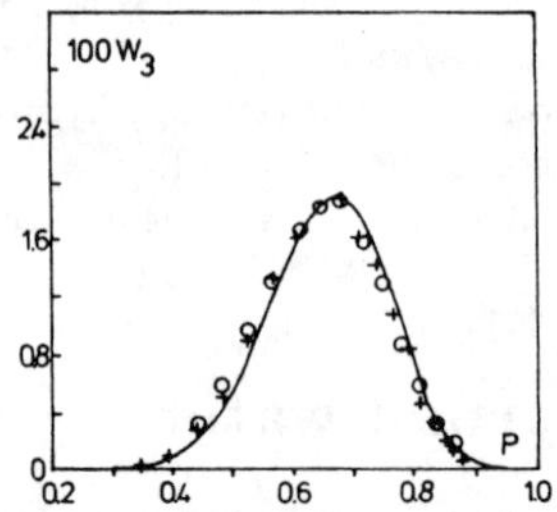

Figure 4: Fraction W_3^c of water molecules, which belong to clusters of three four-bonded molecules.

ice lattice topology is assumed), but close to the percolation point, the average network or cluster size and related properties can be described by critical behavior type equations. As an example, in Fig. 5a the 'spanning length' L of bond networks (and clusters) is shown. L is defined as the difference between the maximum and minimum oxygen coordinates of the networks with respect to some arbitrary coordinate direction. L changes in the vicinity of the percolation threshold n_{HB}^* like

$$L \sim |n_{HB} - n_{HB}^*|^{-\nu}. \tag{3}$$

In Fig. 5b the fraction of molecules, belonging to 'infinite' clusters P_∞ is shown for two slightly different H-bond definitions (see [6]). This 'order parameter' of the percolation transition behaves like

$$P_\infty \sim |n_{HB} - n_{HB}^*|^{\beta}. \tag{4}$$

In accord with lattice percolation theory, we can use for n_{HB}^* the values 1.55 and 3.18 for networks and clusters respectively. The straight lines in Figs. 5a,b have been drawn with a slope

$$\begin{aligned} \nu &= 0.75 \\ \text{and } \beta &= 0.38, \end{aligned}$$

as predicted by lattice percolation theory [6]. The very good fit of the simulation data is obvious.

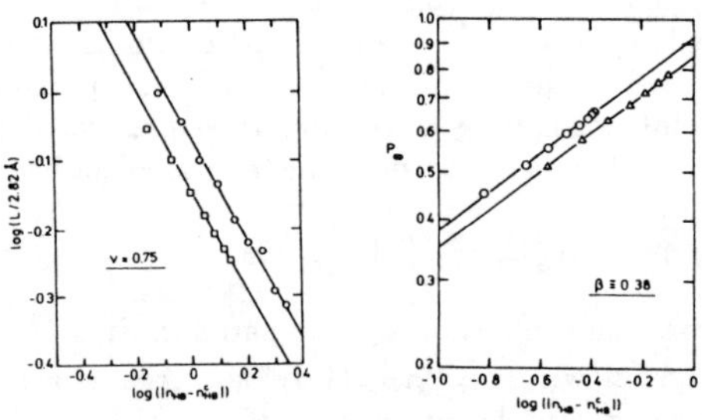

Figure 5: (a) Average length of bond networks and clusters. (b) Fraction of molecules which belong to spanning clusters.

The ramified networks close to the percolation threshold can be described as fractals, which means that the mass distribution in a percolating network is such that the number of particles N, which are found within a sphere of radius r around some reference particle is

$$N(r) \sim r^{d_f}. \tag{5}$$

A so called 'hyperscaling hypothesis' suggests for the fractal dimension d_f a value which is determined by the discussed critical exponents (D=3 is the Euclidian dimension) [13]:

$$d_f = D - \beta/\nu. \tag{6}$$

With the values given above, we expect $d_f = 2.49$. To determine d_f via eq. (5), systems of 216 particles are not sufficient. Therefore we simulated a system of 1728 water molecules, but even with this size only a hint for the appropriateness of eq. (6) can be given [26].

In Fig. 6a, the derivative of N(r) viz. the average number of molecules c(r) within spherical shells is shown for networks of size 850 to 900 close to the bond percolation threshold $n^c_{HB} = 1.55$. To remedy the limitations of our finite size system, we determine the slope $d_f - 1$ for various network sizes N and plot d_f vs. N^{-1}. Fig. 6b reveals that our results from a finite system simulation at least do not contradict the prediction of eq. (6).

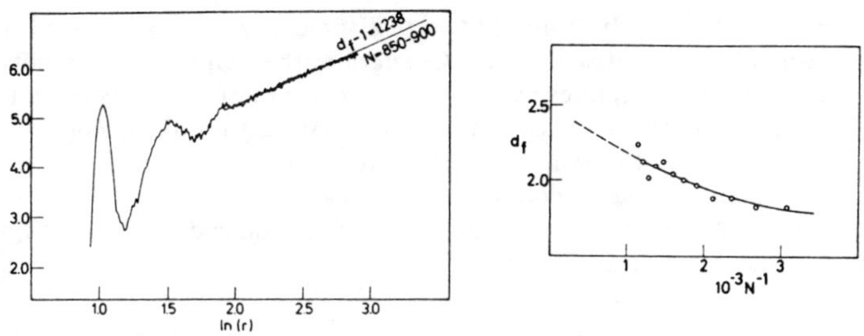

Figure 6: (a) Radial distribution c(r) of water molecules, which belong to same network. (b) Fractal dimension: slope $d_f - 1$ for various network sizes N.

5 Temperature and density dependence of hydrogen-bond network topology.

Fig. 4a revealed very small, but systematic deviations between the prediction from combination theory and the MD results. Now we discuss possible reasons for these deviations. As an example we plot in Fig. 7 these differences for the previously discussed clusters of size three:

$$\Delta W^c_3(p) = W^c_3(p) - W^c_{3,MD}(p) \tag{7}$$

where $W^c_3(p)$ is given by eq. (2) and $W^c_{3,MD}(p)$ are the MD values.

An important difference between the pure temperature effect at constant density (Fig. 7a), and the pure density effect at constant temperature (Fig. 7b) can be seen. The influence of the temperature can be scaled out: by plotting vs. the average number of H-bonds, all difference curves fall on top of each other. In contrast, the difference ΔW_n diminishes with decreasing density and seems to vanish over the total range of p within the given numerical accuracy at $\varrho \approx 0.8 g/cm^3$. Of course, the average number of H-bonds changes (increases) with decreasing temperature and density [11,12], and by that also the occurance probability of the networks, but apparently this effect can be scaled out in one case, but not in the other.

This can be explained as follows: the observed differences occure, because the topology of the H-bond network in the liquid is not the ice lattice topology, and in particular, there are deviations

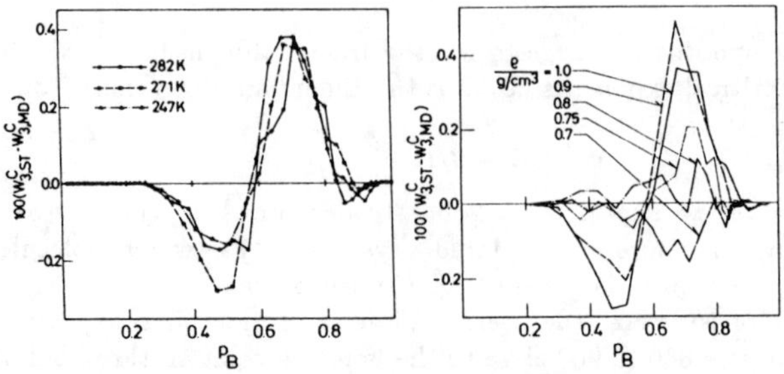

Figure 7: Difference between cluster size distribution in MD simulation and prediction of percolation theory (a) temperature dependence, (b) density dependence.

from the perfect local fourbondedness. If medium to weak H-bonds are also considered, there may be molecules with five H-bonds, as discussed before. Changes of the temperature do not primarily change the topology, but only the thermal excitation of molecular motions. This leads to changes in p and W_n at fixed V_{HB}, which can be scaled out: by just changing the definition value V_{HB} to bring p back to its old value, exactly the same network picture appears.

In contrast to that, the applied density decrease also changes the topology. As one can see in Fig. 8, the mole fraction of molecules which have **more** than four bonds decreases with decreasing density, even when very weak H-bonds are allowed.

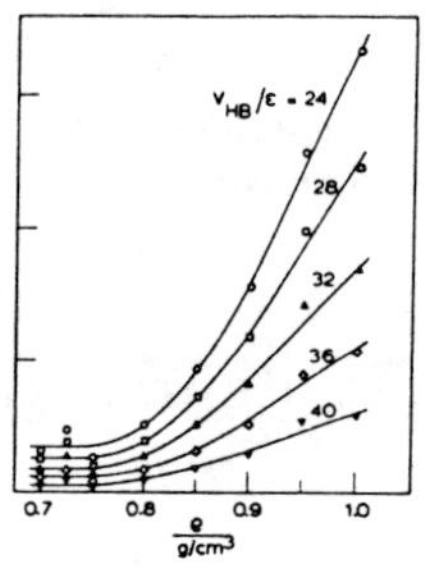

Figure 8: Mole fraction of water molecules with **more** than four H-bonds (for various H-bond definitions).

The local connectivity approaches a more perfect tetrafunctional local arrangement (as assumed for the derivation of the W_n [7]).

6 Structure of low density metastable water

In the previous section, only topological properties of the H-bond network have been discussed. We now consider the structural (geometrical) changes, which are observed as density decreases. This is of special interest, because of a conjectured limit of stability (spinodal) of 'stretched' water [15], which was estimated to be at a negative pressure of roughly -200 MPa and a density of $\varrho \approx 0.8 g/cm^3$ for the temperature of our simulation series ($T \approx 273K$).

In fact, there are a number of observations which indicate the disintegration of the bulk water system at the above given conditions. Fig. 10 gives the virial, calculated from the pair interaction forces (without cutoff correction). A decrease of the pressure at high densities ends at a pressure of about -200 MPa, when the density $0.75 g/cm^3$ is reached. Although these numbers can only be

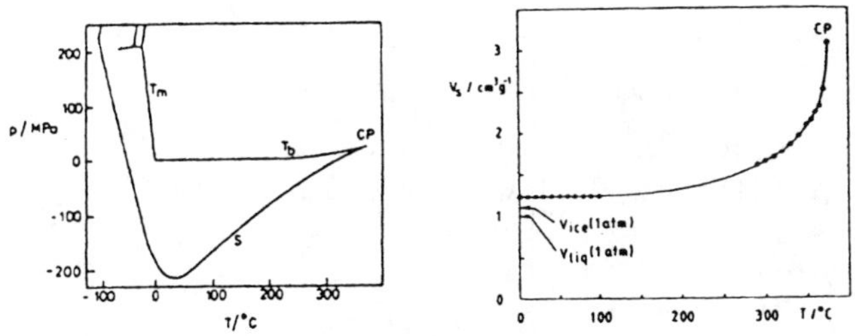

Figure 9: (a) Line of mechanical stability limit for metastable water, (b) specific volume at the limit of stability. Both according to Speedy [15].

considered as rough estimates, they are consistent with Speedy's conjecture [15].

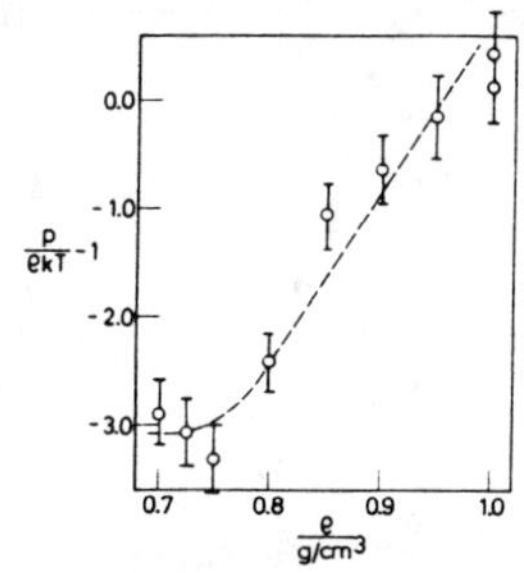

Figure 10: Decrease of pressure virial with density.

Fig. 11a gives a graphical description of the local density fluctuations in water at different global densities ϱ. To get these pictures, the simulation box is cross-sectioned by an arbitrarily positioned, finely meshed plane. Then, each mesh point, which is closer than $2.5Å$ to any water oxygen in the system is marked by a cross. Thus the empty white areas in Fig. 11a indicate cavities in the water. As one can see, between 1.0 and $0.8g/cm^3$ the decreasing global density produces more, but not larger cavities. But below $0.8g/cm^3$ very large cavities occure, indicating the tendency for decomposition. The presence of these large scale density fluctuations can be proven by calculating the static structure factor

$$S(\mathbf{Q}) = \frac{1}{N} < \sum_{i,j} exp(i\mathbf{Q}\mathbf{R}_{ij}) > \tag{8}$$

where $\mathbf{R}_{ij}$ is the oxygen distance vector between molecules i and j. For scattering vectors with $|\mathbf{Q}| \geq 1.2Å^{-1}$, S(Q) has been calculated by Fourier-transforming the oxygen pair correlation function $g_{OO}(r)$ (see Fig. 13a). For smaller Q-values S(Q) is calculated directly from the sum in eq. (8). The smallest accessible wave vector is $Q_1 = 2\pi/L$, where L is the simulation box size, which varies between 18.6 and 21.0 $Å^{-1}$ for the different densities. As one can see from Fig. 11b, for densities below $0.8g/cm^3$ a strong increase at small Q-values indicates the presence of large scale density fluctuations.

The transition point from a homogeneous to a decomposing system with large density fluctuations can be localized by considering the Voronoi polyhedra [17,16] which are obtained by using the oxygens to describe the positions of the water molecules. Such Voronoi constructions are spacefilling and uniquely partition the total volume between the molecules. Fig. 12a gives the distributions

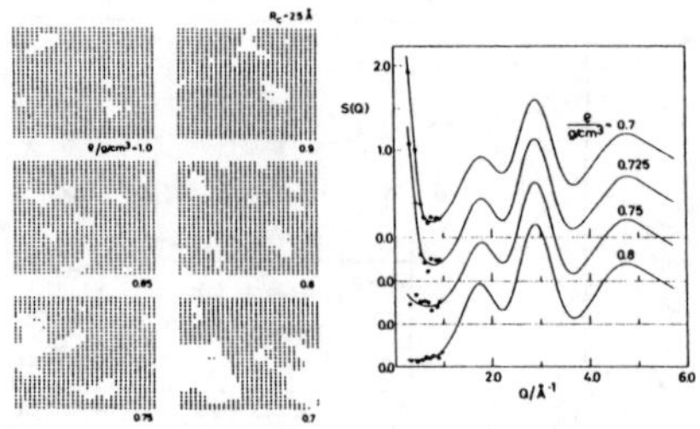

Figure 11: (a) Picture of the density fluctuations in low density water. (b) Structure factor S(Q).

of the Voronoi cell volumes at various densities. For higher densities these bell shaped curves are symmetric. The maxima coincide with the averages (first moments) of the distributions and as the Voronoi construction is spacefilling, the average shifts proportional to the global density change (see Fig. 12b, crosses and circles for averages and maxima positions, respectively).

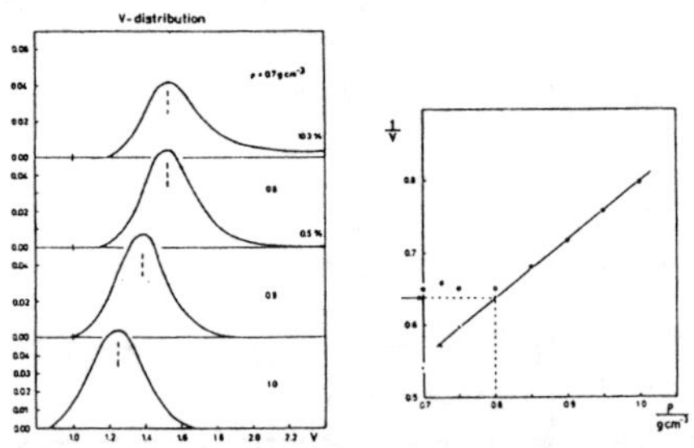

Figure 12: (a) Volume distribution of Voronoi polyhedra. (b) Maximum position (o) and mean value (+) of Voronoi distribution.

Below $0.8g/cm^3$ the volume distribution becomes very unsymmetric, because polyhedra of large size are occuring. These belong to molecules which are in the neighborhood of the large cavities. At these low densities, a global density decrease only enlarges these cavities, but leaves the density of the 'bulk' unchanged. Therefore the maximum does not shift further. Thus, the divergence of first moment and maximum position indicates the decomposition of the system into a 'bulk' phase and cavities [18]. This begins at $\varrho \approx 0.8g/cm^3$, as one can see from Fig. 12b.

To understand the mechanisms that lead to the decomposition at this particular density of $\varrho \approx 0.8g/cm^3$, we determine the number of first neighbors $N(R_1)$ by integrating over the first peak of the oxygen-oxygen pair distribution function (Fig. 13a). Note, that the change of $g_{OO}(r)$ with density is very unusual: the oscillations of $g_{OO}(r)$ become more pronounced and in particular the height of the first peak increases with decreasing density. This demonstrates the development of a more ordered structure with decreasing density. Also, the slight shift of the second peak to larger distances indicates the formation of more linear hydrogen bonds.

It is well known that in the liquid at $1.0g/cm^3$ the water molecule has more than four nearest neighbors (of course, the actual number depends on the upper integration limit). The shape of $N(R_1)$, as shown in Fig. 13b, shows the same density dependence as the maximum positions of the Voronoi volume distribution (Fig. 12b) and the mole fraction of water molecules with more than four H-bonds (Fig. 8). Above and below $0.8g/cm^3$ the system reacts with two very different mechanisms upon an imposed density decrease. Above $0.8g/cm^3$ a global density decrease is achieved by reducing the number of nearest neighbors. At about $0.8g/cm^3$ a more ordered

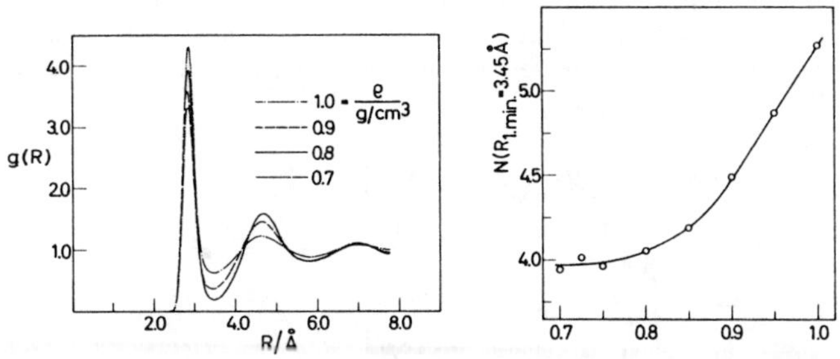

Figure 13: (a) Oxygen-Oxygen pair distribution function in low density water, (b) number of nearest neighbors.

structure with four nearest neighbors per water molecule is reached (see also the discussion of Fig. 7b). A further reduction of the density is then achieved by the formation of large holes. This is energetically more favourable than to loosen the hydrogen-bonding interaction to one of the four remaining nearest neighbors: By adopting a 'straddling' orientation towards the cavity [12] (as this is also known from hydrophobic hydration shells [19]), the water molecules are able to keep four H-bonded neighbors, while reducing the density globally by the formation of cavities.

It is quite comprehensible that at this stage the system becomes very unstable with respect to density fluctuations. This is a simple explanation for the location of the mechanical stability limit at $\varrho = 0.8 g/cm^3$, as suggested by Speedy [15].

In Fig. 14 the oxygen-hydrogen and hydrogen-hydrogen pair correlation functions $g_{OH}(r)$ and $g_{HH}(r)$ are shown. The prominent peaks are characteristic for water and are produced by the mutual arrangement of H-bonded molecules [9]. With decreasing density we observe again a more pronounced structural order. In particular the first peak of $g_{OH}(r)$, which corresponds to the distance between the hydrogen in the H-bond and its acceptor oxygen, increases very rapidly and the following minimum is lowered accordingly.

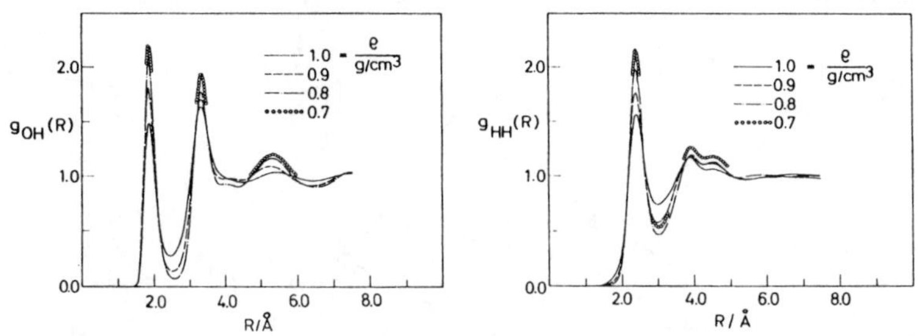

Figure 14: Intermolecular pair distribution functions in low density water: (a) oxygen-hydrogen, (b) hydrogen-hydrogen.

With neutron diffraction experiments on D_2O, a weighted sum of the shown pair correlation functions can be determined (assuming structural equivalence of H and D):

$$g_{tot}(r) = 0.092 g_{OO}(r) + 0.422 g_{OH}(r) + 0.486 g_{HH}(r). \qquad (9)$$

These sums are shown in Fig. 15 and compared with a neutron scattering result for amorphous D_2O of Chowdhury et al. [21]. It turns out that the changes caused by the expansion of the water are comparable to those observed in n-scattering experiments on undercooled water [20]: In those regions, where $g_{tot}(r)$ of the amorphous ice is lower than in the liquid (for example at about $3 \mathring{A}$), stretching or cooling causes a decrease, in other regions an increase. On the whole, an approach to the amorphous ice structure is occuring. The first two peaks even show an 'overshooting', probably due to a well-known defect of the ST2 model, which overemphasizes the water structure.

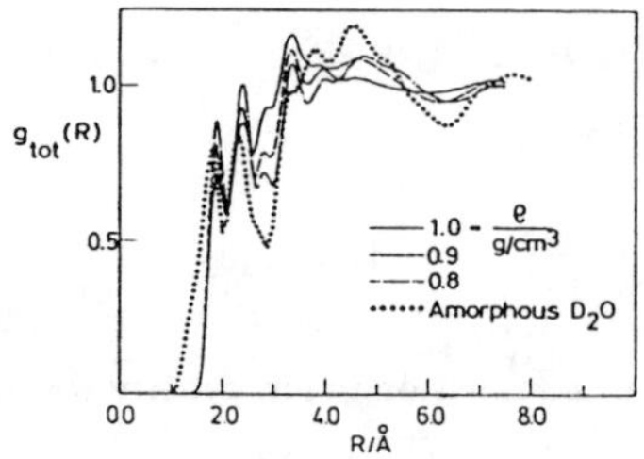

Figure 15: Weighted sum of pair correlation functions from MD compared with neutron-scattering results of amorphous ice.

Following a proposition of Giguère [22], which had been taken up by Dore to interpret the n-scattering results [20], a very simple and grafic explanation for the discussed changes can be given. At higher densities and temperatures an appreciable amount of local arrangements occure, where the hydroxyl group is pointing between the oxygens of two neighbouring water molecules as drawn in Fig. 16 [22]. This transient arrangement, which may be called a bifurcated H-bond, can be fitted in the overall tetrahedral network and corresponds to O-H distances of 2.3 rather than 1.85$\AA$. Accordingly, the increase of the first peak of $g_{OH}(r)$ and the simultaneous depletion of the following minimum indicates a strong reduction of bifurcated H-bonds and is in perfect agreement with our previous observations like the disappearance of five-bonded water molecules (Fig. 8).

It has to be stressed that the term 'bifurcated H-bond' is used here in the above described energetic or geometric sense. It seems that using a hydrogen-bond definition which incorporates also the lifetime [14], such arrangements can no longer be addressed as bonds. Nevertheless, such configurations, even though only transient, are characteristic for liquid water and its structural and dynamical distinction from amorphous ice.

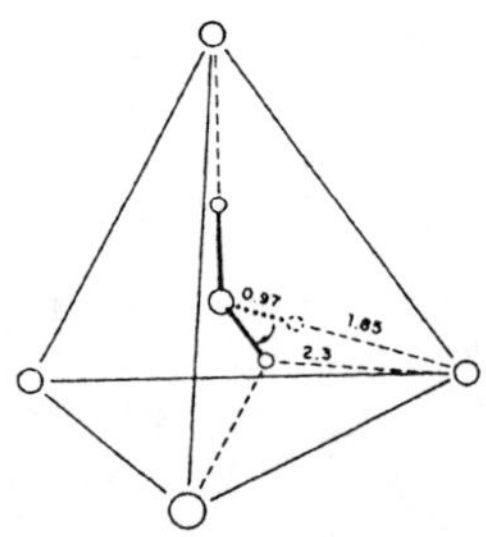

Figure 16: 'Bifurcated H-bond' arrangement. According to Giguère [22].

7 Hydrogen-bonding and microdynamics

The dynamics of the water molecules in the expanded liquid also reveals a density dependence which is completely diverse to what is expected from 'normal' liquids. Usually, expansion increases the molecular mobility, due to reduced steric hindrance. In contrast to this, in water a drastic reduction of molecular mobility is observed, even at small density decreases (and at constant temperature).

In ref. [12] we discussed the mean square displacements $< \Delta r^2(t) >$ and the corresponding self diffusion coefficients D: A density decrease of 20 % (from 1.0 to $0.8g/cm^3$) produces a reduction of D by nearly a factor of 10! An analogueous behavior can be observed when considering the reorientational motion of the molecules. Here we calculate the time correlation functions

$$\mathrm{T}_l(t) =< P_l(\cos\theta(t)) >,$$

where P_l is the Legendre Polynomial of degree l and

$$\cos\theta(t) = \widehat{\mathbf{l}}_i(0) \cdot \widehat{\mathbf{l}}_i(t)$$

describes the rotation of a molecule fixed unit vector $\widehat{\mathbf{l}}_i$. The correlation time τ_l is defined as

$$\tau_l = \int_0^\infty \mathrm{T}_l(t)\, dt.$$

As an example, the density dependence of $\mathrm{T}_2(t)$ and τ_2 of the intramolecular H-H-vector is shown in Fig. 17a. For the integration to obtain τ_l, $\mathrm{T}_l(t)$ is extrapolated to infinity by an exponential function. The very rapid initial decay of $\mathrm{T}_l(t)$ is due to the well-known librational (hindered rotational) motion of the water molecules. With decreasing density, the librational amplitude is slightly diminished and the following exponential decay is slowed down considerably. Accordingly the reorientation times τ_l increase until a plateau is reached at about $0.8g/cm^3$ (Fig. 17b).

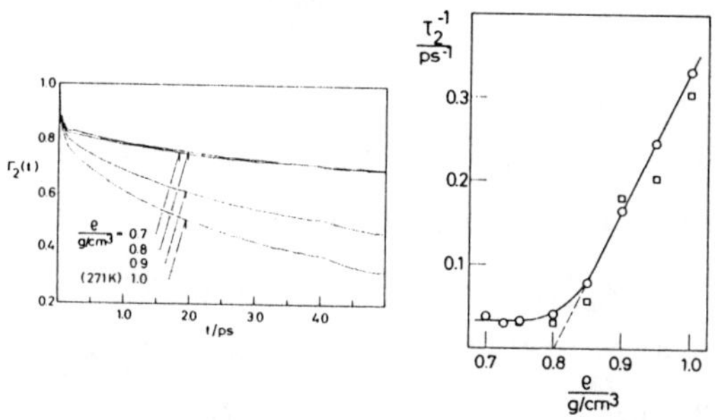

Figure 17: Density dependence of reorientational correlation functions and corresponding correlation times.

A mechanism, which allows us to understand this loss of mobility, has been proposed by Naberukhin [23] in correspondence with the structural considerations of Giguère [22], which we discussed above. Accordingly, the mobility of the water molecules is strongly influenced by the presence of 'bifurcated H-bonds'. Structures which comprise many local arrangements of the kind discussed in Fig. 16, reflect a facilitated switching from one 'regular' H-bond arrangement to another, across lowered energy barrieres. The structural changes observed in Figs. 8 and 14 reflect the disappearance of such 'bifurcated H-bonds' (*defects* of the tetra-functional bond network) and explain the observed loss of mobility.

8 Conclusions

We have shown that the *connectivity* of the hydrogen-bond network can be described quite well by random bond percolation on a lattice. It has also been shown that the *structure* of amorphous ice can be represented by so called continuous random network models, developped for amorphous semiconductors [3,20,24]. A close inspection of the stretched liquid revealed however that the deviations of the structure from the 'ideal' tetra-functional network with four neighbors and four bonding-possibilities per molecule are very important for the properties of water. The ideal tetrahedral network structure is attained only at the mechanical stability limit, which thus can be identified with a new kind of order transition (with the quantities shown in Figs. 8 and 13b as order parameter).

In retrospect, the high mobility of the water molecules seems to be in contradiction to the results of the percolation study, that at any instance a spanning network of strong H-bonds ($V_{ij} \leq -18$ kJ/mole) is present. In view of the fact that this interaction energy is large compared to the thermal energy kT, one should expect water to be gel-like, at least highly viscous. In fact this can be observed in the simulation at low densities.

Therefore, the deviations from a perfect tetrahedral network are of decisive importance for the molecular dynamics. Random network models which focus on distorted tetrahedral local order without considering explicitely arrangements of five nearest neighbors habe to be re-examined under this point of view.

There are experimental observations, which perfectly support our findings regarding low density water: the hydrophobic hydration effect. It is well-known that inert solutes produce increased structural order and decreased mobility in its hydration shell [19,25]. This can easily be understood by realizing that the main effect of the inert solute is to lower *locally* the density of water and to prevent the 'straddling' tetrahedral arrangement (see [19]) of the water molecules in the first hydration shell from being disturbed by the approach of 'fifth neighbors'.

References

[1] F.H. STILLINGER, Science **209**, 451 (1980)

[2] H.E. STANLEY, J. TEIXEIRA, J. Chem. Phys. **73**, 3404 (1980)

[3] S.A. RICE, M.G. SCEATS, J. Phys. Chem. **85**, 1108 (1981)

[4] A. RAHMAN, F.H. STILLINGER, J. Am. Chem. Soc. **95**, 7943 (1973)

[5] A. GEIGER, F.H. STILLINGER, A. RAHMAN, J. Chem. Phys. **70**, 4185 (1979)

[6] A. GEIGER, H.E. STANLEY, Phys. Rev. Lett. **49**, 1895 (1982)

[7] R.L. BLUMBERG, H.E. STANLEY, A. GEIGER, P. MAUSBACH, J. Chem. Phys. **80**, 5230 (1984)

[8] M.P. ALLEN, D.J. TILDESLEY, "Computer Simulation of Liquids", Clarendon Press, Oxford 1987

[9] F.H. STILLINGER, A. RAHMAN, J. Chem. Phys. **60**, 1545 (1974)

[10] O. STEINHAUSER, Mol. Phys. **45**, 335 (1982)

[11] A. GEIGER, P. MAUSBACH, J. SCHNITKER, R.L. BLUMBERG, H.E. STANLEY, J. Physique Colloq. **45**, C7-13 (1984)

[12] A. GEIGER, P. MAUSBACH, J. SCHNITKER, in "Water and Aqueous Solutions", ed. G.W. NEILSON and J.E. ENDERBY: Adam Hilger, Bristol 1986

[13] D. STAUFFER, "Introduction to Percolation Theory", Taylor and Francis, London 1985

[14] F. SCIORTINO, S.L. FORNILI, J. Chem. Phys. **90**, 2786 (1989)

[15] R.J. SPEEDY, J. Phys. Chem. **86**, 982 (1982)

[16] N.N. Medvedev, V.P. Voloshin, Yu.I. Naberukhin, J. Phys. A: Math. Gen. **21**, L247 (1988)

[17] N.N. Medvedev, J. Computat. Phys. **67**, 223 (1986)

[18] N.N. Medvedev, A. Geiger, to be published

[19] A. Geiger, A. Rahman, F.H. Stillinger, J. Chem. Phys. **70**, 263 (1979)

[20] J.C. Dore, in "Water Science Reviews", Vol. 1, ed. F. Franks, University Press, Cambridge 1985

[21] M.R. Chowdhury, J.C. Dore, J.T. Wenzel, J. Non-Cryst. Solids **53**, 247 (1982)

[22] P.A. Giguère, J. Chem. Phys. **87**, 4835 (1987)

[23] Yu.I. Naberukhin, J. Struct. Chem. **25**, 223 (1984)

[24] M. Popescu, J. Non-Cryst. Solids **75**, 483 (1985)

[25] A. Geiger, Ber. Bunsenges. Phys. Chemie **85**, 52 (1981)

[26] P. Mausbach, A. Geiger, Z. angew. Math. Mech. **69**, T55 (1989)

A MOLECULAR MODEL FOR AQUEOUS SOLUTIONS

D. BRATKO* and L. BLUM
Department of Physics
University of Puerto Rico
Rio Piedras, Puerto Rico 00931
U.S.A.

ABSTRACT. Simulation studies of aqueous systems in which water molecules are represented as dipolar hard spheres with anisotropic adhesion that mimicks hydrogen bonding are reviewed. The collected results including new data for ion hydration show that the essential features of liquid water and aqueous solutions are reproduced by this simple model designed for use in theories of ionic solutions and electrical double layer. An acceptable agreement with the results of neutron diffraction experiment and with predictions of established models for water is observed. A plausible interpretation of structural or solvation forces between hydrophobic solutes is given.

1. INTRODUCTION

In the past decade various pair potentials of water have been proposed for use in molecular theories of aqueous solutions. They mainly belong to the interaction site models in which the intermolecular potential is a sum of interactions between a number of sites fixed in each molecule, or to the central force models which represent water as a mixture of hydrogen and oxygen atoms with appropriate pair potentials [1,2]. Improvements accounting for internal degrees of freedom [3,4] , polarizability [5,6], and quantum effects [4,7,8] are also studied. In view of the complexity of the available models computer simulation has established itself as a method of choice in studies of liquid water [9]. In typical solutions, however, the number of water molecules that should be taken into account together with a statistically significant amount of the solute often exceeds the practical limits of simulation. Integral equation theories being an inviting alternative in treatments of pure solvent remain the only suitable approach in these cases [9]. Numerical solution of RISM equations for modified MCY [10], SPC [11], and TIPS [12] model has been reported as well as the solution of HNC [13,14] and RHNC [14] equation for central force water [15] and its solutions. LHNC and HNC approaches to the systems where water was modelled by hard spheres with embedded electrostatic multipoles were also described [16-19].

*Permanent address: Department of Chemistry, University E. Kardelj, Murnikova 6, Ljubljana, Yugoslavia

J.C. Dore and J. Teixeira (eds.), Hydrogen-Bonded Liquids, 185–196.

Considering nontrivial numerical procedures used in these works, however, studies of simpler models that would be analytically tractable are also important. So far, an exact analytical solution of Ornstein-Zernike (OZ) equation for spatial correlation functions [20] has been obtained using the model which represented the ionic solution as a mixture of hard ions and hard spherical dipoles with no account of hydrogen bonding [21-24]. The properties of such a system are still far from those of a real solution [25]. In this lecture we consider a simple intermolecular potential that should improve the properties of the hard dipole model for water while remaining suitable for the analytical solution of the appertaining Ornstein-Zernike equation. When dealing with anisotropic interactions, the molecular OZ equation

$$h_{ij}(\vec{X}_i,\vec{X}_j)=c_{ij}(\vec{X}_i,\vec{X}_j)+(1/8\pi^2)\sum_k \rho_k \int c_{ik}(\vec{X}_i,\vec{X}_k)\, h_{jk}(\vec{X}_j,\vec{X}_k)\, d\vec{X}_k \qquad (1)$$

where $\vec{X}_i=(\vec{R}_i,\vec{\Omega}_i)$ denotes the position $\vec{R}_i=(x_i,y_i,z_i)$ and orientation $\vec{\Omega}_i=(\alpha_i,\beta_i,\gamma_i)$ of the particle i, is usually solved by the method of invariant expansion [20,26-29]. The total and the direct correlation functions, h_{ij} and c_{ij}, are expanded in terms of generalized spherical harmonics depending on the orientations $\vec{\Omega}_i$, $\vec{\Omega}_j$, and $\tilde{R}_{ij}=(\theta_{ij},\phi_{ij})$. Here, $\vec{R}_{ij}$ specifies the direction from the first to the second molecule. The resulting systems are well suited for analytical manipulation [26] if the intermolecular potential $u_{ij}(\vec{X}_i,\vec{X}_j)$ or the corresponding Mayer function $f_{ij}=\exp(-u_{ij}/kT)-1$ is also expandable in spherical harmonics. In particular, this approach was shown to yield an analytic solution to the Percus-Yevick (PY) Ornstein-Zernike equation for the fluid of hard spheres with orientation dependent 'adhesion' [31], thus 'the potential that may mimick the short ranged anisotropic interactions encountered in hydrogen bonded fluids' [30]. Our objective is to explore the applicability of this class of potentials in modelling aqueous systems. The proposed intermolecular potential for water consists in adding to the hard spheres with a point dipole a narrow potential well with a tetrahedral octupole of the form $v(r)\sin\theta \sin 2\theta \cos 2\phi$ [32]. While an analytic treatment is being dealt with in a separate study [33], in this work we use computer simulations to review the properties of the model and to compare its predictions with neutron diffraction measurements and with some results of known models for water. The outline of the article is as follows. In the next section, a detailed description of the model and the water-water potential is given. In section 3, the simulation results for the structure of the pure liquid and for ion hydration are discussed and compared with the results from other sources. A Grand Canonical Ensemble Monte Carlo (GCEMC) study of hydrophobic interaction is also described. Finally, the main conclusions are drawn and future work is briefly outlined.

2. DESCRIPTION OF THE MODEL

Water molecules are pictured as impenetrable spheres of diameter σ, each with a point dipole μ in its centre. Analogously, the ions of valence z_i are hard spheres of diameter σ_i and charge $z_i e_o$. The configuration of the system is specified by the coordinates $\vec{X}_j=(\vec{R}_j,\vec{\Omega}_j)$ that comprise the positions and the orientations of anisotropic water molecules, and the positions $\vec{R}_i$ for the ions. The coordinates of the hydrogen and the oxygen nuclei in a water molecule at x=y=z=0 and with orientation $\alpha=\beta=\gamma=0$ are:

$$
\begin{array}{llll}
H_1: & x_1=0, & y_1=0.075\ \text{nm}, & z_1=0.053\ \text{nm} \\
H_2: & x_2=0, & y_2=-0.075\ \text{nm}, & z_2=0.053\ \text{nm} \\
O: & x_O=0, & y_O=0, & z_O=-0.00654\ \text{nm}
\end{array} \quad (2)
$$

This way, the centre of mass coincides with the centre of the molecule, the directions from the centre to the protons form the angle 109.5^o, and the correct OH distance 0.0957 nm is obtained. To incorporate the effects of hydrogen bonding, the pair potential among water molecules contains short ranged forces that will impose tetrahedral structure to to the first coordination shell. The diffraction studies of water show that the nearest neighbour coordination is tetrahedral but the structure is not found to persist in further shells [34-36] due to the free rotation around the hydrogen bond. The proposed potential therefore depends on the direction from the first molecule to its neighbour but not on the relative orientation around the bond. The angle dependence of the short ranged potential includes the lowest spherical harmonic of tetrahedral symmetry which is the octupole $Y^e_{23}(\theta,\phi)$, where θ and ϕ are the polar angles of either of the molecules in the frame of its neighbour.

The radial projection of the short ranged potential is described by a narrow potential well corresponding closely to the analytically tractable [30] sticky potential of Baxter [31]. The total intermolecular potential among water molecules consists of three terms:

$$u_{ij} = u^{hs}_{ij} + u^{\mu}_{ij} + u^{sp}_{ij} \quad (3)$$

u^{hs}_{ij} is the hard core potential

$$u^{hs}_{ij}(r_{ij}) = \infty \quad \text{if } r_{ij} < (\sigma_i + \sigma_j)/2 \text{ and zero otherwise} \quad (4)$$

u^{μ}_{ij} is the interaction of two point dipoles located in the centres of molecules i and j at given orientations and distance, and u^{sp}_{ij} is the anisotropic adhesive or sticky potential

$$u^{sp}_{ij} = v_{ij}(r_{ij})\ w_{ij}(\vec{\Omega}_i,\vec{\Omega}_j,\tilde{R}_{ij}) \quad (5)$$

In integral equation treatments [30,33] the distance dependent term v_{ij} is described in terms of Dirac δ function while the form of a narrow potential well

$$v_{ij}(r) = a - b\ (\sigma+\Delta/2-r^2) \quad \text{at} \quad \sigma \leq r \leq \sigma+\Delta \quad \text{and zero elsewhere} \quad (6)$$

is preferred in simulations. Above, $b=4a/\Delta^2$, and a and Δ are constants that determine the depth and the width of the potential well.

The orientation dependent part

$$w_{ij} = Y^e_{23}(\theta'_{ij},\phi'_{ij}) + Y^e_{23}(\theta'_{ji},\phi'_{ji}) \quad (7)$$

involves the tesseral harmonics $Y^e_{23}(\theta,\phi)=(15/2)\sin\theta \sin 2\theta \cos 2\phi$ [37]. θ'_{ij} and ϕ'_{ij} are the polar angles of molecule j in the system with molecule i at the origin and with $\alpha_i=\beta_i=\gamma_i=0$. An identical term with molecule j in the centre is added for symmetry reasons.

As illustrated in Fig. 1, the desired coordination is produced by the proposed potential. However, the four tetrahedral sites corresponding to the directions of two protons and two lone electron pairs of the oxygen atom are equivalent. The dipole-dipole interaction brings in the distinction between the four apparently equal binding sites preventing

Figure 1. Angle dependence of short ranged (u_{ij}^{sp}) and total (u_{ij}) pair potential for water molecules with parallel dipoles at the distance $r_{ij}=\sigma+\Delta/2$.

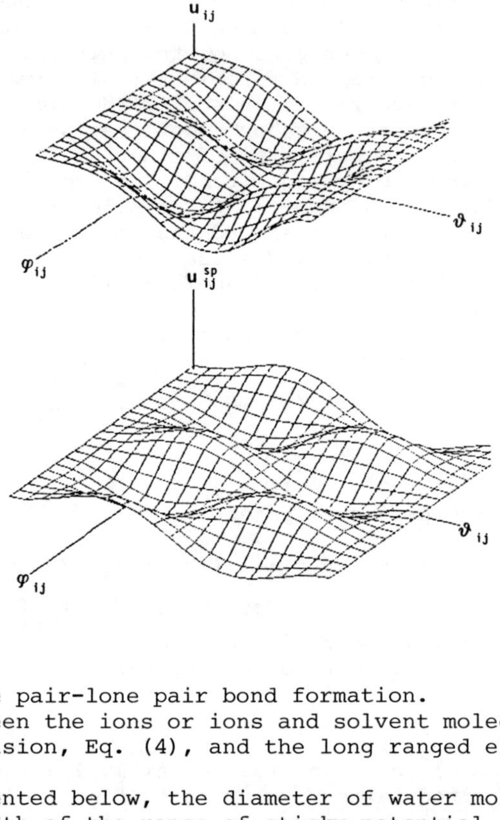

the hydrogen-hydrogen or lone pair-lone pair bond formation.

The pair potentials between the ions or ions and solvent molecules consist of the hard core repulsion, Eq. (4), and the long ranged electrostatic interaction.

In all calculations presented below, the diameter of water molecules σ was 0.275 nm and the width of the range of sticky potential Δ was 0.03 nm. The potential constant a of Eq. (6) was 0.87 kJ/mol. This corresponds to the maximum energy of the hydrogen bond of 10.0 kJ/mol. The temperature T was 298 K and the background permittivity of vacuum was presumed in calculations of the electrostatic interactions.

3. RESULTS AND DISCUSSION

In this section, the simulation results for the structure of pure water, for the hydration of simple ions, and for hydrophobic solvation determined by the above model are collected. Standard Metropolis sampling was used for a system that usually contained 70 water molecules in a cubic box of sides L=1.28 nm. Periodic boundary conditions and the spherical cutoff of the intermolecular potential at the distance L/2 proved satisfactory [32]. When studying the hydrophobic solvation, an open system of varying size was treated by the Grand Canonical Ensemble Monte Carlo (GCEMC) method [38]. The number of generated configurations was typically $5 \cdot 10^5$ in equilibration and $2\text{-}3 \cdot 10^6$ in production runs. References [33] and [38] contain further technical details.

In Fig. 2, the atom-atom correlation functions g_{OO}, g_{OH}, and g_{HH} for our model are presented. In spite of the simple form of the applied potential the main features of liquid water are reproduced. The position of the first peak of g_{OO} is, of course, adjusted by the choice of the molecular size σ, but the first and the second peak of g_{OH} are also of about the right magnitude and are located at the same distances as observed by neutron diffraction [34-36]. The distances at which the peak of g_{HH} and the second peak of g_{OO} are found are somewhat too large but on the whole a surprisingly good agreement with the experimental profiles is observed. With regards to the first peak of g_{OO} it is particularly interesting to study the coordination number $n_{OO}(r)$ defined as the number of nearest neighbours within a specified distance from the central atom or molecule

$$n_{ij}(r) = 4\pi\rho_j \int_0^r g_{ij}(r_{ij}) r_{ij}^2 \, dr_{ij} \quad (8)$$

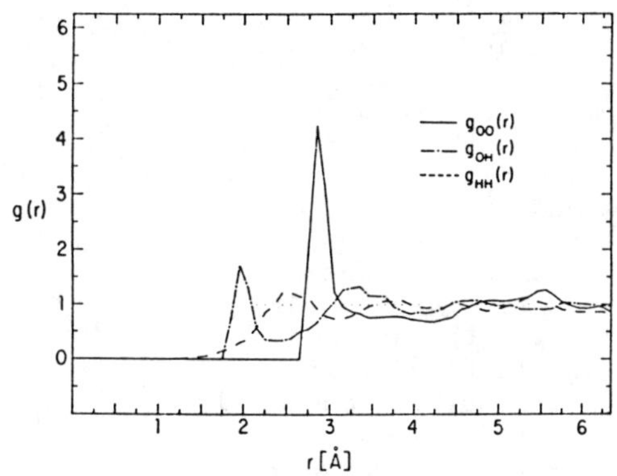

Figure 2. Intermolecular atom-atom correlations in liquid water.

In Table 1. we compare the oxygen-oxygen coordination numbers of our model with those determined by neutron scattering experiment [34-36] or by Monte Carlo and the RISM integral equation (IE) [9] for MCY [10] and TIPS [12] water.

Table 1. Coordination numbers n_{OO} in liquid water

System:	EXPT	SP-MC	MCY-MC	MCY-IE	TIPS-MC	TIPS-IE
Distance r						
0.35 nm	5.0	5.3	4.8	5.7	5.1	5.7
0.38 nm	6.6	6.6	6.5	7.4	6.9	7.4

If compared with the experiment (EXPT), the coordination numbers are overestimated in the integral equation (IE) approach while all the three models including the SP potential of this work give essentially correct results in simulation.

Having tested the model in preditcing the structure of pure water we turn attention to simple ionic solutions. In Figs. 3-5 the results of the simulation of ionic solvation are presented in the form of ion-oxygen and ion-hydrogen radial distribution functions $g_{ij}(r)$ and running coordination numbers $n_{ij}(r)$ in the hydration shells of the ions. The results for di- and trivalent ions are new while those for monovalent ions are from ref. [39] where technical details can also be found. The model for water is the same as above and the ions are represented by charged hard spheres. The ionic diameters and valences corresponding roughly to Na^+, F^-, Ca^{++}, S^{--}, and La^{+++} ion are collected in Table 2, where a comparison of some of our results with those from other sources is made. The distribution functions and the running coordination numbers shown in Figs. 3-5 compare favourably with the reported [40,41] results for Lennard-Jones ions in ST2 [42] water. With all ions, a pronounced maximum of g_{iO} located at or close to the ion-water contact distance is observed. The peaks of g_{iH} in the hydration shells of anions are broader while they are rather smeared with the cations. In all cases, the positions of the peaks indicate a roughly normal orientation of water dipoles with respect to the ion in the first hydration layer. The peaks in ion-oxygen distribution functions are somewhat higher

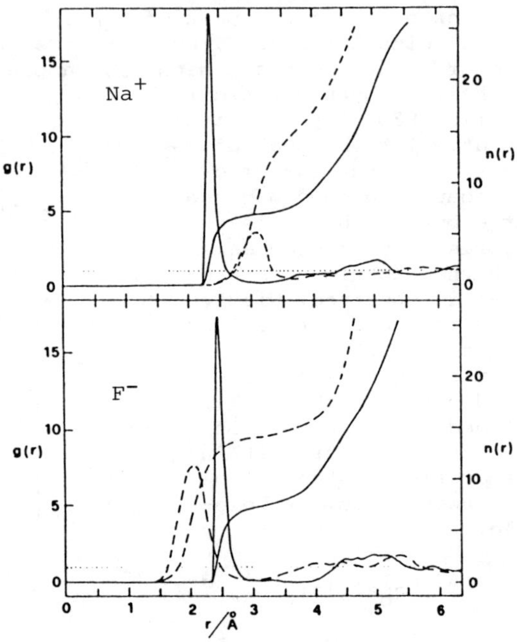

Figure 3. Ion-oxygen (——) and ion-hydrogen (- - -) correlation functions g_{ij} and running coordination numbers n_{ij} in the hydration shells of Na^+ and F^-ion.

Table 2. Characteristic parameters and hydration numbers

Ion:	Na^+	F^-	Ca^{++}	S^{--}	La^{+++}	
Charge:	1	-1	2	-2	3	
σ / Å :	2.0	2.0	2.4	2.4	2.7	
n_h (SP):	7.0	7.5	10.2	9.6	11.4	
n_h (ST2):	6.5	6.8	9.2	-	9-13	[40,41]

and narrower than observed in interactions of soft ions with ST2 water [40,41]. This deficiency reflects the approximate, hard core ion-water repulsion in the model.

The quantity of particular importance is the ion hydration number n_h defined as the plateau value of the running ion-oxygen coordination number $n_{iO}(r)$ read at the position of the first minimum of $g_{iO}(r)$. In Table 2 the calculated hydration numbers n_h (SP) are compared with the values determined for the same ions in ST2 water [40,41]. Although the hydration numbers are slightly higher in sticky water of the present work, a very good agreement between the two sets of results is found.

Finally, we describe an application of the model [38] in a study of structural effects exhibited by the molecular solvent in the form of solvation forces [43] between solute particles. Among these effects, special attention is paid to hydrophobic interaction. To obtain an estimate for the hydrophobic force between colloidal entities, we consider the system consisting of two parallel smooth hard plates immersed in water. The pressure exerted on the walls by the confined fluid varies with separation h between them. Since only the hard core interaction between the plates and the molecules is assumed, the pressure on the wall is proportional to the wall-fluid contact number density ρ^c. The net force F acting on either of the plates is then given by the following relation:

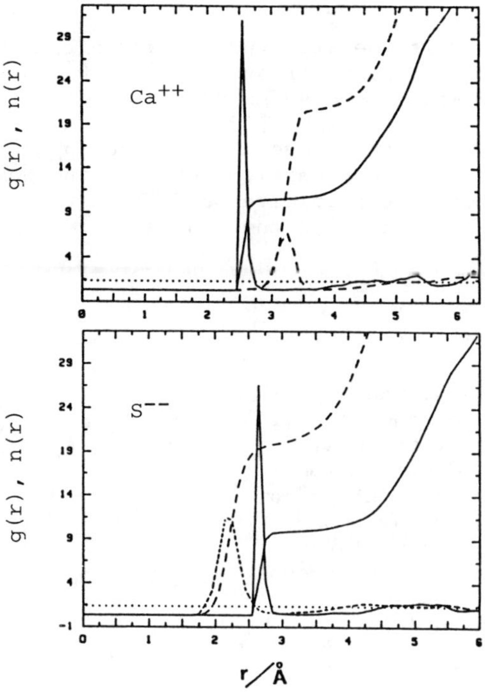

Figure 4. As in Figure 3 but for Ca^{++} and S^{--} ion.

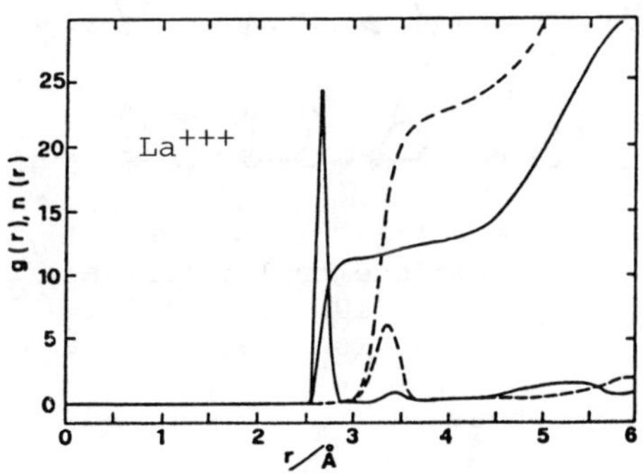

Figure 5. As in Figures 3 and 4 but for La^{+++} ion.

$$\frac{F}{S} = (\rho^c_{in} - \rho^c_{out})\, kT \tag{9}$$

where S is the area and ρ^c_{in} and ρ^c_{out} the wall-fluid contact number densities on the inner and on the outer side of the plates, respectively [44]. Fig. 6 shows the variation of ρ^c_{in} with the separation between the plates h determined [38] by the Grand Canonical Ensemble Monte Carlo method for the above model. Since ρ^c_{out} is equal to ρ^c_{in} at large separations h, we can easily recognize the domains with $\rho^c_{in} > \rho^c_{out}$, where the net force is repulsive, and those with $\rho^c_{in} < \rho^c_{out}$ that correspond to net attraction between the plates.

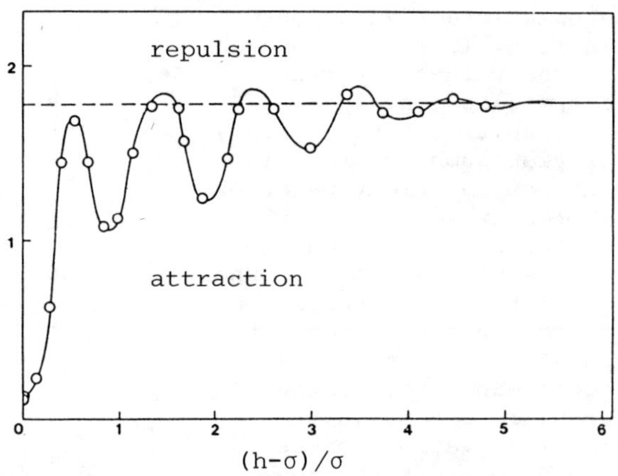

Figure 6. The dependence of the contact value of the wall-fluid distribution function $g^c(h)=\rho^c_{in}/\bar{\rho}$ on the separation between the plates. The solvation force varies as $[g^c(h)-g^c(\infty)]\,\bar{\rho}kT$.

Formally, the fluid density profile in the gap can be expressed in terms of graphical expansions [20,45]

$\ln(\rho/Z) =$  (10)

$\ln(\rho/Z) =$ 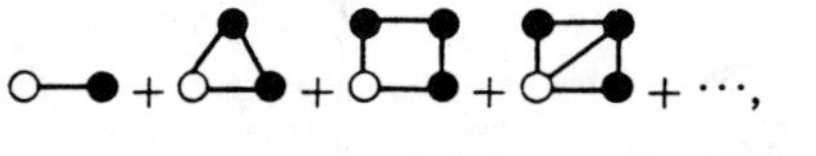 (11)

where ρ is the local number density, $Z=\bar{\rho}\exp(\mu_{ex}/kT)$ the fluid fugacity, μ_{ex} the excess chemical potential, and the black circles carry the factor Z in Eq. (10) or ρ in Eq. (11). In sufficiently narrow pores, all the graphs vanish [38,45,46] and the limiting law

$$\rho \to Z \quad \text{when} \quad h \to \sigma \tag{12}$$

applies. With dense hard sphere fluids this gives rise to very high limiting values of ρ. In liquids with sufficiently attractive interactions such as water, however, $\mu_{ex} < 0$ and the limiting density in the gap is lower than in the bulk phase. In our case, the GCEMC result for

$\mu_{ex}/kT=-2.0$ [38] gives

$$\lim \rho = \bar{\rho} \exp(\mu_{ex}/kT) = 0.135 \bar{\rho} \quad \text{at } h \to \sigma \tag{13}$$

As shown in Fig. 6, the limiting law, Eqs. (12) and (13), is confirmed in the simulation. Considering this result, it is not surprising that the attractive force dominates at small separations h even in the absence of any direct attraction between the plates or the plates and the liquid.

Another interesting feature are distinct but rapidly decaying oscillations in the solvation force F(h) and $\rho^c_{in}(h)$ with the period noticeably smaller than would be expected with the hard core fluid of the same molecular diameter [45]. This result reflects the characteristic structure of water and is in a substantial agreement with celebrated measurements of hydration forces by Israelachvili and coworkers [47].

In view of the above examples we conclude that the introduction of the proposed orientation dependent adhesion improves essentially the properties of the hard dipole model for water and aqueous solutions. It appears that the main features of hydrogen bonding can be captured by this simple intermolecular potential or its variations [32,33]. Analytical treatments of the model are being considered. Extensions of the Percus-Yevick Ornstein-Zernike study [30] of the fluid with infinitesimaly short ranged [31] quadrupolar adhesion by adding higher multipoles and electrostatic terms will be reported. Applications in theory of ionic solvation and electrical double layer are planned.

ACKNOWLEDGMENT. Collaboration and valuable discussions with Dr. A. Luzar, Prof. P. T. Cummings, and Dr. D. Q. Wei are acknowledged. This work was supported by the National Science Foundation through the Department of Physics, University of Puerto Rico, and through the U. S. - Yugoslav Fund for Scientific cooperation.

REFERENCES

[1] Dogonadze, R. R., Kalman, E., Kornyshev, A. A. and Ulstrup, J. (eds.) (1985) The Chemical Physics of Solvation, Part A, Theory of Solvation, Elsevier, Amsterdam.

[2] Kleeberg, H.(ed.) (1987) Interactions of Water in Ionic and Nonionic Hydrates, Springer, Berlin-Heidelberg.

[3] Teleman, O., Jönsson, B. (1987) "A molecular dynamics simulation of water model with intramolecular degrees of freedom", Mol. Phys. 60, 193-203.

[4] Kuharsky, R. A. and Rossky, P. J. (1985) "The contribution of intramolecular vibrations to observed structure of liquid water", J. Chem. Phys. 82, 5284-5291.

[5] Watanabe, K. and Klein, M. (1989) "Effective pair potentials and the properties of water", Chem. Phys. 131, 157-167, and references therein.

[6] Finney, J. L. (1982) "Towards a molecular picture of liquid water", in F. Franks (ed.), Biophysics of Water, Wiley, New York, pp. 73-95.

[7] Kuharsky, R. A. and Rossky, P. J. (1985) "A quantum mechanical study of structure in liquid H_2O and D_2O", J. Chem. Phys. 82, 5164-5177.

[8] Rossky, P. J. (1989) "Quantum effects in hydrogen bonded liquids", J. Dore and J. Teixeira (eds.), Proc. NATO ASI on Hydrogen-Bonded Liquids, Cargese 1989, in press.

[9] Pettitt, B. M. and Rossky, P. J. (1982) "Integral equation predictions of liquid state structure for water-like potentials", J. Chem. Phys. 77, 1451-1457.

[10] Matsuoka, M., Clementi, E. and Yoshimine, M. (1976) "A study of the water dimer potential surface", J. Chem. Phys. 64, 1351-1361.

[11] Berendsen, H. J. C., Postma, J. P. M., Gunsteren, W. F. and Hermans J. (1981), in B. Pullman (ed.), Intermolecular Forces, Reidel, Dordrecht, p. 331.

[12] Jorgensen, W. (1981) "Transferable intermolecular potential functions for water, alcohol, and ethers. Application to liquid water" J. Am. Chem. Soc. 103, 335-340.

[13] Thuraisingham R. A. and Friedman, H. L. (1983) "HNC solution for the central force model for liquid water, J. Chem. Phys. 78, 5772-5775.

[14] Ichiye, T. and Haymet, A. D. J. (1988) "Accurate integral equation theory for the central force model of liquid water and ionic solutions", J. Chem. Phys. 89

[15] Stillinger, F. H. and Rahman, A. (1978) "Revised central force potentials for water", J. Chem. Phys. 68 (1978) 666-670.

[16] Carnie, S. L. and Patey, G. N. (1982) "Fluids of polarizable hard spheres with dipoles and tetrahedral quadrupoles. Integral equation results with application to liquid water", Mol. Phys. 47, 1129-1151.

[17] Fries, P. H. and Patey, G. N. (1985) "The solution of the hypernetted chain approximation for fluids of nonspherical particles. A general method with application to dipolar hard spheres", J. Chem. Phys. 82, 429-440.

[18] Caillol, J. M., Levesque, D., Weis, J. J., Kusalik, P. G. and Patey, G. N. (1987) "A comparison between computer simulations and theoretical results for ionic solutions", Mol. Phys. 62, 461-473.

[19] Kusalik, P. G. and Patey, G. N. (1988) "The solution of the reference hypernetted chain approximation for water-like models", Mol. Phys. 65, 1105-1119.

[20] Friedman, H. L. (1985), A Course in Statistical Mechanics, Prentice-Hall, Englewood Cliffs.

[21] Wertheim, M. S. (1979) "Equilibrium statistical mechanics of polar fluids", Ann. Rev. Phys. Chem. 30, 471-501.

[22] Adelman, S. A. and Deutch, J. M. (1974) "Exact solution of the mean spherical model for strong electrolytes in polar solvents", J. Chem. Phys. 60, 3935-3949.

[23] Blum, L. (1974) "Solution of a model for the solvent-electrolyte interactions in the mean spherical approximation", J. Chem. Phys. 61, 2129-2133.

[24] Blum L. (1978) "Solution of the mean spherical approximation for hard ions and dipoles of arbitrary size", J. Stat. Phys. 18, 451-474.

[25] Chan, K. Y., Gubbins, K. E., Henderson, D. and Blum, L. (1989) "Monte Carlo and simple theoretical calculations for ion-dipole mixtures", Mol. Phys. 66, 299-316.

[26] Blum, L. and Torruella, A. J. (1972) "Invariant expansion for two body correlations: Thermodynamic functions, scattering and the Ornstein-Zernike equation", J. Chem. Phys. 56, 303-310.

[27] Blum, L. (1972) "Invariant expansion II. The Ornstein-Zernike equation for nonspherical molecules and an extended solution to the mean spherical model" , J. Chem. Phys. 57, 1862-1869.

[28] Blum, L. (1973) "Invariant expansion III. The general solution of the mean spherical model for neutral spheres with electrostatic interactions", J. Chem. Phys. 58, 3295-3303.

[29] Blum, L. and Torruella, A. J. (1988) "Invariant expansion IV. The exponentials of tensorial expressions", J. Chem. Phys. 89, 4976-4980.

[30] Cummings, P. T. and Blum, L. (1986) "Analytic solution of the molecular Ornstein-Zernike equation for nonspherical molecules. Spheres with anisotropic surface adhesion" J. Chem. Phys. 84, 1833-1842.

[31] Baxter, R. J. (1968) "Percus-Yevick equation for hard spheres with surface adhesion", J. Chem. Phys. 49, 2770-2774.

[32] Bratko, D., Blum, L. and Luzar, A. (1985) "A simple model for intermolecular potential of water", J. Chem. Phys. 83, 6367-6370.

[33] Blum, L., Cummings, P. T. and Bratko, D., to be published.

[34] Narten, A. H. and Levy, H. A. (1971) "Liquid water: Molecular correlation functions from X-ray diffraction", J. Chem. Phys. 55, 2263-2269.

[35] Narten, A. H., Thiessen, W. E. and Blum, L. (1982) "Atom pair distribution functions of liquid water at 25°C fron neutron diffraction", Science, 217, 1033-1034.

[36] Thiessen, W. E. and Narten, A. H. (1982) "Neutron diffraction study of light and heavy water mixtures at 25°C", J. Chem. Phys. 72, 2656-2662.

[37] Morse, P. and Feshbach, H. (1953), Methods of Theoretical Physics, Part II, McGraw-Hill, New York, p. 1264.

[38] Luzar, A., Bratko, D. and Blum, L. (1987) "Monte Carlo simulation of hydrophobic interaction", J. Chem. Phys. 86, 2955-2959.

[39] Bratko, D., Blum. L. and Wei, D. Q. (1987) "A model of ion hydration", in ref. [2], pp.27-31.

[40] Heinzinger, K. and Palinkas, G. (1985) "Computer simulations of ion-solvent systems", in ref.[1], pp. 313-353.

[41] Heinzinger, K. and Palinkas, G. (1987) "Interactions of water in ionic hydrates", in ref. [2], pp. 1-22.

[42] Stillinger, H. F. and Rahman, A. (1974) "An improved simulation of liquid water by molecular dynamics", J. Chem. Phys. 60, 1545-1557.

[43] Ninham, B. W. (1980) "Long range vs. short range forces. The present state of play", J. Phys. Chem. 84, 1423-1430.

[44] Evans, R. (1989) "Microscopic theories of simple fluids and their interfaces", in Proc. NATO ASI on Liquids at Interfaces, Les Houches 1988, in press.

[45] Wertheim, M. S., Blum., L. and Bratko, D. (1989) "Statistical mechanics of confined systems: The force between smooth parallel plates", in S. H. Chen and R. Rajagopalan (eds.), Micellar Solutions and Microemulsions, Springer, New York.

[46] Bratko, D., Blum, L. and Wertheim, M. S. (1989) "Structure of hard sphere fluids in narrow cylindrical pores", J. Chem. Phys. 90, 2752-2757.

[47] Israelachvili, J. N. (1985) "Measurements of hydration force between macroscopic surfaces", Chem. Scripta 25, 7-14.

A MODEL FOR HYDROGEN-BONDING EFFECTS AT AQUEOUS INTERFACES

A. LUZAR
Institute of Microbiology, Medical Faculty
Zaloška 4, 61105 Ljubljana and "J. Stefan"
Institute, E. Kardelj University, Ljubljana,
Yugoslavia

ABSTRACT. An attempt is made to derive a model for thermodynamic properties of water and water mixtures that incorporates the effect of strong orientation dependent hydrogen bonding interactions in the mean-field approximation. The theoretical analysis presented permits us to describe the physically important features of the liquid/vapour interfaces of aqueous mixtures with the aid of four phenomenological parameters, E^{11}, E^{12} and ΔS^{11}, ΔS^{12}, determined in bulk solution. The parameters E^{11} and E^{12} are the energies of hydrogen bond formation between water molecules or water molecules and a second component. ΔS^{11} and ΔS^{12} are corresponding entropies of bond formation. The theory provides relations for the contribution of hydrogen bonding to bulk and surface equilibrium properties: surface free energy of water, excess energies and free energies of mixing and excess surface free energy of the mixture. An accurate description of experimental data for the system water-dimethylsulphoxide is obtained.

1. Introduction

In this article we review our work on hydrogen-bonding effects at aqueous interfaces. A detailed exposition may be found in Refs. [1-5].

Aqueous interfaces are of considerable importance in many biological and chemical systems. A key quantity in the characterization of interfaces is the surface tension.

Theoretical methods that were very useful for the study of simple fluids at the vapour/liquid interface, such as perturbation theory, density functional approach and the integral equation theories [6] have not been yet successful in accounting for the properties of aqueous interfaces. The nonspherical nature of the water molecule, combined with large intermolecular attractive energy, have made it very difficult to extend these theoretical methods to the study of aqueous interfaces.

In considering specific types of interaction that occur in the aqueous phase, by far the most important is the hydrogen bond. The bottom line of this lecture is the hypothesis that the local environment of a water molecule is correlated with its degree of bondedness. Spatial

J.C. Dore and J. Teixeira (eds.), Hydrogen-Bonded Liquids, 197–210.

confinement (propinquity of the surface) and/or foreign components (added solute to the solvent; aqueous mixtures) locally modify the structure so that bonding is perturbed from the bulk average. It is therefore of interest to examine the surface-induced perturbation in hydrogen bonding of water [1,2] and its mixtures [3-5] near the interface. We consider hydrogen bonding characteristics and orientational distribution and present a molecular picture of the interfacial region based on these results.

2. Contribution of hydrogen bonds to the surface tension of water

The physical basis of the difference between water and other fluids can be explained by the fact that a water molecule can form hydrogen bonds to four other near-neighbouring water molecules leading to an open tetrahedral network structure in the liquid. This idea has been used within the context of the mixture [7], continuum [8], and random network models [9] to account for the properties of water. We focus on a simple two-state hydrogen bonding model for water [1,2,10] as it apparently gives a valuable insight into the relation between bulk and interfacial water structure.

In the model, water is treated as an ensemble of a large number of O—H⋯O_{H2O} pairs, that are able or unable to form a bond, due to orientational restrictions. In the former case, there is an equilibrium between formed and broken bonds. This is a crude description of the concept that a hydrogen bond consists of an interaction between a positively charged proton on one molecule and a negatively charged lone electron pair in the valence shell of the oxygen atom on the other. The total number density of water-water (11) pairs, n_{11}, is then equal to the number density of water protons, $2n_1$, where n_1 is the number density of water molecules. The phase space of the 11 pair is formally divided into two subvolumes, V_1^{1i}(i=1) for hydrogen bonded and V_0^{1i}(i=1) for nonhydrogen bonded pairs. There are only two energy levels for 11 pairs: E^{1i}(i=1) < 0 for formed bonds and zero otherwise. Only a fraction of all possible bonds ω^{1i} is formed:

$$\omega^{1i} = \frac{V_1^{1i}\, e^{-\beta E^{1i}}}{V_1^{1i}\, e^{-\beta E^{1i}} + V_0^{1i}} = \frac{1}{1 + \alpha^{1i} e^{\beta E^{1i}}} \quad ; \quad i=1 \qquad (1)$$

where $\beta = 1/kT$, and kT has its usual meaning. $\alpha^{1i}=V_0^{1i}/V_1^{1i}$ is the degeneracy ratio.

The free energy change due to the formation of a pair 11 able to form a bond is given by

$$\Delta A^{1i} = -\, kT \ln q^{1i} \quad ; \quad i=1 \qquad (2)$$

where q^{1i} denotes the partition function of a single hydrogen bonding pair relative to a pair unable to form a bond. It is expressed as a sum of two terms:

$$q^{1i} = \frac{\alpha^{1i}}{\alpha^{1i}+1} + \frac{e^{-\beta E^{1i}}}{\alpha^{1i}+1} \qquad ; \quad i=1 \qquad (3)$$

The first term corresponds to the non-bonded state and the second to the bonded state. Parameters of the model, the energy E^{11} and the entropy $\Delta S^{11} = -kT \ln\alpha^{11}$ of the hydrogen bond formation are obtained according to the best agreement of the theoretical temperature dependence of the fraction of formed bonds, ω^{11} (eq.1), with the spectroscopic data [11,12]. The value of $E^{11} = -13.4$ kJ/mol [1] is in good agreement with the hydrogen bond energy obtained in recent experiments [13] and $\alpha^{11}=28$ [1].

In the spirit of the work of Stanley et. al. [14], liquid water is considered to be composed of molecules with up to four hydrogen bonds. This means that in bulk solution there are four other near-neighbouring water molecules located at the correct distance and angle required for hydrogen bonding. Close to the interface, however, the number of hydrogen bonds between the water molecules decreases, simply for geometrical reasons: the number of neighbouring molecules at exactly located sites and with appropriate orientation for the formation of hydrogen bonds is decreased within the surface layer.

In the interfacial layer of the thickness, equal to the length of the hydrogen bond, $d_{O-H\cdots O} = 0.3$ nm, the number density of water-water pairs becomes a function of the distance from the interface, $n_{11}(x)$. A function $0 \le h^{11}(x) \le 1$ is introduced, which is a relative measure of the average number of possible bonds per molecule at the distance x from the interface, compared to the bulk phase, therefore $h^{11}(b)=1$;

$$h^{11}(x) = \frac{1}{4} \sum_{l=1}^{4} l \; Y_l^{11}(x) \qquad (4)$$

$Y_l^{11}(x)$ represent the fractions of water molecules able to form at most l bonds ($1 \le l \le 4$) with neighbouring water molecules at the distance x from the vapour/liquid interface, which is here defined as a plane impenetrable for oxygen atoms. The detailed procedure of the calculation of $Y_l^{11}(x)$ in which Boltzmann statistics was used to account for the increase in the probability of the orientations with higher number of bonds is found in Refs. [1,5]. The sum of $Y_l^{11}(x)$ in eq.4 is presented in Fig. 1 (curve a). It turns out that the maximum local decrease in the average number of hydrogen bonds for water molecules is only ≅ 30% [3,5] instead of 50% decrease found if all orientations of surface molecules were equally probable [1]. The orientations where the surface molecule may participate in forming three hydrogen bonds are most probable due to the assumed tendency of the liquid to maximize the number of hydrogen bonds. This result is in essential agreement with molecular dynamics simulation of ST2 water [Fig.5 of Ref.15], as well as with theoretical predictions of Marčelja et. al. [16]. The thickness of the interval where the average number of hydrogen bonds per molecule is lower than the bulk value is equal to 0.3 nm. This result is due to the approximation of the model where only the nearest neighbour interactions among water molecules are taken into account. It is gratifying to note that this

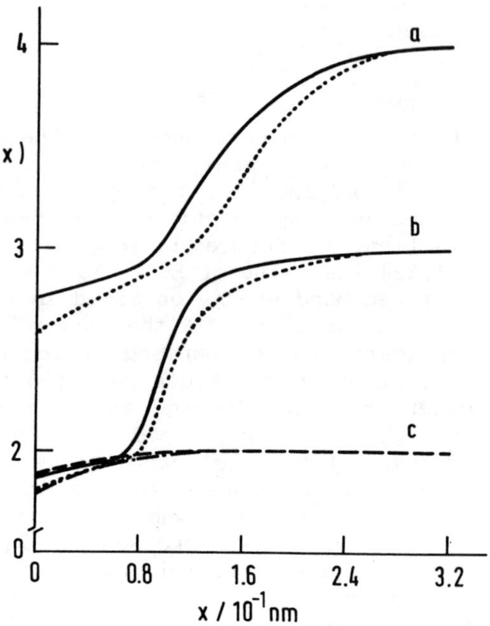

Figure 1. Average number of possible bonds per molecule as a function of the distance from the interface for T = 25°C (——) and T = 60°C (········); a: i=1, 0 ≤ l ≤ 4; b: i=2, 0≤ l ≤ 3; c: i=2, 0≤ l ≤ 2.

thickness agrees with the simulation results of Lee et. al [15] obtained by considering a detailed molecular structure. The correlations with distant water molecules appear to be less important for the dependence of the average number of hydrogen bonds on the distance from the interface. This dependence is further used in the calculations of the free energy density function of the system.

We write the free energy density function of the system as:

$$f(x) = n_{11}(x)\ \Delta A^{11} \tag{5}$$

Considering that $n_{11}(x)=2n_1 h^{11}(x)$, $h^{11}(b)=1$ and assuming that the number density of water molecules in the interfacial layer is equal to the bulk value, the increase in the free energy density function close to the interface is given by:

$$f(x) - f(b) = 2n_1\ \Delta A^{11}\ [h^{11}(x) - 1] \tag{6}$$

The surface tension γ of a pure liquid is equal to the specific excess surface free energy A^{σ} [17]. The contribution of hydrogen bonds to the total surface tension, γ_1^{HB}, can be expressed in terms of the free energy function of eq.6, integrated over the interval $0 \le x \le d = 0.3$ nm. Below, we give a few values of γ_1^{HB} (in mN/m) calculated at different temperatures:

T [°C]	0	25	60	100
	28.1	27.7	26.5	24.6

It is worth while to mention that γ_1^{HB} is overestimated by approximately one half if we describe the hydrogen bonding as random [1]. In addition, much weaker temperature dependence of hydrogen bond contribution to the surface tension of water is observed if weighted orientational probabilities are taken into account.

3. Contribution of hydrogen bonds to the surface tension of aqueous mixtures: water-DMSO

The above approach can be extended to binary systems to study interfacial properties of water mixtures. DMSO has been chosen as the second component in the system because of its high ability to participate in hydrogen bonding [18] and because of its interesting physiological properties [19].

The unique capability of DMSO to penetrate living tissues without causing a significant damage is most probably related to its relatively polar nature, its capacity to accept hydrogen bonds, and its relatively small and compact nature. It has been proposed that DMSO, acting as a proton acceptor, can alter the structure of water. This behaviour may also account for its cryoprotective properties. DMSO is best formulated as a resonance hybrid of the canonical structure having a polarized S—O bond and that having a $(p \rightarrow d)_\pi$ SO double bond [18]:

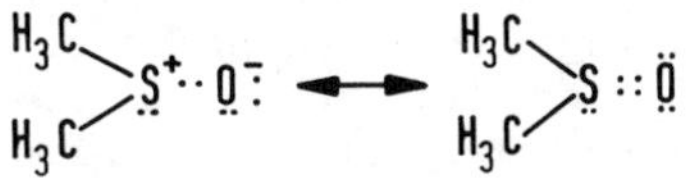

(7)

This representation agrees best with the physical properties and the reactivity of the compound. DMSO contains two acceptor sites: the oxygen atom, which is the position of highest electron density, and the sulfur atom, which, in spite of its partially positive character, still possesses nucleophilic properties because of its lone electron pair. Nevertheless, the only possible point of association for the proton is the O atom of DMSO [18]. Considering eq.7 the number of lone electron pairs in the valence shell of the oxygen atom in DMSO, m, is either 2 or 3. There are no hydrogen bonds between DMSO molecules [18]. According to the previous model for pure water, two kinds of pairs are considered in the mixture: 11 pair, able to form a bond $(O\text{—}H\cdots O_{H2O})$ and water-DMSO (12) pair able to form a bond $(O\text{—}H\cdots O_{DMSO})$. The total number density of pairs $n_{11} + n_{12}$ remains equal to the number density of water protons in the mixture, $2n_1$. The probability of the formation of the bond between a proton and a lone electron pair in the valence shell of the oxygen atom in water or in DMSO is determined by Boltzmann distribution.

In the calculation of the free energies of the bonding pair formation, we follow the approach described in Ch.2 for pure water. All the relations (eqs.1-3) retain their form, except for the superscript i which has now two values: i=1 for 11 pair and i=2 for 12 pair.

The parameters E^{12} and α^{12} for 12 pair in eqs.1-3 are determined [4] from the experimental data for the enthalpies of mixing, ΔH^{ex}_{mix}, and the Gibbs free energies of mixing, ΔG^{ex}_{mix}, at $T = 25^0C$ [20,21].

In the derivation of expressions of these excess thermodynamic functions, the small differences between the energies and the enthalpies as well as those between the free energies and the free enthalpies of mixing are neglected [24].

Thermodynamic effects of mixing are interpreted in terms of the changes in n_{11} and n_{12} and of the energy and free energy of the formation and the decomposition of a pair able to form a hydrogen bond,

E^{1i} and ΔA^{1i}:

$$\Delta A^{ex} = [N_A/(n_1+n_2)]\; n_{12}\; (\Delta A^{12} - \Delta A^{11}) \tag{8}$$

$$\Delta E^{ex} = [N_A/(n_1+n_2)]\; n_{12}\; (E^{12}\omega^{12} - E^{11}\omega^{11}) \tag{9}$$

Above ΔA^{ex} and ΔE^{ex} are the excess free energy and the excess energy of mixing per mole of the solution and N_A the Avogadro number. For the values of the parameters [4] see text of Fig.2.

The number densities of 11 and 12 pairs in eqs.8,9 are obtained in the following way [14]: an imaginary binding process is presumed during which 11 and 12 pairs are being formed. The probability of a proton to enter a pair 11 or 12 is taken to be proportional to the corresponding Boltzmann factor $\exp(-\beta\Delta A^{1i})$ and to the instantaneous number densities of free electron pairs in the valence shell of the oxygen atoms in DMSO or in water, $mn_2 - n'_{12}$ or $2n_1 - n'_{11}$, respectively. The number of free electron pairs decreases during the above process. In the calculation of the instantaneous probabilities for the formation of the given bond species only the number densities of unoccupied electron pairs are considered:

$$\frac{dn'_{12}}{d(n'_{11}+n'_{12})} = \frac{(mn_2 - n'_{12})\exp(-\beta\Delta A^{12})}{(2n_1-n'_{11})\exp(-\beta\Delta A^{11}) + (mn_2-n'_{12})\exp(-\beta\Delta A^{12})} \tag{10}$$

Integrating eq 10 from $n'_{11}+n'_{12} = 0$ to $n'_{11}+n'_{12} = 2n_1$, the following solution is obtained:

$$\exp(-\beta\Delta A^{11})\ln\left(1 - \frac{n_{12}}{mn_2}\right) - \exp(-\beta\Delta A^{12})\ln\left(\frac{n_{12}}{2n_1}\right) = 0 \tag{11}$$

A comparison between theoretical and experimental values of the excess enthalpy and excess free energy of mixing at different temperatures is given in Fig.2. E^{12} and α^{12} are treated [4] as temperature independent parameters determined according to the best agreement of calculated curves with experimental data [20,21]. Following the concept of non-random mixing, all the theoretical curves exhibit the experimentally observed skewness with respect to the mole fraction, which has not been the case in the study, where a random formation of interacting pairs was presumed [3].

According to the model, the observed thermodynamic effects are the consequence of the difference in the strength of the hydrogen bond between 11 and 12 pairs, rather than the result of any considerable change in the total numbers of hydrogen bonds in the system. In Fig. 3 the average numbers of hydrogen bonds per water molecule for 1i pairs are presented as functions of the mole fraction of water. When $X_1(b)$ is small, each water molecule is surrounded by DMSO molecules and forms

Figure 2. Excess enthalpy and free enthalpy of mixing for water-DMSO solution as functions of the mole fraction of water at T=25^{0}C (—) and 70^{0}C (·······); calculated values are depicted with dashed (m=2; E^{12}= -16.95 kJ/mol, α^{12}= 48 [4]) or with solid (m=3; E^{12}= -16.25 kJ/mol, α^{12}= 41 [4]) curves whereas experimental values are shown using open (ΔH^{ex} at 25^{0}C; [20,21]), full (ΔG^{ex} at 70^{0}C; [22]) circles or open (ΔG^{ex} at 25^{0}C; [22]) triangles.

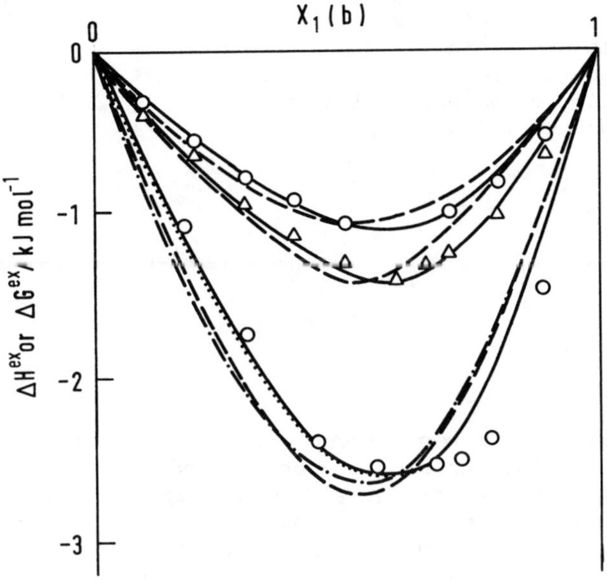

hydrogen bonds with it. Accordingly $\langle n_{HB}^{12} \rangle$ decreases and $\langle n_{HB}^{11} \rangle$ increases with the mole fraction of water. In water rich solutions, water molecules may form hydrogen bonds with other molecules of water as well as with DMSO molecules. The total number of hydrogen bonds in the system at any value of the mole fraction of water is roughly the same before and after mixing, reflecting the fact that the fractions of formed bonds for 11 and 12 pairs do not differ significantly, i. e. $\omega^{11} \cong \omega^{12}$.

The hydrogen bonding to DMSO leads to a decrease in the number of hydrogen bonds among water molecules (Fig. 3). At the interface, there is an additional effect on hydrogen bonding: some protons on the water molecule or lone electron pairs in the valence shell of the oxygen atom in water or in DMSO are oriented towards the surface, therefore they are not able to participate in forming a hydrogen bond.

In order to take into account the orientational constraints imposed by the surface, simple structural models of water and DMSO molecule are used [5]. For water, a tetrahedral coordination is presumed again. The shape of the DMSO crystal structure is pyramidal, the corners being occupied by the two C atoms, the O atom and the S atom [23]. According to eq.7, there may be up to three free electron pairs in the valence shell of the oxygen atom in DMSO. If we assume that these electron pairs point in the directions of three corners of an imaginary tetrahedron with the O atom in the center and the S atom in the remaining corner, a tetrahedral coordination of the O atom of DMSO in the hydrogen bonded water-DMSO complex is obtained [5]. In its centre, there is the oxygen nucleus of a DMSO molecule and the vertices of the tetrahedron are occupied by the sulfur atom and two (m=2) or three (m=3) oxygen atoms of the neighbouring, hydrogen bonded water molecules. Regardless of the value of m, there are always three disposable sites where the oxygen

Figure 3. Average number of hydrogen bonds per water molecule, $\langle n_{HB}^{1i} \rangle = n_{1i}\,\omega^{1i}/n_1$ (b), as functions of the mole fraction of water; full lines (m=3), dashed lines (m=2).

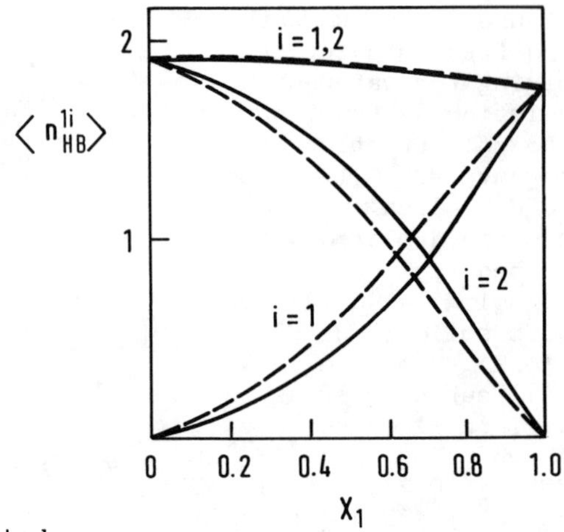

atoms of water molecules can be located.

The surface is approximated by a smooth plane, impenetrable to the hydrophilic sulfur and oxygen atoms. Approaching the surface, only the orientations of the complex with the sulfur and oxygen atom of DMSO located in the liquid phase are taken into account. The orientations with the hydrophobic methyl groups pointing into the vapour phase are allowed [17] in this approach. The number of possible hydrogen bonds at the given orientation is then equal to the number of those sites available for the oxygen atoms of hydrogen bonded water in the complex which are located in the liquid phase.

The thermodynamic properties of the mixture in the interfacial layer are treated by the methods, developed for bulk mixture, but with the number of lone electron pairs in the valence shell of the oxygen atom multiplied by $h^{1i}(x)$, where $h^{11}(x)$ is defined in eq. 4 and

$$h^{12}(x) = \frac{1}{m} \sum_{l=0}^{m} l\, Y_l^{12}(x) \tag{12}$$

$Y_l^{12}(x)$ are proportional to the fractions of the orientational space within which the central DMSO molecule can form l hydrogen bonds ($0 \le l \le m=2$, or $0 \le l \le m=3$), multiplied by the Boltzmann factors corresponding to the specified number of bonds l. Detailed calculation is described in Ref. [5]. The sums $\Sigma l Y_l^{12}(x)$ are presented in Fig. 1 (curve b and c). From curve c we can see that if there are two waters per one DMSO, practically no decrease in $\Sigma l Y_l^{12}$ is observed, which indicates that DMSO molecules at the surface mainly orient in such a way that they can retain two hydrogen bonds. This result is in accord with the expectation that DMSO molecules positively adsorb on the surface with the orientational bias that favours more hydrogen bonds.

Applying eq. 10 to the layer at the distance x from the interface, we have to consider, that the effective numbers of protons and of free electron pairs per water molecule become $2h^{11}(x)$ and that of electron pairs on DMSO are now $mh^{12}(x)$. Eq. 11 takes the form:

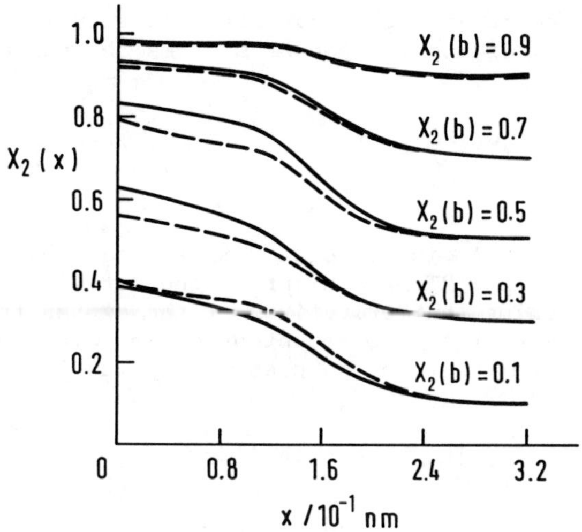

Figure 4. Calculated profiles for DMSO as functions of the distance from the interface for various compositions of the bulk mixture; full lines (m=3), dashed lines (m=2).

$$\exp(-\beta\Delta A^{11})\ln\left[1 - \frac{n_{12}(x)}{mn_2(x)h_{12}(x)}\right] - \exp(-\beta\Delta A^{12})\ln\left[\frac{n_{12}(x)}{2n_1(x)h_{11}(x)}\right] = 0 \qquad (13)$$

where $n_i(x)=X_i(x)/\Sigma X_i(x)V_i$ are the local number densities of the corresponding species. Considering small volume effects of mixing [24], partial molar volumes of the two components are approximated by the molar volumes of pure components in calculations. Here, $X_i(x)$ denotes the local mole fraction of i-th component.

The equilibrium density profiles, $X_i(x)$ are determined by minimizing the free energy of the system with respect to the local mole fraction of water at constant volume of the mixture. The rigorous procedure is described elsewhere [5]. Calculated profiles $X_2(x)$ are presented in Fig. 4. In the mixture, the number of free electron pairs on the oxygen atoms exceeds that of the protons available for the hydrogen bond formation. If some of the electron pairs point to the surface, this may not lead to the reduction of the total number of bonds in the system. This number is, however, affected if proton donating water molecules are brought to the surface layer and some of the protons are oriented towards the surface. This explains the adsorption of DMSO at the liquid/vapor interface, shown in Fig. 4.

Fig. 5 illustrates the combined effects of mixing and of steric restrictions at the liquid/vapour interface on the number densities of 11 and 12 bonding pairs.

The free energy density function for the mixture is derived [5] by adding the ideal configurational entropy term to the hydrogen bond terms as defined in eq. 5:

$$f(x) = \sum_i n_{1i}(x)\,\Delta A^{1i} + kT\,\ln \sum_i n_i(x)\,\ln X_i(x) \qquad i=1,2 \qquad (14)$$

and in analogous way for bulk solution.

In multicomponent systems the surface tension is determined as [16]:

$$\gamma = A^{\sigma} - \sum_i \mu_i \Gamma_i^{\sigma} \tag{15}$$

where A^{σ} is calculated in the same way as for pure water. Further, $\mu_i = \mu_i^0 + RT\ln X_i + RT\ln f_i$; the activity coefficients are determined by taking the derivatives of the excess free energy of the mixture with respect to the amount of given component while the values of μ_i^0 are irrelevant in the present application [5]. Γ^{σ} is the surface excess

$$\Gamma_i^{\sigma} = \int_0^d [n_i(x) - n_i(b)]\, dx \tag{16}$$

$n_i(x)$ and $n_i(b)$ are the number densities at x or in bulk solution.

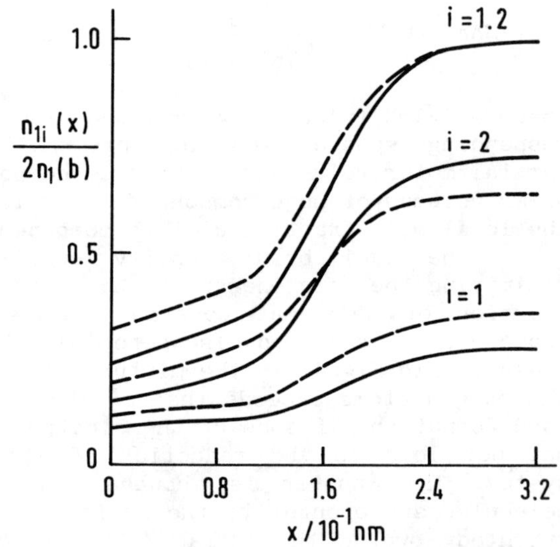

Figure 5. Number densities of 11 and 12 pairs relative to the total number density of hydrogen bonds in bulk solution at 25°C and $X_1(b)=0.5$, as functions of the distance from the interface, calculated for m=3 (full lines) and for m=2 (dashed lines).

The contribution of the perturbations of hydrogen bonding at the interface on the surface tension of the mixture is studied in terms of the excess surface tension $\gamma^{ex} = \gamma(\text{observed}) - (\gamma_1 X_1 + \gamma_2 X_2)$. X_1 and X_2 are the bulk liquid mole fractions of the two species and γ_1 and γ_2 the surface tensions of pure components at the same temperature. γ^{ex} has no fundamental thermodynamic significance and is merely a useful quantity to consider when comparing theory with experiment. It is our belief that a 'good' theory of the interface should accurately describe the measured results for γ^{ex}.

Considering only the hydrogen bond contribution to γ^{ex}, the approximate relation $\gamma^{ex} \cong \gamma^{HB} - X_1 \gamma_1^{HB} - X_2 \gamma_2^{HB}$ is applied. γ^{HB} is the surface tension estimated using the above approach, γ_1^{HB} is the hydrogen bond contribution to the surface tension of pure water (see Ch.2), and $\gamma_2^{HB}=0$, since no significant hydrogen bond effects have been observed in pure DMSO [18]. In Fig. 6 the calculated γ^{ex} is compared with the experimental results for various compositions of the mixture. The experimental excess surface tension has been found [25] to be temperature independent within experimental uncertainties. This observation is in accordance with the theory (Fig. 6). We conclude that the "microscopic" model gives a good account of the excess surface tension of the binary mixture while using the bulk model parameters given previously.

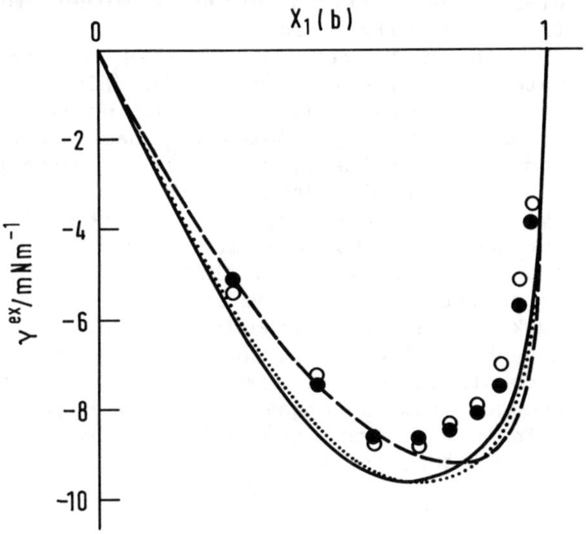

Figure 6. Excess surface tension as a function of the mole fraction of water. Full lines: m=3 and T=25°C (——) or 60°C (·········). Dashed line: m=2 and T=25°C. Open ($T = 25^{0}C$) and full ($T = 60^{0}C$) circles are experimental values for water-DMSO mixture [25].

4. Conclusions

The present theory represents an effort to include in a single, unified theoretical framework, the effect of hydrogen bonding on the excess bulk <u>and</u> surface thermodynamic properties of water and aqueous mixtures. We have adopted a thermodynamic approach in which we introduce a phenomenological expression for the free energy that incorporates the essential physical ingredients of aqueous interfaces. From the free energy other thermodynamic functions of the mixture can be calculated. As a result, we have put the theory of aqueous mixtures in the same category as existing mean-field type theories (density functional approximation) of simple binary fluids, but with two improvements: 1) our approach implies a step beyond Flory-Huggins treatment, where random mixing is presumed 2) the short-range anisotropy of the solvent molecules is taken into account.

The four adjustable parameters of the model presented for bulk mixture are determined from independent physico-chemical measurements.

Using the above parameters, the surface properties of aqueous mixtures are described satisfactorily without invoking any additional parameter. The model introduced emphasizes both hydrogen bonding and the orientational degrees of freedom of interfacial molecules and offers the possibility of predicting the surface properties from the bulk ones. The calculation of the surface tension represents a stringent test of the proposed model because of the important role of the geometry of the hydrogen bonding complex. The structure of the complex is determined by the strong directional character of the hydrogen bonds in contrast to some other possible types of intermolecular interactions. It is already clear, that the inclusion of other interactions would not yield a tractable analytic solution.

The model reviewed in this lecture describes the interactions in water and aqueous mixtures in only a schematic way. Nevertheless, it yields several useful conclusions concerning role of hydrogen bonding in the surface phenomena of these systems. It is not easy to make a comparison of the present approach with computer simulations, since the contribution of hydrogen bond effects has not been explicitly determined by the latter method. It is, however, obvious that simulations of surface thermodynamics are presently far less accurate than those for the bulk phase [26-28]. The reason for this lies in the use of effective potentials in which the pair interactions are adjusted [29] to fit the bulk properties without incorporating polarization [30] and other many-body effects. It should be emphasized that no such problem has been encountered in the present study in which surface thermodynamics is deduced from the behaviour of the bulk. This may indicate that many-body effects are less important in hydrogen bond interaction if compared to other intermolecular forces.

5. Acknowledgements

I would like to thank John Dore and Jose Teixeira, who organized the ASI School on "Hydrogen-bonded liquids" held at the Centre de Physique Theorique, Cargese, France (1989), for giving me the opportunity to present this work. The support of the Research Council of Slovenia is also acknowledged.

6. References

1. Luzar, A., Svetina, S., and Zeks, B. (1983) 'The contribution of hydrogen bonds to the surface tension of water', Chem. Phys. Letters 96, 485-490.
2. Luzar, A., Svetina, S., and Zeks, B. (1985) 'Consideration of the spontaneous polarization of water at the solid/liquid interface', J. Chem. Phys. 82, 5146-5154.
3. Luzar, A. (1987) 'A simple approach to hydrogen bond effects in Me_2SO-water mixtures', Period. biol. 89, 349-351.
4. Luzar, A. (1987) 'Thermodynamics of DMSO hydration', in H. Kleeberg (ed.), Interactions of Water in Ionic and Nonionic Hydrates,

Springer-Verlag, Berlin, Heidelberg, pp. 125-129.

5. Luzar, A. (1989) 'The contribution of hydrogen bonds to bulk and surface thermodynamic properties of dimethylsulfoxide-water mixtures', J. Chem. Phys. 91, in press.
6. For an excellent review on this subject see: Evans, E. 'Microscopic theories of simple fluids and their interfaces', in J. Charvolin and J. F. Joanny (eds.), NATO ASI Series volume on Liquids at Interfaces, Les Houches, France, 1988.
7. Davis, C. M. and Jarzynski, J. (1972) 'Mixture models of water', in A. Horne (ed.), Water and Aqueous Solutions, Wiley, New York, pp. 377-423.
8. Kell, G. S. (1972) 'Continuum theories of liquid water', in Ref. 7, pp. 331-376.
9. Rice, S. A. and Screats, M. (1981) 'A random network model for water', J. Phys. Chem. 85, 1108-1119.
10. Angell, C. A. (1971) 'Two-state thermodynamics and transport properties for water as zeroth-order results of a "Bond Lattice Model"', J. Phys. Chem. 75, 3698-3705.
11. Luck, W. A. P. (1967) 'Spectroscopic studies concerning the structure and thermodynamic behaviour of H_2O, CH_3OH and C_2H_5OH', Discuss. Faraday Soc. 43, 115-132.
12. Walrafen, G. E. (1971) 'Raman and infrared spectral investigations of water structure in water', in F. Frank (ed.), Water, A Comprehensive Treatise, Plenum, New York, vol. 1, pp. 151-214.
13. Walrafen, G. E., Hokmabedi, M. S., and Yang, W. H. (1986) 'Raman isosbestic points from liquid water', J. Chem. Phys. 85, 6964-6969.
14. Stanley, H. E., Blumberg, R. L., Geiger, A., Mausbach, P., and Teixeira, J. (1984) 'The "locally-structured transient gel" model of water structure' J. de Physique C7, 3-12.
15. Lee, C. Y., McCammon, J. A., and Rossky, P. J. (1984) 'The structure of liquid water at an extended hydrophobic surface', J. Chem. Phys. 80, 4448-4455.
16. Marcelja, S., Mitchell, D. J., and Ninham, B. W. (1977) 'Role of solvent structure in solution theory', J. Chem. Soc.Faraday Trans. II 73, 630-648.
17. Adamson, A. W. (1982) Physical Chemistry of Surfaces, Wiley, New York, 4th edition.
18. Martin, D., Weise, A., and Niclas, H. J. (1967) 'Dimethylsulfoxide', Angew. Chem. 6, 318-334.
19. For a review on this subject see: (1975) 'Biological Actions of dimethylsulfoxide', Ann. N. Y. Acad. Sci. 243.
20. Cowie, J. M. G. and Toporowski, P. M. (1964) 'Association in the binary liquid system dimethylsulfoxide-water', Can. J. Chem. 39, 2240-2243.
21. Clever, H. L. and Pigott, S. P. (1971) 'Enthalpies of mixing of dimethlysulfoxide with water and with several ketones at 298.15 K', J. Chem. Thermodynamics 3, 221-225.
22. Kenttamaa, J. and Lindberg, J. J. (1960) 'The thermodynamic excess functions of the system Dimethylsulphoxide-water', Suomen Kemistilehti B33, 89-100.
23. Horrocks, W. D.,JR and Cotton, F. A. (1961) Infrared and Raman

spectra and normal co-ordinate analysis of dimethylsulfoxide and dimethyl sulfoxide-d_6', Spectrochim. Acta 17, 134-147.

24. de Visser, C., Heuvelsland, W. J. M., Dunn, L. A., and Sumsen, G. (1978) 'Some properties of binary aquoeus liquid mixtures', Chem. Soc. Faraday Trans. I 74, 1159-1169.

25. Tommila, E. and Pajunen, A. (1969) 'The dielectric constant and surface tensions of dimethylsulfoxide-water mixtures', Suomen Kemistilehti B41, 172-176.

26. Borstnik, B., Janezic, D., and Azman, A. (1979) 'Surface tension of water. The Fowler model', Z. Naturforsch. 34, 1239-1242.

27. Lee, C. Y. and Scott, H. L. (1980) 'The surface tension of water: a Monte Carlo calculation using an umbrella sampling algorithm', J. Chem. Phys. 73, 4591-4596.

28. Wilson, M. A., Pohorille, A., and Pratt, L. R. (1988) 'Surface potential of the water liquid-vapour interface', J. Chem. Phys. 88, 3281-3285.

29. Heizinger, K., Palinkas, G. (1985) 'Computer simulation of ion-solvent systems', in R. R. Dogonadze, E. Kalman, A. Kornyshev, J. Ulstrup (eds.), The Chemical Physics of Solvation, Part A, Theory of Solvation, Elsevier, Amsterdam, pp. 313-353.

30. Finney, J. L. (1982) 'Towards a molecular picture of liquid water', in F. Franks (ed.), Biophysics of Water, Wiley, New York, pp. 73-95.

A COLLECTIVE APPROACH TO THE DYNAMICS OF WATER

Emilio Del Giudice
INFN Sezione di Milano

Giuliano Preparata
Dipartimento di Fisica - Universita' di Milano
INFN Sezione di Milano

ABSTRACT.
The usually neglected interaction between the electric dipole of the water molecule and the quantized electromagnetic (em) radiation field is shown to give rise to a collective (superradiant) dynamics, where two phases coexist: one in which molecules rotate coherently in phase, the other, gas-like, produced by thermal fluctuations. Such dynamics occurs in microdomains (the coherence domains) whose size is of the order of hundreds microns. Hydrogen bonding appears to be a consequence of the coherent rotation of the molecules.

The generally accepted approach to water in the liquid state is formulated in terms of short range interactions, such as the hydrogen bond, which appears to be highly directional and ineffective beyond a few A's. However, as clearly discussed, for instance, in Ref. 1 the electron density of the isolated molecule is almost isotropic, leaving no space for "hooks" or "rabbit's ears": the well known electron lone-pairs picture. Quoting Buckingham[*]: << it appears that the hydrogen bond can be understood in terms of an electrostatic attraction, implying that distortion of the electronic structure is not important >>. However, it must be pointed out that the idea of an electrostatic attraction among water molecules, to account for cohesion, must face the fatal difficulty of the disorder brought about by thermal agitation and the uncorrelated rotation of the water molecules. This idea, which is the only one to survive, so far, serious theoretical analysis, could be salvaged if one succeeded in the task of, so to say, keeping the water molecules rotating in an orderly fashion, or in phase.

In this paper we propose an approach to this problem, where such correlation among water molecules is generated by a "superradiance" phenomenon, which may emerge[2] when a large number of elementary systems interact through their own e.m. radiative field. Let us emphasize that the argument that leads to the usual neglect of this interaction for its trifling contribution to the two-body force between

(*) See the discussion on page 13 in Ref. 1.

J.C. Dore and J. Teixeira (eds.), Hydrogen-Bonded Liquids, 211–220.

the components, is seen to be untenable when a large number of systems at high density is at play[3]. Indeed, we shall show that the "superradiant" interaction dominates the dynamics of water molecules in the liquid state.

Our model for water is then as follows (notice that the "natural" units $\hbar=c=1$ are used throughout):

a) the water molecule is a rigid rotator of average moment of inertia

$$I \simeq 2m_H d^2 = 6.\ 10^{-3}\ \text{cm} \quad , \tag{1}$$

b) the rotator is endowed with a permanent electric dipole $\vec{D}=D\vec{u}$, with

$$D=2e\delta=1.85\ \text{Debye}\ (\delta=.2\mathring{A}) \ ,$$

or in natural units

$$D=1.2\ 10^{-9}\ \text{cm.} \quad , \tag{2}$$

c) the density of molecules is

$$\frac{N}{V}=\left(\frac{1}{d_o}\right)^3=3.7\ 10^{22}\ \text{cm}^{-3} \ , \ (d_o \simeq 3\mathring{A}); \tag{3}$$

d) the single molecule Hamiltonian is

$$H^{sm} = \frac{1}{2I} L^2_{\vec{u}} + \vec{A}.\dot{\vec{D}} + \frac{\vec{P}^2}{2M} \quad , \tag{4}$$

where $\vec{L}$ is the angular momentum operator, $\vec{A}$ the e.m. potential (in the gauge $A_o=\nabla.\vec{A}=0$) and $\vec{P}^2/2M$ is the center of mass kinetic energy;

e) the complete Hamiltonian is obtained by summing up all the single molecule Hamiltonians and adding to them the Hamiltonian for the e.m. radiation field

$$H_{em}=\frac{1}{2}\int_V d^3x\ (\vec{E}^2+\vec{B}^2) \ .$$

We quantize the radiation field by writing, as usual,

$$\vec{A}(\vec{x},t)= \sum_{\vec{k}r} \frac{1}{(2\omega_{\vec{k}}V)^{1/2}} [a_{\vec{k}r}(t)\ \vec{\varepsilon}_{\vec{k}r}\ e^{-i(\omega_{\vec{k}}t-\vec{k}\vec{x})} \tag{5}$$

$$+ a^+_{\vec{k}r}(t)\ \vec{\varepsilon}^{\,*}_{\vec{k}r}\ e^{i(\omega_{\vec{k}}t-\vec{k}\vec{x})} \quad] \ ,$$

where $\varepsilon_{\vec{k}r}$ is the polarization vector corresponding to the mode with momentum $\vec{k}$ and polarization r, and the creation and annihilation operators $a^+_{\vec{k}r}$ and $a_{\vec{k}r}$ obey the equal time commutation relations

$$[a_{\vec{k}r}\ (t),\ a^+_{\vec{k}'r'}\ (t)]=\delta_{\vec{k}\vec{k}'}\ \delta_{rr'} \ . \tag{6}$$

We associate to the water molecules a "wave-field" ψ:

$$\psi(\vec{x},\vec{u};t)= \sum_{\vec{p},lm} a_{\vec{p}lm}(t)\ \phi_{\vec{p}lm}(\vec{x},\vec{u}) \quad , \tag{7}$$

where

$$H^{sm}\ \phi_{\vec{p},lm}(\vec{x},\vec{u})= \left(\frac{\vec{p}^2}{2M} + \frac{l(l+1)}{2I}\right)\phi_{\vec{p},lm}(\vec{x},\vec{u}) \ , \tag{8}$$

and the operators $a_{\vec{p},lm}$ ($a^+_{\vec{p},lm}$) are annihilation (creation) operators, with

$$[a_{\vec{p},lm}\ a^+_{\vec{p}',l'm'}]=\delta_{\vec{p}\vec{p}'}\delta_{ll'}\delta_{mm'} \ , \tag{9}$$

which imply

$$[\psi(\vec{x},\vec{u},t),\psi^+(\vec{x}',\vec{u}',t)]=\delta^2(\vec{u}-\vec{u}')\delta^3(\vec{x}-\vec{x}'). \tag{10}$$

In terms of the above fields, the total Hamiltonian H can be written as the sum of a matter Hamiltonian H_m, a field Hamiltonian H_f and a possible short range term H_{sr}:

$$H=H_m+H_f+H_{sr} \quad , \tag{11}$$

where

$$H_m=\int d^3x\ d\Omega_{\vec{u}}\ \psi^+\left[-\ \frac{\nabla^2_x}{2M} + \frac{\vec{L}^2_{\vec{u}}}{2I}\right]\psi\ + \tag{11'}$$

$$+ \sum_{r\vec{k}} \left\{\frac{a_{\vec{k}r}}{(2\omega_{\vec{k}}V)^{1/2}}\ e^{-i\omega_{\vec{k}}t}\int d^3x\ d\Omega_{\vec{u}}\ e^{i\vec{k}\vec{x}}\psi^+\ \vec{\varepsilon}_{\vec{k}r}\cdot\dot{\vec{D}}\psi + h.c\right\}$$

and

$$H_f= \sum_{r\vec{k}} \left[\omega_{\vec{k}}\ a^+_{\vec{k}r}\ a_{\vec{k}r}+i/2\ (a^+_{\vec{k}r}\ \dot{a}_{\vec{k}r} - \dot{a}^+_{\vec{k}r}\ a_{\vec{k}r}) + \frac{1}{2\omega_{\vec{k}}}\ \dot{a}^+_{\vec{k}r}\ \dot{a}_{\vec{k}r}\right] \tag{11''}$$

The associated Lagrangian L is then:

$$L(t)=\int d^3x d\Omega_{\vec{u}}\ \psi^+\ \left[i\ \partial/\partial t + \frac{\vec{\nabla}^2_x}{2M} - \frac{\vec{L}^2_{\vec{u}}}{2I}\ \right]\psi \tag{12}$$

$$+ \sum_{\vec{k}r} \left[ia^+_{\vec{k}r}\ \dot{a}_{\vec{k}r} + \frac{1}{2\omega_{\vec{k}}}\ \dot{a}^+_{\vec{k}r}\ \dot{a}_{\vec{k}r}\right]$$

$$+ \sum_{\vec{k}r} \left[\frac{a_{\vec{k}r}}{(2\omega_{\vec{k}}V)^{1/2}}\ e^{-i\omega_{\vec{k}}t}\ \int d^3x\ d\Omega_{\vec{u}}\ e^{i\vec{k}\vec{x}}\ \psi^+\ \vec{\varepsilon}_{\vec{k}r}\cdot\dot{\vec{D}}\psi + h.c.\right] - H_{sr} \ ,$$

and the action:

$$I_{fi}=\int_{t_i}^{t_f}dt\ L(t)=I_{fi}[\psi,\ \psi^+,\ a_{\vec{k}r},\ a^+_{\vec{k}r};\ D]\ . \tag{13}$$

The transition amplitude from the initial state i to the final state f is given by the Feynman path integral (5):

$$\langle f|i\rangle = Z_{fi} = \int_i^f [d\psi\ d\psi^*\ dada^*]\ \ \exp \frac{i}{\hbar} I_{fi} \tag{14}$$

and the Noether's theorem guarantees that:

$$\int d^3x d\Omega_{\vec{u}}\ \psi^+\psi = N \tag{15}$$

is a constant of motion. It is interesting to realize that in Eq. (12) the first two terms in the r.h.s. are bilinear terms in the fields, whereas the one which contains the coupling constant D is trilinear. As a consequence if we introduce the new fields ϕ and α:

$$\begin{aligned} \psi&=N^{1/2}\phi\ , \\ a&=N^{1/2}\alpha\ , \end{aligned} \tag{16}$$

we get

$$Z_{fi}\ \ \alpha\int_i^f [d\phi\ d\phi^*\ d\alpha\ d\alpha^*]\ \ \exp\ i\ \frac{N}{\hbar}\ I_N\ , \tag{17}$$

where

$$I_N=I[\phi,\ \phi^*,\ \alpha,\ \alpha^*;\ N^{1/2}D] \tag{18}$$

Eqs. (17) and (18) have an interesting physical meaning since

a) the large value of N in eq. (17) mimicks a vanishingly small $\hbar$, so that a large N implies "classically looking" solutions, obeying

$$\delta\ I_N=0 \tag{19}$$

as in classical physics;

b) the rescaled action I_N looks just like the original action I, but for the very important fact that the <u>small</u> coupling constant D becomes the <u>big</u> coupling constant $N^{1/2}D$.

In order to extract the "classically looking" contribution, we split the fields ϕ and α as:

$$\phi=\xi+\eta\quad ,\quad \alpha=a+\beta \tag{20}$$

where ξ and a are O(1) and η and β are $O(N^{-1/2})$, so that:

$$\delta I(\xi,\ \xi^*,\ a,\ a^*;\ N^{1/2}D)=0 \tag{19'}$$

and the quantum fluctuations of the system are governed by the Lagrangian L^{fl}:

$$L^{fl}=\int d^3x\ d\Omega_{\vec{u}}\ \eta^*\left[i\frac{\partial}{\partial t}+\frac{\vec{\nabla}^2}{2M}-\frac{L^2_{\vec{u}}}{2I}\right]\eta+$$

$$+\left(\frac{N}{V}\right)^{1/2}\left\{\sum_{\vec{k}r} a_{\vec{k}r}\frac{e^{-i\omega_{\vec{k}}t}}{(2\omega_k)^{1/2}}\int d^3x d\Omega_{\vec{u}}\ e^{i\vec{k}.\vec{x}}\ \eta^*(\vec{\varepsilon}_{\vec{k},r}.\dot{\vec{D}})\eta+h.c.\right\}+$$

$$+\sum_{\vec{k},r}(i\dot{\beta}^*_{\vec{k},r}\ \dot{\beta}_{\vec{k},r}+\frac{1}{2\omega_k}\ \dot{\beta}^*_{\vec{k},r}\ \dot{\beta}_{\vec{k},r})-H_{sr} \tag{21}$$

In order to derive the evolution equations for the fields, we notice that if a single mode $\omega_{\vec{k}}$ dominates the interaction between radiation and matter, then the ground state wave function ξ will be (almost) constant within a sphere of radius R, where

$$\int d\Omega_{\vec{u}}\ e^{i\vec{k}.\vec{x}}\Big|_{|\vec{x}|<R}\sim j_o(\omega_k|\vec{x}|)\sim \text{constant}$$

which implies $R\sim\pi/\omega_k=\lambda/2$.
In this domain all the dipoles are coupled to the e.m. field in the same way. By limiting ourselves to the volume V of the above domain, we get from eq. (19′) the evolution equations for the fields ξ and a (we obviously neglect the $\vec{x}$ dependence):

$$i\frac{\partial\xi}{\partial t}=\frac{L^2_{\vec{u}}}{2I}\ \xi-\left(\frac{N}{V}\right)^{1/2}D\sum_{\vec{k},r}\frac{\vec{\varepsilon}_{\vec{k}r}.\vec{u}}{(2\omega_k)^{1/2}}\ .$$

$$.[(\dot{a}_{\vec{k}r}-i\omega_{\vec{k}}a_{\vec{k}r})\ e^{-i\omega_k t}+(\dot{a}^*_{\vec{k}r}+i\omega_{\vec{k}}a^*_{\vec{k}r})e^{i\omega_{\vec{k}}t}]\xi \tag{22}$$

and

$$i\dot{a}_{\vec{k}r}-\frac{1}{2\omega_{\vec{k}}}\ \ddot{a}_{\vec{k},r}=\left(\frac{N}{V}\right)^{1/2}\frac{D}{(2\omega_{\vec{k}})^{1/2}}\ \vec{\varepsilon}_{\vec{k},r}\ e^{i\omega_k t}\ \frac{d}{dt}\int d\Omega_{\vec{u}}\xi^*\vec{u}\xi \tag{23}$$

As noted in Ref. 4, Eqs. (22) and (23) are different from Eqs. (14) and (15) of Ref. 3 since in the present treatment we have dropped the "slowly varying envelope approximation" used there, which in the case of the real liquid is unphysical. In order to extract the relevant information from Eqs. (22) and (23) we expand

$$\xi(\vec{u},\ t)=\Sigma_{lm}c_{lm}(t)Y^m_l(\vec{u}) \tag{24}$$

in spherical harmonics and keep only the first two levels, l=0 and l=1, whose energy difference is $\omega_o=1/I \sim 170\text{cm}^{-1}$ (remember that in our "natural" units, $\hbar=c=1$, $1\text{cm}^{-1}=1/2\pi$ wave numbers), so that the size of the domain considered in Eqs. (22) , (23) is ~ 180 μm. Let us keep now only the resonating e.m. mode $\omega_{\vec{k}}=\omega_o$ and define an e.m. field amplitude averaged over the polarizations:

$$a_m=\int d\Omega_{\vec{k}}\ a_{\vec{k}r}\ \varepsilon_{\vec{k}r,m}. \tag{25}$$

We set

$$b_o(t)=c_{oo}(t)$$

$$b_m(t)=\sqrt{3}\ c_{1m}(t)\ e^{-i\omega_o t}\quad , \tag{26}$$

introduce the "dimensionless time" $\tau=\omega_o t$ and realize that rotational invariance implies that dependence upon m of the fields a_m and b_m must be dropped. After defining the constant

$$g=\left(\frac{N}{2\omega_o V}\right)^{1/2}\frac{D}{3}\sqrt{8\pi} \tag{27}$$

and
$$a=\sqrt{\frac{g}{8\pi}}\ a_m$$

we finally get the equations:

$$\dot{b}_o=-g\ (a^*-i\dot{a}^*)\ b_1 \tag{28}$$

$$\dot{b}_1=g\ (a+i\dot{a})\ b_o \tag{29}$$

$$\dot{a}+\frac{i}{2}\ \ddot{a}=-g\ (1-i\frac{d}{d\tau})\ (b_o^*\ b_1) \tag{30}$$

with the constants of motion:

$$|b_o|^2+|b_1|^2=1 \tag{31}$$

$$Q=\frac{1}{2}\ (|b_1|^2-|b_o|^2)\ +\ |a|^2\ +\ \frac{1}{2}(a^*\dot{a}-\dot{a}^*a)\ -ig(ab_1^*b_o-a^*b_1b_o^*) \tag{32}$$

Since we assume as initial condition the distribution at thermal equilibrium, setting $|b_1(0)|^2-|b_o(0)|^2=\Delta$ and $a(0)=0$, we have $Q=\Delta/2$ where Δ is the difference of the populations of the levels l=0 and l=1 at t=0. We now show that this initial condition may be unstable. For $\tau\simeq0$, after differentiation and insertion of the constants of motion Eq.(30) becomes

$$i\frac{\dddot{a}}{2}+\ddot{a}\ (1+g^2\Delta)\ -\ 2i\dot{a}g^2\Delta\ -\ ag^2\Delta=0. \tag{33}$$

This equation is controlled by the parameter $g^2\Delta$, which depends through

Eq. (27) on the density of molecules. It is easy to see that for a large range of $g^2\Delta$ values Eq. (33) admits exponentially exploding solution $e^{\mu\tau}$ ($\mu>0$) (runaways) that push the system far from the gas-like thermal situation at a=0. For the system (28) ÷ (30) it is possible to derive (4) the stationary solutions:

$$b_1=\cos\frac{\theta}{2}\ e^{i\alpha_1} \tag{34}$$

$$b_o=\mathrm{sen}\frac{\theta}{2}\ e^{i\alpha_o} \tag{35}$$

$$a=2gAe^{i\beta} \tag{36}$$

where:

$$\theta=\frac{\pi}{2}+\frac{1}{(4g^4\Delta)^{1/3}} \tag{37}$$

$$A=\left(\frac{4\Delta}{g^2}\right)^{1/3} \tag{38}$$

$$\alpha_1-\alpha_o-\beta=\frac{\pi}{2} \tag{39}$$

$$1-\dot{\beta}=2\ \frac{\sin\ \theta}{A}\ \sim\ -\frac{2}{A} \tag{40}$$

in the limit g>>1. As a matter of fact, Eq. (27) gives g~17 for liquid water. By substituting the above solutions in the total Hamiltonian, for the energy of the ground state, we get

$$E=N\omega_o\left(\frac{3}{2}+\frac{9}{8}\ g^2\right) \tag{41}$$

which, being positive, shows that the system could not reach this situation without an external pump. However, this is not the whole story.

Let's look at the contribution of the fluctuations, which are governed by the Lagrangian of Eq. (21). The total number N of molecules comprises the number N_s of molecules participating to the "superradiant" dynamics [Eqs. (34)÷(36)] and the number $N_f=N-N_s$ of molecules fluctuating according to the fields η_m (m=0,1,2,3) defined in Eq. (20). Eq. (21) becomes

$$L^{fl}=\int d^3x\left\{\eta_m^*(\vec{x},t)\left(i\frac{\partial}{\partial\tau}-\frac{L^2}{2I}\right)\eta_m(\vec{x},t)-\eta_m^*\ M_{mn}\ \eta_n\right\} \tag{42}$$

By diagonalizing the matrix M as in Ref. 4, we get negative eigenvalues corresponding to one out of four orthogonal new "fluctuation fields". An instability around η~0 then arises which compels us to define a new fluctuating ground state, by adding to the Lagrangian a positive term proportional to $|\eta|^4$, which corresponds to the very short range repulsion among molecules. This is a standard procedure in quantum field theory when particles are introduced with an imaginary mass (5). After some manipulations described in Ref. 4, we get the energy for a ground state where a fraction $p_s=N_s/N$ obeys to the "superradiant dynamics" and a fraction $1-p_s$ fluctuates:

$$E_{gs}=N\omega_o\left(\frac{3}{2}\,p_s + \frac{9}{8}\,g^2p_s^2 - g^2p_s\,(1-p_s)\right)\,, \tag{43}$$

which has a negative (!) minimun for $p_s=4/17$, with a gap against the fluctuations,

$$\Gamma=g^2\omega_o 4/17 \tag{44}$$

which for liquid water turns out to be $\Gamma\sim 7kT$! The ground state energy $E_{GS}=-N\omega_o g^2 2/17$ being negative, implies that this state can be reached spontaneously by the system in the gaseous phase, when an equilibrium between the superradiant molecules and fluctuations is established. It is interesting that the relevant thermodynamical quantities depend on $g^2\omega_o$, which is independent of $\omega_o=1/I$, so that we have no isotope effect!

From our analysis of liquid water a two-fluid picture emerges, which reminds us of the models used by chemists before 1933, when the paper of Bernal and Fowler (6) convinced everybody that, since all molecules were alike, there was no reason for a two-phase water. Evidence for a two-phase water was given in more recent years by Raman spectroscopy (7) and its relevance has been discussed in Ref. 8. We point out here that since the superradiant phase is completely ordered it must be considered at T=0, while all the thermal agitation is in the fluctuating phase, whose presence accounts for the high thermal capacity of water. The two phases are spatially superimposed within the coherence domain defined above, so that density gradients are not expected. The whole picture shares some features with the two-phase model of superfluid helium.

The fact that water molecules in the superradiant phase rotate coherently allows a non zero electrostatic attraction among them. Consider two molecules rotating in phase at an angular speed Ω around an axis passing through the oxygens. If $\vec{x}$, $\vec{y}$, $\vec{z}$ are the orthogonal unit vectors, we describe the two electric dipoles rotating around the z -axis as

$$\vec{d}_1=D[\sin\beta\cos(\Omega t+\phi_1)\vec{x}+\sin\beta\sin(\Omega t+\phi_1)\vec{y}+\cos\beta\,\vec{z}] \tag{45}$$

$$\vec{d}_2=D[\sin\theta\cos(\Omega t+\phi_2)\vec{x}+\sin\theta\sin(\Omega t+\phi_2)\vec{y}+\cos\theta\,\vec{z}] \tag{46}$$

It is clear that in order to minimize the electrostatic energy the first requirement is to have an H of one molecule, pointing toward the O of the other molecule while keeping its rotation. This means that the z-axis must be aligned with an OH direction of the first molecule, so that β=52°. The electrostatic energy E of the pair of dipoles is:

$$E=\frac{\Sigma_{i,j}\,d_{1i}\,d_{2j}\,(\delta_{ij}-3\delta_{i3}\,\delta_{j3})}{R^3}= \tag{47}$$

$$= \frac{D^2}{R^3} [\sin\beta \sin\theta \cos(\phi_1-\phi_2) - 2\cos\beta \cos\theta]$$

The minimum occurs when $\phi_1-\phi_2=\pi$ and tg $\theta=1/2$tg β, namely $\theta\sim32°$, a configuration compatible with that shown on page 254 in Ref. 1, which has been derived from the spectroscopical data of the dimer. Moreover the dipoles attract themselves until the distance R reaches the value R~3Å where the repulsion comes into play. At that point the H-bond is established and, as predicted by Buckingham, the electrostatic attraction has done the job. The above mechanism works among the molecules of the superradiant phase and is absent of course in the fluctuating phase. We have then a number of H-bonds in the bulk water smaller than that expected on geometrical grounds. This result allows to reinterpret in terms of the two-fluid picture the results obtained in terms of "intact" and "broken" H-bonds in Refs. 9 and 10.

The large degree of coherence existing in the superradiant phase makes it possible to store a large amount of electrostatic energy in liquid water. It can be shown that one gets as a consequence a strong polarizability of the system. Actually, in the absence of externally applied fields, the expectation value of the electric polarization vector in the ground state is:

$$P_k(t)=\frac{N}{V}D\langle u_k\rangle_{GS}=\frac{N}{V}D\left[p_s\frac{1}{3}(b_o b_k^* + b_o^* b_k) + (1-p_s)\frac{1}{3}\langle\eta_-|\sigma_k|\eta_-\rangle\right]=$$

$$=\frac{ND}{3V}(2p_s-1)\cos\left[2\left(\frac{g^2}{4\Delta}\right)^{1/2}\omega_o t-\phi_k\right] \quad (48)$$

where σ_k is a Pauli matrix. Eq.(48) gives a zero time average polarization. If we apply, for instance in the direction z, a weak static electric field E, as shown in Ref. 3, the states $|0,0\rangle$ and $|1,0\rangle$ get mixed.

$$|0'\rangle= \sin\alpha|0,0\rangle + \cos\alpha|1,0\rangle$$

$$|1'\rangle= -\cos\alpha|0,0\rangle + \sin\alpha|1,0\rangle \quad (49)$$

where

$$\text{tg}\alpha=\frac{\omega_o}{2V}\left[1-\left(1+\frac{4V^2}{\omega_o^2}\right)^{1/2}\right]\sim\frac{V}{\omega_o} \;;\; V = -Du_zE \quad (50)$$

The polarization in the z-direction becomes:

$$P_z(t)=\frac{ND}{3V}\left[\frac{\sin 2\alpha}{\sqrt{3}} + (2p_s-1)\cos2\alpha \cos\left(2\left(\frac{g^2}{4\Delta}\right)^{1/3}\omega_o t-f_k\right]\right. \quad (51)$$

The mixing induced by the external perturbation produces then a non vanishing permanent polarization along the z-axis. If we take the static part of the r.h.s. of Eq.(51) in the limit $E\to0$ and use Eq.(50), we get a very large static dielectric constant ε (11) $[P=(\varepsilon-1)E]$:

$$\varepsilon=1+\frac{2}{3\sqrt{3}}\frac{N}{V}\frac{D^2}{\omega_o} \tag{52}$$

which gives a value not far from 100. A more refined treatment would require a more realistic description of the single molecule states. The appearance of a sizeable permanent polarization in the aqueous surroundings of giant polar biomolecules has been suggested (12) to be the origin of a coherent dynamics in living matter.

REFERENCES

1. "Water and Aqueous Solutions" (eds G.W. Neilson and J.E. Enderby). Colston Papers n. 37-Hilger, Bristol and Boston (1985). See in particular the contribution of A.D. Buckingham and the discussion at pages 11, 12 and 13;

2. R.H. Dicke, Phys. Rev. 93, 99 (1954);

3. E. Del Giudice, G. Preparata and G. Vitiello, Phys. Rev. Lett. 61, 1085 (1988);

4. G. Preparata, Quantum Field Theory of Superradiance, Preprint MITH 89/5 (1988). Milano;

5. See for instance L.D. Fadeev and A.A. Slavnov, Gauge fields, introduction to Quantum Theory. The Benjamin Cummins Co. Reading, Mass. 1980;

6. J.D. Bernal and R.H. Fowler, J. Chem. Phys. 1, 515 (1933);

7. G.E. Walrafen, J. Chem. Phys. 48, 244 (1968);

8. H.S. Frank, Science 169, 635 (1970);

9. H.E. Stanley and J. Teixeira, J. Chem. Phys. 73, 3404 (1980);

10. D. Bertolini, M. Cassettari, M. Ferrario, P. Grigolini, G. Salvetti and A. Tani - J. Chem. Phys. - in press (1989);

11. E. Del Giudice and G. Preparata "Some remarks on the electrostatics of water". Milano preprint (1988);

12. E. Del Giudice, S. Doglia, M. Milani and G. Vitiello, Nucl. Phys. 251B [FS13] 375 (1985); Nucl. Phys. 275B [FS17] 185 (1986).

Editors note: During the discussion of this paper several participants expressed critical reservations about the theoretical methods used. The paper is reproduced here in its submitted form and it is anticipated that the controversy will be resolved by future debate.

PROPERTIES OF H-BONDING IN THE INFRARED SPECTRAL RANGE

S. BRATOS[1], H. RATAJCZAK[2] and P. VIOT[1]
[1]*Laboratoire de Physique Théorique des Liquides**
Université Pierre et Marie Curie
4, place Jussieu 75252 Paris Cedex 05
France
[2]*Institute of Chemistry, University of Wroclaw*
ul. Joliot-Curie 14, 50383 Wroclaw
Poland

ABSTRACT. A review is presented covering the experimental and theoretical aspects of infrared spectroscopy of hydrogen bonded liquids. The discussion is mainly limited to the study of the hydrogen stretching bands perturbed by this interaction. It is shown that their profile is generated by complex relaxation processes in which several oscillators, coupled by anharmonic forces, are simultanously relaxing. These processes are very fast, and the liquid acts, in essence, as a static disordered medium.

1 Introduction

A considerable effort has been made in the last fifty years to understand the properties of the ν_{AH} infrared bands of systems containing hydrogen bonds. The main reason for this endeavour is that the ν_{AH} band represents the most easily recognizable signature of hydrogen bonding. Moreover, and this is the second reason for the scientific action, spectral properties of the ν_{AH} bond are highly spectacular and unusual. Its intensity may exceed that of the free band by an order of magnitude, and its half-width by two! No perturbation, even one much stronger, has such far reaching spectral consequences as hydrogen bonding.

Unfortunately, its study has turned out to be exceptionally difficult. About thirty five years, 1935-1970, were required before a consensus was reached that the anharmonicity of the $AH \cdots B$ system and the disorder in the fluid phases are at

* *Unité de Recherche Associée au CNRS.*

J.C. Dore and J. Teixeira (eds.), Hydrogen-Bonded Liquids, 221–235.

the origin of this complex behaviour. Fifteen additionnal years, after 1975, were needed to produce semi-quantitative theories, but a really quantitative description is still missing. The source of the difficulties is now well localized. The ν_{AH} band profiles are generated by complex relaxation processes in which several oscillators, coupled by anharmonic forces, are simultanously relaxing. As these processes take place in time scales short compared to the molecular dynamics of the liquid, the latter acts, in essence, as a static disordered medium.

The purpose of the present paper is to give a brief review of this field. Both experimental and theoretical aspects will be covered, but the accent will be put on the theory. The relation between spectral anomalies in infrared and the processes governing the acid-base equilibria is emphasized. For text-books and monographs in this domain, see Refs[1-4].

2 Phenomenological description

a) Generalities. The prominent role of the ν_{AH} band in hydrogen bond research was perceived as early as in 1935. It was also rapidly understood that spectral effects strongly depend on the $AH \cdots B$ distance (Fig.1). The weak, medium strong and strong H-bonds will be treated separately in this paper. The subsequent discussion is limited to the $OH \cdots O$ bonds; the characteristics of other $AH \cdots B$ systems are qualitatively similar.
b) Weak H-bonds[6,7]. It is convenient to start with weakly bonded systems for which the enthalpy for H-bond formation is of the order of, or smaller than, 5-6 Kcal/mole and the $OH \cdots O$ distance between 2.8 and 2.9Å; water is a well known example of this kind(Figs 2a and b). (i) The ν_{OH} bands of this class undergo a low frequency shift $\Delta\nu$ of $100 - 300\,cm^{-1}$ with respect to the position they have when H-bonding is absent, i.e. in the $3600\,cm^{-1}$ region. Simultaneously, their integrated intensity A increases with respect to the integrated intensity A_f of the free band by a factor of the order of 5 to 10. (ii) These bands are continous, intrinsically structureless, and broad. Their half-width $\Delta\nu_{\frac{1}{2}}$ is of the order of $100 - 300cm^{-1}$, exceeding the free ν_{OH} bands about $10 - 30$ times. (iii) The half width increases with temperature roughly as $\sqrt{T}$. It should be noted that vibrational bands do not all behave in this way. (iv) On deuteration, the frequency ν_H, the intensity A and the half-width $\Delta\nu_{\frac{1}{2}}$ decrease by a factor close to $\sqrt{2}$. This ratio is characteristic of a harmonic vibrator.
c) Medium strong H-bonds[7,8]. The next important class of systems is that for which the enthalpy of the H-bond formation is between 6 and 8 Kcal/mole, the distance $OH \cdots O$ being between 2.6 and 2.8Å; the formic acid dimer is a representative example (Fig. 2c). (i) The ν_{OH} bands exhibits a low frequency shift of the order of $300 - 1000cm^{-1}$ and $A/A_f \sim 10 - 15$; the size of these effects is

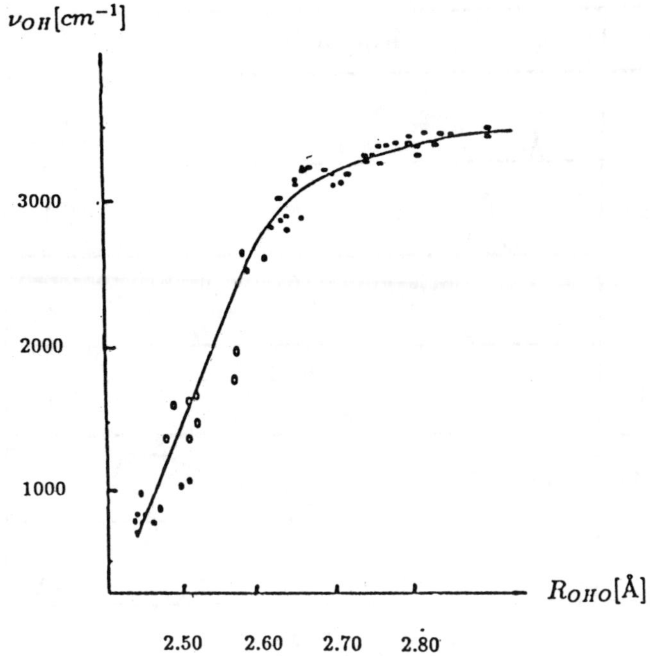

Figure 1: Experimental relation between the ν_{OH} frequency and the H-bond length $R_{OH \cdots O}$[5]. This relation is of basic importance for the theory.

surprising. (ii) These bands are irregular and very broad. The profiles, if they are sufficiently well defined, are asymmetrical. (iii) The ν_{OH} bands of this class exibit an appearently erratic fine structure. After a long lasting controversy, it was assigned to various overtone and combination levels. (iv) The spectral resolution improves at low temperatures, but a complete separation is rarely achieved and, if so, never in the liquid phase. (v) The D-substitution produces low frequency shifts and intensity variations which do not follow usual harmonic rules. One finds $\nu_H/\nu_D \sim 1.3$ and $A_H/A_D \sim 2$. Moreover, these changes are generally accompanied by a profound alteration of the band structure; no obvious correlation exists between spectral details of the ν_{OH} and ν_{OD} bands.

d) Strong H-bonds[9,10]. The systems of this class are characterized by a large variety of highly unusual and spectacular ν_{OH} bands. They were discovered and described in the late fifty's. These H-bonds may be symmetrical or asymmetrical; the two types of bonds exibit very different spectral characteristics. The asymmetric bonds correspond to enthalpies of formation of the order of 9-15 kcal/mol and to $OH \cdots O$ bond lengths between 2.5 and 2.6$\AA$; the complex between the trichloroacetic acid and sulfoxyde is a representative example. Other data are as

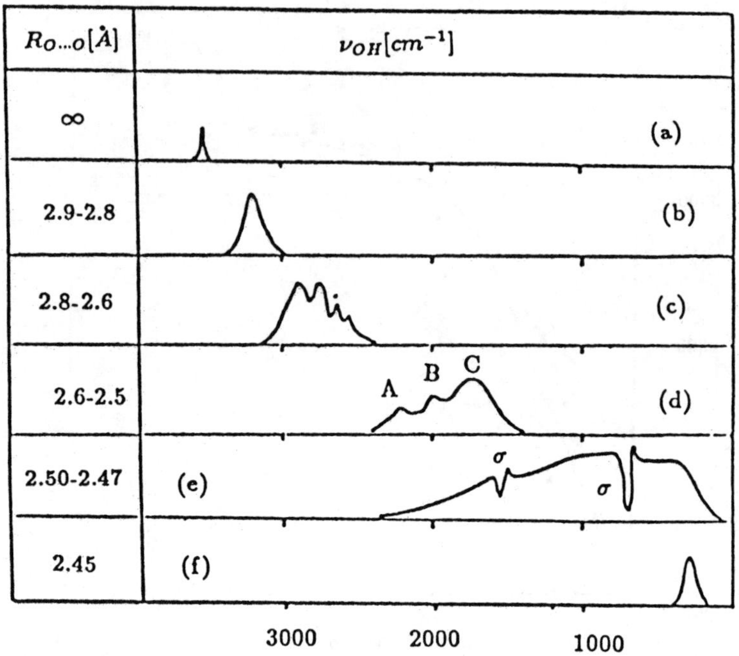

Figure 2: The evolution of the ν_{OH} band as a function of the H-bond strength. The profiles refer to a free molecule(a), and to weak (b), medium strong (c), strong asymmetrical (d), strong nearly symmetrical (e) and to truly symmetrical H-bonds (f).

follows (Fig. 2d). (i) The ν_{OH} band is located in the $1500-2000cm^{-1}$ region and is extremely intense. Although the presence of superposed bands precludes any quantitative study, the ratio A/A_f is believed to be as high as 30 ! (ii) The profiles ν_{OH} are extremely broad and extend over $1000-1500cm^{-1}$; for comparison, it should be recalled that $\Delta\nu_{\frac{1}{2}} \sim 10cm^{-1}$ for a free ν_{OH} bond. (iii) The profiles exhibit a characteristic triplet structure; the sub-bands are generally called the A, B, C bands. This structure survives the D-substitution although the intensity of sub-bands is distributed differently. (iv) The isotope effect is highly anomalous (Fig. 3). One finds $\nu_H/\nu_D \sim 1.0-1.2$ and $A_H/A_D \sim 3-4$; unfortunately, the values cited for A/A_f represent only a rough estimate for reasons mentioned above. It results from this description that the complexity of the spectral behaviour attains its maximum for this class of hydrogen bonds.

The nearly symmetrical and symmetrical H-bonds may readily be found as positive or negative ions in aqueous acid or base solutions together with other H-bonded species. The $[F\cdots H\cdots F]^-$ is the best known example of this sort. Although isolated complexes may be available in the solid, they are rarely found in

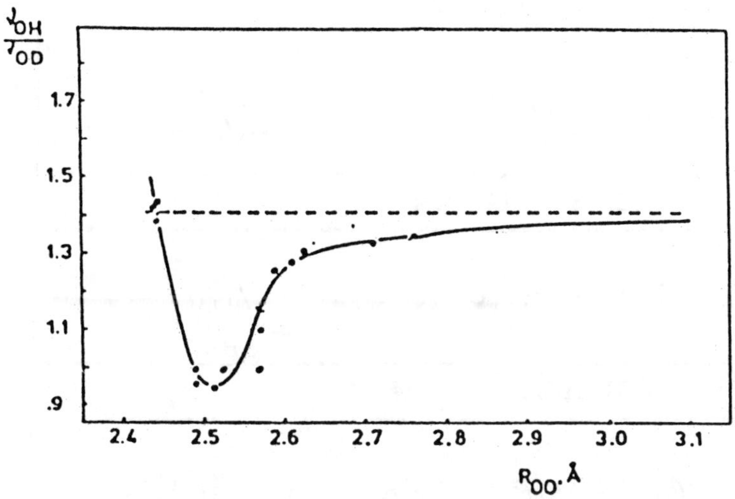

Figure 3: The isotope effect on the ν_{OH} band. The curve illustrates the dependence of the ν_H/ν_D ratio on the $R_{O\cdots O}$ distance

the liquid phase. The enthalpy of the bond formation is of the order of, or larger than, 20 Kcal/mole and the $OH \cdots O$ bond length shorter than 2.5Å The spectral data are as follows (Figs 2e and f). (i) The ν_{OH} bands of this type are located in the $1500 - 600cm^{-1}$ region; one recalls that the free ν_{OH} bands are usually centered at $3600cm^{-1}$. (ii)The profiles are so broad that they substantially cover the medium IR spectra region; half-widths of the order of $2000cm^{-1}$ are quite common. These profiles are separated by sharp transmission windows called, according to the author who first described them, Evans holes. (iii) When the H-bond distance decreases from 2.50 to 2.45Å, the IR spectra change dramatically; $\Delta\nu_{\frac{1}{2}}$ drops from 2000 to $200cm^{-1}$; A/A_f diminishes by a factor or the order of 10, while the isotope effect remains anomalous (Fig. 3). Any acceptable theory must be able to explain these data.

3 Search for band shaping mechanisms

The spectacular behaviour of the ν_{AH} bands was in the center of interest of spectroscopists for several decades. The difficulties were aggravated by a complete absence of statistical theories of IR and Raman band shapes of liquids. The ambition of the scientific community at that time was thus limited to the search for band shaping mechanisms. Even then, 35 years of effort and endless controversies were required before a consensus was reached!

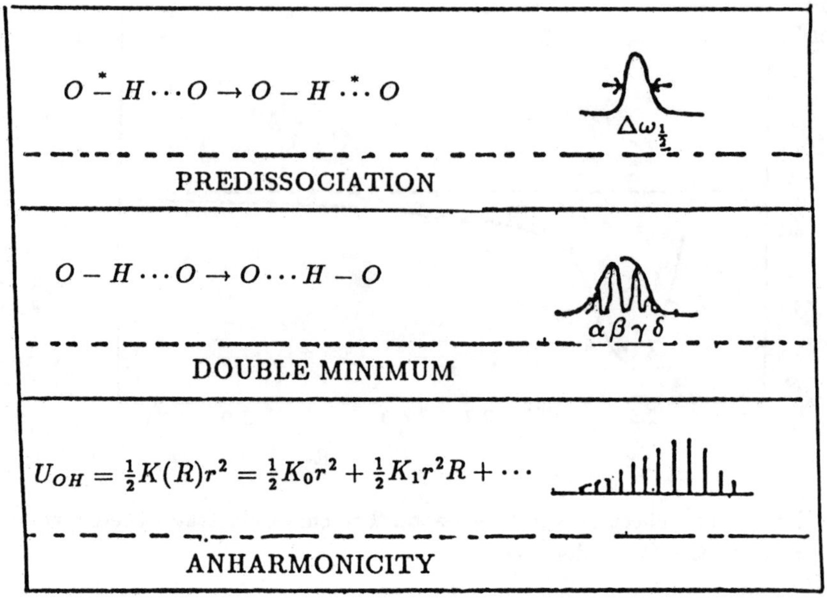

Figure 4: Illustration of the three main mechanisms proposed to explain spectral anomalies of the ν_{OH} band.

Leaving a number of rapidly abandoned proposals aside, the following three mechanisms merit a close examination (Fig. 4). (i) The vibrational excitation can migrate from AH to the $H \cdots B$ bonds. If $h\nu_{AH} > U_{AH \cdots B}$, the energy transfer is followed by a rupture of the H-bond which produces a life-time broadening[11] (Fig. 4a). This so called predissociation mechanism is highly inefficient and was thus abandoned. (ii) The potential energy curve of an $AH \cdots B$ system has two minima which correspond to the configurations $AH \cdots B$ and $A \cdots HB$. The low lying vibrational levels are thus split and the number of allowed spectral transitions increases correspondingly. If the latter are sufficiently blurred by molecular ineractions, a broad ν_{AH} band results (Fig. 4b); see e.g. Ref[12]. The major objection against the double-minimum mechanism is that it does not provide a satisfactory correlation between the wave numbers needed to reproduce the fine structure of the ν_{AH} bond and those observed in the far IR. (iii) The vibrations of the $AH \cdots B$ group are strongly anharmonic. The most important coupling is between the AH and $H \cdots B$ stretching motions. This can be conjectured by simply assuming that the force constant K associated with $r_{AH} = r$ depends on $r_{H \cdots B} = R$. One then finds readily :

$$\begin{aligned}\frac{1}{2}K(R)r^2 &= \frac{1}{2}(K_{rr} + K_{rrR}R + \ldots)r^2 \\ &= \frac{1}{2}(K_{rr}r^2 + K_{rrR}r^2R + \ldots)\end{aligned} \quad (1)$$

There thus results a progression in the $\nu_{AH\cdots B}$ mode and hence a broad band (Fig. 4c). Implicit in the early fluctuation theory [13], this mechanism was only formulated explicitely in 1957[14]. It is now generally accepted and is the basis of all modern band shape theories of liquids and gases.

4 Statistical theory of the ν_{AH} band shapes

a) Generalities. The modern theories of IR band shapes of H-bonded liquids are based on the well known Gordon formula for infrared absorption coefficient $\alpha(\omega)$:

$$\alpha(\omega) = \frac{2\pi}{3c\hbar nV}\omega\left(1 - e^{-\frac{\hbar\omega}{k_BT}}\right)\int_{-\infty}^{+\infty} dte^{-i\omega t}\left\langle\vec{M}(0)\vec{M}(t)\right\rangle \quad (2)$$

where $\vec{M}(t)$ is the dipole moment of the liquid sample[15]. The important limitation of Eqn(2) in the present context is that it neglects the difference between the external and the internal Maxwellian electric field in the liquid sample. It can thus be applied only to solutions, which is a severe restriction. The low frequency shift $\Delta\nu$ and the variation of the ratio A/A_f will not be studied in this chapter. These effects are of an electronic origin and will be discussed in another part of this book.

b) Weak H-bonds. The theory refers to a single H-bonded system $AH \cdots B$ isolated in an inert solvent. This system is considered as a semi-rigid unit, incapable of large amplitude internal motions, and is pictured as two coupled oscillators ν_{AH} and $\nu_{AH\cdots B}$. Other details of the model are as follows. (i) The ν_{AH} vibrations are described by a normal coordinate n obeying the laws of quantum mechanics. (ii) The $\nu_{AH\cdots B}$ vibrations are described by a classical stochastic function $N(t)$. (iii) The Hamiltonian for internal vibrations $H(n,t)$ and the corresponding dipole moment $\vec{M}(n,t)$ are

$$H(n,t) = \frac{1}{2}p^2 + \frac{1}{2}\omega_0^2n^2 + \frac{1}{2}K_{11I}n^2N(t) \quad (3)$$

$$\vec{M}(n,t) - \vec{M}_0 = \vec{M}_1n + \vec{M}_{1I}nN(t) \quad (4)$$

where ω_0 is the non-perturbed ν_{AH} frequency. The model thus considers two band shaping mechanisms described by the terms $\frac{1}{2}K_{11I}n^2N(t)$ and $\vec{M}_{1I}nN(t)$; they express the effects of the mechanical and electrical anharmonicities of the ν_{AH} and

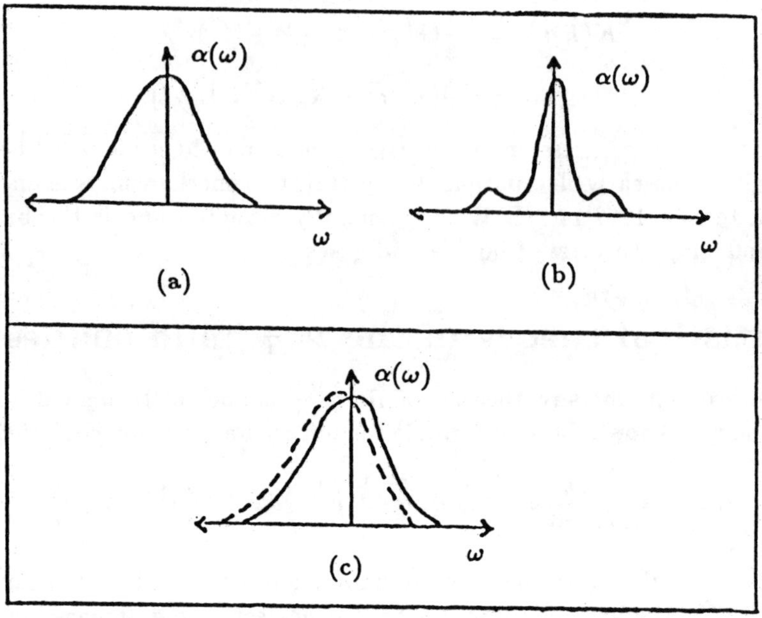

Figure 5: Theoretical shapes of the ν_{AH} bands characteristic of weak H-bonds. The mechanical n, N coupling produces a Gaussian (a) in the high, and the profile (b) in the small friction limit. The electrical n, N coupling induces an overall shift of the ν_{AH} band to the higher or lower frequencies (c).

$\nu_{AH\cdots B}$ motions. (iv) The dynamics of the external $N(t)$ mode is described on different levels of approximation. In the simplest analysis, its time evolution is wholly suppressed by noticing that the time scale of interest, $1/\Delta\nu_{\frac{1}{2}}$, is much shorter than times characteristic of molecular motions of liquids[16]. Alternatively, the dynamics of the N-motion is considered explicitely either in the frame of the simple[17], or the generalized Langevin theory[18]. One has, eitheir $N(t) \rightarrow N$ or

$$\ddot{N} + \gamma\dot{N} + \Omega^2 N = f(t) \quad (5)$$

$$\ddot{N} + \int_0^t d\tau\gamma(t-\tau)\dot{N}(\tau) + \Omega^2 N = f(t) \quad (6)$$

where γ is the friction coefficient, $\gamma(t)$ the corresponding memory function, $f(t)$ the random force and Ω the frequency of the bridge vibration. The precision of the description is improved by going from Eqn(5) to Eqn(6).

If the above model is adopted, the ν_{AH} band profile can be determined by solving Eqn(3) in the adiabatic approximation; the random force f(t) is considered as a

Markovian-Gaussian process. The following conclusions can be reached from the theory. (i) The basic band shaping mechanism is the n, N coupling. It produces a broad Gaussian; however, if the friction is weak enough, a progression in the N-mode appears in the spectrum (Fig. 5a and b). The link with the gas phase theories may be established in this way[19]. (ii) The electrical anharmonicity does not affect the ν_{AH} band very much. It is responsible, however, for the non-coincidence of infrared and Raman bands[20] (Fig.5c). (iii) The isotopic effect on the position and the intensity of the ν_{AH} is normal.

In more recent work, the assumption of semi-rigidity of the H-bonded complexes has been relaxed, and the effect of internal rotations of the proton donor and acceptor molecules with respect to each other examined[21]. This study is still in a preliminary stage, but certainly merits being pursued; the experimental data seem to confirm theoretical predictions[22].

c) Medium strong H-bonds. The H-bonded complex is considered, here again, as a semi-rigid unit and is pictured by three oscillators ν_{AH}, $\nu_{AH\cdots B}$ and δ_{AH}. The last of them refers to the bending motion of AH, or to some other degree of freedom perturbed by H-bonding. Other details of the model are as follows. (i) The ν_{AH} and δ_{AH} vibrations are quantum mechanical whereas the $\nu_{AH\cdots B}$ vibrations are classical. The former two are represented by the normal coordinates n_1, n_2 and the latter by a stochastic function $N(t)$. (ii) The Hamiltonian $H(n_1, n_2, t)$ and the dipole moment $\vec{M}(n_1, t)$ are

$$H(n_1,n_2,t) = \frac{1}{2}p_1^2 + \frac{1}{2}p_2^2 + \frac{1}{2}\omega_1^2 n_1^2 + \frac{1}{2}\omega_2^2 n_2^2 + \frac{1}{2}K_{11I}n_1^2 N(t) + \frac{1}{2}K_{122}n_1 n_2^2 \quad (7)$$

$$\vec{M}(n_1,t) - \vec{M}_0 = \vec{M}_1 n_1 + \vec{M}_{1I} n_1 N(t) \quad (8)$$

where ω_1 and ω_2 represent the non-perturbed frequencies of ν_{AH} and δ_{AH}. This model thus involves three, rather than two, band shaping mechanisms. Two of them express the effect of mechanical anharmonicity and are described by the terms $\frac{1}{2}K_{11I}n^2 N(t)$ and $\frac{1}{2}K_{122}n_1 n_2^2$; the last mechanism is the electrical anharmonicity and is pictured by the term $\vec{M}_{1I}n_1 N(t)$. It should be pointed out that the term $\frac{1}{2}K_{122}n_1 n_2^2$ generates a Fermi resonance mixing of the modes n_1, n_2, but no variation of the integrated intensity A. (iii) The $N(t)$ motions are described by different levels of approximation. Either the time dependence of $N(t)$ is suppressed altogether[16], or is treated in terms of a one-dimensional Smoluchowski, or Fokker-Planck equation[23]. One has, either $N(t) \to N$ or

$$\frac{\partial W(N,t)}{\partial t} = \gamma\left[\frac{\partial}{\partial N}N + \langle N^2\rangle \frac{\partial^2}{\partial N^2}\right] W(N,t) \quad (9)$$

$$\frac{\partial W(P,N,t)}{\partial t} = \left[\Omega^2 N \frac{\partial}{\partial P} - P\frac{\partial}{\partial N} + \gamma\left(\frac{\partial}{\partial P}P + \langle P^2\rangle \frac{\partial^2}{\partial P^2}\right)\right] W(P,N,t) \quad (10)$$

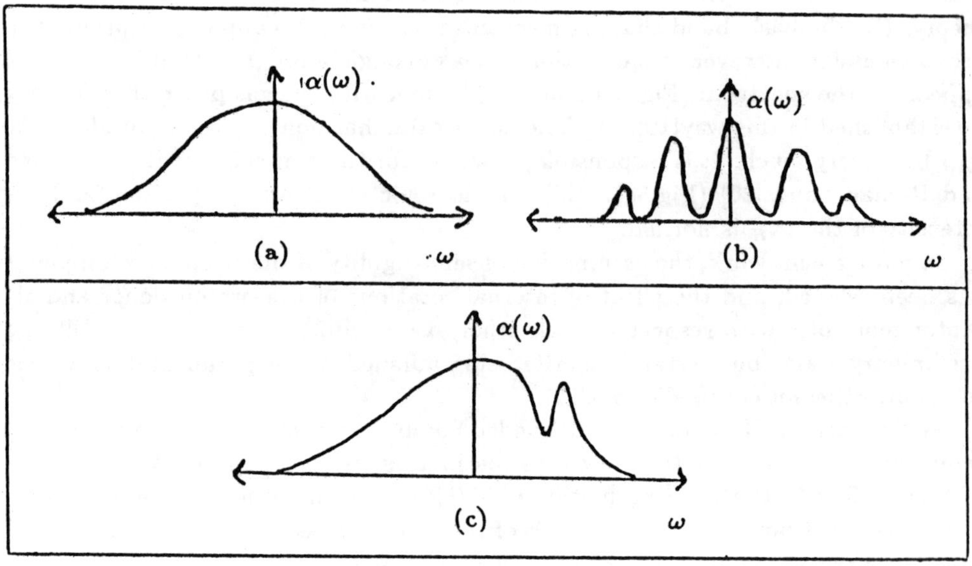

Figure 6: Theoretical band shapes of the ν_{AH} of systems containing medium strong H-bonds. The mechanical n, N coupling produces a Gaussian (a) in the high, and a progression (b) in the $\nu_{OH\cdots O}$ mode in the low friction limit. The Fermi resonance with some other degree of freedom of the system creates an Evans transmission window in the ν_{AH} band profile (c).

where $P = \dot{N}$, γ is the friction coefficient, Ω the frequency of the bridge vibration and $W(N,t)$ the probability of the variable N at time t; $W(P,N,t)$ has a similar meaning. The quality of the description improves when going from Eqn(9) to Eqn(10).

The profiles may be calculated either by treating $H(n_1, n_2, t)$ in an approximation in which only the excited $1\nu_{AH}$ and $2\delta_{AH}$ levels are mixed by the Fermi resonance, or by building the Fokker-Planck techniques into the bandshape theory. The following results have been reached in this way. (i) The anharmonic n_1, N coupling remains the main shaping mechanism. It produces either a broad Gaussian or a progression of the N-bands if $\gamma \to 0$ (Figs 6a and b); this progression has never been observed, however, in the liquid phase. (ii) The n_1, n_2 Fermi resonance coupling generates narrow and deep, or broad and shallow Evans holes. The details of transmission windows reflect molecular dynamics of liquids and can only be

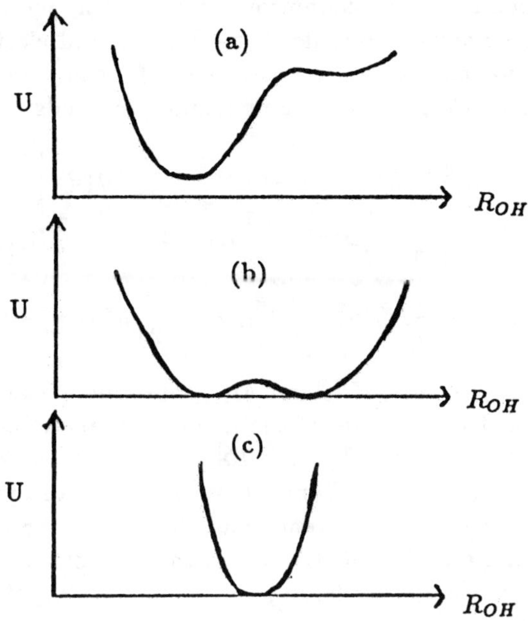

Figure 7: Potential energy curves $U = U(R_{OH})$ for strong H-bonds. Figs(a,b,c) refer to asymmetrical, nearly symmetrical and truly symmetrical H-bonds, respectively. The rapid variation of their form with R_{OO} is noteworthy.

imperfectly reproduced in the static limit (Fig.6c). (iii) The electrical anharmonicity slightly deforms band shapes, but leaves the position of transmission windows unchanged[24]. The experimental data fit well this description.

As far as the recent work is concerned, a special technique to eliminate spectral effects of the Fermi resonance merits attention. This "peeling off" procedure consists in transforming the observed band shape into the one that one would have if the n, N coupling were present alone; the results are sometimes spectacular[25]. Another question, recurrent in the theory of H-bonding and reexamined again, is whether spectral anomalies in the ν_{AH} are due to the breakdown of the Born-Oppenheimer separation between electrons and nuclei. It was conjectured that the ring structure of carboxylic acid dimers may favour a "quasi-Jahn-Teller" effect[26]. For the moment, the evidence for its existence is still incomplete. Finally, a recent attempt to transpose the Heller theory of photodissociation processes[27] to the present problem should be mentionned[28]; the feasability of this approach is not obvious and should be checked carefully.

d) Strong H-bonds. The strong H-bonds may be treated much in the same way as the medium strong bonds. The assumptions (i-iii) still apply except that the anharmonicity of the ν_{AH} normal mode itself, mechanical and electrical, must be added to those coupling n_1, n_2 and $N(t)$. Eqs(3) and (4) then transform and take the following form, adapted to asymmetric and symmetric bonds

$$\begin{aligned} H(n_1,n_2,t) &= \left[\frac{1}{2}p_1^2 + \frac{1}{2}\lambda_{11}n_1^2 + \frac{1}{6}\lambda_{111}n_1^3 + \frac{1}{24}\lambda_{1111}n_1^4\right] + \\ &+ \left[\frac{1}{2}p_2^2 + \frac{1}{2}\omega_2^2 n_2^2\right] + \frac{1}{2}K_{11I}n_1^2 N(t) + \frac{1}{2}K_{122}n_1 n_2^2 \end{aligned} \tag{11}$$

$$\vec{M}(n_1,t) - \vec{M}_0 = \vec{M}_1 n_1 + \frac{1}{2}\vec{M}_{11}n_1^2 + \frac{1}{6}\vec{M}_{111}n_1^3 + \frac{1}{24}\vec{M}_{1111}n_1^4 + \vec{M}_{1I}n_1 N(t) \tag{12}$$

where $\lambda_{11} = \omega_1^2$ and $\lambda_{111}, \lambda_{1111}, \vec{M}_{11}, \vec{M}_{111}$ and $\vec{M}_{1111}$ are different anharmonic coupling coefficients. Another basic element of the theory is that the potential energy of the n_1 mode changes very rapidly when the $R_{OH\cdots O}$ distance varies from 2.50 to 2.45Å (Fig. 7); the system transforms from a loosely bonded complex into a true molecule. The theory involves four different band shaping mechanisms and leads to the following results[29]; see also [30,31]. (i) In the case of a strong asymmetric H-bond, the n_1 motion has a very large amplitude. The n_1, N coupling term $\frac{1}{2}K_{11I}n_1^2 N$ is thus very large and generates a Gaussian of an exceptional width. In turn, the Fermi resonance of the $1\nu_{OH}$ and the $2\delta_{OH}$, $2\gamma_{OH}$ levels produces the A,B,C structure (Fig. 8a). The D-substitution effect is totally anomalous. (ii) In the case of a strong nearly symmetrical H-bond, the amplitude of the n_1 mode attains its extreme value, a consequence of the flat-bottomed potential energy surface characteristic of this bond. The n, N coupling then produces a profile of an extreme width; its lack of symmetry is due to the induced emission at low frequencies(Fig. 8b). The calculated isotope effect remains highly anomalous. (iii) If the H-bond is strong and truly symmetrical, the potential energy surface becomes harmonic again; the amplitude of the ν_{OH} vibration, and the n_1, N coupling term $K_{11I}n_1^2 N$ are comparatively small. This mechanism thus produces a Gaussian of moderate width (Fig. 8c), and a nearly normal D-substitution effect. These results all indicate that the anomalous ν_{OH} band width is due to the large amplitude of the hydrogen stretching vibration along the $AH\cdots B$ bond. The onset of the dissociation of the acid proton manisfests itself in this way.

5 Conclusions

It results from the present review that the modern theory of band shapes attained a level permitting a qualitative or semi-quantitative description of spectral effects of the H-bonding in diluted solutions. To go further and construct a quantitative theory, computer simulations should be applied. A limited number of papers have

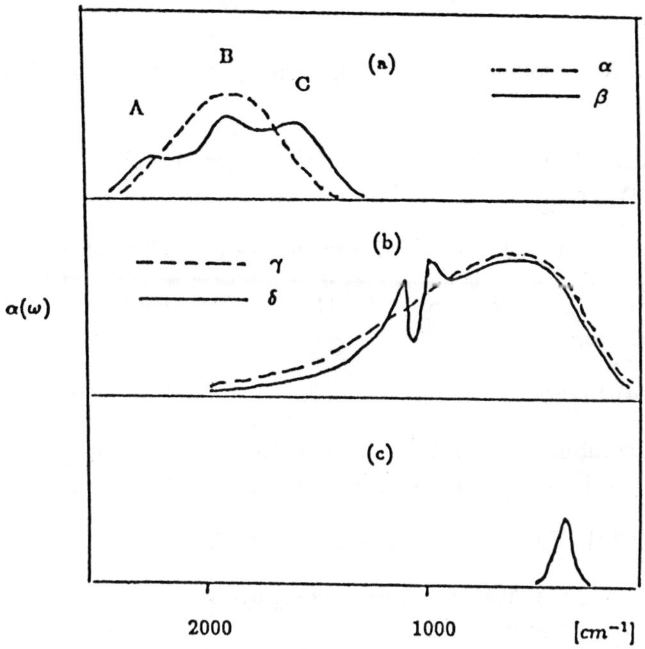

Figure 8: Theoretical band shapes of the ν_{AH} band of systems containing strong H-bonds. (a) In the case of an asymmetrical H-bond, the n_1, N mechanical coupling produces a broad Gaussian α and the n_1, n_2 coupling a profile β exhibiting the A, B, C structure. (b) In the case of nearly symmetrical H-bond, the n_1, N mechanical coupling gives rise to an asymmetric profile γ, whereas the coupling of n_1 with a skeletal mode n_2 generates a profile δ containing a narrow Evans hole. (c) In the case of a truly symmetrical H-bond, the n_1 ,N mechanical coupling produces a relatively narrow Gaussian.

already been published in this direction[32-36]; however, one should avoid duplicating well known and firmly established results. At the moment, the evolution in this direction is hindered by an insufficient experience in simulating chemical equilibria. Another unsolved problem is that of the internal field effect in pure H-bonded liquids. No treatment of intermolecular vibrational couplings, and of their effect on infrared band band shapes is envisageable unless this arduous and ungrateful problem is solved. Finally, the present knowledge of intermolecular potentials, as well as of their dependence on internal coordinates, should be improved to insure meaningful computer simulations. Still, the recent progress is impressive enough to envisage the future with optimism!

References

1 Pimentel, G.C., Mc Clellan, A.L. (1960) 'The Hydrogen Bond', Freeman, San Francisco.

2 Vinogradov, S.N., Linnel, R.H. (1971) 'Hydrogen Bonding', Van Nostrand-Reinhold, New York.

3 Joesten, M.J., Schaad, L.J. (1974), 'Hydrogen Bonding', Dekker, New York.

4 Schuster, P., Zundel, G., Sandorfy, C. (1976), 'The Hydrogen Bond', North-Holland, Amsterdam.

5 Novak, A. (1974) Struct. Bonding, **18**, 177.

6 Bratos, S., Lascombe, J., Novak, A. (1980) in 'Molecular Interactions' Vol 1, Ratajczak, H., Orville-Thomas, W.J., editors, Wiley, NewYork, p. 301.

7 Sandorfy, C. (1984) 'Topics in Current Chemistry', **120**, 41.

8 Maréchal, Y. (1987) in 'Vibrational Spectroscopy and Structure', vol. 16, Durig, J. editor, Elsevier, Amsterdam, p. 311.

9 Hadzi, D. (1972) Chemia **26**, 7.

10 Zundel, G. (1976) in 'The Hydrogen Bond', Schuster, P., Zundel, G., Sandorfy, C. editors, Amsterdam, North Holland, p. 685.

11 Stepanov, B.I. (1945) Zhur. Fiz. Khim. **19**, 507.

12 Blinc, R., Hadzi, D. (1958), J. Mol. Phys., **1**, 397.

13 Landsberg, G.S., Baryshanskaia, F.S. (1946), Izv. Akad. Nauk. SSSR, Ser. Fiz, **10**, 509.

14 Bratos, S., Hadzi, D. (1957), J. Chem. Phys. **27**, 991.

15 Gordon, R.G. (1968) Adv. Mag. Res. **3**, 1.

16 Bratos, S. (1975), J. Chem. Phys. **63**, 3499.

17 Robertson, G.N., Yarwood, Y. (1978), Chem. Phys. **32**, 267.

18 Sakun, V.P. (1985), Chem. Phys. **99**, 457.

19 Maréchal,Y., Witkowski, A. (1968), J. Chem. Phys. **48**, 3697.

20 Maréchal, E., Bratos, S. (1978), J. Chem. Phys. **68**, 1825.

21 Abramczyk, H. (1985), Chem. Phys. **94**, 91.

22 Abramczyk, H. (1987), Chem. Phys. **116**, 249.

23 Johnson, W.G., Oxtoby, D.W. (1987), J. Chem. Phys. **87**, 781.

24 Maréchal, E., Ratajczak, H. (1986), Chem. Phys. **110**, 103.

25 Maréchal,Y. (1983), Chem. Phys. **79**, 69.

26 Maréchal,Y. (1987), J. Chem. Phys. **87**, 6344.

27 Heller, E.J. (1978), J. Chem. Phys. **68**, 2066.

28 Filliaux, F. (1983), Chem. Phys. **74**, 395.

29 Bratos, S., Ratajczak, H. (1982), J. Chem. Phys. **76**, 77.

30 Romanowski, H., Sobczyk, L. (1977), Chem. Phys. **19**, 361.

31 Librovich, N. B., Sakun, V.P., Sokolov, N.D. (1978) **39**, 351.

32 Belch, A.C., Rice, S.A. (1983), J. Chem. Phys. **78**, 4817.

33 Rice, S.A., Bergren, M.S., Belch, A.C., Nielson, G. (1983), J. Phys. Chem. **87**, 4295.

34 Reimers, J.R., Watts, R.O., (1984), Chem. Phys. **91**, 201.

35 Postma, J.P.M., Berendsen, H.J.C., Straatsma, T.P. (1984), J. Phys. (Paris) **C7**, 31.

36 Bansil, R., Berger, T., Toukan, K., Ricci, M.A., Chen, S.H. (1986), Chem. Phys. Letters, **132**, 165.

INFRARED SPECTRA OF H-BONDED MOLECULES

Y. MARECHAL
DRF/SPh/PCM -CEN Grenoble
85X
F 38041 Grenoble-Cedex
FRANCE

ABSTRACT

ν_s bands(A-$\vec{H}$...B) which appear in vibrational spectra of H-bonded molecules exhibit unusual shapes and intensities.In order to quantitatively understand these features ,two types of models have been studied:single H-bonded species of the type ClH...$O(CH_3)_2$ and cyclic dimers such as found in carboxylic acids.The evolution of the shapes of ν_s in these systems when varying temperature and mass of the proton(deuteration) has allowed to elaborate precise descriptions of the mechanics of vibrations of H-atoms in H-bonds,which allow to precisely understand the peculiarities of the shapes of these bands.If the dramatic enhancement of the intensity of ν_s as compared to the corresponding ν_{OH} band with no H-bond is also understood,the origin of isotopic anomalies of these intensities in H-bonded cyclic dimers still remains obscure.It makes us think that IR spectroscopy might be an interesting tool to study transfers of protons inside such cyclic structures.These transfers of protons are thought to be important in water,and specially in the majority of biological reactions at the molecular level.The advent of recent techniques to obtain reliable IR spectra of water seems then most promisefull.

J.C. Dore and J. Teixeira (eds.), Hydrogen-Bonded Liquids, 237–260.

1. INTRODUCTION

Water is well-known to exhibit exceptional physical,chemical and biological properties which are not found in any other species.The origin of all these singular properties is to be found in its exceptional density of H-bonds,which is so high that water may be characterized as the unique species where the number of such intermolecular bonds is roughly equal to the number of covalent bonds.H-bonds have two properties which will give water its crucial properties [1]:first,they are linear.As a consequence H-bonds will be at the origin of well defined molecular edifices.Secondly,their energies of formation are of some kT at 300K.It implies that these molecular edifices will be stable at room temperature,but will at the same time require a small amount of energy only to be modified,which thermal fluctuations can provide.In other words H-bonds will be at the origin of stable but nevertheless evolutive molecular edifices.This is one of the reason why it plays a central role in biology.This is also the origin of all exceptional properties of water.

The concept of H-bonds is now more than 80 years old[2]. Various techniques,such as heat measurements,X-Ray or neutron diffraction,spectroscopy,etc...,have been used to study these particuliar intermolecular bonds.We may then think that everything that concerns these bonds is now clear.This is not completely true,and we shall see,in the course of this lecture,that there still exists points which are not fully understood and make us suspect that they play an even greater role in aqueous media than that one which we normally attribute them.These obscure points are to be found in vibrational spectroscopy,which encompasses such various techniques as IR,Raman and more recently INS(inelastic neutron scattering).Using these techniques we may detect H-bonds ,as in other techniques such as NMR,microwave or UV spectra,by spectral shifts due to the fact that transitions which we are looking at may be (slightly) affected by the presence of H-bonds.We also may detect H-bonds by some specific manifestations ,such as *intermonomer modes*,that is intermolecular vibrations of the two molecules linked by such bonds one with respect to the other which appear in the low frequency part of vibrational spectra,in the region of 100 cm^{-1} or so.The most spectacular manifestation of H-bonds,however,where there still exist obscure points,is found in that part of IR(Raman) spectra which corresponds to the stretching vibration of the H atom of an H-bond,parallel to its covalent bond.In Fig.1 we illustrate the dramatic effect H-bonds have in this part of vibrational spectra taking as an example spectra of D3-acetic acids in the gas phase,where monomers CD_3COOH are in

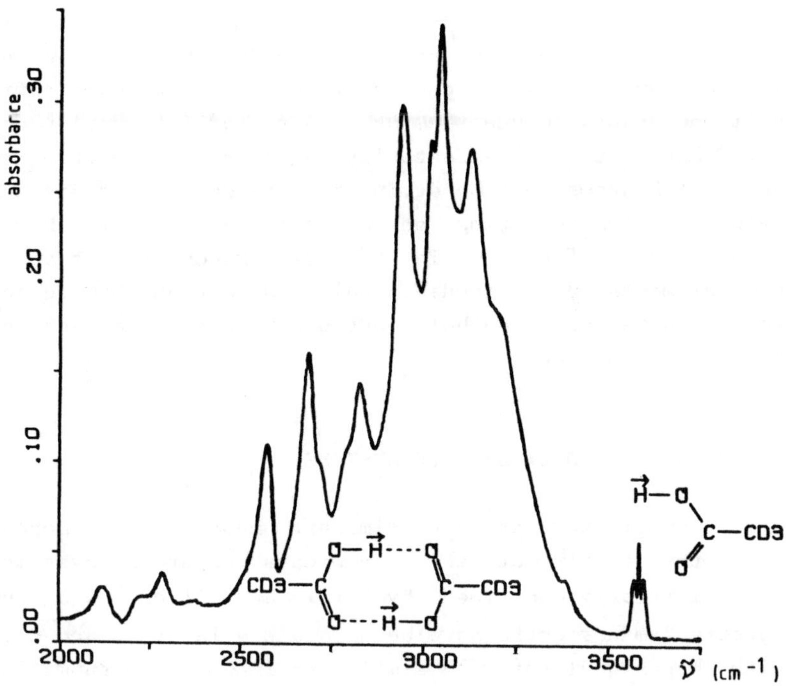

Figure 1. IR spectrum of D3-acetic acid vapor,illustrating the dramatic change H-bonds have in the strectching vibration of H-atoms.

equilibrium with H-bonded cyclic dimers $(CD_3COOH)_2$.In monomers with no H-bonds the stretching vibration ν_{OH} of the H-atom is the quite ordinary band falling at 3600cm^{-1}.Its counterpart ν_s (we shall use this denomination throughout this paper) in H-bonded dimers is the most intense,broad and structured band extending from 2000 to 3500cm^{-1}.When we know that the conditions for recording the spectrum in Fig.1 were such that there were as many monomers as dimers,we realize the big enhancement of intensity that H-bonds induce in those IR bands.This is not the case for Raman spectra in which ν_s bands are in general less intense than corresponding ν_{OH} bands.It makes Raman spectroscopy a less powerfull technique for the study of H-bonds than IR spectra,for which the analysis of the intensity of these bands will be shown to reveal interesting features on the dynamics of H-bonds.The particiuliarly unuasual shape of ν_s is found in both Raman and IR spectra.

It is clear that such conspicuous effects H-bonds exhibit in IR and,in a less marked way in Raman spectra,are worth being quantitatively understood if one wishes to use them as a source of information on the dynamics of H-bonds.It implies establishing a precise description of the mechanics of H-atoms in H-bonds.In order to do so physicists have been striving for manipulating on molecular edifices which may be considered as models of H-bonds.In fact two kinds of such edifices have been experimentally defined.We shall comment on them selecting experimental results that can help establishing such a description which we shall then outline.

2. HYDROCHLORIC ACID ...DIETHER TYPE SYSTEMS

These may be considered as the simplest models of H-bonds.The archetype of the family of these nowadays numerous systems is $ClH...O(CH_3)_2$ initially discribed by Bertie and Millen[3] who studied their IR spectra.Raman spectra have been obtained later[4] and display ν_s bands with shapes qualitatively similar to IR ones.The submaxima of ν_s bands of these systems are equally spaced by some $100cm^{-1}$ and Bertie and Millen have interpretated this spacing as being equal to the frequency of the intermonomer stretching vibration.This is a precious point to start defining the Hamiltonian $\mathcal{H}$ which governs ν_s vibrations.The simplest form which we can think of for it is that of an Hamiltonian harmonic in *q(the coordinate of* ν_s)whose force constant is modulated by *Q(the coordinate of the stretching intermonomer mode)*.The main argument for doing so is,beside simplicity,the fact that $0 \rightarrow n$ transitions in ν_s are very weak for $n>1$ in all H-bonded systems.It constitutes a mathematical description of the "band-shaping" mechanism proposed by Bratos and Hadzi[5].We may then write,assuming that the potential ruling Q is also harmonic:

$$\mathcal{H} = \frac{p^2}{2m} + \frac{P^2}{2M} + \frac{1}{2} k(Q)\, q^2 + \frac{1}{2} M\, \Omega^2\, Q^2 \tag{1}$$

$$p = -i\hbar \frac{\partial}{\partial q} \qquad P = -i\hbar \frac{\partial}{\partial Q}$$

The wavenumber ω at which q vibrates and which is such that

$$k(Q) = m\, \omega^2(Q) \tag{2}$$

is of the order of $2500 cm^{-1}$,whereas the wavenumber Ω for the intermonomer mode Q is of the order of $100cm^{-1}$.We can then perform an adiabatic separation between q and Q.Its physical justification is that q is supposed to be enough rapid to adapt itself immediatally to any change in Q.Mathematically we write the total wavefunction $\Phi(q,Q)$ as:

$$\Phi(q,Q) = \varphi_n(q,Q)\, \alpha_n^m(Q) \tag{3}$$

with $\varphi_n(q,Q)$ being the nth function describing the q vibration which parametrically depends on Q,and α_n^m the mth vibrational function for Q,when q vibrates in its nth level.Choosing φ_n to be an eigenfunction of $\frac{p^2}{2m} + \frac{1}{2} m\, \omega^2(Q) q^2$ shows that in this approximation,slow vibrations Q experience an averaged(on q) potential $\frac{1}{2} M\, \Omega^2 Q^2 + W_n(Q) = \frac{1}{2} M\, \Omega^2\, Q^2 + (n+\frac{1}{2})\, \hbar\omega(Q)$.
This approximation mathematically consists in neglecting such terms as $(P|\varphi_n\rangle)\, P\alpha_n^m$,which hinders slow Q vibrations to induce transitions between different states φ_n and φ_m of the rapid q vibration.Hence the denomination "adiabatic".We may then write:

$$\mathcal{H}|\varphi_n(q,Q)\rangle_q = |\varphi_n(q,Q)\rangle_q\, \mathbb{H}_n \tag{4}$$

$$\mathbb{H}_n = \frac{P^2}{2M} + \frac{1}{2} M\, \Omega^2 Q^2 + (n+\frac{1}{2})\hbar\omega(Q) \tag{5}$$

$\mathbb{H}_n$ is the Hamiltonian governing intermonomer mode Q when ν_s is in its nth state.Its eigenfunctions are the α_n^m . The various transitions composing the ν_s bands are then transitions between levels α_0^m of $\mathbb{H}_0$ and $\alpha_1^{m'}$ of $\mathbb{H}_1$,which are illustrated in Fig.2.They are "Franck-Condon" transitions which are well known in spectra of electronic transitions,and give lines separated by Ω,which give a fair description of experimental spectra[6].More elaborated simulations of experimental spectra starting on equivalent premisses have been performed recently[7].

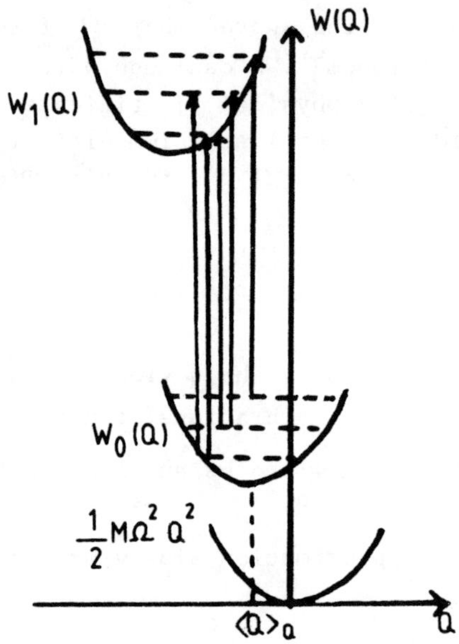

Figure 2. Energetics Scheme of the various transitions occuring in the vibrational spectra defined by Eq.5

Such simple models as $ClH...O(CH_3)_2$ are most interesting,and they have been the object of important developments.Their vibrational spectra have been reviewed by Lassègues and Lascombe [8] .Their interest,however,extends far beyond this type of spectroscopy,and they are presently studied more by microwave spectroscopy,which leads to precise determinations of their geometrical as well as spectroscopical parameters.They constitute a precious experimental basis from which theoreticians strive for giving a detailed description of the electronic structure of H-bonds,which leads to a physically simple image for the origin of the important intensity of ν_s bands[9] (Fig.3).All these recent developments have been recently reviewed by

Legon and Millen [10].These systems suffer,however,from a lack of flexibility in their manipulation which makes,for instance,variations of temperature available on a restricted range only.Also spectra of

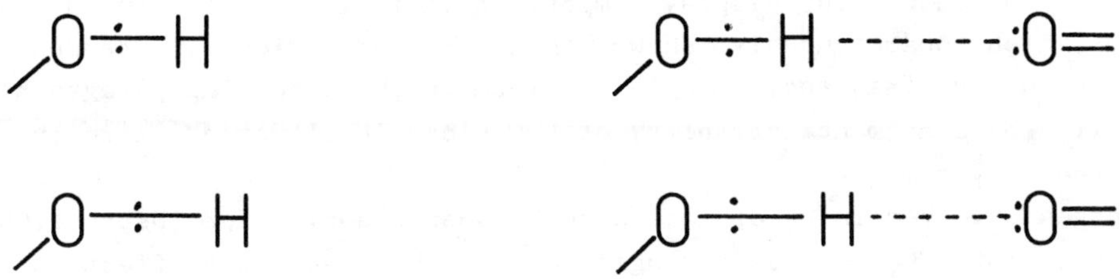

Figure 3. Intensity of the ν_s band:In an O-H group with no H-bond displacement of the proton parallel to the valence bond provokes a displacement of the average position of the two electrons of this bond(left).When this O-H group is part of an H-bond(right) the presence of the two electrons in the non-bonding orbital of the atom,to which this group is H-bonded,prevents this displacement of the average position of valence electrons.The enhancement of the dipole moment is then much bigger.

their intermonomer modes,in the far IR region,is difficult to obtain accurately.These are difficulties which are not encountered with the second type of model.

3. H-BONDED CYCLIC DIMERS

3.1. EXPERIMENTAL RESULTS:ESTABLISHMENT OF THE VIBRATIONAL HAMILTONIAN

These cyclic dimers are found in vapours or crystals of carboxylic or dicarboxylic acids.They have the well known planar structure displaid in Fig.1.Their advantages and disadvantages,as compared to preceding sytems have been recently examined [11].As a disadvantage they loose some elements of simplicity.They gain,however,in flexibility of manipulation,which enables for instance to study ordered H-bonds in monocrystals,using polarized IR beams.More important,the range of temperature where variations of ν_s may be observed is more extended.In the case of adipic acids COOH-$(CH_2)_4$-COOH,for instance,the shape of ν_s has been shown to display important variations [12] between 0 and 400K,which unambiguously shows that the modulation by intermonomer modes of the frequency of ν_s,as defined in the preceding paragraph,is still the fundamental property of the vibrational potential ruling ν_s in these systems.

Changes observed in ν_s of acetic acids when changing,for instance,CH_3 groups into CD_3 groups strongly suggest that resonance effects,often referred to as Fermi resonances,have as well important responsabilities in the appearance of submaxima[13].The analysis which we shall make of the evolution of the width of ν_s with temperature will confirm this point of view.Under this denomination of Fermi resonances lies the influence on the first excited state of ν_s of other vibrational levels which are almost degenerate with it,but have a very weak probability to be directly attained from the ground vibrational level by absorption of an IR photon.Typical candidates for such resonance levels are binary combinations of other vibrations of the whole dimeric system.One of their typical manifestation is the dedoubling of narrow bands.They constitute a complication which should be eliminated.

Finally the presence of two identical H-bonds in these dimers is at the origin of an interaction term between the two ν_s vibrations,which we shall take as harmonic so as to keep $0\longrightarrow n$ transitions weak for $n>1$.The magnitude V_0 of this interaction may be determined if one can compare spectra of mixed dimers that is cyclic dimers having one H-bond and one D-bond with spectra of ordinary dimers[14].

All these considerations lead then to writing the vibrational Hamiltonian $\mathcal{H}$ for these cyclic dimers as:

$$\mathcal{H} = T + \frac{1}{2}\sum_{i=1,2}\left(m\omega^2(Q_i)q_i{}^2 + M\Omega^2 Q_i^2\right) + \frac{1}{2}\sum_{\delta,i=1,2} m_\delta\omega_\delta^2\, q_{i,\delta}^2$$

$$+\sum_{i=1,2}\sum_{\delta,\delta'} F_{\delta,\delta'}\, q_i\, q_{i,\delta}\, q_{i,\delta'} + \tilde{V}_0\, q_1\, q_2 \qquad (6)$$

where T is the kinetic energy for all modes, $F_{\delta,\delta'}$ is the coefficient of the resonant (Fermi)coupling between ν_s and the combination of modes δ and δ'. Indices $\delta(\delta')$ represent any vibrational mode of the dimer other than ν_s and intermonomer modes. These δ modes vibrate at wavenumber ω_δ and have a reduced mass m_δ. $\tilde{V}_0$ is the coefficient of the interaction term between the two ν_s modes. Finally we have omitted, for simplicity, terms in Q_1Q_2 which may be present in the first summation. Their effect is only to change the decomposition of normal modes on local vibrations.

3.2. IR TRANSITIONS

Before calculating bandshapes that vibrations governed by $\mathcal{H}$ give in IR spectra, let us recall some well-established results which concern treatment of transitions, which may easily be generalized beyond IR transitions and which we shall be using subsequently. In optics the experimentally measured quantity, absorbance, is proportional to $\tilde{\nu}.s(\tilde{\nu})$, where $\tilde{\nu}$ represent wavenumbers and $s(\tilde{\nu})$ may be called the shapefunction of the considered band. In the linear response theory, where an external field $\vec{E}=E\vec{\mathcal{E}}$(an electric field for instance; $\vec{\mathcal{E}}$ is a unit vector)provokes, in the considered molecular system, the appearance of a conjugated moment $\mathcal{M}=\varepsilon''E$(an electric dipole moment in the case of an electric field), $s(\tilde{\nu})$ is proportional to the imaginary part ε'' of the susceptibility(in our case the dielectric constant)of the molecular system. Fluctuation-dissipation theorem[15] then states that at wavenumber $\tilde{\nu}$ much greater than $kT/\hbar$, $\varepsilon''(\tilde{\nu})$ is the Fourier transform of the autocorrelation function $\langle\langle\mathcal{M}(t)\mathcal{M}(0)\rangle\rangle$ of the component along $\vec{\mathcal{E}}$ of the conjugated moment $\mathcal{M}$, where $\langle\langle\ \rangle\rangle$ means thermal average on all quantum levels of the molecular system. This is strictly equivalent to the Fermi Golden Rule[16]. Because of its flexibility we shall rather use this formalism. It implies that the Fourier Transform $s(t)$ of $s(\tilde{\nu})$ is proportional to the autocorrelation function $\langle\langle\mathcal{M}(t)\mathcal{M}(0)\rangle\rangle$. An immediate interest of this is that development of $s(t)$ in powers of it has coefficients M_n equal to the nth moment of $s(\tilde{\nu})$ [16]:

$$s(t)=\sum_{n=0}^{\infty}\frac{(it)^n}{n!}M_n \qquad M_n=\int_{-\infty}^{+\infty}\tilde{\nu}^{\,n}\ s(\tilde{\nu})\ d\tilde{\nu} \tag{7}$$

From these formula we may obtain the firsts coefficients $\bar{\omega}$ and μ of the cumulant expansion of s[17]

$$s(t) = M_0 \, e^{i\bar{\omega}t} \, e^{-\mu^2 t^2/2} \quad \ldots. \tag{8}$$

as:

$$\bar{\omega} = M_1/M_0 \qquad \mu^2 = M_2/M_0 - (M_1/M_0)^2 \tag{9}$$

$\bar{\omega}$ is the centre of intensity of the band,while μ represents its width(μ^2 is called its variance).

In case of IR spectra $\mathcal{M}$ is equal to:

$$\mathcal{M} = e\left(-\sum_i \vec{r}_i + \sum_n Z_n \vec{R}_n\right) . \vec{\mathcal{E}} \tag{10}$$

where e is the absolute value of the charge of the electron.The position of the ith electron is $\vec{r_i}$ while that of the nth nucleus is $\vec{R_n}$ (its charge is $Z_n e$) .Within the Born-Oppenheimer approximation (separation of motions of electrons and nuclei) the sum over positions of electrons is to be replaced by its quantum mean value $\bar{r}$ in the ground electronic state of the molecular system.We may write:

$$\mathcal{M} = \mathcal{M}(0) + \sum_n R_n \frac{\partial \mathcal{M}}{\partial R_n} + \ldots = \mathcal{M}(0) + (q_1 - q_2)\left(\frac{\partial \mathcal{M}}{\partial q}\right) + \ldots \tag{11}$$

$$= \mathcal{M}(0) + (q_1 - q_2)\; e\; (1 - \frac{\partial \bar{r}}{\partial q}) + \ldots$$

where $\mathcal{M}(0)$ is the static dipole moment of the molecular system which has no role in optical transitions.In the case of our centrosymmetric cyclic dimers it is equal to zero.In the last two members of Eqs.(11) we only kept those terms in the summation over n which concern ν_s .We may recognize with this formula that only the antisymmetric mode $q_1 - q_2$ with respect to the centre of inversion appears in IR spectroscopy.Its physical meaning is that the symmetric mode $q_1 + q_2$ creates no dipole moment.It nevertheless appears in Raman spectroscopy whereas $q_1 - q_2$ is not visible as an antisymmetric mode does not modify polarizability.

3.3. FIRST MOMENTS OF ν_s

We have seen that s(t) is proportional to $\langle\langle \mathcal{M}(t)\mathcal{M}(0)\rangle\rangle$. Assuming the Born-Oppenheimer decoupling of electrons and nuclei to be valid, we have to take the average value symbolized by $\langle\langle\ \rangle\rangle$ on q and Q. With this condition we may write, with the help of Eq.(11) and assuming $\frac{\partial \bar{r}}{\partial q}$ to be independent of Q(approximation of electrical harmonicity):

$$\langle\langle \mathcal{M}(t)\mathcal{M}(0)\rangle\rangle = \left(\frac{\partial \mathcal{M}}{\partial q}\right)^2 \langle\langle e^{-i\mathcal{H}t/\hbar}(q_1 - q_2)e^{i\mathcal{H}t/\hbar}(q_1 - q_2)\rangle\rangle_{q,Q} \tag{12}$$

Because frequencies of vibrations of q and q_δ modes are always greater than $1500\text{cm}^{-1} \simeq 2500\text{K}$, we have to consider ground states only of these modes in Eq.(12). An adiabatic decoupling between these modes in their ground state and Q modes is then possible. It allows to define an Hamiltonian $\mathbb{H}_0$ for Q as:

$$\mathbb{H}_0 = \sum_{i=1,2} \left(\frac{P_i^{\,2}}{2M} + \frac{1}{2}M\,\Omega^{\,2}Q_i^{\,2} + \frac{1}{2}\hbar\omega(Q_i)\right) + \sum_{\delta} \hbar\omega_\delta \tag{13}$$

This Hamiltonian is a generalization of that found in the case of a single H-bond(Eq.(5) with n=0). In contradistinction with this simpler case of a single H-bond with no Fermi resonances, an adiabatic decoupling is no longer possible when ν_s is in one of its first excited state, as these states are degenerate(the excitation may be either on q_1 or on q_2) and is also quasidegenerate with resonant states of various combination modes δ,δ'. We may then always define such an Hamiltonian as $\mathbb{H}_1$, but it no longer rules Q states when ν_s is in one of its first excited states. Taking advantage of this adiabatic decoupling when ν_s is in its ground state, we may then write:

$$\langle\langle \mathcal{M}(t)\mathcal{M}(0)\rangle\rangle = \left(\frac{\partial \mathcal{M}}{\partial q}\right)^2 \langle e^{-i\mathbb{H}_0 t/\hbar}\,\mathcal{O}^{+}(t)\rangle_Q \tag{14}$$

where the average $< >_Q$ means thermal average($< A >_Q = \sum <l e^{-\mathbb{H}_0/kT} A l>_Q \Big/ \sum <l e^{-\mathbb{H}_0/kT} l>_Q$ where $\sum$ extends over all eigenstates $l>_Q$ of $\mathbb{H}_0$ .It is often written as $< A >_Q = \mathrm{Tr}(e^{-\mathbb{H}_0/kT} A) \,/\, \mathrm{Tr}\, e^{-\mathbb{H}_0/kT}$).The Q dependent operator is simply equal to :

$$\mathcal{O}^{+}(t) = <(q_1 - q_2) \quad e^{i\mathcal{H}t/\hbar} \, (q_1 - q_2) >_q \quad (15)$$

where the integration on q is performed on the ground state in q_1 and q_2 .Developping Eqs.14 and 15 in powers of it rapidly gives the first integral parameters of ν_s as defined in Eqs.(7)-(9):

$$M_0 \approx \left(\frac{\partial \mathcal{M}}{\partial q}\right)^2 < \frac{\hbar}{m\omega(Q)} >_Q \simeq \left(\frac{\partial \mathcal{M}}{\partial q}\right)^2 \frac{\hbar}{m<\omega(Q)>_Q} \qquad (16)$$

$$\bar{\omega} = < \omega(Q) >_Q - < \tilde{V}_0 \frac{\hbar}{m\omega(Q)} >_Q$$

$$\mu^2 = \sigma_F^2 + \sigma^2 \qquad \text{with} \quad \sigma^2 = < (\omega(Q) - \bar{\omega})^2 >_Q$$

$$\sigma_F^2 = \frac{3}{2} \sum_{\delta,\delta'} \sigma_{\delta,\delta'}^2 \qquad \text{and} \quad \sigma_{\delta,\delta'} = F_{\delta,\delta'} \sqrt{\frac{\hbar}{m\omega} \frac{\hbar}{m_\delta \omega_\delta} \frac{\hbar}{m_{\delta'} \omega_{\delta'}}}$$

These quantities are those which describe ν_s in a first approach.Let us note that it clearly appears that Fermi resonances do not modify intensity and centre of intensity of ν_s .They first appear in the variance of ν_s as an additive contribution σ_F^2 to the variance σ^2 due to the modulation of ν_s by intermonomer modes.Integral quantities defined in Eq.(16) are related either to the integrated intensity of this band(M_0) or to its shape($\bar{\omega}$, σ and higher order cumulants) which are two independent kinds of quantities(in Eq.(9) M_0 acts as a normalization factor only).Let us first examine the shape of ν_s .

3.4. THE ν_s BANDSHAPE

In the expression giving the value of the center of band $\bar{\omega}$ as defined in Eq.(16) we made no supposition as to whether the interaction coefficient $\tilde{V}_0$,defined in Eq.(6),depends or not on intermonomer distances Q.Comparisons of the evolution with temperature of the $\bar{\omega}$'s of ν_s bands in imidazole monocrystals with polarizations of the IR beam either along their $\vec{c}$ axes,or along their $\vec{b}$ axes,or in isotopically mixed crystals,indicate that in these species $\tilde{V}_0$ does not depend on Q[18].Along $\vec{c}$ axis antisymmetric vibrations in q_1 and q_2 are seen and $\bar{\omega}$ is then given by the same formula as in Eq.(16).Along $\vec{b}$ axis symmetric vibrations are seen whose $\bar{\omega}$ is given by Eq.(16) changing however the sign before the term in $\tilde{V}_0$ .The difference in the $\bar{\omega}$'s in these two different polarization does not vary with temperature,indicating that $\tilde{V}_0$ does not depend on Q.This is corroborated by the $\bar{\omega}$'s of the isotopically mixed crystals,which fall half-way between the $\bar{\omega}$'s of these previous polarizations.In these isotopically mixed crystals, $\bar{\omega}$ is given by Eq.(16) with $\tilde{V}_0=0$ .With these conditions which we shall suppose valid for carboxylic acids we may then write :

$$\bar{\omega} = \langle\omega(Q)\rangle_Q - V_0 \qquad\qquad V_0 = \tilde{V}_0 \frac{\hbar}{m\omega} \tag{17}$$

A comparison of the centres of intensity of ν_s between normal H-bonded or D-bonded dimers of carboxylic acids and mixed dimers,where Eq.(6) still holds with $V_0=0$ gives then values for V_0 of some $50cm^{-1}$ in H-bonded acids and of some $30cm^{-1}$ in D-bonded acids[14] .As a consequence of Eq.(17),the variation of $\bar{\omega}$ with temperature is then equal to:

$$\frac{d\bar{\omega}}{dT} = \frac{d\,\langle\omega(Q)\rangle_Q}{dT} \tag{18}$$

Developping $\omega(Q)$ in a Taylor series in Q,which leads to writing :

$$\langle \omega(Q) \rangle_Q = \omega(\langle Q \rangle_Q) + \frac{1}{2}\frac{d^2\omega}{dQ^2} < (Q - \langle Q \rangle_Q)^2 >_Q + \; .... \qquad (19)$$

gives interesting information on the kind of intermonomer modes that are responsible for this shift of $\overline{\omega}$ with temperature:as the variation of $\langle Q \rangle_Q$ with temperature (the ordinary dilatation) is too small to give values for $\frac{d\overline{\omega}}{dT}$ in agreement with experimental ones[14][19] shifts of $\overline{\omega}$ with T are due to the increasing amplitudes of vibrations of intermonomer modes with temperature(last term in Eq.(19)).It implies that $\frac{d^2\omega}{dQ^2}$ is positive which is only possible with bending intermonomer modes,as seen in Fig.(4).

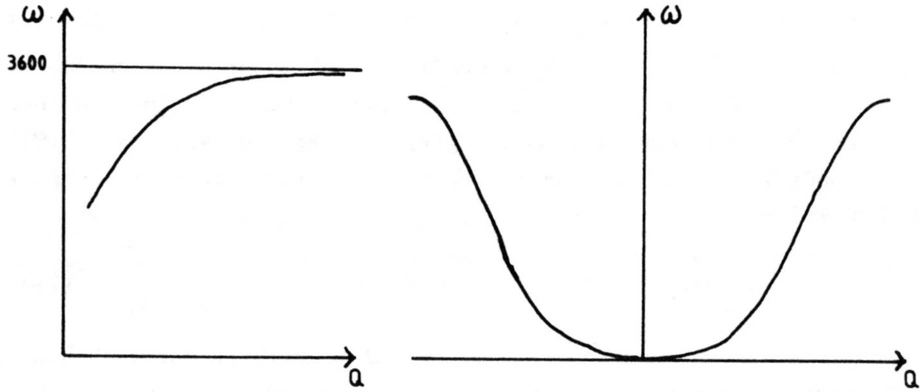

Figure 4. Schematic dependence of wavenumber ω of ν_s on intermonomer stretching(left) or bending(right)modes.Only bending intermonomer modes can exhibit a positive curvature $d^2\omega/dQ^2$.

It also implies that the relative variation of σ(Eq.(16)) with temperature is important.The fact that experimentally μ shows no important relative variation with temperature shows that with the T-independent term σ_F Fermi resonances have a predominant responsability in the width of ν_s ,which has been an argument for the existence of these resonances[14].

With the formalism of Eqs.(14)(15) we may calculate with more details the shape of ν_s .This has been already done in the case of H-bonds diluted in an inert liquid [20] where one can then consider intermonomer modes as ruled by classical mechanics.In Eqs.(14)(15) these modes are considered as obeying quantum mechanics.It may then be shown[14][21] that the operator $\mathcal{O}^+(t)$,which for simplicity we shall take in the case of a single H-bond,obeys the set of coupled equations:

$$-i\frac{d\mathcal{O}^+}{dt} = (\mathbb{H}_0/\hbar + \omega(Q))\,\mathcal{O}^+(t) + \sum_{\delta,\delta'} f_{\delta,\delta'}\,\mathcal{R}^+_{\delta,\delta'}(t) \qquad (20)$$

$$-i\frac{d\,\mathcal{R}^+_{\delta,\delta'}(t)}{dt} = (\omega_\delta + \omega_{\delta'} + \mathbb{H}_0/\hbar)\,\mathcal{R}^+_{\delta,\delta'}(t) + f_{\delta,\delta'}\,\mathcal{O}^+(t)$$

$$\text{with}\quad \hbar f_{\delta,\delta'} = \frac{1}{2\sqrt{2}}\,\sigma_{\delta,\delta'} \quad \text{(see Eq.(16))}.$$

This set of coupled equations simply describes the jump of the excitation in the ν_s vibrational system after absorption of an IR photon at time t=0,which makes ν_s to transit from it ground state to its first excited state.This is reflected in the the fact that $\mathcal{O}^+(t=0)$ equals 1 and all $\mathcal{R}^+_{\delta,\delta'}(t=0)$ equal 0.After some time this excitation will jump back and forth on Fermi levels δ,δ' which will make $\mathcal{R}^+_{\delta,\delta'}(t)$ to take some finite value[14].The difficulty for solving this set of coupled equations is that $\mathbb{H}_0$ is an operator which does not commute with $\omega(Q)$.It rules the excitation when it is on a Fermi level(second equation in (20)),while when this excitation is on ν_s it is ruled by $\mathbb{H}_0+\hbar\omega(Q)$.A great simplification may be achieved if one uses a short time approximation,corresponding to a low resolution for $s(\tilde{\nu})$ where one is then allowed to replace all levels in Q corresponding to the excitation on Fermi levels by their average value[21].One is then able to "peel-off" ν_s bands,that is calculate the shape it would have when Fermi resonance are suppressed.An example is given in Fig.5 for $(CD_3COOH)_2$ dimers.Peeled-off spectra are akeen with spectra of simple H-bonds of $ClH...O(CH_3)_2$ type.They may be analyzed in terms of

Franck-Condon types transitions, that is a transition in ν_s accompanied by various transitions in intermonomer modes[22]. It allows to measure the values of anharmonic coupling parameters $\frac{\partial\omega}{\partial Q}$ (Q standing here for bending as well as stretching type intermonomer modes). These may be interesting to know as they are at the origin of the influence of temperature on ν_s and probably beyond on sensibility of properties of H-bonds to temperature.

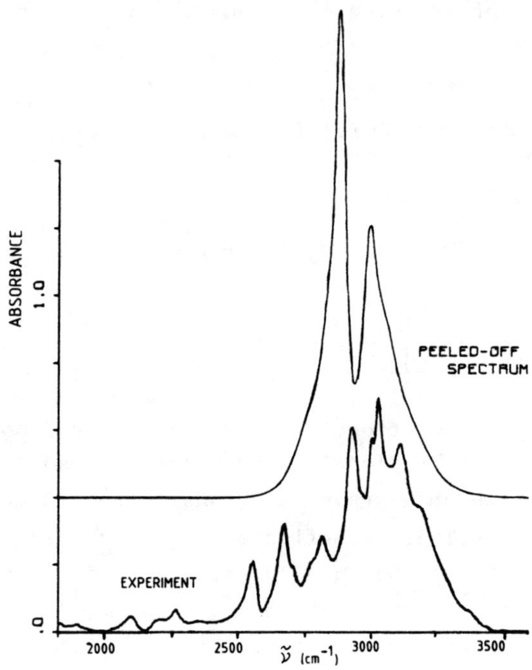

Figure 5. Experimental(bottom) and peeled-off ν_s band of $(CD_3COOH)_2$ drawn at the same scale.

The obtention of peeled-off spectra allows precise determinations of the relative contributions of the two anharmonic couplings ν_s establishes with intermonomer vibrations, through modulation of ω by these modes, on one hand, and Fermi resonances on the other hand. In the case of adipic acids, for instance, the values of widths σ and σ_F (Eq.(16)) due to these kinds of couplings are[21] σ_F = 185cm^{-1} , σ=44cm^{-1} at OK for H-adipic acid and σ_F=100cm^{-1} , σ=25cm^{-1} at OK for D-adipic acid. It clearly shows the predominance of Fermi resonances in the width of ν_s at OK. Upon raising temperature the σ's rapidly increase so as to become comparable at 400K to the σ_F's which are temperature independent. Similar results have been determined for simple carboxylic acids in the gas phase(300K)[22].

We may conclude this paragraph by stating that the so particuliar shape ν_s exhibits in vibrational spectra is satisfactorily understood. Arriving at this point required the elaboration of a quantitative description of vibrations of H-atoms in H-bonds which followed several specific measurements. We are tempted to no longer consider it a problem. As we are to see it, it does not mean that ν_s no longer contains unclear points.

3.5. INTENSITY OF ν_s

Very few measurements of intensities of vibrational bands have been performed, if one excludes simple molecules . This is because intensities are delicate to measure quantities, which are most sensitive to the presence of impurities and that one should be most careful to obtain reproducible results. The experimental quantities corresponding to integrated intensities of absorption bands are[22]:

$$\mathcal{P}_{band} = \int_{band} \log_{10}\left(\frac{I_0}{I}\right)\frac{d\tilde{\nu}}{\tilde{\nu}} \qquad (21)$$

The corresponding SI unit($m^2.mole^{-1}$) is:

$$\Gamma_{band} = \frac{Ln(10)}{\rho\mathbb{L}}\,\mathcal{P}_{band} \qquad (22)$$

where $\mathbb{L}$ is the length of the absorbing cell and ρ the number per unit volume of moles of the species at the origin of the considered band. By varying pressure at a defined temperature we have shown that we could measure ρ's of monomers and of dimers in a vapour of carboxylic acids and thus obtain values of Γ's of various bands in monomers and in dimers[22] at that temperature. One noticeable result, even if already announced, which concerns values of these Γ's is the dramatic enhancement of intensity of the stretching vibration of the H-atom induced by the establishment of the H-bond: between one O-H group and one O-H...O group, the ratios of the Γ's in ν_s and ν_{OH} falls between 20 and 25 for the various carboxylic acids.

The most puzzling result, however, comes from the comparison of the Γ's of corresponding bands between H-bonds and D-bonds: for all carboxylic acids having these cyclic H-bonded dimers the ratios of these quantities is greater than expected. In the case of ν_s it is found equal to 2. This value is greater than the value 1.3 of the ratio of reduced masses one expects when applying results for M_0 (Eq. (16)). For intense $\nu_{C=O}$ bands it is found equal to 1.3 instead of the expected value 1. For all other bands it is greater than the expected value by some 10%. This most surprising result has still not received any precise explanation. It most probably is the manifestation of a cooperative interaction between the two H-bonds due to this particuliar cyclic structure of dimers of carboxylic acids. The most probable candidate is then the transfer of protons that this cyclic structure allows to achieve. It is represented in Fig.6 which shows that passing from tautomer A to tautomer B is possible with such a transfer of protons.

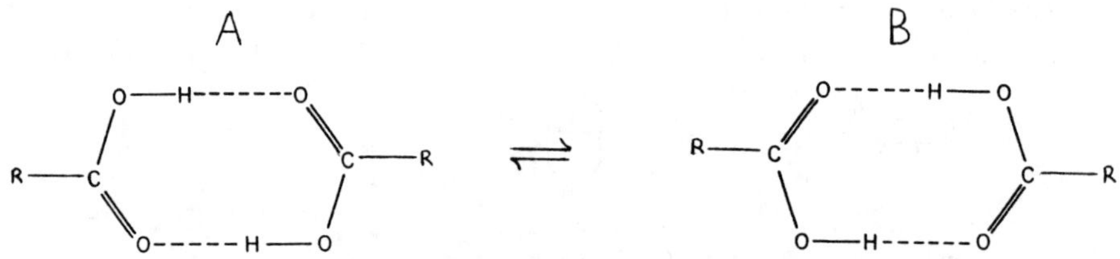

Figure 6. The two tautomers of H-bonded cyclic dimers.

Such transfers have been seen by Hayashi and Coworkers who studied $\nu_{C=O}$ stretching bands($\overrightarrow{C}{=}\overleftarrow{O}$ vibrations) in crystals of benzoic acids

C_6H_6COOH(23)(24),where molecules form the same cyclic dimers as in the gas phase.In crystals the two tautomers A and B of Fig.6 have not the same energy.At temperatures of some K,only one type of tautomers,say A type,are present,and the polarization of the $\nu_{C=O}$ band is unique and directed along C=O bonds of tautomers A.As one raises temperature a second polarization for this band appears at 120° from the preceding one.It is the direction of C=O bonds in tautomer B.From NMR results the frequency of these transfers has been shown to be of the order of $10^{10} s^{-1}$ at room temperature.It has be shown to be of some 10^8 s^{-1} at 2K by multiquanta transitions measurements(25) .They are presently the object of studies in the FIR region where quantum levels corresponding to such transfers are looked at in our group.

Such transfers of protons are most probably present in cyclic dimers in gases.As can be seen in the case of formic acid(14) they induce distorsions in the positions of other nuclei which should also be moved when passing from tautomer A to tautomer B.In consequence they are certainly faster in these light carboxylic acids found in gas phase than in rigid benzoic acid in crystals.It is therefore not surprising that with frequencies higher than 10^{12} s^{-1} they have an effect in IR spectra.The origin of their appearance is however still unclear,because such simultaneous transfers of protons affect the symmetric motion of protons described by coordinate q_1+q_2 .As we have seen the ν_s IR band corresponds to motions in the antisymmetric coordinate q_1-q_2 .In opposition to well spread ideas the enhancement of the intensity of ν_s by such a transfer of protons can not therefore be the direct consequence of a double-minimum potential,which would only concern q_1+q_2 modes.It is then more likely to be a consequence of non-adiabatic separability between electrons and nuclei in these systems.Calculating the integral intensity of ν_s would require then going beyond the Born-Oppenheimer approximation,which we have supposed to be valid for establishing Eq.(12).We expect its non applicability to primarily affect M_0 and much less higher order cumulants(Eq.(16)) where,as already seen, M_0 acts like a renormalization factor.This practically signifies that the analysis which we have performed on the shape of ν_s is likely to remain valid even in the still hypothetical presence of this non-adiabatic effect.The physical image which we may give of such a breakdown of the Born-Oppenheimer approximation is that during such a transfer of protons electrons do not follow protons immediately but require some delay to adapt themselves to new positions of nuclei.This delay may magnify any dipole moment due to the presence of an antisymmetric type distorsion which will be seen in IR spectra as an enhancement of intensities of transitions.

We see that this point of an anomalous enhancement of intensity when passing from an H-bond to a D-bond is still far from being completely clear,and will certainly require new experimental data to be fully understood.One point worth noting is that it certainly much owes to the particuliar cyclic structure of these dimers.We may then hope to have,with measurements of IR intensities,a mean to detect such cyclic structures,and also transfers of protons for which such cyclic structures seem necessary.These resonance-type transfers of protons,as opposed to relaxation-type transfers which for instance follow an electronic excitation due to absorption of a visible or UV photon,certainly play a fundamental role in the majority of biological reactions when considered at the molecular level [1] .In particular even if they have never been really seen,they are almost always invoked in any biomechanism considered at the molecular level,to lower barrier potentials of the numerous intermediate states that enzymatic reactions require.It must then be stressed that with the flexibility that its H-bond network confers it,water is particularly suited to establish such cyclic structures and consequently permit such reactions to occur.In the case of pure water,for instance,six membered rings are well known configurations inside which these resonance transfers of protons may occur.Because of the lack of experimental data,they have,however,scarcely been considered.This is a reason why applications of these previous methods of quantitative IR spectroscopy to water seem most promising.

4. INFRARED SPECTROSCOPY OF WATER

The first point to be stressed is that most surprisingly quantitative spectra of water where such theoretical analyses as presented precedingly may be applied are quite recent,even in the conventional region 500-5000 cm^{-1},where most vibrations appear.This is because the density of strong absorbing H-bonds in water is so high that water is in practice opaque to IR radiation.It is consequently out of the question to obtain precise reliable measurements from absorption spectroscopy,as previously described in this paper.This is the reason why up to now Raman techniques,where specific bands of H-bonds are not intense,have been the main technique of vibrational spectroscopy for water,in opposition to the situation that prevails for H-bonds in general.Fortunately recent advances in ATR (attenuated total reflection) techniques have allowed to obtain quantitative IR spectra of water.With such techniques the IR beam does not go through water,but travels in a crystal of high refractive indice.The geometry of this crystal is such that once inside the crystal the IR beam suffers total reflections before having the possibility of getting out of the crystal.The outcoming beam is strongly modified if,instead of a non absorbing medium,this crystal is plunged in an absorbing medium such as water.In principle the knowledge of the indice of the crystal in the IR region allows,by application of Fresnel Equations at each "total" reflection of the IR beam,to determine from these modifications of the outcoming beam the optical constants,or equivalently the complex dielectric constant of this absorbing medium.The physical ground of this technique is that we allow the medium around the crystal to absorb a part only of the IR beam which is represented by the evanescent wave going out of the crystal after a "total" reflection is achieved.The amplitude of this evanescent wave exponentially decreases with the distance from the surface of the crystal,even in the case of a transparent surrounding medium.As a consequence it falls to zero at a distance from this surface of the order of one micron only.In practice commercial apparatus for performing ATR measurements in liquids have recently appeared.They require a particuliar optical set-up to be arranged around a crystal having a cylindrical geometry.The feasibility of the determination of optical constants from ATR measurements has been demonstrated by Bertie and Eysel [26] .It requires using iteration procedures and Kramers-Kronig integral calculations which are now easily replaced by double Fourier transforms.

We show in Fig.7 $\varepsilon''(\tilde{\nu})$ of water and heavy water at various temperatures,as obtained from such ATR measurements.As already mentioned $\varepsilon''(\tilde{\nu})$is proprtional to $s(\tilde{\nu})$.With such spectra we are then in a good way to apply quantitative analysis to water in the same way as precedingly described for carboxylic acids.IR spectroscopy is,in my

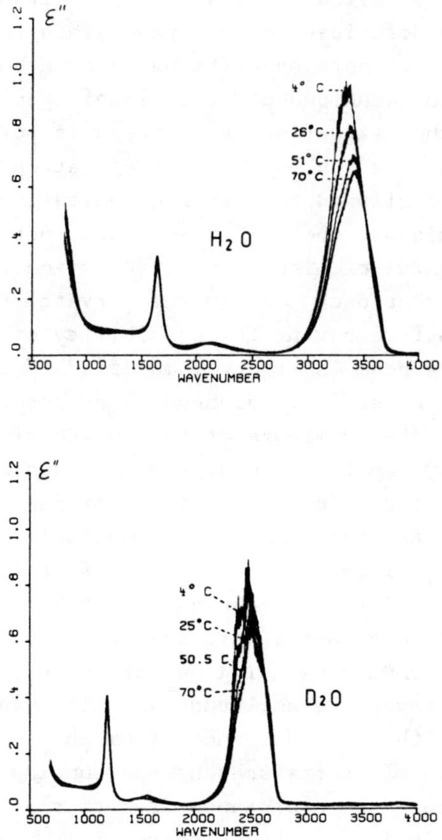

Figure 7. IR spectrum of water and heavy water deduced from ATR measurements using Bertie-Eysel iterative method[26].

opinion, one of the most powerfull technique,mainly because,as may appear from preceding paragraphs, it is backed up by a precise theory.It is then our hope that analysis of quantitative spectra of water will bring some new insight on the properties of water at molecular level,and more particularly on resonance-type transfers of protons inside ring structures which we suspect,as already mentioned,to be a fundamental mechanism especially in enzymatic reactions in biology.

REFERENCES

(1) Marechal, Y. (1989) 'La Liaison Hydrogène', La Recherche, 20, p. 480. .

(2) Lippert, E. (1976)'Introduction' , in P. Schuster, G. Zundel and C. Sandorfy (eds.),*The Hydrogen Bond:Recent Developments in Theory and Experiments*, North-Holland, Amsterdam, pp.5-6.

(3) Bertie, J.E. and Millen, D.J. (1965), J. Chem. Soc.,p. 497 and 514.

(4) Desbat, B. and Lassègues, J.C. (1979), J. Chem. Phys. 70, p. 1824.

(5) Bratos, S. and Hadzi, D. (1957), 27, p. 991.

(6) Bouteiller, Y. and Maréchal, E. (1976), Mol. Phys. 32,p. 277.

(7) Bernadet, F. and Fillaux, F. (1986), Chem. Phys. 106,p. 447.

(8) Lassègues, J.C. and Lascombe, J. (1982), 'Vibrational Spectra of gaseous Hydrogen-bonded compounds', in J.R. Durig (ed.), *Vibrational Spectra and Structure*,Elsevier,Amsterdam, **Vol.11**, p 51.

(9) Szczesniak, M.M. ,Kurnig, Ingrid J. and Scheiner,Steve (1988), J. Chem. Phys., 89,p. 3131.

(10) Legon, A.C. and Millen, D.J. (1986), Chem. Rev., 86,p. 635.

(11) Maréchal, Y. (1988), J. Mol. Struct. 189, p. 55.

(12) Auvert, G. (1979), Thesis at the Université Scientifique et Médicale de Grenoble.

(13) Haurie, M. and Novak, A. (1965), J. Chim. Phys. 62,p. 146.

(14) Maréchal, Y. (1987), 'Infrared Spectra of Cyclic Dimers of Carboxylic Acids:the Mechanics of H-bonds and Related Problems'in J.R. Durig (ed.), *Vibrational Spectra and Structure*, **Vol.16**,Elsevier,Amsterdam, p. 311.

(15) Callen, H.B. (1962),'The Fluctuation-Dissipation Theorem and Irreversible Thermodynamics', in ter Haar (ed), *Fluctuation, Relaxation and Resonance in Magnetic Systems*,Oliver and Boyd, London,p. 15.

(16) Gordon, R. (1965), J. Chem. Phys., 43 p 1307.

(17) Kubo, R. (1962), 'A Stochastic Theory of Line-shape and Relaxation',in ter Haar (ed), *Fluctuation,Relaxation and Resonance in Magnetic Systems*,Oliver and Boyd,London,p. 23.

(18) Excoffon, P. and Maréchal Y. (1980), Chem. Phys. 53, p.23.

(19) Excoffon, P. and Maréchal Y. (1980), Chem. Phys. 52, p.245.

(20) Bratos, S. (1975), J. Chem. Phys. 63, p. 3499.

(21) Maréchal, Y. (1983), Chem. Phys. 79, p. 69 and 85.

(22) Maréchal, Y. (1987), J. Chem. Phys. 87, p. 6344.

(23) Hayashi S. and Umemura J. (1974), J. Chem. Phys. 60, p. 2630.

(24) Nagaoka S., Terao T., Imashiro F., Saika A., Hirota N. and Hayashi S. (1983), J. Chem. Phys. 79, p. 4694.

(25) Hochstrasser, R.M. and Trommsdorff, H.P. (1987), Chem. Phys. 115 p. 1.

(26) Bertie, J.E. and Eysel, H.H. (1985), Applied Spectroscopy 39, p. 39

Vibrational and Collision-Induced Raman Scattering from Water and Aqueous Solutions

G. E. Walrafen, M. S. Hokmabadi, and Y. C. Chu
Chemistry Department
Howard University
Washington, D. C. 20059

ABSTRACT.

Vibrational and collision-induced (CI) Raman spectra were obtained from liquid water, and from aqueous solutions of $NaClO_4$ (7.4 M), NaI (8.2 M), and sucrose (2.3 M). The inter- and intra-molecular vibrational spectra were found to lie above the collision-induced (CI) Raman baselines which were approximated, above 20 cm^{-1}, by $d\ln I/d\ln w = -m$, with $m \approx 1.2-1.3$ (I = intensity, w = vibrational frequency in cm^{-1}). Significant vibrational effects above the CI baselines, e.g., shape changes in the restricted translational, librational, H_2O bending, and OH-stretching bands, including marked changes in the depolarization ratios, were observed for the solutions, compared to pure water.

The pure isotropic Raman intensities of the restricted translational band near 155 cm^{-1}, and of the OH-stretching band near 3480 cm^{-1}, were enhanced significantly for the NaI solution, compared to pure water. These modes, which refer to restricted translations of first-sphere H_2O molecules against the I^- ion, and vibrations of the proton along the $I^-\cdots H-O$ hydrogen bond direction, respectively, both involve increases of the isotropic polarizability derivatives because of the high polarizability of the I^- anion, compared to H_2O.

Increased Raman polarization and sharpening of the hydrogen bond $O\cdots H-O$ stretching band from the sucrose solution near ≈200 cm^{-1} was also observed, along with a prominent 3210 cm^{-1} OH-stretching component from the concentrated sucrose-water glass. These observations suggest an ordering of the hydrogen bonds via the sucrose structure.

Increased Raman polarization in the restricted translational region from the $NaClO_4$ solution indicates strong interactions between ClO_4^- and H_2O, which are also evident from the large librational intensity enhancements. The perchlorate-water interactions may involve bifurcation between pairs of perchlorate oxygen atoms, and protons of first-sphere H_2O molecules.

1. INTRODUCTION.

The CI Raman baseline scattering, occurring underneath the inter- and intra-molecular vibrational bands, was recently described for liquid water and aqueous solutions [1,2]. CI processes in water and aqueous solutions are of considerable

J.C. Dore and J. Teixeira (eds.), Hydrogen-Bonded Liquids, 261–282.

interest, of course, but we shall be concerned here, primarily, with results which arise from use of the CI baselines, that is, with the shapes of the inter-and intra-molecular vibrational features resulting after the CI baseline intensity has been removed, or otherwise considered.

Failure to recognize the magnitude and frequency dependence of the CI baseline intensity, especially in the inter-molecular vibrational region, can lead to erroneous conclusions about the effects of temperature, or of solute addition, on water, and thus correction for the CI Raman baseline intensity is especially important [1,2].

In this chapter we shall describe the effects of $NaClO_4$, NaI, and sucrose on water structure, and we will attempt to arrive at mechanisms leading to these effects.

$NaClO_4$ has long been known to produce marked effects on the water structure, similar to those of temperature rise in the inter- and intra-molecular vibrational regions. For example, the bands from water near 170 cm^{-1} and between 2800 and 3800 cm^{-1} are strongly affected by the ClO_4^- anion [3,4,5]. Moreover, the effects of ClO_4^- on water are similar to those of PF_6^- or BF_4^- [5]. The ClO_4^-–H_2O interactions must clearly be very strong because of large librational intensity enhancements observed in this work. Further, because of the similarities of the perchlorate spectra to spectra of water at elevated temperatures, the possibility of bifurcated interactions, e.g., between pairs of O atoms on the perchlorate ion, and protons of the surrounding H_2O molecules exists [6,7].

NaI, in contrast, produces different effects, spectroscopically, from those of temperature rise on pure water. I^- produces a marked decrease in the Raman depolarization ratio near 155 cm^{-1}, compared to pure water, and an intensified, highly-polarized, OH-stretching component near 3480 cm^{-1} (the peak depolarization ratio is about 0.3). Because these changes occur at a stoichiometric $[H_2O]$ to $[I^-]$ molarity ratio of 4.6, it is probable that most of the H_2O molecules occur in the first hydration sphere of the I^- anion. Hence, it is reasonable to treat the I^-–H_2O interactions as fundamental vibrations of an $I^-(H_2O)_n$ structure, where n is 4, 5, or 6, but there also may be sharing of first sphere water between I^- anions. For example, the stoichiometry is consistent with one second-sphere H_2O molecule between two 4-coordinated anionic structures, or one shared first-sphere H_2O between 4- and 6-coordinated structures, or three shared H_2O's between two 6-coordinated structures, i.e., no single unique structure, even ignoring the Na^+ coordination.

Highly concentrated sucrose, like NaI, is almost certainly fully hydrogen bonded to the surrounding H_2O molecules, and also intramolecularly. The sucrose spectra show a marked sharpening near 200 cm^{-1}, not seen for NaI, and this may indicate a significant ordering of the hydrogen bonds.

2. RAMAN DATA.

2.1. Pure Liquid Water.

Depolarized, X(ZX)Y, and polarized, X(ZZ)Y, Raman spectra from pure liquid water are shown in the (a) (upper) sections of Figs. 1 and 2, respectively. The (a) spectra of these two figures were corrected for the Bose-Einstein thermal population factor [1,2], for the fourth-power dependence of the scattering intensity upon the actual scattering frequency, and for the instrumental sensitivity, which was very accurately measured [2]. The CI Raman intensity of Figs. 1 and 2 corresponds to the area under the curved (solid) line shown in (a), but this baseline intensity was determined as described next.

The Raman intensities, corrected for all factors <u>except</u> for the Bose-Einstein factors, were plotted on lnI versus lnw (natural logarithm) plots in the (c) (lower) sections of Figs. 1 and 2. From such plots, straight (solid) lines were determined as described in [1,2]. The slopes of these straight lines are given by $d\ln I/d\ln w = -m \approx -1.2$ to -1.3, and the area under these straight lines constitutes the CI Raman intensity.

Plots corresponding to the (c) plots, but with the Bose-Einstein correction included, were then computed, and they are shown in the (b) (middle) sections of Figs. 1 and 2. Note that the straight lines of the (c) plots are now curved downward, particularly at low w values.

The curved (solid) ln-ln CI baselines were converted into the corresponding I versus w baselines, which are shown in the (a) plots of the figures. The intensities under the curved baselines of the (a) plots of Figs. 1 and 2 are considered to arise from CI Raman processes alone, whereas the intensities above the curved lines refer to inter- and intra-molecular vibrational Raman scattering (albeit possibly perturbed by CI processes, but not readily separable from such processes).

A great improvement in the (a) presentation of Figs. 1 and 2 can be achieved by subtracting the CI Raman intensity. (This was accomplished with digital computers, as all of the other presentations were accomplished.) The result of the CI baseline subtraction is shown in the (b) (lower) section of Fig. 3 for the X(ZX)Y or depolarized Raman spectrum, which includes Bose-Einstein correction as well. The (b) spectrum constitutes a very nice spectral display of the complete inter- and intra-molecular vibrational spectrum, which is amenable to further computerization as described in [2]. That is, if the frequency dependent coupling coefficient, C(w), is known accurately, multiplication of the intensities, Fig. 3 (b), by w/C(w), should yield the vibrational density of states, free from the effects of the CI baseline contribution.

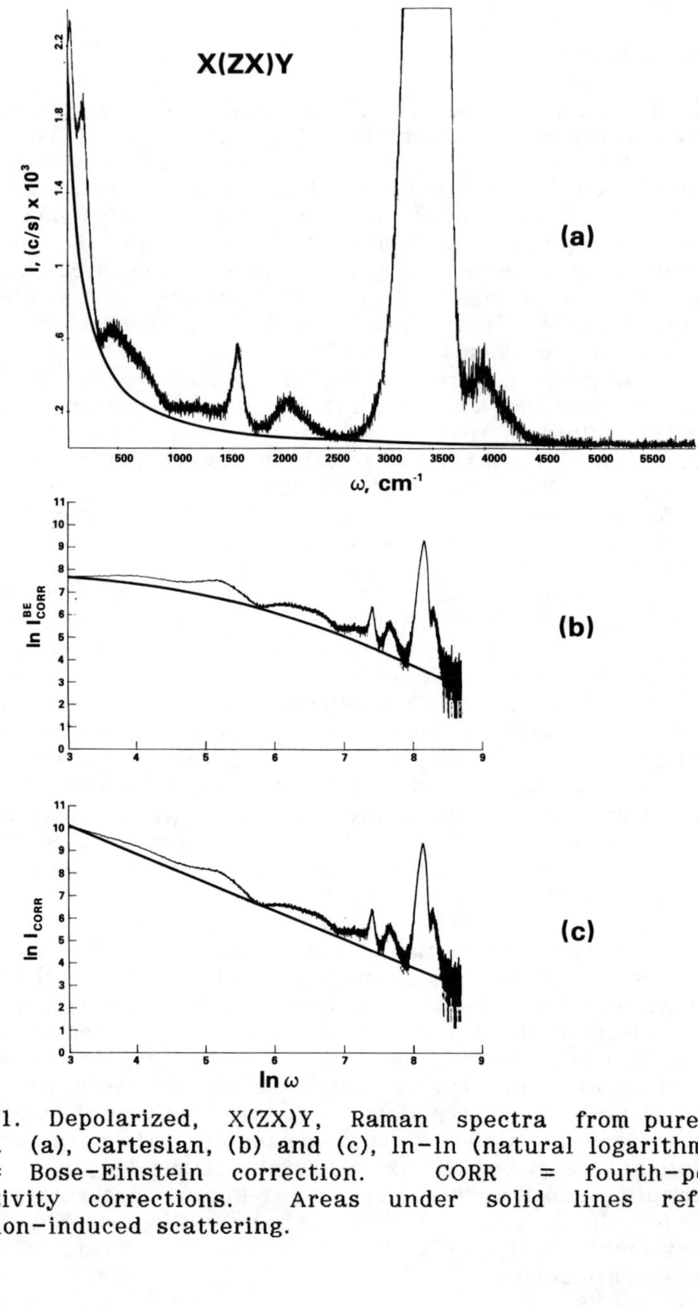

Fig. 1. Depolarized, X(ZX)Y, Raman spectra from pure liquid water. (a), Cartesian, (b) and (c), ln-ln (natural logarithms). BE = Bose-Einstein correction. CORR = fourth-power and sensitivity corrections. Areas under solid lines refer to collision-induced scattering.

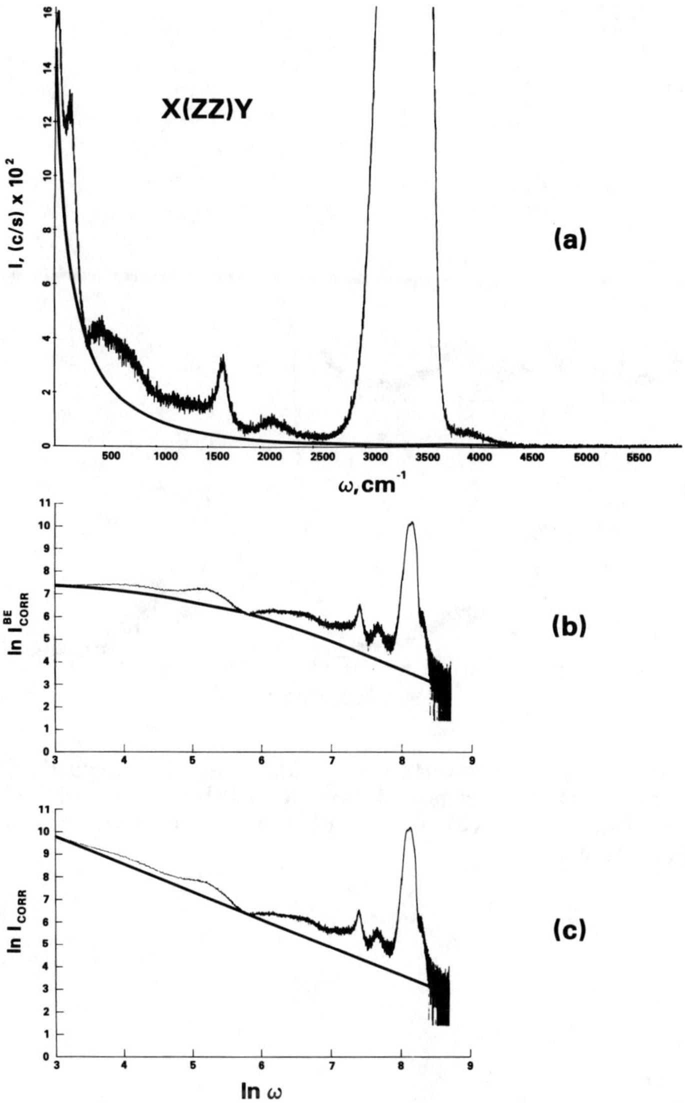

Fig. 2. Polarized, X(ZZ)Y, Raman spectra from water. See Fig. 1 caption for details.

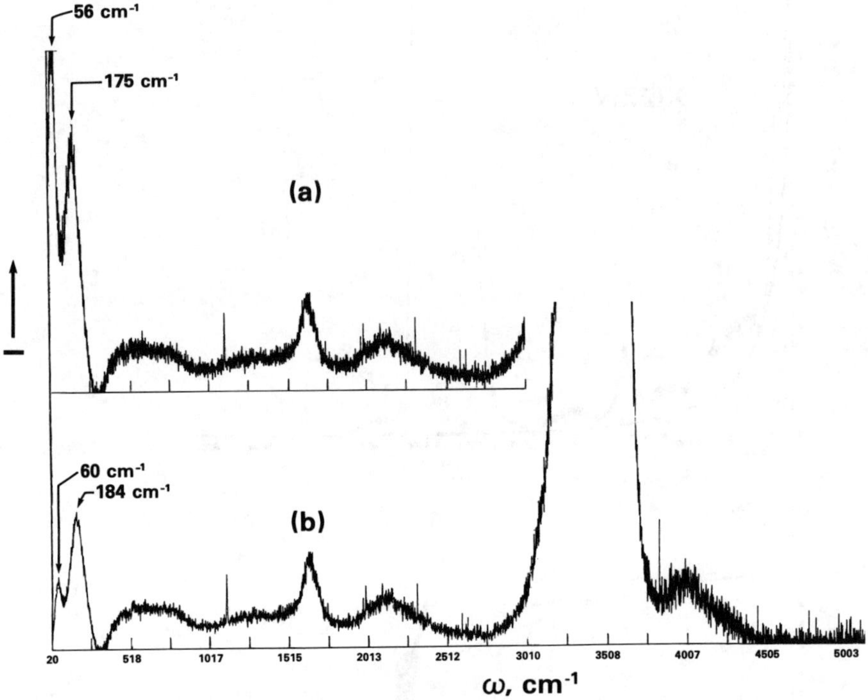

Fig. 3. Raman spectra from water with Gaussian-ln-ln collision-induced baseline scattering removed, see Ref. [2] for details of Gaussian-ln-ln function. (b), Bose-Einstein corrected, (a), no Bose-Einstein correction.

A depolarized I versus w presentation, with the CI baseline subtracted, but <u>without</u> Bose–Einstein correction, is also shown in (a) of Fig. 3. This (a) spectrum clearly indicates that it is possible to demonstrate the existence of the <u>peaks</u> at ≈56 cm^{-1} and ≈175 cm^{-1} without Bose–Einstein correction, provided that the CI baseline contribution has been removed. The corresponding peaks in the (b) spectrum are slightly shifted upward to 60 cm^{-1} and 184 cm^{-1}.

Both Raman spectra of Fig. 3, of course, refer to the pure <u>anisotropic</u> spectra. Pure <u>isotropic</u> spectra, with or without Bose–Einstein correction, but otherwise fully corrected, and with the CI baseline scattering subtracted, could be accomplished from I[X(ZZ)Y] – (4/3) I[X(ZX)Y]. This procedure is slightly different from obtaining the fully-corrected, pure isotropic spectrum <u>without</u> CI baseline subtraction--a procedure which requires less computation. The CI baseline Raman scattering is approximately depolarized. However, there may be a very slight amount of polarization at low frequencies, roughly below 1000 cm^{-1}, see Table III of [2]. This slight polarization, which seems from repeated measurements beyond those reported in [2], to be real, means that the isotropic procedure alone (no CI subtraction) does not remove all of the CI baseline effect.

2.2. $NaClO_4$ Solution (7.4 M).

Ln–ln Raman spectra, (a) X(ZZ)Y, and (b) X(ZX)Y, are shown without Bose–Einstein correction, but otherwise fully corrected, in Fig. 4 for a 7.4 M (molar) $NaClO_4$ solution. The librational areas are shown by the dashed lines in (a) and (b), above the CI contribution (solid) straight line. A considerable change in shape has occurred, compared to water, and a downward shifting of the librational frequencies (three in pure, cold water) has also occurred. These large effects of ClO_4^- are not easily seen in the ordinary I versus w presentation, partly because of the interference of the deformation modes of the anion. They are, however, readily seen on the ln–ln presentation, and in addition the CI contribution is easily determined from such a presentation. The ClO_4^- anion is a strong structure-breaker as detailed in [3,4,5].

2.3. NaI Solution (saturated, 8.2 M).

Ln–ln Raman spectra, (a) X(ZZ)Y, and (b) X(ZX)Y, fully-corrected except for the Bose–Einstein correction, are shown in Fig. 5 for an 8.2 M NaI solution. Again large spectral changes, compared to pure water, are made readily evident by the ln–ln presentation, and by the linear ln–ln CI baseline. Large intensity enhancements are evident for the librational bands, shown between about 5.5 and 6.8 on the lnw scale. However, these bands, like those from pure water, are almost completely

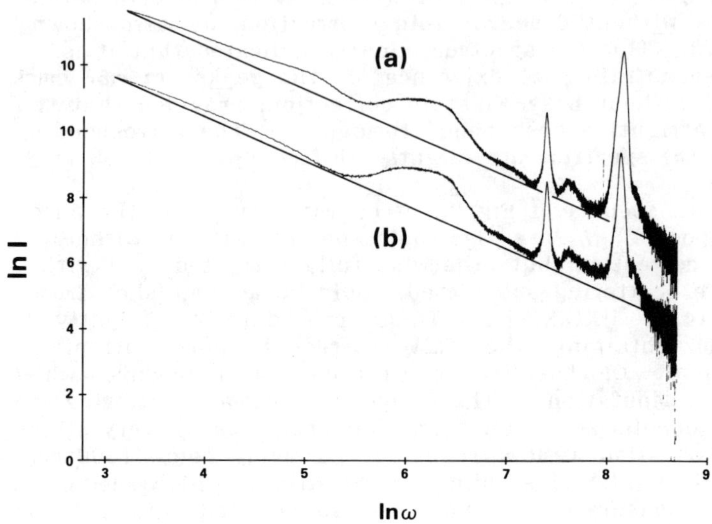

Fig. 4. Ln-ln Raman spectra from an aqueous 7.4 M $NaClO_4$ solution. No Bose-Einstein correction. (From Ref. 2)

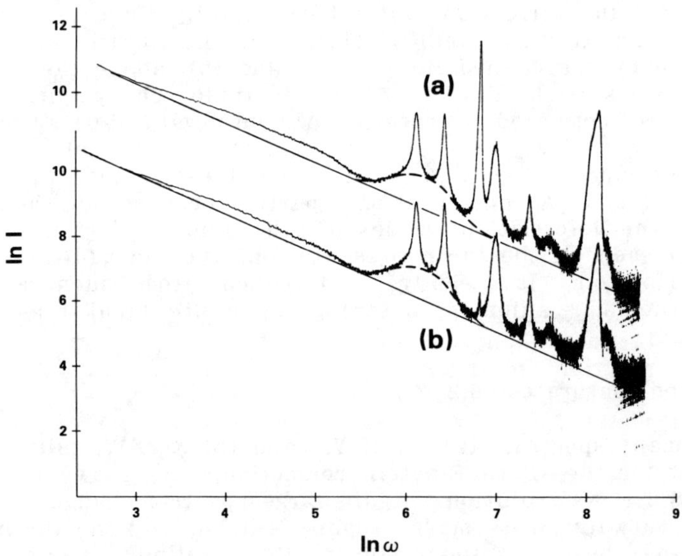

Fig. 5. Ln-ln Raman spectra from an aqueous 8.2 M NaI solution. No Bose-Einstein correction. (From Ref. 2)

depolarized, that is, the depolarization ratio is equal to, or just a little smaller than 3/4. Large intensity enhancements also occur in the H_2O bending region, near 7.4 on the lnw scale, and in the OH-stretching region, near 8.15 on the lnw scale. In particular the depolarization ratio at the peak frequency of 3480 cm^{-1} is about 0.3. Also, the hydrogen-bond stretching band, near 5 on the (a) lnw scale, and near 4.8 on the (b) lnw scale, is now seen to be polarized from visual inspection alone, when the linear ln-ln CI baseline is present. This obvious polarization constitutes a large change from pure water, compare Figs. 1 and 2, where the depolarization ratio near 170 cm^{-1} is just slightly smaller than 3/4, i.e., weakly polarized. (Some polarization, however weak, is required for a symmetric intermolecular mode of this type.)

The (a) and (b) spectra from NaI were corrected for the Bose-Einstein factors, and the corresponding pure isotropic spectra were computed. These isotropic spectra are shown in Fig. 6, where the corresponding spectra from pure water are included for comparison (below). No CI subtraction was accomplished in obtaining the Fig. 6 spectra.

From Fig. 6, it is evident that a particularly large isotropic intensity enhancement has occurred for the 155 cm^{-1} component, compared to pure water. In analogy with the assignments for pure water, the 45 and 155 cm^{-1} features from the NaI solution are assigned to "bending" and stretching motions, respectively, of the $I^-\cdots H-O$ hydrogen bonds. Here, "bending" actually refers to $O-I^--O$ bending, just as the term hydrogen bond "bending" refers to O-O-O bending for pure water, that is, H_2O translation, as opposed to proton motion, is involved.

2.4. Sucrose Solution (2.3 M).

Ln-ln Raman spectra, X(ZZ)Y (a), and X(ZX)Y (b), are shown in Fig. 7 for a 2.3 M aqueous sucrose solution. In this case, it is virtually impossible to determine the shape of the librational contour, accurately, because of the numerous interfering vibrations of the sucrose. Nevertheless, a lower limit to the librational contour intensity was estimated, and it is shown by the dashed lines between about 5.5 and 6.8 on the lnw scale, and above the CI contribution (solid) line.

The lower librational intensity limit shows little increase, compared to pure water. But the sucrose vibrations are difficult to subtract, and conventional I versus w spectra suggest sizeable enhancement [3]. However, the hydrogen-bond O-O stretching peak near 5.3 on the lnw scale is clearly enhanced, and sharply peaked as well, compared to pure water. In addition, this O-O stretching peak is obviously polarized, as seen visually, from comparisons between the (a) and (b) spectra.

Evidence to follow indicates that a saturated sucrose solution is extensively hydrogen-bonded, that is, extensively

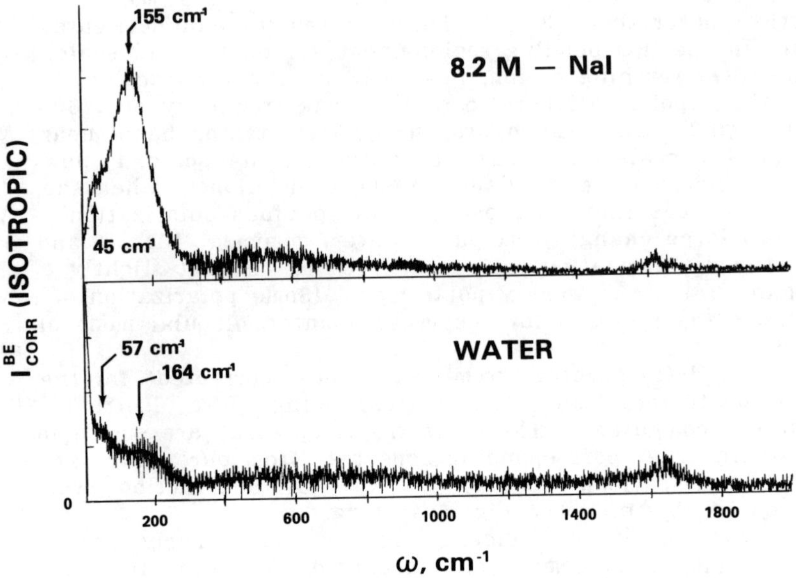

Fig. 6. Isotropic, I[X(ZZ)Y] − (4/3)I[X(ZX)Y], Bose-Einstein corrected, Raman spectra from 8.2 M NaI in water (upper), compared to pure water (lower).

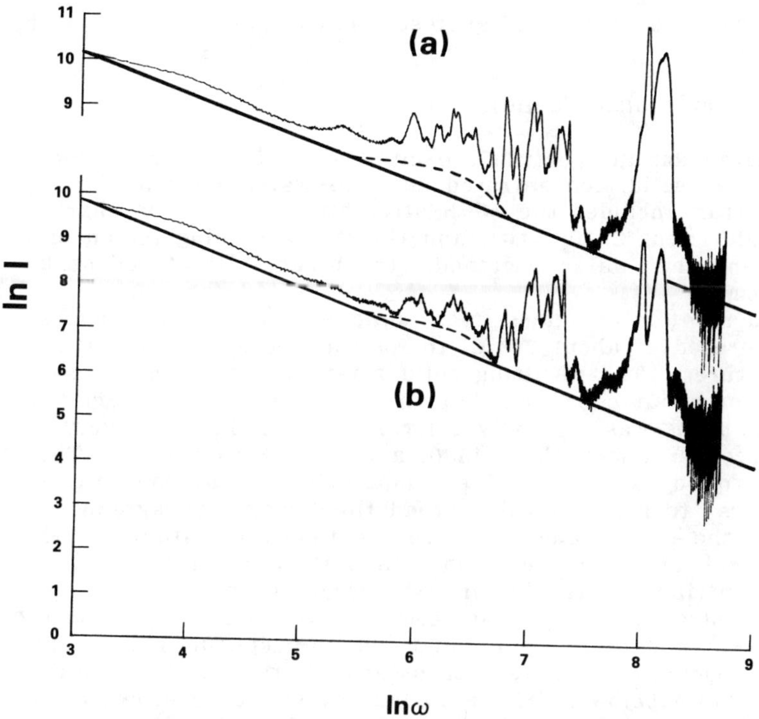

Fig. 7. Ln-ln Raman spectra from an aqueous 2.3 M sucrose solution. No Bose-Einstein correction. The (a) spectrum refers to X(ZZ)Y, and the (b) spectrum to X(ZX)Y, but these conditions refer to the electric vector as the beam first enters the sample, and to the polaroid orientation in front of the slit. We estimate that the rotation of the input electric vector is no more than 5 degrees, and the rotation of the scattered electric vector somewhat less than this. Nevertheless, the polarization data cannot be considered to be quantitative, although they clearly show considerable polarization for several bands.

hydrogen-bonded in terms of sucrose-sucrose plus sucrose-water hydrogen bonds.

2.5. Sucrose-Water Glass (Solid).

A polarized Raman spectrum, excited with 5145 Å radiation, is shown for a solid, concentrated, sucrose-water glass in Fig. 8. This spectrum includes the C-H stretching region, the sharp peak, off-scale near 2910 cm^{-1}, and the OH-stretching region of the sucrose and the water combined, the broad, structured peak near ≈3400 cm^{-1}.

The sucrose-water glass was prepared by solidifying a solution, saturated at about 70° C, to form a colorless glass at room temperature. The resulting solid glass was stable at rooom temperature for about one week, but then it slowly crystallized. Although this glass was visually colorless, it produced a very broad fluorescence under 4579, 4880, and 5145 Å excitation. The sloping background with negative slope shown on the Fig. 8 spectrum refers to part of this broad fluorescence background. The shape of the fluorescence was very extensively studied with the three exciting lines mentioned, and thus the fluorescence background contribution was known with great accuracy.

The C-H stretching peak was removed by means of a computer from the OH-stretching contour, and the OH-amplitudes above the fluorescence background were transferred to a horizontal baseline, as shown below in Fig. 8. This transferred contour, as well as all of the original contours obtained with the three exciting lines used, indicated two-component structure, visually. The transferred contour was decomposed into two Gaussian components with the du Pont 310 analog computer, and the results of this two-Gaussian decomposition are shown in Fig. 8.

One of the Gaussian components is centered near 3210 ± 10 cm^{-1}, and the other component is centered near 3425 ± 10 cm^{-1}. These values are essentially identical with the values reported from Gaussian decompositions of pure water [8,9,10]. On the basis of previous assignments [8], the 3210 cm^{-1} component is assigned to OH-oscillators between several OH-groups which vibrate in phase (five H_2O molecules in the case of pure water), whereas the 3425 cm^{-1} component is assigned to out-of-phase OH-stretching between several OH-groups.

A further key observation resulting from Fig. 8 is that no additional Gaussian components at higher frequencies were required to fit the OH-stretching contour, that is, no components, whatever, near 3500 and 3620 cm^{-1} were required [8,9,10]. The higher frequency components arise when hydrogen bonds are broken, e.g., bifurcated [6,7], and their absence in the Fig. 8 spectrum indicates that the very concentrated sucrose-water glass is fully hydrogen bonded. The sucrose concentration of the glass, of course, was much greater than 2.3 M.

It is probable that the 2.3 M sucrose solution was

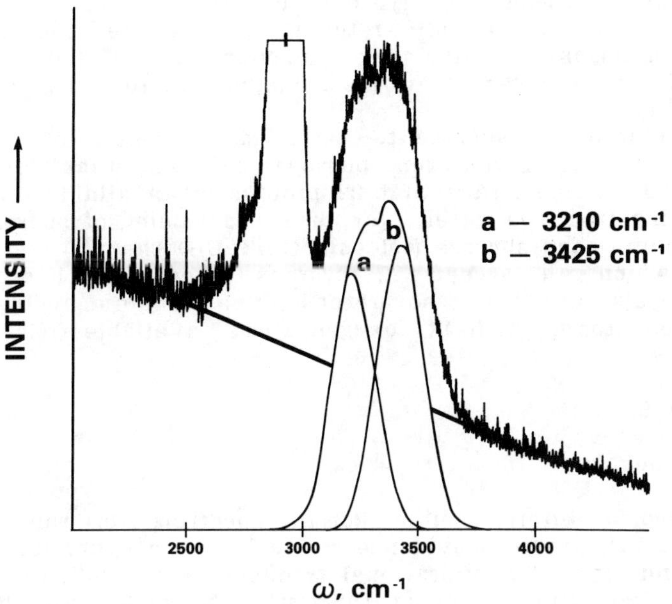

Fig. 8. Raman spectra from a concentrated (>> 2.3 M) sucrose-water glass (solid). The heavy, solid, curved baseline results from fluorescence, due to an impurity. The lower spectrum refers to the upper average signal minus the baseline, and also minus the 2910 cm^{-1} feature, whose center is shown by the vertical bar (┼). (a) and (b), Gaussian components obtained with a computer.

extensively (but not completely) hydrogen bonded, because only about 4 % of the total OH-stretching contour area could be attributed (from computer analysis) to components in the 3500-3650 cm^{-1} region. However, it is also instructive to consider the stoichiometry.

The stoichiometric sucrose-to-water molarity ratio of the 2.3 M solution is 11.6. For present purposes consider that this is essentially 11. This means that 22 protons are available for hydrogen bonding from the water, for each molecule of sucrose present. But on each sucrose molecule ($C_{12}H_{22}O_{11}$) there are 11 oxygen atoms which can accept 22 protons when fully hydrogen bonded. There also are 8 sucrose protons which can form hydrogen bonds to oxygen atoms, with 11 oxygen atoms available from the water alone.

3. CONCLUSIONS.

3.1 NaI Solutions.

The isotropic intermolecular Raman spectrum from the NaI solution, Fig. 6 (a), provides evidence related to the hydration structure around I^-. Two vibrational frequencies are indicated from the 8.2 M NaI solution, one at 155 cm^{-1}, the peak, and the other, very roughly near 45 cm^{-1}, the shoulder. These two features may be assigned, in very broad terms, to $I^- \cdots H-O$ stretching and "bending", respectively, but further detailed treatments yielded more accurate and extensive data. The Raman spectra corresponding to these more involved treatments will not be shown here (for brevity). But, the results derived from them are described in detail.

An isotropic Raman spectrum was calculated, after Bose-Einstein correction, and with the CI baselines subtracted. The isotropic intensity is given by, I(iso) = [l/(1 + n)]{I[X(ZZ)Y]-CI[IX(ZZ)Y]} - [(4/3)/(1 + n)]{I[X(ZX)Y] - CI[IX(ZX)Y]}, where CI refers to the collision-induced baseline intensity, and 1/(1 + n) is the Bose-Einstein factor. (Note that the polarized and depolarized CI intensity contributions are not the same.) This spectrum indicated an intense, broad peak near 159 cm^{-1}. Only a very small amount of intensity remained near 45 cm^{-1}. This result indicated that the 159 cm^{-1} feature is strongly polarized.

Next, the two spectra used to calculate the isotropic spectrum, that is, the polarized spectrum minus its CI baseline, and the depolarized spectrum minus its CI baseline, both multiplied by 1/(1 + n), were decomposed with the duPont 310 analog computer using Gaussian components.

The polarized spectrum (CI subtracted, Bose-Einstein corrected), yielded three components centered near 43 cm^{-1} (peak, PH = 1440 cps, FWHH = 38 cm^{-1}), 153 cm^{-1} (peak, PH = 2150 cps, FWHH = 80 cm^{-1}), and a central component at 84 cm^{-1} (PH = 1750 cps, FWHH = 84 cm^{-1}). (PH, peak height; FWHH, full-width, half height.

The depolarized spectrum (same corrections) also yielded three components near 30 cm^{-1} (peak, PH = 1010 cps, FWHH = 36 cm^{-1}), 80 cm^{-1} (PH = 1195 cps, FWHH = 88 cm^{-1}), and a weak component (PH = 205 cps, FWHH = 60 cm^{-1}) centered, in this case, near 129 cm^{-1}, instead of at 153 cm^{-1}.

The depolarization ratios of the three major components were: 0.70 (at 30–43 cm^{-1}), 0.68 (at 80–84 cm^{-1}), and 0.04 (at 153 cm^{-1}). Gaussian computer decomposition of the isotropic spectrum also indicated a significant component near 129 cm^{-1}. Therefore, the 129 cm^{-1} component must be polarized. However, it was not feasible to carry out four-Gaussian decompositions of the polarized and depolarized spectra to determine its depolarization ratio. Nevertheless, from computer decompositions of the isotropic and depolarized spectra we were able to determine that the depolarization ratio of the 129 cm^{-1} component was roughly 0.4.

The two depolarized components (now quoted in terms of their average positions) at 37 cm^{-1} and at 82 cm^{-1}, plus the moderately polarized, but weak component at 129 cm^{-1}, and the intense, strongly polarized component at 153 cm^{-1} suggest approximate interpretation in terms of a tetrahedral XY_4 model, in which the H_2O molecules are treated as single Y units (mass 18), with X = I^- (mass 126.9). (Here we shall treat the 129 cm^{-1} component as if it were depolarized. Its moderate polarization may result from interactions, e.g., with neighboring I^- ions, i.e., from the sharing of H_2O molecules by I^- ions, etc.)

We applied the central force method to the tetrahedral XY_4 model using the equations given by Herzberg [11]. (The valence force method gave poor results.) We employed v_1A_1 = 153, v_2E = 36.5, v_3F_2 = 129 and v_4F_2 = 82 cm^{-1}. These average values yielded excellent results.

We obtained values of k_2 = $1.8_6 \times 10^4$ dyne/cm (k_2 is the stretching force constant), k_1 = $1.5_5 \times 10^3$ dyne/cm (k_1 is the bending force constant) and k' = −135 dyne/cm (k' is the interaction force constant). (This negative k' value was negligibly small, but k' was sensitive to the input data. For example, in one test calculation, k' changed to +3 dyne/cm for v_1 = 151.9 cm^{-1}, v_2 = 39 cm^{-1}, v_3 = 129 cm^{-1}, and v_4 = 82.4 cm^{-1}. Nevertheless, the LHS = RHS test, see Herzberg [11], gave good results, namely, LHS = 3.81 and RHS = 4.64.) Our stretching force constant, $1.8_6 \times 10^4$ dyne/cm, is just below the value of 2.18×10^4 dyne/cm calculated simply from I^- versus H_2O. Moreover it falls between the value of 1.6×10^4 dyne/cm for liquid water, and the value for a strong, linear O···H−O hydrogen bond, e.g., ice, 2.4×10^5 dyne/cm.

We next tested our results further by use of a second model, namely, a linear XYZ model with X = I^-, Y = H, and Z = O. We used a value of 1.82 Å for the I^-···H distance, and a value of

0.97 Å for the H–O distance. We assigned values of 3480 cm^{-1} to the O–H stretch, and 153 cm^{-1} to the I^-···H stretch. For the bend we used Gaussian librational component values from the 7.8 M iodide solution of 382 cm^{-1} and 566 cm^{-1}. These involve large proton amplitudes, and the proton amplitudes of the true bending mode (not the O–I^-–O bending) must also be large because of momentum conservation involving the heavier I^- and O atoms.

Again using equations given by Herzberg [11], we obtained stretching force constants of $1.9_9 \times 10^4$ dyne/cm for the I^-···H stretch, and $6.9_4 \times 10^5$ dyne/cm for the O–H stretch. The 382 cm^{-1} libration, which corresponds to the largest moment of inertia, yielded a reasonable bending force constant of $1.9_1 \times 10^3$ dyne/cm, but the higher librational frequency yielded a high value for the force constant, 4.2×10^3 dyne/cm, which we rejected. (If Z = OH, instead of O, one obtains, $K_1 = 6.93 \times 10^5$, $K_2 = 2.10 \times 10^4$, and K(bend) = 1.91×10^3 dyne/cm.) The stretching force constant of $1.9_9 \times 10^4$ dyne/cm compares favorably with the value of $1.8_6 \times 10^4$ dyne/cm obtained from the T_d XY_4 model. Moreover, the comparison between the XYZ bending force constant, 1.9×10^3 dyne/cm, and that from the XY_4 model of $1.5_5 \times 10^3$ dyne/cm is also reasonably good.

The force constant of $6.9_4 \times 10^5$ dyne/cm obtained for the OH-stretching mode of the H_2O molecule hydrated to I^- in the XYZ model is also in reasonably good agreement with a stretching force constant of 6.4×10^5 dyne/cm which we estimated for the OH-stretching force constant using Raman data from liquid water.

Now apart from the approximate nature of the above models, one caveat should be made, namely, a perturbed six-coordinated O_h model might also fit the present intermolecular vibrational frequencies reasonably well. Certainly, it is not unreasonable to assume that the $I^-(H_2O)_6$ structure is significant, as described above.

To illustrate this caveat, consider the vibrational frequencies of the octahedral $[BiBr_6]^{3-}$ ion [12], an ion which was chosen solely because of the similarities between its vibrational frequencies and the frequencies presently observed from the iodide solution. (We do not imply that there are other similarities, such as the nature of the bonding, etc.) The six vibrational frequencies of this O_h ion, from ν_1 to ν_6, respectively, are, 156, 130, 128, 75, 62, and 44 cm^{-1} [12]. Now, consider that these frequencies all become Raman active by virtue of some perturbation affecting three bromine atoms differently from the other three, in analogy to the sharing of three H_2O molecules between a six-coordinated I^- ion and a neighboring I^- ion to give a stoichiometric water to iodide ratio of 4.5. Further, consider that this sharing produces splittings and/or broadenings, leading to unresolved features. Accordingly, one might expect to see unresolved features at 156, 129, 69, and 44 cm^{-1} in surprisingly close agreement with the frequencies seen here from the iodide solution.

One might also question a T_d assignment in which ν_3 occurs at a lower frequency than ν_1. In regard to this, we examined vibrational frequencies from 140 T_d molecules and ions [12]. In 16 cases the frequency of ν_3 was found to be less than ν_1, and in each of these 16 cases the mass of X was greater than the mass of Y, which is the situation in our model. (Of course, $\nu_3 > \nu_1$ occurs in many cases where the $M_X > M_Y$.)

There can be no doubt, however, that our T_d model reproduces the salient features of the I^-–H_2O hydration, and is more probable than the O_h model. Much more detailed examinations of the Raman spectra from a wide range of NaI concentrations will be required before more refined calculations can be justified.

Finally, the XYZ model seems appropriate with regard to justifying the intensifications observed in the isotropic Raman spectra from the iodide solution at 155 cm^{-1}, Fig. 6(a) and in the polarized ln–ln Raman spectra of Fig. 5(a), e.g, in the librational region, in the H_2O bending region, and in the OH-stretching region near 3480 cm^{-1}.

The two stretching modes of the XYZ model involve motion of the proton back and forth between the I^- ion and the O atom. This motion would clearly increase the isotropic polarizability derivatives of both the OH-stretching and the I^-–O stretching modes, because of the large polarizability of the I^- ion. Similarly, the motion of the proton in the bending mode would be expected to intensify the librational mode for the same reason. The H_2O bending intensity enhancement, of course, requires a more detailed model, but again the large polarizability of the I^- ion is almost certainly involved.

Ionic polarizabilities, from molar refraction, for F^- to I^- are, respectively, 0.84, 3.00, 4.13, and 6.18 $Å^3$ [13]. The polarizability of H_2O is 1.48 $Å^3$, which places it between F^- and Cl^-. (Ionic radii in Å for the series F^- to I^- are, respectively, 1.36. 1.81, 1.95, and 2.16 [14].) From these polarizability values the F^- might conceivably decrease the polarized intensity, and the depolarized intensity, accordingly, to yield a depolarization ratio of 3/4 for the $F^- \cdots H$–O stretch (the $O \cdots H$–O value for water is equal to, or very slightly below 3/4) whereas continuously larger Raman polarization (smaller depolarization ratios) would be expected for Cl^- to I^- [15].

3.2 Sucrose, Solution and Glass.

The data of this work indicate that the 2.3 M sucrose solution is extensively hydrogen-bonded, and further that the more concentrated sucrose-water glass is fully hydrogen bonded. Solutes such as sucrose and I^- form hydrogen bonds to H_2O molecules, which at high solute concentrations, necessarily replace the hydrogen bonds that occur between H_2O molecules. High concentrations of these two solutes replace the water structure with their own specific solute-solvent hydration or hydrogen bonding. Moreover, a concomitant effect seems to be that the concentration of broken hydrogen bonds, as deduced from the low intensity in the 3650 cm^{-1} region, also declines at high solute concentrations [3]. This suggests that high concentrations of solutes such as I^- and sucrose interact strongly with <u>all</u> structures that could possibly give rise to the production of hydrogen bonds.

A further observation relates to the enhanced polarization, intensification, and sharpening of the peak in the Raman spectrum from the sucrose solution near ≈200 cm^{-1}, and to the very pronounced nature of the Raman feature near 3210 cm^{-1} from the sucrose-water glass. Both of these effects indicate that a specific ordering has occurred, which is not seen for the NaI solutions, for example. In regard to this, it is probable that the hydrogen bonds are coupled in their motions, that is, more of the O-O stretching motions between neighboring hydrogen bonds, as well as more of the OH-streching motions between neighboring H_2O molecules, occur in-phase.

We suggest that hydrogen bonding to the sugar backbone enhances the chances for in-phase coupling. Such in-phase coupling is present in cold water, but it decreases as hydrogen bonds are disrupted by rising temperature [8,9,10,15]. But the sucrose backbone is always present as a structuring element, and, in fact, it is probable that the sucrose-water glass can sustain locallized phonon modes.

3.3 $NaClO_4$ Solution.

The pronounced intensification of the librational region from the perchlorate solution, compared to water, Fig. 4, indicates that the perchlorate-water interactions must be strong. This intensification is thought not to have been seen before the ln-ln presentation was used [1,2], and it is in the opposite direction to the effects of temperature rise on pure water. However, many other effects of ClO_4^- are similar to the effects of temperature rise on pure water. For example, difference spectra between perchlorate solutions and pure water [3] indicate effects in the OH-stretching region very similar to those of temperature rise on pure water alone. Moreover, a marked decrease in the intensity near 175 cm^{-1} has been reported for ClO_4^- in water [3].

The effects of temperature rise on pure water have recently been explained by Giguère in terms of bifurcated hydrogen bonds [6,7]. In regard to such bifurcated hydrogen bonds, we wish to emphasize that the perchlorate ion is an ideal structure for the formation of six bifurcated hydrogen bonds, one for each of the six tetrahedral edges. Such a fully-bifurcated 6-coordinated perchlorate hydration structure would be perfectly octahedral in terms of the protons giving rise to the bifurcation. On the other hand, a fully-trifurcated 4-coordinated structure would be tetrahedral in terms of the protons of the surrounding H_2O molecules. Either of these bifurcated or trifurcated interactions would be predominantly electrostatic, but strong. Each would exert a sizeable influence on the librational modes of H_2O, in accord with the observed large librational intensifications.

A visual maximum (broad) at about lnw = 4.25 in Fig. 4 corresponds to ≈70 cm^{-1}, that is, a rise of about 10 cm^{-1} seems to have occurred, relative to pure water. The value of 70 cm^{-1} is the same value as that assigned to the bifurcated hydrogen bond in liquid water [2]. Accordingly, it is not unreasonable to attempt an analysis of the intermolecular Raman frequencies in terms of a bifurcated, 6-coordinated hydration structure around the perchlorate anion. In fact, the data to follow seem, surprisingly, to demand such an analysis, although we emphasize once more that several coordinations may be involved.

It should be noted also that the stoichiometric [H_2O] to [ClO_4^-] molarity ratio for the 7.4 M sodium perchlorate solution is roughly 4.4. Hence, if this stoichiometry were to result from 6-coordinated structures alone, three of the first-shell H_2O molecules in these 6-coordinated structures would have to be shared with neighboring perchlorate ions to give a ratio of 4.5. Accordingly, the selection rules for octahedral symmetry would not be obeyed, and this is what our data indicate.

(It is also interesting to note that the rate at which the solubility of sodium perchlorate rises with temperature is constant down to an [H_2O]/[ClO_4^-] ratio of about 3.0 to 2.6, after which it declines with rising temperature, and decreasing [H_2O]/[ClO_4^-] ratio [16]. A ratio below 3 in terms of 6-coordination means that more than three first-shell H_2O molecules are shared.)

As before, we corrected the Raman intensity data for all factors including the Bose-Einstein factors, and then we subtracted the appropriate CI baselines, that is, depolarized or polarized. Component data are as follows: (1) Polarized spectrum: 54 cm^{-1} (peak, PH = 566 cps, FWHH = 54 cm^{-1}), 95 cm^{-1} (PH = 437 cps, FWHH = 73 cm^{-1}), and 156 cm^{-1} (PH = 434 cps, FWHH = 94 cm^{-1}), depolarized spectrum: 52 cm^{-1} (peak, PH = 413 cps, FWHH = 50 cm^{-1}), 98 cm^{-1} (PH = 256 cps, FWHH = 73 cm^{-1}), and 159 cm^{-1} (PH = 212 cps, 93 cm^{-1}), isotropic spectrum: 70 cm^{-1} (PH = 84 cps, FWHH = 72 cm^{-1}), 120 cm^{-1} (PH = 125 cps, FWHH = 72 cm^{-1}), and 170 cm^{-1} (PH = 111, FWHH = 85 cm^{-1}). The isotropic components at frequency values of 170, 120, and 70 cm^{-1} must be polarized (not measured). The measured depolarization ratios for components at average frequency values of 158 cm^{-1}, 96 cm^{-1}, and 53 cm^{-1} were, respectively, 0.49, 0.59, and 0.73. Note that only the 53 cm^{-1} component (out of the six) is depolarized.

The 170 cm^{-1} component was assigned as ν_1, and the 158 cm^{-1} component was assigned as ν_3. The remaining components at 120, 96, 70, and 53 cm^{-1}, respectively, were assigned as ν_2, ν_4, ν_5, and ν_6. We used the G and F matrix elements for O_h symmetry given by Nakamoto [12]. We obtained the following force constants: (1), $f_r = 1.99 \times 10^4$, (2) $f_{rr} = 2.56 \times 10^3$, (3) $f'_{rr} = 5.4 \times 10^2$, (4) $f_\alpha = 2.37 \times 10^3$, (5) $f_{\alpha\alpha} = 4.4 \times 10^2$, and $f'_{\alpha\alpha} = 5.4 \times 10^2$, all in dyne/cm.

The stretching force constant, f_r, for the perchlorate O_h case, 1.99 x 10^4 dyne/cm, agrees well with the stretching force constant, k_2, obtained for the iodide T_d case of $1.8_6 \times 10^4$ dyne/cm. The bending force constant, f_α, for the perchlorate O_h case, 2.37 x 10^3 dyne/cm, is also of about the same magnitude as the bending force constant, k_1, $1.5_5 \times 10^3$ dyne/ cm, for the previous iodide T_d case. These agreements are quite satisfactory, in view of the uncertainties in the structures, and of the approximate nature of the models employed.

We emphasize, nevertheless, that all of the spectral contours (and Gaussian components) of this work are very broad, and that the models employed refer to the structures that give rise, specifically, to the frequencies of the Gaussian component centers. Obviously, a wide range of bond distances, angles, and force constants must exist for each of the structures involved.

3.4 Precautions.

Two precautions relative to obtaining accurate lnI versus lnw Raman spectra should be very strongly emphasized.

First, when obtaining ln-ln Raman spectra, it is absolutely imperative to remove all traces of fluorescence by purification methods. (For example, many electrolyte solutions which are stable at elevated temperatures may be cleaned by prolonged stirring [24 hours] at roughly 70° C with activated charcoal [Darco G-60], followed by removal of the charcoal by prefiltration, and then by further repeated [10 or more] filtrations through extremely fine filters, e.g., 0.2 μm or less.) If more than a few photon counts per second arise from fluorescence in the 3000-6000 cm^{-1} region due to the tail of a far-red fluorescence, the linear ln-ln relation, or the more accurate Gaussian ln-ln relation, will appear to fail badly. A false rise in the total scattering toward the red region will occur.

Second, the solutions studied must be as free as possible of fine particles, such as dust. The presence of fine particles will cause the stray light in the spectrometer to rise because of Tyndall scattering, and the stray light will produce an apparent increase in the low-frequency Raman scattering, which again will cause the linear ln-ln (or the Gaussian ln-ln) relation to appear to fail. Pure water, for example, can be considered to be sufficiently free of particles, if only one particle is observed to fall through the 2-cm path of a focussed, horizontal laser beam in 5 minutes [2]. Unfortunately, this criterion may be too stringent, i.e., nearly impossible to achieve, for concentrated aqueous solutions, but the number of particles falling through the beam should not exceed 2-3 per minute.

Finally, in regard to scattering from fine particles, it should be made clear that no multiple monochromator, however ideal, can obviate the need for rigorous filtration. The Tyndall scattering reflects light at the exciting frequency, and the rejection exactly at the exciting frequency is nil for any (even multiple) monochromator. Hence, the failure to remove fine particles will always increase the stray light at sufficiently low Raman frequencies. Of course it is possible to use a multi-pass Fabry-Perot in front of a multiple monochromator, but that approach is difficult, and has not been exploited extensively.

4. REFERENCES.

1. Walrafen, G. E., Hokmabadi, M. S., and Yang, W.-H. (1988), "Reply to comment on: a new method of determining the Stokes base line shape from 20-4600 cm^{-1}", J. Chem. Phys.,88, 4555-4556.
2. Walrafen, G. E., Hokmabadi, M. S., Yang, W.-H., Chu, Y. C., and Monosmith, B. (1989), "Collision-induced Raman scattering from water and aqueous solutions", J. Phys. Chem., 93, 2909-2917.
3. Walrafen, G. E., and Fisher, M. R. (1986), "Low-frequency Raman scattering from water and aqueous solutions: a direct measure of hydrogen-bonding", in Biomembranes, a volume of Methods of Enzymology, edited by L. Packer, (Academic, New York), 91-105.
4. Walrafen, G. E. (1970), "Raman spectral studies of the effects of perchlorate ion on water structure", J. Chem. Phys., 52, 4176-4198.
5. Walrafen, G. E. (1971), "Raman spectral studies of the effects of solutes and pressure on water structure," J. Chem. Phys., 55, 768-792.
6. Giguère, P. A. (1987), "The bifurcated-hydrogen-bond model of water and amorphous ice", J. Chem. Phys., 87, 4835.
7. Giguère, P. A. (1988), "An electrostatic model for hydrogen bonds in alcohols", J. Solution Chem., 1007-1014.
8. Monosmith, B. and Walrafen, G. E. (1984), "Temperature dependence of the Raman OH-stretching overtone from liquid water", J. Chem. Phys., 81, 669-674.
9. Walrafen, G. E. (1974), "Spontaneous and stimulated Raman spectra from water and aqueous solutions", in Structure of water and aqueous solutions, edited by W. A. P. Luck, (Verlag Chemie, Weinheim), 301-321.
10. Walrafen, G. E. (1967), "Raman spectral studies of the effects of temperature on water structure", J. Chem. Phys., 47,

114–126.

11. Herzberg, G. (1945), "Infrared and Raman spectra of polyatomic molecules. Vol. II. Molecular spectra and Molecular structure", (Van Nostrand, Princeton, New Jersey).

12. Nakamoto, K. (1978), "Infrared and Raman spectra of inorganic coordination compounds", (Wiley, New York).

13. Jolly, W. L. (1984), "Modern Inorganic Chemistry", (McGraw–Hill, New York), 287 (see table 11.8), and 281 (see table 11.3).

14. Atkins, P. W. (1986), "Physical chemistry", (Freeman, New York), 579 (see table 24.1).

15. Walrafen, G. E. (1962), "Raman spectral studies of the effects of electrolytes on water", J. Chem. Phys., 36, 1035–1042; and, Walrafen, G. E. (1966), "Raman spectral studies of the effects of temperature on water and electrolyte solutions", J. Chem. Phys., 44, 1546–1558.

16. Seidell, A. (1940), "Solubilities of inorganic and metal organic compounds", (Van Nostrand, New York).

Acoustic S- and P-Wave Character of the Intermolecular 60 and 175 cm^{-1} Raman Bands from Liquid Water

G. E. Walrafen
Chemistry Department
Howard University
Washington, D. C. 20059

ABSTRACT.

The acoustic character of the intermolecular 60 cm^{-1} and 175 cm^{-1} Raman bands from liquid water, that is, their respective S-wave (transverse acoustic, shear) and P-wave (longitudinal acoustic, dilatational) nature seems hitherto unappreciated. Consideration of these two intermolecular vibrations as acoustic modes indicates that both are collective, involving second- and third-nearest-neighbor interactions.

1. INTRODUCTION.

Since the discovery of the ≈175 cm^{-1} Raman maximum from water by Segré in 1931 [1], and the subsequent discovery of the weaker ≈60 cm^{-1} Raman feature by Bolla in 1932 [2], it has been apparent that these two intermolecular vibrations are of major importance in understanding the structure of liquid water. (The 175 cm^{-1} band is infrared active [3], and evidence for infrared activity of the 60 cm^{-1} band has been reported very recently [4].)

The 175 cm^{-1} vibration is thought to arise from the restricted translation of two H_2O molecules against each other along the O−H···O (parallel) direction [5], whereas the 60 cm^{-1} vibration may be the perpendicular restricted translation, i.e., the O−O−O bending mode [5].

The purpose of this article is to point out that clear evidence also exists for the acoustic S- and P-wave character [6] of the 60 and 175 cm^{-1} modes, i.e., their (shear) transverse, and (primary) dilatational, longitudinal character. This evidence does not negate the previous assignments--it augments them.

2. HIGH-FREQUENCY-LIMITING MODULI, ACOUSTIC WAVELENGTH, AND RAMAN FREQUENCIES.

Slie et al. [7] report values of the high-frequency-limiting (HFL) compressional or bulk modulus, K_{inf}, and shear modulus, G_{inf}, for liquid water. The HFL longitudinal modulus, M_{inf}, may be calculated from these bulk and shear moduli, and is given by $M_{inf} = K_{inf} + (4/3)G_{inf}$. At 25°C, $M_{inf} \approx 7.19 \times 10^{10}$ dyne-cm^{-2}. Accordingly, the HFL longitudinal P-wave velocity, V_p, is calculated to be $\approx 2.68 \times 10^5$ cm-sec^{-1}. At 0°C V_p is $\approx 2.81 \times 10^5$ cm-sec^{-1} [7].

J.C. Dore and J. Teixeira (eds.), Hydrogen-Bonded Liquids, 283–287.

The HFL longitudinal and shear moduli are considered to provide a reasonably good measure of the elasticity of the short-range hydrogen-bonded water structure. Accordingly, the acoustic wavelength, λ_p, of the 175 cm^{-1} feature considered as a P-mode was calculated using the longitudinal HFL velocity, $V_p = 2.68 \times 10^5$ $cm\text{-}sec^{-1}$. The Bose-Einstein (BE) corrected X(ZX)Y Raman shift at 25°C is about 173 cm^{-1} [8], and the collision-induced (CI) corrected X(ZX)Y value without BE correction is about 175 cm^{-1} [8]. The values of 173 cm^{-1} and 175 cm^{-1} correspond to 5.19 and 5.25 THz, respectively. The corresponding λ_p values are thus 5.2 Å and 5.1 Å.

Of course, a value of ≈2.9 Å could be calculated using 173 cm^{-1} and the 4.4 GHz sound velocity of 1494 $m\text{-}sec^{-1}$ at 25°C [9], in agreement with the nearest-neighbor x-ray RDF O-O peak at 2.85 Å [10]. However, the P-mode wavelength should be, very roughly, twice the nearest-neighbor distance.

The calculated P-mode wavelength of ≈5 Å is sufficiently close to the second-nearest-neighbor peak in the X-ray RDF, namely, 4.5 Å at 20°C [10], to identify the very broad P-mode [5] with the second-nearest-neighbor O-O distance. This agreement means that the assignment as a restricted translation between two hydrogen-bonded H_2O molecules [5] should be modified to involve (at least) three hydrogen-bonded H_2O molecules. But the O-O-O angle must be ≈109.5 degrees, for 4.5 Å, which requires a T_d structure.

P-wave particle motions are radially symmetric. Four hydrogen bonded H_2O molecules breathing against a fifth, central H_2O molecule of a T_d structure in terms of O atoms (separated by 2.85 Å) corresponds to radial motion [11], but an acoustic phonon wavelength of ≈5 Å requires breathing of second-nearest-neighbor molecules against the central H_2O molecule. This collective mode may refer more closely to the P-wave particle motions.

Polycrystalline ice provides an interesting comparison. At 0°C the longitudinal velocity, averaged from C_{11} and C_{33} [12], is 3.837×10^5 $cm\text{-}sec^{-1}$, or polycrystalline, −20°C [13], is $\approx 3.84 \times 10^5$ $cm\text{-}sec^{-1}$. The peak Raman shift is about 220 cm^{-1}, but a prominent shoulder also occurs near 275 cm^{-1}, this work. The corresponding longitudinal wavelength range is thus ≈4.5 Å to ≈5.8 Å. The transverse velocity (from C_{44} at 0°C, [12]) is $\approx 1.75 \times 10^5$ $cm\text{-}sec^{-1}$, and the Raman shear peak occurs near 62 cm^{-1}, this work. The transverse wavelength is thus ≈10 Å. Note also that the HFL longitudinal velocity calculated from M_{inf} lies between the ordinary sound velocities of water and ice [9,12,13].

The possibility also exists that the P- or longitudinal acoustic (LA) mode may couple mechanically with one of the ≈3215 cm^{-1} or ≈3390 cm^{-1} OH-stretching (optical) modes [14], although the vibrational periods of these OH-stretches are admittedly very short, $\approx 10^{-14}$ sec, compared to the P-mode, $\approx 2 \times 10^{-13}$ sec. The evidence for this coupling comes from the frequency difference, namely, 3390 − 3215 cm^{-1} = 175 cm^{-1}.

The $\approx$3215 cm^{-1} vibration has been assigned to in-phase breathing of all OH units in the five-molecule hydrogen-bonded structure [14,15], and the 3390 cm^{-1} vibration to symmetric intramolecular stretching, but out-of-phase between the five molecules [14,15].

The $\approx$3215 cm^{-1} mode is more highly polarized, depolarization ratio, $\approx$0.04, than the $\approx$3390 cm^{-1} mode, 0.1_3 [14]. The 175 cm^{-1} band must be polarized because it appears weakly in the isotropic spectrum [8]. Its very weak polarization is sufficient to indicate A_1 symmetry for a weak, predominantly electrostatic intermolecular interaction [11]. Hence, polarization for both the 175 cm^{-1} and 3215 cm^{-1} vibrations agrees with in-phase radial particle motions, that is, breathing.

It should also be noted that the correspondence between the P-mode wavelength and the second-nearest-neighbor O-O distance means that there is a connection between the P-mode wavelength and a major density-related O-O spacing. This is reasonable because the P-wave changes the volume, adiabatically, as it propagates--hence its velocity involves the adiabatic bulk modulus, K_s, as well as the shear modulus, G.

The 60 cm^{-1} feature is assigned to the S-wave in agreement with its smaller Raman shift. This mode is almost completely depolarized [8]. Slie et al. [7] report a value for the HFL shear modulus, $G_{inf} = 1.68 \times 10^{10}$ dyne-cm^{-2}. $V_s^2 = G_{inf}/\rho$. Hence, V_s is 1.296×10^5 cm-sec^{-1}. The wavelength, λ_s, at the measured peak value of 55 cm^{-1} [5] is thus $\approx$7.9 Å.

The high-distance tail of the 7.0 Å (last) peak in the X-ray RDF (20°C) fades away at 7.8 Å [10], i.e., $G(r) \approx 1$ at 7.8Å. The agreement between 7.8 Å and the shear wavelength (7.9 Å) is good, considering that G_{inf} was obtained by extrapolation [7]. It means that the shear wave is damped-out when no significant correlation remains--an important required condition linking the longest phonon wavelength to the structural correlation length.

The 7.0 Å peak corresponds to the O-third-O distance [10]. Hence, it is evident that the 60 cm^{-1} Raman feature may refer to a more extensive collective motion than the 175 cm^{-1} mode.

3. CAVEATS.

Caution should be exercized when comparing variations in the frequencies of the 60 and 175 cm^{-1} modes with variations in the low-frequency sound velocities, or even GHz sound velocities in view of the small velocity dispersion [9,12,13]. Velocities calculated from the HFL bulk and shear moduli should be used, if available. (This was not done for ice, because 0°C values were needed for comparison with water, and when it was not done for water, the value of $\approx$2.9 Å resulted, which was too small.)

At 30°C the low-frequency sound velocity rises about 11% as the pressure increases from 1 to ≈1000 atm [16, 17]. However, high-pressure Raman measurements at 27°C (1967) indicated constancy of the 60 and 175 cm^{-1} peak positions from 1 to 1200 atm, within experimental errors of ±5 and ±3%, respectively [18].

The low-frequency sound velocity increases to a maximum with rising temperature--the low-frequency velocity is 1543.11 $m\text{-}sec^{-1}$ at 100°C, whereas the value at 20°C is 1482.34 $m\text{-}sec^{-1}$. [13]. However, the position of the 175 cm^{-1} Raman peak decreases to 100°C.

The above complications involving pressure or temperature changes emphasize some unusual physical properties of liquid water, but they do not correspond to the Raman data, because the sound velocities involved do not correspond to the HFL elastic moduli. In contrast, M_{inf} and G_{inf} [7], and the 60 and 175 cm^{-1} Raman peak positions [5], all decrease with rising temperature.

The fundamental difficulty with using sound velocities measured even at GHz frequencies, is that the period is so long, ≈2 x 10^{-10} sec, that the local structure has formed, broken down, and reformed many times, with the result that the corresponding modulus is too small to represent the true elasticity of the short-range hydrogen-bonded structure.

4. REFERENCES

1. E. Segré, Rend. Lincei 13, 929(1931).
2. G. Bolla, Nuovo Cimento 9, 290(1932);10, 101(1933);12, 243(1935).
3. G. E. Walrafen, J. Chem. Phys. 44, 1546(1966).
4. J. B. Hasted, S. K. Husain, F. A. M. Frescura, and J. R. Birch, Chem. Phys. Lett. 118, 622(1985).
5. R. Bansil, J. Wiafe-Akenten, and S. Krishnamurthy, chapter in, "Lasers as Reactants and Probes in Chemistry", edited by W. M. Jackson and A. B. Harvey, Howard University Press, Washington, 1985. G. E. Walrafen, M. R. Fisher, M. S. Hokmabadi, and W.-H. Yang, J. Chem. Phys. 85, 6970(1986). G. E. Walrafen, M. S. Hokmabadi, W.-H. Yang, Y. C. Chu, and B. Monosmith, J. Phys. Chem. 93, 2909(1989).
6. The terms S-wave and P-wave were used, purposely, instead of transverse acoustic (TA) and longitudinal acoustic (LA) to emphasize the connection to the stress-wave theory of isotropic elastic solids, as opposed to terminology related to phonons in crystals. Liquid water behaves like a glass at THz sound frequencies. Hence, it may be treated as an isotropic elastic solid at such frequencies.

7. W. M. Slie, A. R. Danfor, Jr., and T. A. Litovitz, J. Chem. Phys. 44, 3712(1966).
8. G. E. Walrafen, M. S. Hokmabadi, and Y. C. Chu, chapter in, "Hydrogen-Bonded Liquids", NATO ASI, edited by J. C. Dore and J. Teixeira, Kluwer Academic, Dordrecht, 1990. Also G. E. Walrafen, M. S. Hokmabadi, and Y. C. Chu, unpublished work.
9. D. P. Eastman, A. Hollinger, J. Kenemuth, and D. H. Rank, J. Chem. Phys. 50, 1567(1969).
10. A. H. Narten and H. A. Levy, chapter in, "Water, A Comprehensive Treatise, Vol. 1. The Physics and Physical Chemistry of Water", edited by F. Franks, Plenum, New York, 1972.
11. G. E. Walrafen, J. Chem. Phys. 40, 3249(1964).
12. G. Dantl, in "Physics of Ice", edited by N. Riehl, B. Bullemer, and H. Engelhardt, Plenum, New York, 1969.
13. G. W. C. Kaye and T. H. Laby, "Tables of Physical and Chemical Constants", (15th), Longman, London, 1986, pgs. 74, 76.
14. G. E. Walrafen, chapter in, "Structure of Water and Aqueous Solutions", edited by W. A. P. Luck, Verlag Chemie, Physik Verlag, Weinheim/Bergstr., 1974.
15. W. B. Monosmith and G. E. Walrafen, J. Chem. Phys. 81, 669(1984).
16. A. H. Smith and A. W. Lawson, J. Chem. Phys. 22, 351(1954).
17. T. A. Litovitz and E. H. Carnevale, J. Appl. Phys. 26, 816(1955).
18. Unpublished high-pressure Raman measurements carried out by G. E. Walrafen at A. T. and T. Bell Laboratories in 1967.

5. NOTE ADDED IN PROOF:

After this chapter was submitted, Professor E. Duval kindly sent an article to G. E. Walrafen in which the 50 cm^{-1} band from water was assigned as a transverse acoustic mode. That article by J. L. Rousset, E. Duval, and A. Boukenter is titled: "Dynamic Structure of Water: Low Frequency Raman Scattering from a Disordered Network and Aggregates", and it is scheduled to appear in 1990 in the Journal of Chemical Physics. Rousset et al. were not aware of the work of Walrafen, who, in turn, was unaware of the work of Rousset et al.

Editors note: The paper by Duval and colleagues is published in J.Chem.Phys. (Feb.1990).

The work reported in the two papers was done over the same period without prior knowledge and the difference in publication times has arisen solely from the time delay in the production of the ASI book.

QUASI-ELASTIC AND INELASTIC NEUTRON SCATTERING AND MOLECULAR DYNAMICS OF WATER AT SUPERCOOLED TEMPERATURE

Sow-Hsin Chen
Nuclear Engineering Department and
Center for Materials Science and Engineering
Massachusetts Institute of Technology
Cambridge, Massachusetts 02139, USA

ABSTRACT. Recent results of incoherent neutron scattering investigations of water (H_2O) at various temperatures, ranging from above the room temperature down to $-20^{\circ}C$, are summarized. Analysis of these results in terms of various models are discussed and compared with the available computer molecular dynamics (CMD) simulations in the literature. For the quasi-elastic scattering (QENS) region, the self dynamic structure factor has been analysed in terms of a convolution of translation and rotation of individual water molecules. The translational and the rotational parts were modeled respectively by a jump diffusion and rotational diffusion. Comparisons are then made with recent CMD results to assess the validity of the model description. For the inelastic scattering (INS) region, the analysis focused mainly on the intramolecular vibrations in the liquid state. A satelite peak which appeared beyond the stretch vibrational band was analysed in terms of a quantum mechanical process of exciting the stretch vibrational mode with a simultaneous breaking of the hydrogen bond. The absolute cross section in the stretch vibrational region was compared quantitatively with a CMD result as a function of temperature from $80^{\circ}C$ down to $-15^{\circ}C$.

J.C. Dore and J. Teixeira (eds.), Hydrogen-Bonded Liquids, 289–332.

1. INTRODUCTION

Inelastic neutron scattering studies of water began in 1958 when Brockhouse [1] discovered a strong quasi-elastic component in the inelastic scattering spectra of H_2O which is energy broadened proportional to Q^2 (Q is the magnitude of wave-vector transfer in a scattering experiment). The origin of the broadening was promptly identified with the diffusion of water molecules in liquid state in accord with the theory of Van Hove [2]. In 1962, Sakamoto, Brockhouse, Johnson and Pope [3] published the first systematic study of the quasi-elastic scattering line shape of H_2O using a rotating crystal time-of-flight spectrometer having a sufficient energy resolution (ΔE = 200μeV) to allow them to extract the half-width of the quasi-elastic peak $\Gamma(Q)$ as a function of Q^2. A time Fourier transform of the self dynamic structure factor combined with a Gaussian approximation of the intermediate scattering function also allowed them to extract the mean square distance of travel in time t, $\langle r^2(t)\rangle$, of a water molecule as a function of time at 25°C and 75°C. Both these quantities can be related to the self-diffusion constant of the water molecule as measured, for example, by the nuclear magnetic resonance technique [4]. Both experimental techniques gave consistent results. Ever since this pioneering study, there have been numerous neutron scattering studies of water, both quasi-elastic and inelastic, mainly above 0°C. A comprehensive summary of these data, together with his own contribution, was made by von Blanckenhagen in 1972 [5]. Blanckenhagen noticed that there were substantial disagreement in $\Gamma(Q)$ vs. Q^2 data at 25°C among all six experiments surveyed in his paper [5]. Analyses of these early experimental data were largely made without considering the rotational motions of water molecules. The rotational motions, however, turned out to be important in treating the self diffusion of water molecules because of the extensive hydrogen bond network existing in water at and below room temperature. More recent neutron scattering results were discussed in a review article by Chen and Teixeira [6] and also in a summary by Teixeira et al. [7]. In this article I briefly review the material covered in [6] and [7], and present some new results obtained during the period 1986-88. The main focus of the present review is to discuss these relatively new results in light of recent computer molecular dynamic sumulations. In this way one can correlate the results of the experimental analyses with the microscopic picture of hydrogen bond dynamics of water obtained by CMD at various temperatures including supercooled temperatures.

I shall first discuss the interpretation of the incoherent quasi-elastic scattering line shape of H_2O in section II. Then I go on to the analyses of the incoherent inelastic spectra of H_2O at high and low temperatures in section III. In this section I shall focus attention mainly on the stretch vibration band and its satelite peak, and shall describe a quantum mechanical calculation of the intensity of the satelite peak. In section IV, I will make a brief summary of the impact of neutron scattering work so far on understanding of the temperature dependence of hydrogen bonding dynamics and touch upon the prospect of using the low frequency inter-molecular translational and librational modes as a probe for the H-bond dynamics.

2. INTERPRETATION OF THE RESULTS FROM QUASI-ELASTIC NEUTRON SCATTERING

The differential scattering cross section per water molecule per unit solid angle, per unit energy transfer, is given in standard notation by [8]:

$$\sigma_{inc}(E_o, E', \theta) = \frac{n_H \sigma_H}{4\pi h} \left(\frac{E'}{E_o}\right)^{1/2} S_s(Q, \omega) \tag{1}$$

where n_H is the number of hydrogen atoms per molecule, σ_H the incoherent bound atom cross section of hydrogen, E_o the incident neutron energy, E' the scattered neutron energy and $S_s(Q,\omega)$ the self part of the dynamic structure factor [2]. The energy transfer and wave vector transfer in the scattering process are defined by: $E = \hbar\omega = E_o - E'$ and $\underline{Q} = \underline{k}_o - \underline{k}'$. For an isotropic liquid like water, the dynamic structure factor is a fuction of the magnitude of $\underline{Q}$ only and the latter is given in an energy loss experiment by

$$Q(\text{Å}^{-1}) = 0.693\ [2E_o - E - 2\sqrt{E_o(E_o - E)}\ \text{Cos}\theta]^{1/2}\ , \tag{2}$$

where θ is the scattering angle and energies are measured in unit of meV. In particular, for a quasi elastic scattering, Eq. (2) simplifies to

$$Q = \frac{4\pi}{\lambda} \sin(\theta/2) \tag{3}$$

where λ is the wave length of neutrons.

The dynamic structure factor is given in terms of an intermediate scattering function $F_s(Q,t)$ by a Fourier transform

$$S_s(Q,\omega) = \frac{1}{\pi}\int_0^{\infty} dt\ \cos\omega t\ F_s(Q,t). \tag{4}$$

To calculate the intermediate scattering fraction, the usual procedure is to decompose the position vector of the hydrogen atoms $\vec{r}_H$ into a sum of three independent vectors:

$$\vec{r}_H = \vec{R} + \vec{b} + \vec{u}. \tag{5}$$

$\vec{R}$ is the position of the center of mass of the water molecule, which is very close to the oxygen atom, $\vec{b}$ the equilibrium position of the hydrogen atom measured from the center of mass, and $\vec{u}$ the instantaneous position of hydrogen atom measured from its equilibrium position. One then makes a decoupling approximation by assuming that three kinds of motions, namely, the translation, rotation and vibration are independent. One can thus writes the total intermediate scattering function as a product of the three component functions:

$$F_s(Q,t) = F_T(Q,t)\cdot F_R(Q,t)\cdot F_V(Q,t) \ . \tag{6}$$

For a small energy transfer experiment like a quasi-elastic scattering, it is a good appromation to retain in the vibrational intermediate scattering function just the Debye-Waller factor [8],

$$F_V(Q,t) = \langle e^{-i\underline{Q}\cdot\underline{u}} e^{-i\underline{Q}\cdot\underline{u}(t)} \rangle \cong \exp[-\tfrac{1}{3}Q^2\langle u^2\rangle] \tag{7}$$

where $\langle u^2\rangle$ is the mean square vibrational amplitude of the hydrogen atom.

The rotational intermediate scattering function can generally be written as an expansion in various partial waves each corresponding to a rotational angular momentum state ℓ of a water molecule, [9] (see appendix A for the derivation of the partial wave expansion):

$$F_R(Q,t) = j_0^2(Qb) + \sum_{\ell=1}^{\infty} (2\ell+1) j_\ell^2(Qb) F_\ell(t). \tag{8}$$

The amplitude factor for each partial wave, $S_\ell = (2\ell+1)j_\ell^2(Qb)$, is plotted in Fig. 1, for ℓ = 0, 1, 2 and 3 as a function of Qb. Since b $\cong$ 1.0 Å for a water molecule, one immediately see that for the range $0 \leq Q \leq 2.5$ Å^{-1}, which is the practical range from an experimental point of view, one needs only to keep $\ell \leq 2$ terms in the expansion. The time-dependent factor, $F_\ell(t)$, is the ℓ-th order rotational correlation function which is defined by

$$F_\ell(t) = \langle P_\ell(\hat{b}\cdot\hat{b}(t)) \rangle \ . \tag{9}$$

It depends on the cosine of the angle between two unit vectors,

$\hat{b}\cdot\hat{b}(t) = \cos\alpha(t)$, which changes in time as the molecule rotate around the center of mass. $F_0(t) = \langle P_0(\cos\alpha(t))\rangle$ is equal to unity at all times. $F_1(t) = \langle P_1(\cos\alpha(t))\rangle$ is the single-molecule dipole autocorrelation function. It can be obtained, for example, from a Fourier transform of an infrared absorption spectrum centered at the vibrational frequency [10,11]. $F_2(t) = \langle P_2(\cos\alpha(t))\rangle$ is also related to the depolarized component of a rotational Raman band or the depolarized part of a vibration-rotation Raman band [10,11]. In the case of water, it turns out that both $F_1(t)$ and $F_2(t)$ decay exponentially, for time longer than 0.15 psec [12,13] with time constants τ_1 and τ_2 respectively. In this case it is more convenient to obtain τ_1 from the dielectric relaxation time τ_D [14,15] and τ_2 from the NMR quadrupole relaxation time T_1 of deuterium or oxygen-17 nuclei in a water molecule [16].

For the center-of-mass translation, we shall assume that $F_T(Q,t)$ also decays exponentially in time for a sufficiently long time probed by the quasi-elastic scattering, with a Q-dependent decay constant $\Gamma(Q)$,

$$F_T(Q,t) = \exp[-\Gamma(Q)t] \ . \tag{10}$$

From the above equations we see that for $Qb < 1$, the dynamic structure factor is simply given by

$$S_s(Q,\omega) = \exp[-\tfrac{1}{3}Q^2\langle u^2\rangle]\ j_o^2(Qb)S_T(Q,\omega) \tag{11}$$

which contains only the translational motion of the center of mass.

But in the range $1 < Qb < 2.5$, the dynamic structure structure factor becomes a convolution of the translational and rotational factors, i.e.,

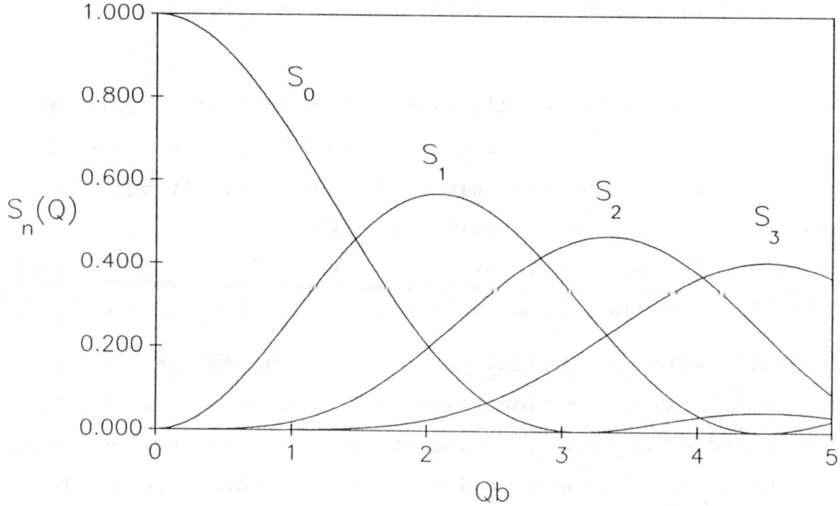

Fig. 1: The partial wave amplitude factor $S_\ell(Q) = (2\ell+1)j_\ell^2(Qb)$ vs Qb plots for ℓ = 0, 1, 2, 3. Note the experimentally accessible Qb range is between 0.2 to 2.1. At Qb = 2.2 the contribution from $S_3(Q)$ is only 0.04.

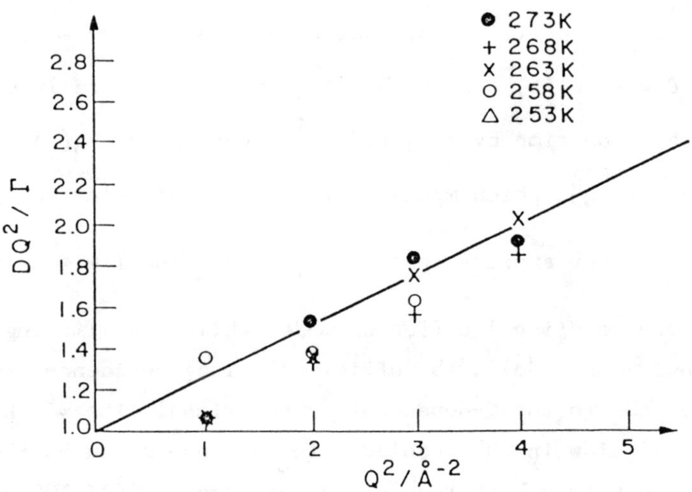

Fig. 2: $DQ^2/\Gamma(Q)$ vs Q^2 plot for five temperatures ranging from T = 273°K down to 253°K. It is seen that the experimental point suggests the empirical relation $1 + Q^2\ell^2$.

$$S_s(Q,\omega) = \exp[-\frac{1}{3}Q^2\langle u^2\rangle]\, S_T(Q,\omega) \otimes S_R(Q,\omega) \tag{12}$$

It is worth noting that for all experiments done up to 1972 on quasi-elastic scattering of water, no account has been taken of the rotational effect on the line width [5]. Essentially Eq. (11) was used for analyses of data for the entire Q-range.

From the CMD work of Impey, Madden and McDonald [12] and Lie and Clementi [13], it was well established that both $F_1(t)$ and $F_2(t)$ decay exponentially after an initial period of libration lasting for 0.15 psec. The librational motion gives rise to a prominent "hinder rotation peak" situated at 500 cm^{-1}, in the spectrum of the hydrogen atom velocity auto-correlation function. It is therefore reasonable to take a simple rotational diffusion model for $F_\ell(t)$ as a first approximation [10],

$$F_\ell(t) = \exp[-\ell(\ell+1)D_R t] \; . \tag{13}$$

D_R in Eq. (13) is the rotational diffusion constant of the water molecule. For $\ell = 1$, $F_1(t) = \exp[-2D_r t]$. One then naturally define a rotational correlation time by $\tau_1 = (2D_r)^{-1}$. For $\ell = 2$, $F_2(t) = \exp[-6D_R t] = \exp[-t/\tau_2]$, which means that one has the relation $\tau_2 = \tau_1/3$. In practice, one expects that $1 \leq \frac{\tau_1}{\tau_2} \leq 3$. The limiting value 3 corresponds to the rotational diffusion model while the limiting value 1 to the rotational jump model with sufficiently long residence time [10].

Finally we come to the Q-dependent translational line width $\Gamma(Q)$. Since diffusion is slow in supercooled water, if one measures the quasi-elastic line with a high resolution instrument ($\Delta E < 100\mu ev$), in the low Q region one essentially obtains only the translational line because the rotational line is so much broader and can be subtracted off as a D. C. base line. Chen, Teixeira and Nicklow [17] realized this point in 1982 and they used an ice background as the D. C. base line.

Fig. 2 shows the plot of $DQ^2/\Gamma(Q)$ as a function of Q^2 for temperatures between 273° and 253°K. From the straight line behavior of these experimental points it was inferred that one may take the empirical form $\Gamma(Q) = DQ^2/(1 + Q^2\ell^2)$ to fit these data points. This form of P(Q) is analogous to the well-known expression for a jump diffusion model [18], which is normally written as

$$\Gamma(Q) = \frac{DQ^2}{1 + DQ^2\tau_0} . \qquad (14)$$

In Eq. (14) D is the self diffusion constant of water and τ_0 is called the residence time, which is the time interval between each diffusional jump. One can then introduce the mean square jump length by the relation

$$\langle \ell^2 \rangle = 6D\tau_0 . \qquad (15)$$

Equation (6), (7), (8), (10), (13) and (14) taken together completely define the dynamic structure factor through Eq. (4) and hence the cross section through Eq. (1). In practice, one knows the self diffusion coefficient D as a function of temperature from the room temperature down to supercooled temperatures [19,20]. For example, recently Prielmeier, Lang, Speedy and Lüdemann gave a convenient parametric form for D applicable to the temperature range, 242°K < T < 333°K [21],

$$D = 1.67 \times 10^{-4} \left(\frac{T}{223} - 1\right)^{1.82} . \qquad \mathrm{cm^2/sec} \qquad (16)$$

Therefore, the theory has three fitting parameters: $\langle u^2 \rangle$, τ_0 and τ_1. Fig. 3 shows the line shape of three quasi-elastic peaks at Q = 1.05Å^{-1}, 1.44Å^{-1} and 2.03Å^{-1} measured at a temperature of -5°C, by Teixeira, Bellissent-Funel, Chen and Dianoux [22]. The theoretical dynamic structure factors have been convolved with a Gaussian resolution function ($\Delta E \sim 100\mu ev$) of the time-of-flight instrument (IN6, ILL) and compared with the experimental line shape. The simple theory was able

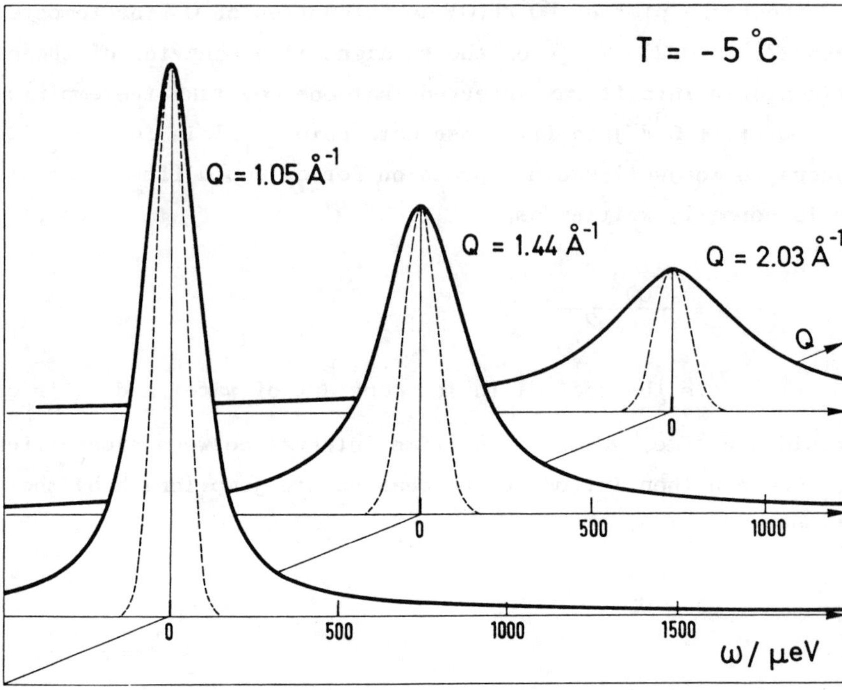

Fig. 3: Experimentally measured quasi-elastic peaks at -5°C for three values of Q = $1.05Å^{-1}$, $1.44Å^{-1}$, $2.03Å^{-1}$. It is noted that the peak heights decreases rapidly as Q gets larger. In practice it is difficult to measure the quasi elastic scattering beyond Q = $2.2Å^{-1}$. The dotted lines are the shape of the resolution functions of the time-of-flight instrument (IN6, ILL).

to fit to all available data set (337 peaks per temperature) to within 2 ~ 3% accuracy. The extracted $\Gamma(Q)$ plotted as a function of Q^2 are shown in Fig. 4, for the nine temperatures studied, ranging from 20° down to −20°C. Solid lines are the theoretical calculation according to the jump diffusion model of Eq. (14). The jump diffusion model predicts that at sufficiently large Q such that $DQ^2\tau_0 >> 1$, the line width $\Gamma(Q)$ approaches $1/\tau_0$ independent of Q. One can clearly see this type of behavior for data below − 15°C in Fig. 4. That is, as far as the translational motion of the center of mass is concerned, the short time diffusion of water molecules is well approximated by the picture of jump diffusion at supercooled temperatures. The three parameters extracted from the experimental data are: the root mean square vibrational amplitude of the hydrogen atom, $\sqrt{\langle u^2 \rangle} = 0.48$Å, the temperature dependence of the jump diffusion residence time τ_0, and the rotational correlation time τ_1. Numerical values of τ_0 and τ_1 are given in Fig. 5. It is readily seen from Fig. 5 that the temperature dependence of τ_0 is strongly non-Arrhenius, while that of τ_1 is approximately Arrhenius, with an activation energy $E_A = 1.85$ Kcal/mol (931°K). The root mean square vibrational amplitude was measured once before by Franks, Ravenhill, Egelstaff and Page [23] at 21°C. The value given was $\sqrt{\langle u^2 \rangle} = 0.58$Å. In this paper of Franks et al. [23], they also showed an analysis of $\Gamma(Q)$ vs. Q^2 curve at 21°C by the jump diffusion model. The experimental data agreed well with the model, giving $\tau_0 = 1.2$ ps (21°C), which compares well with the result of Teixeira et al. [22], $\tau_0 = 1.25$ ps (20°C).

It is instructive to compare the rotational relaxation times τ_1 and τ_2 measured by other techniques. In water, many experimental investigations showed that, at micro-wave frequencies, the dispersion of the complex permittivity ϵ^* of water can be described with sufficient accuracy by the Debye equation [14,15],

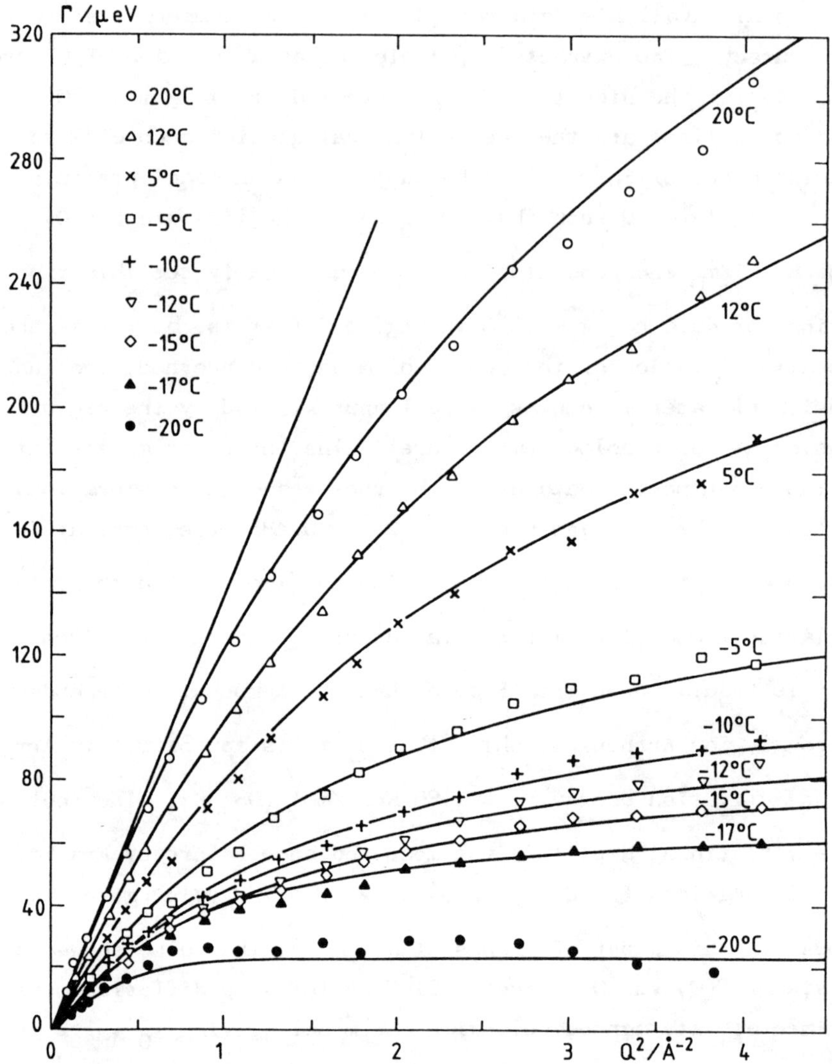

Fig. 4: $\Gamma(Q)$ vs Q^2 plots for a series of 9 temperatures ranging from 20° down to -20°. The temperature measurement of the last one (-20°) was uncertain. It might have been -25°C. The solid lines are the model prediction of Eq. (14). The experimental value of D was used to evaluate Eq. (14). It is noteworthy that $\Gamma(Q)$ tends to a constant value at large Q for temperatures below -15°C.

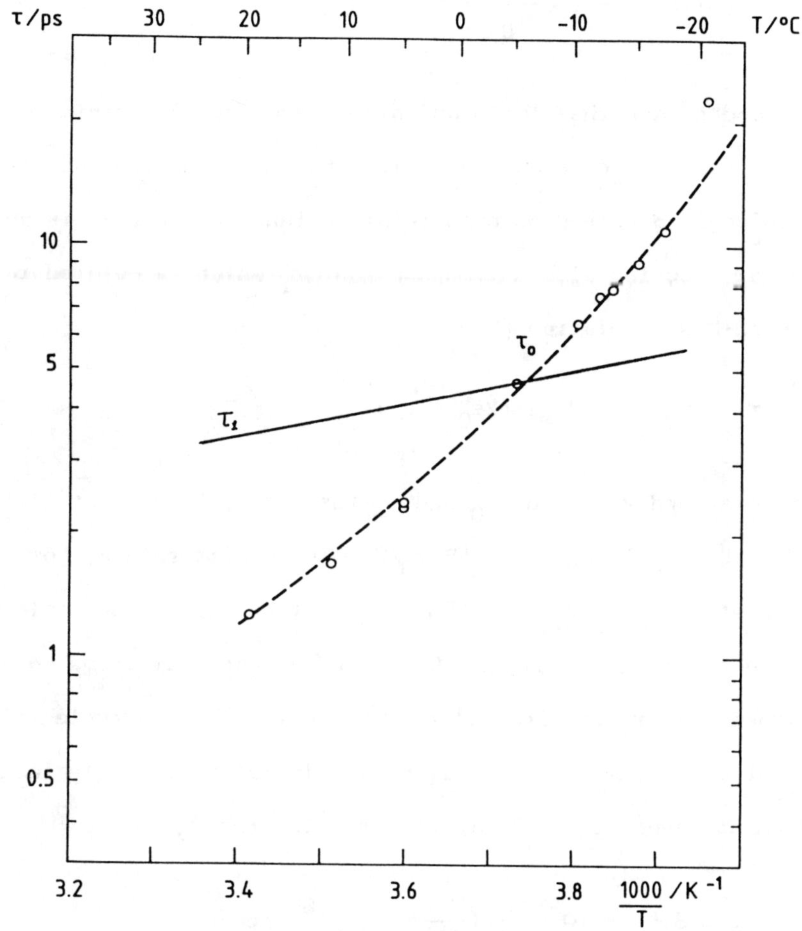

Fig. 5: τ_0 and τ_1 vs 1000/T plots. Notice the non-Arrhenius behavior of τ_0 in particular. It is about 1 ps at room temperature and increases rapidly to more than 10 ps at -20°C.

$$\epsilon^*(\omega) = \epsilon_\infty + \frac{\epsilon_0 - \epsilon_\infty}{1-i\omega\tau_D} \tag{17}$$

where ϵ_0 and ϵ_∞ are the static and high frequency dielectric constants and τ_D the dielectric relaxation time. Eq. (17) implies that the single-molecule dipole-dipole correlation function, $F_1(t)$, is an exponential function with a time constant τ_1 which is related to τ_D by the Fatusso-Mason relation [24]

$$\tau_1 = \tau_D(\epsilon_0 + 2\epsilon_\infty)/(2\epsilon_0 + \epsilon_\infty) \ . \tag{18}$$

From the measured values of ϵ_0 and ϵ_∞ for water, Nobokov et al. [15] concluded that $\epsilon_0 \gg \epsilon_\infty$ and $\tau_1 \cong \tau_D/2$. In the literature, however, it was often stated that $\tau_1 \cong \tau_D$ [25,14], so there is an uncertainty of factor two relating τ_1 and τ_D. On the other hand, it seems more firmly established that τ_2 is directly related to the NMR quadruple relaxation time T_1 of deuterium or Oxygen-17 nuclei in water [26]. In fact Lang and Lüdemann gave [16] an empirical relation for τ_2:

$$\tau_2 = 3.42 \times 10^{-13} \cdot \left(\frac{T}{223} - 1\right)^{-1.89} \text{ sec} \tag{19}$$

which is applicable from 473°K down to the homogeneous nucleation temperature. Furthermore, very recently, Kivelson and Kivelson [27] argued that in the supercooled liquids the ratio $(\tau_1/\tau_2) = r$ can decrease from 3 to 1 upon approach to the glass transition. In view of these available data, it is probably of interest to reanalyse the neutron data of Teixeira et al. [22] using the two known relations Eq. (16) for D and Eq. (19) for τ_2, and take the three fitting parameters as

$\sqrt{\langle u^2 \rangle}$, τ_0 and r. It is intriguing to see whether the ratio r would indeed start from 3 at room temperature and gradually decrease toward unity upon deep supercooling.

Table 1. Relaxation Times Extracted from QENS

$T(^{\circ}K)$	τ_0(ps)	τ_1(ps)	τ_2(ps)	p	F_M	τ_{NMR}(ps)	$100\tau_0 F_M$
293	1.25	3.51	1.17	0.628	0.0192	3.06	2.39
285	1.66	3.81	1.27	0.660	0.0134	3.84	2.22
278	2.33	4.14	1.38	0.688	0.00948	4.82	2.21
268	4.66	4.71	1.57	0.728	0.00547	7.04	2.55
263	6.47	5.01	1.67	0.748	0.00403	8.80	2.61
261	7.63	5.16	1.72	0.756	0.00354	9.70	2.70
258	8.90	5.37	1.79	0.768	0.00289	11.3	2.58
256	10.8	5.52	1.84	0.776	0.00251	12.7	2.72
248	22.7	5.76	1.92	0.808	0.00136	15.2	3.08

Table I lists the temperature dependence of τ_0, τ_1 and τ_2 taken from the analysis of Teixeira et al. [22]. In this analysis the rotational diffusion model was used, so $\tau_1 = 3\tau_2$. Also listed are τ_{NMR} calculated from Eq. (19) which should be compared with τ_2 from the neutron data. One sees that the agreement is poor. This may be because of the assumption of rotational diffusion which is not valid in water. It is perhaps more appropriate to take a rotational jump model for which $\tau_1 = \tau_2 = \tau_{NMR}$. In fact the neutron data may not be sensitive to the particular rotational model because the accessible Q range is limited, and the line shape is dominated by τ_0. It is therefore of interest to look into the origin of the non-Arrhenius behavior of τ_0. It should be noted that τ_0 is a time constant that no other technique except neutron scattering can get access to.

Stanley and Teixeira [28] have introduced a percolation model to interpret the anomalous thermodynamic behavior of supercooled water. Central to their arguments is the probability p of formation of an intact hydrogen bond per water molecule. The probability p is given a temperature dependence, $p = 1.8 - 0.004T$ for H_2O. For example, the average number of hydrogen bonds per water molecule is calculable in terms of p as:

$$n_{HB} = \sum_{j=1}^{4} j \binom{4}{j} p^j (1-p)^{4-j} = 4p . \tag{20}$$

According to Eq. (20), $n_{HB} = 2.51$ at $T = 293^{\circ}K$ and $n_{HB} = 3.47$ at $T = 233^{\circ}K$. The probability that a water molecule is bonded by j hydrogen bonds is given by $f_j = \binom{4}{j} p^j (1-p)^{4-j}$. Stanley and Teixeira [28] introduced the concept of a mobile fraction of water molecules, F_M, which is the fraction that can participate in diffusion at a given time. It was assumed that $F_M = f_0 + f_1$. This ansatz was used later by Bertolini et al. [14] to interpret the temperature dependence of their data on dielectric relaxation time τ_D of supercooled water. I found, however, it is more reasonable to use $F_M = f_0 = (1-p)^4$ for interpretation of the temperature dependence of τ_0. In the last column of Table I, I listed the product $100\tau_0 F_M$ as a function of T. It is seen that except the last point ($t = -20^{\circ}C$, this temperature was uncertain in the experiment), the products are near constant and equal to 2.5 ± 0.2. This means that the jump rate τ_0^{-1} is proportional to the mobile fraction F_M which is a very reasonable result.

It has been known in the literature that the temperature dependence of transport coefficients, such as viscosity η and self diffusion constant D, is noticeably non-Arrhenius below the melting temperature for water. It has, however, been also established that the following relations are well obeyed: $\eta D/T = C_1$, $\tau_D T/\eta = C_2$ and hence $D\tau_D = C_3$. Prupacher [19], in particular, gave a value for $C_3 = 2.11 \times 10^{-16} \mathrm{cm}^2$, which is interpreted by Sposito [25] to mean that

$$D\tau_D = a^2 \tag{21}$$

where a = 1.45Å is the effective hydrodynamic radius of a water molecule. Since τ_D is related to the rotational correlation time τ_1 by Eq. (18), it might be, at first sight, puzzling to see a relation such as Eq. (21) between the translational diffusion constant of the center-of-mass and the rotational correlation time. Eq. (21) can however be taken as a hint that the translational diffusion of a water molecule is possible only by first activating the rotation which breaks the hydrogen bonds. When sufficient number of hydrogen bonds are broken by the rotation, the water molecule is free to move to the adjacent available cavity and be trapped again by another set of hydrogen bonds. In between these trapped states there is a rapid translational motion of the molecule as a whole, which appears on the average as a diffusional jump. The residence time τ_0 in this picture thus corresponds to the time interval required to break sufficient number of hydrogen bonds by rotation so that the molecule can begin to translate. It is then understandable that τ_0 is a strong function of temperature, because the average number of hydrogen bonds per molecule n_{HB} (see Eq. (20)) increases rapidly with decreasing temperature.

In summary I would like to stress that the short-time diffusion in a hydrogen bonded liquid such as water is qualitatively very different from that of a simple liquid like argon. In a simple liquid, the atoms make free flights in between short-duration collisions. Thus the short-time diffusion would never be a jump-diffusion-like because there is no valid concept for a residence time in this case. On the other hand, in a hydrogen bonded liquid, a pair of brief free flights are

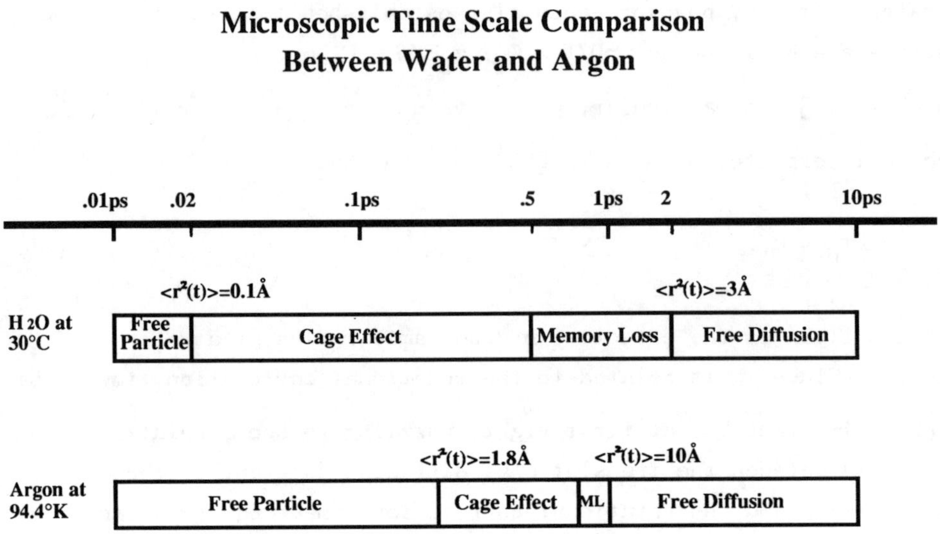

Fig. 6: Comparison of the short-time diffusion between water (a hydrogen bonded liquid at 30°C) and argon (a simple liquid at 94.4°K). The particular feature to be noted is that for a water molecule, between the initial free flight and the final free diffusion there is a lengthy period of cage effect (from 0.02 ps to 2 ps) during which the molecule moves very little. In contrast, for an argon atom, the initial free flight and the final free diffusion are interrupted by only a brief period (from 0.2 ps to 1 ps) of the cage effect.

interrupted by a lengthy period of struggle to break loose from the hydrogen bonds. Thus the concept of a residence time is approximately valid and the jump diffusion model has its validity at low temperatures. Fig. 6 shows a comparison of the sequence of events in molecular time scale for water and argon starting from 0.01 ps to several ps that I reconstructed from looking at the CMD results of Impey et al. [12] and Lie and Clementi [13]. Starting from an initially localized water molecule at the origin, the free flight (actually a free rotation) is interrupted abruptly at about 0.02 ps by the presence of the hydrogen bonds. At this time the hydrogen atom in a water molecule moves only about $\langle r_H^2(t)\rangle = 0.1Å^2$ from its initial position. The molecule feels the presence of the hydrogen bonds and the velocity autocorrelation functions of both the center of mass and the hydrogen atom start to go negative and show the cage effect. This cage effect lasts until 0.5 ps and after that gradually disappears at about 2 ps. At this time the hydrogen atom moves merely $\langle r_H^2(t)\rangle = 3Å^2$ from the origin, largely by rotation around the center-of-mass. After 2 ps, the free diffusional behavior sets-in and the time dependence of the mean square distance of travel of the hydrogen $\langle r_H^2(t)\rangle$ and the oxygen $\langle r_O^2(t)\rangle$ become parallel to each other. Thus it can be said figuratively that at 30°C, where the simulation was made [13], the water molecule takes, on the average, 2 ps to break loose from all the hydrogen bonds by rotation. This time can be taken to be the residence time τ_0 measured by neutrons. On the other hand, for an Argon atom at 94.4°K, the initial free streaming get interrupted by the inter-atomic force only after about 0.2 ps, at which time the argon atom has already traveled $\langle r^2(t)\rangle = 1.8Å$. The cage effect then takes place until 1 ps, at which time the argon atom has moved $\langle r^2(t)\rangle = 10Å^2$. The free diffusion sets in after 1 ps [29]. Between the initial free flight and the free diffusion, there is only a brief period of cage effect, in the case of argon, as compared to the long period of the cage effect, in the case of water.

3. INTERPRETATION OF THE TEMPERATURE DEPENDENCE OF THE INCOHERENT INELASTIC NEUTRON SCATTERING (INS) SPECTRA

In Section II, evidence has been presented which relates the hydrogen-bond dynamics in water to the line widths of the incoherent quasi-elastic scattering spectra. The main conclusion was that the residence time τ_0 deduced from the line-width is a direct measure of the average duration which is required for a water molecule to break loose from its hydrogen bondage by rotational excitation. As such, τ_p^{-1} is proportional to the mobile fraction $(1-p)^4$, which depends on T strongly through various powers up to four.

The incoherent inelastic neutron spectra, on the other hand, is directly related to the Q-dependent density of vibrational states of hydrogen atoms in the molecular system [30]. Since in water a large fraction of hydrogen atoms participate in hydrogen bonding at a given time, and furthermore since this fraction is a strong function of temperature, it is expected that measurements of the hydrogen atom density of states by INS would provide a sensitive means of studying the temperature dependence of hydrogen-bond dynamics.

The basic principle of the method is to extract the Q-dependent density-of-state through the relation (31,32)

$$G_s(Q,\omega) = \frac{\omega^2}{Q^2} S_s(Q,\omega) \quad . \tag{22}$$

$G_s(Q,\omega)$ can then be shown to be given by a Fourier transform of the self part of the longitudinal current correlation function $J_s(Q,t)$,

$$G_s(Q,\omega) = \frac{1}{\pi} \int_0^\infty J_s(Q,t) \cos\omega t dt \tag{23}$$

Since by definition, the $Q \longrightarrow 0$ limit of $J_s(Q,t)$ is the velocity auto-correlation function of the hydrogen atom,

$$\lim_{Q\to 0} J_s(Q,t) = \langle V_H(0)\cdot V_H(t)\rangle \tag{24}$$

We have the very useful relation,

$$\lim_{Q\to 0} G_s(Q,\omega) = f_H(\omega) \cdot \frac{k_B T}{M_{H_2O}} \tag{25}$$

where $f_H(\omega)$ is the true density-of-states of the hydrogen atom. The vibrational density-of-states of a given species of atom in liquid state is defined as the spectral density of the velocity auto-correlation function of that particle. In practice the $Q \longrightarrow 0$ limit is satisfied when the product $Qb < 1$ [30]. Study of $f_H(\omega)$ in liquids is particularly important in their supercooled states where they may approach amorphous or glassy states. For water, b = 1Å and measurement of $f_H(\omega)$ is possible only for $E = h\omega$ less than about 50 meV. However, CMD calculations of both $f_H(\omega)$ and $G_s(Q,\omega)$ are straightforward [32,33]. Thus the strength of INS that it gives directly the proton density-of-states, or its Q-dependent extension, both of which are directly calculable in CMD once the inter- and intra- molecular potentials of water molecules are specified. This feature of INS is in sharp contrast to the traditional optical spectroscopic techniques of infrared and Raman spectroscopies for which CMD calculations are possible only in some approximations [34].

We shall begin by describing the general feature of $G_s(Q,\omega)$ in water. In increasing frequency, after the prominent quasi-elastic peak, the inelastic spectrum shows three inter-molecular bands: A sharp and strong band at about 6meV (48.4 cm^{-1}) called A, a broad and weak band spreading between 20-35 meV (161-282 cm^{-1}) called B, and a strong and broad band from 50-130 meV (403-1049 $cm^{-1)}$ called L. The bands A and B are translational in character and have their counter parts in Raman

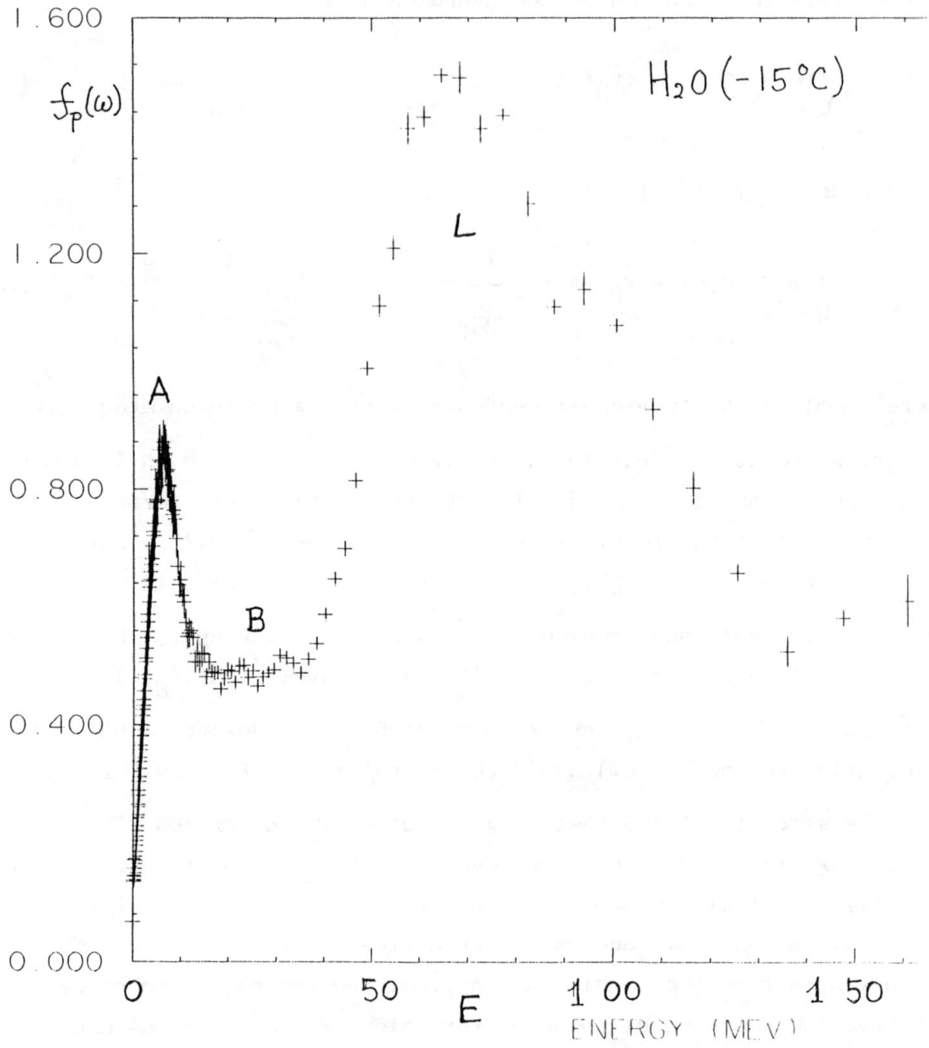

Fig.7: Inter-molecular part of the density-of-states of hydrogen atoms in water at $-15^{\circ}C$ deduced from the INS data. The three bands, A, B and L are visible.

spectrum [35,36,37]. The Raman hand at 60 cm^{-1} corresponding to A, is assigned to be the transverse vibration of water molecule as a whole with aspect to the hydrogen bond direction. This transverse motion induces a large amplitude oscillation of the hydrogen atom perpendicular to the hydrogen bond and is reflected in INS spectrum as a strong peak. The Raman bands at 190 cm^{-1} and 260 cm^{-1}, corresponding to B, are due to O-O stretching modes which induce only small amplitude motions of the H-atom. Thus these bands have weak intensities in INS spectrum. On the other hand, the L band is librational in character. The water molecule rotates around the center of mass (nearly coincident with the oxygen atom) resulting in large amplitude motion of H-atoms. This gives rise to an intense INS spectrum known as the hinder rotation band. In the Raman spectrum, it corresponds to a broad librational band extending from 300 cm^{-1} to 900 cm^{-1} [36]. The reason for such a broad band is that a water molecule has three different moments of inertia and librations around each axes have different frequencies. Fig. 7 shows a proton density of states $f_H(\omega)$ extracted from an INS spectrum for water at -15°C [38]. One readily notice the sharpness of the A band at 6 meV, a very weak B band at around 20-35 meV and an intense L band in the range 50-130 meV.

We now turn our attention to the intra-molecular vibrations. We show in Fig. 8 a Q-dependent proton density-of-state in a two-dimensional plot, as functions of Q and E, obtained for water at -15°C using a high Energy Medium Resolution Chopper Spectrometer (HRMECS) operating at the Intense Pulse Neutron Source (IPNS) of Argonne National Laboratory. Starting from the left-hand corner of the energy scale, the first peak about 80 meV is the L band, the second peak at 220 meV is the OH bending mode (BEN) and the broad peak at around 420 meV is the OH stretch vibration (STR). Actually in a free water molecule there are two stretch vibrations, one is symmetric and the other antisymmetric with the latter at higher frequency. In the condensed phase, INS spectra show that there is only one single broadened peak. Table II list all the observable peaks in INS spectra.

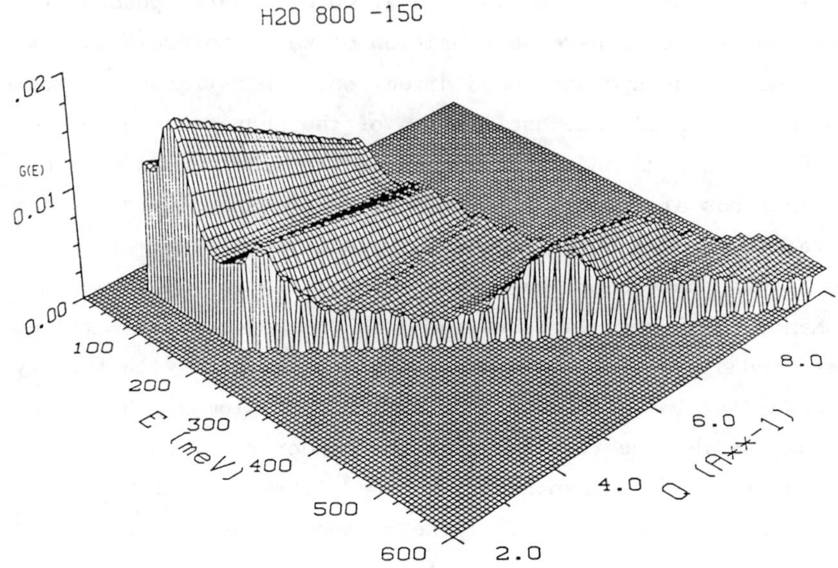

Fig. 8: Q-dependent density-of-states $G_s(Q,\omega)$ for water at -15°C measured by INS technique. From the low frequency side, there are three bands: L, BEN and STR as described in the text.

Table 2. Experimental vibrational energies in water and ice

	T(°C)	E_L(meV)	E_{BEN}(meV)	E_{STR}(meV)
Ice	253	82	207	407 (49[a])
H_2O	258	74	207	418 (49)
H_2O	313	-	207	441 (66)
H_2O	353	-	207	443 (70)
Gas(Raman)	293	-	198	453[b], 466[c]
Q(Å^{-1})		2.6	3.5	6.6

[a]values in parenthesis are bandwidths; [b,c] symmetric and antisymmetric modes

Fig. 9 shows the Q-dependent density-of-states for the three temperatures listed in Table II [32]. We summarize the main features of the experimental spectra as follows: (1) significant temperature dependence is observed for the L-band and STR-band while the BEN-band is nearly temperature independent. The frequency of L-band decreases as the structure changes from ice to water. The opposite behavior is observed for the STR-band. (2) the width of STR-band, which can be more reliably estimated, increases with temperature. The energy resolution at the stretch band is 16 meV. The antisymmetric and symmetric stretch vibrations in gas phase (see Table 2) are split by 13 meV. This means that the symmetric and antisymmetric stretch vibrations in the free molecule, if they persist in the condensed phase, cannot be resolved in the INS experiment.

CMD results of Toukan and Rahman [33] using the SPC flexible water molecule potential have been available for both gas and liquid phases. Careful examination of these results reveals the following two important points: (1) The intra-molecular O-H bond length in liquid is 2% longer than that in the gas phase. This elongation leads to a weakening of the

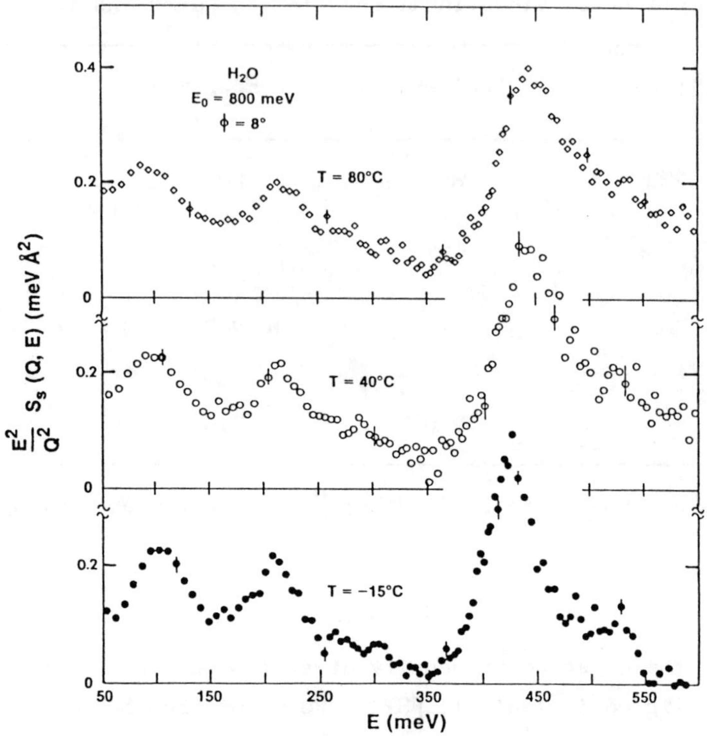

Fig. 9: A Constant scattering angle ($\theta = 8^{\circ}$) cut of $G_s(Q,\omega)$ for water at three temperatures 80°C, 40°C, and −15°C. The data are corrected for multiple scattering.

O-H bonds and hence to mode softening of the stretch vibration. This is observed in the experiment (see Table 2); (2) the formation of hydrogen bonds in the liquid phase leads to further softening of the O-H stretch vibrational mode. As the temperature of the system is decreased, more intact hydrogen bonds are formed, leading to a decrease in the STR-band energy. The experimental results shown in Fig. 9 and listed in Table 2 also support this result. This trend is correlated with an increase in L-band energy (see Table 2).

It is of interest next to compare the so-called density-of-states measured by different spectroscopic techniques. A fair and unambiguous way of comparing the different spectra is to generate them by CMD using the same inter-molecular potential. This can be done by calculating the correlation functions each spectroscopic techniques are supposed to be measuring. In the case of neutron scattering this is straight forward and without ambiguity. In the case of infrared and Raman some additional assumptions are needed to calculate the spectral density functions [34].

For INS, we can write the definition of $J_s(Q,t)$ which appeared in Eq. (23) as

$$J_s(Q,t) = \left\langle \frac{1}{N} \sum_{j=1}^{N} (Q\cdot\vec{V}_j(t+\tau))(Q\cdot\vec{V}(\tau))\exp[iQ\cdot(\vec{R}_j(t+\tau)-\vec{R}_j(\tau))] \right\rangle \qquad (26)$$

where N is the number of H-atoms in the system.

For IR, we can use the bond dipole moment model to write

$$C^{IR}(t) = \left\langle \frac{1}{N} \sum_{j=1}^{N} [(\frac{\delta\mu}{\delta r_j})_0 \ \delta\vec{r}_j(\tau)][(\frac{\delta\mu}{\delta r_j})_0 \ \delta\vec{r}_j(t+\tau)] \right\rangle \qquad (27)$$

where N is the number of O-H bonds. We then obtain the spectrum by

$$I^{IR}(\omega) = \frac{\hbar\omega}{2k_BT} \coth(\frac{\hbar\omega}{2k_BT}) \frac{1}{\pi} \int_0^\infty C^{IR}(t) \cos\omega t \, dt \; . \tag{28}$$

For isotropic Raman scattering, in the bond polarizability model, the corresponding correlation function is

$$C^R_{ISO}(t) = \langle \frac{1}{9N} \sum_{k=1}^{N} \left[\sum_{\alpha=1}^{3} A^k_{\alpha\alpha}(\tau) \cdot \sum_{\alpha=1}^{3} A^k_{\alpha\alpha}(t+\tau) \right] \rangle \tag{29}$$

where the polarizability tensor $A^k_{\alpha\beta}$ is given by

$$A^k_{\alpha\beta}(t) = \{ (\frac{\alpha_\parallel}{\alpha_\perp} - 1)(\vec{r}_k(t)\cdot e_a)(\vec{r}_k(t)\cdot e_\beta) + \delta_{\alpha\beta} \} \, \delta r_k(t) \; . \tag{30}$$

$r_k(t)$ is the unit vector along the direction of the k-th O-H bond and $\delta r_k(t)$ is the change in the k-th O-H bond length from the equilibrium value. The spectrum is then calculated as

$$I^R(\omega) = K(\omega) \frac{1}{\pi} \int_0^\infty C^R(t) \cos \omega t \, dt \tag{31}$$

where $K(\omega)$ is the Bose-Einstein factor

$$k(\omega) = (\omega_L - \omega)^4/[\omega(1-\exp(-\hbar\omega/k_BT))] \; . \tag{32}$$

and ω_L the laser frequency.

Fig. 10 shows the result of CMD calculation [34] using the SPC flexible water molecule potential [33] at T = 250°K. The peaks shown are for the STR-band only. It should be noted that while INS and IR spectra show only one peak, the isotropic Raman spectrum shows a split peak. We can thus conclude that the single broad STR-band one observed in Fig. 9 by INS is not resolution limited but due to the strong damping occurring in the condensed phase. On the other hand the split peak one observed in the experimental Raman spectra [36] should not be taken as evidence for existence of two stretch vibrational modes in the density-of-states of water. The correlation function in Eq. (30) that Raman scattering is observing is in fact quite different from the true density-of-states of water which is defined by Eq. (25).

A nice feature of a time-of-flight instrument such as HRMECS is that it has many detectors at closely spaced angles covering an angular range from about 3° to 130°. It is thus possible to interpolate a two-dimensional representation of $G_s(Q,\omega)$, such as the one shown in Fig. 8, to obtain plots of $G_s(Q,\omega)$ vs ω at constant Q (instead of constant θ). These kinds of plots facilitate comparisons with CMD calculations. Fig. 11 shows two such constant Q plots of $G_s(Q,\omega)$ for H_2O at T = 258°K. Fig. 11(a) covers an energy range appropriate to the L- and BEN-modes and Fig. 11(b) that of the other the STR-mode. In Fig. 11(a) one sees the L-mode peak at about 100 meV and the BEN-mode peak at 230 meV. There is a satelite peak visible at 300 meV. In Fig. 11(b) we can identify the STR-mode peak at 430 meV and a satelite peak at 530 meV. Two aspects of these plots are of particular importance to our discussion here.

3.1 Absolute Intensity of the Stretch Vibrational Band

It is of interest to compare the absolute intensity of an INS spectrum with that generated by CMD. Using the SPC flexible water molecule potential. Toukan et al. [32] have generated $G_s(Q,\omega)$ using the definition of $J_s(Q,t)$ given in Eq. (26) combined with the Fourier transform relation of Eq. (23). The result for H_2O at T = 250°K is shown in Fig. 12. Two features are immediately noticeable from inspection of these graphs: (i) the L-band is considerably broadened by increased Q values; (ii) the BEN-band is moderately affected by Q but the STR-band is hardly affected by it. This latter fact implies that the STR-band represents a well-localized mode of vibration. It is for

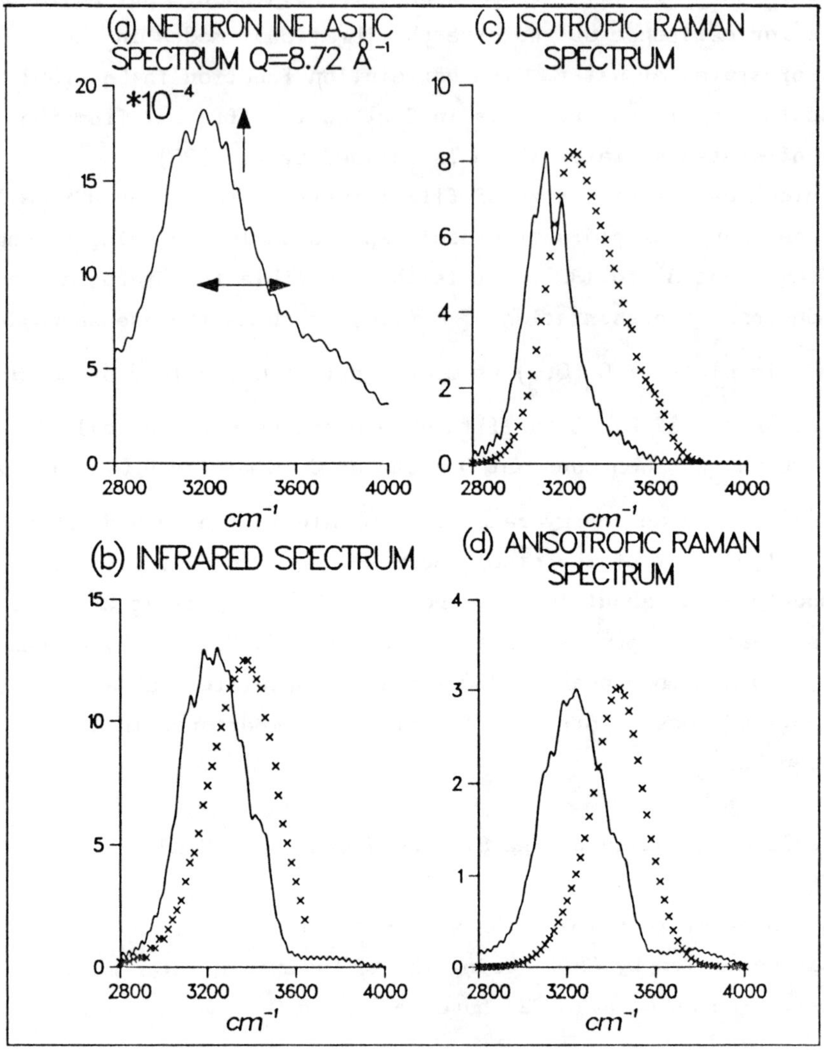

Fig. 10: CMD generated spectra for three different spectroscopic techniques. The crosses are experimental points [34].

this reason that experimentally one can still observe a sharp STR-band at an energy as high as 500 meV and a Q value as large as $10Å^{-1}$. Since it is a localized mode, the mode can be well approximated by a harmonic oscillation.

For a harmonic oscillator (or harmonic lattice vibrations) the quantum mechanical dynamic structure factor can be written as follows [39]:

$$S_s(Q,\omega) = \exp[\hbar\omega/2k_BT]\exp[-\gamma Q^2]S_s^{cl}(Q,\omega) \tag{33}$$

where

$$\gamma = \int_0^\infty d\omega \frac{f_H(\omega)}{\omega^2}\left[\cosh\frac{\hbar\omega}{2k_BT} - 1\right] \tag{34}$$

and $f_H(\omega)$ is the proton density-of-states,

$$f_H(\omega) = \frac{2}{\pi}\int_0^\infty dt \cos\omega t \langle V_H(0)V_H(t)\rangle . \tag{35}$$

$S_s^{cl}(Q,\omega)$ is the classical dynamic structure factor calculated by CMD. To compare the CMD calculation with the experiment, we must multiply the $S_s^{cl}(Q,\omega)$ with the detailed balance factor and the Q-dependent recoil factor given in Eq. (33) and (34). Referring to fig. 11(b), the stretch band peak is at 420 meV. At $T = 298°K$, the detailed balance factor is about 4200. The nominal Q value at the band center is about $8.5Å^{-1}$. The attenuation coefficient γ can be calculated according to Eq. (34) by CMD [40]. At room temperature, for a SPC water, the value of γ is $0.42Å^2$, which gives 1/2000 for the Q-dependent recoil factor. Thus at STR-band center, we have,

$$G_s(Q,\omega)|_{EXP} = 2.1\ G_s^{cl}(Q,\omega)|_{CMD} . \tag{36}$$

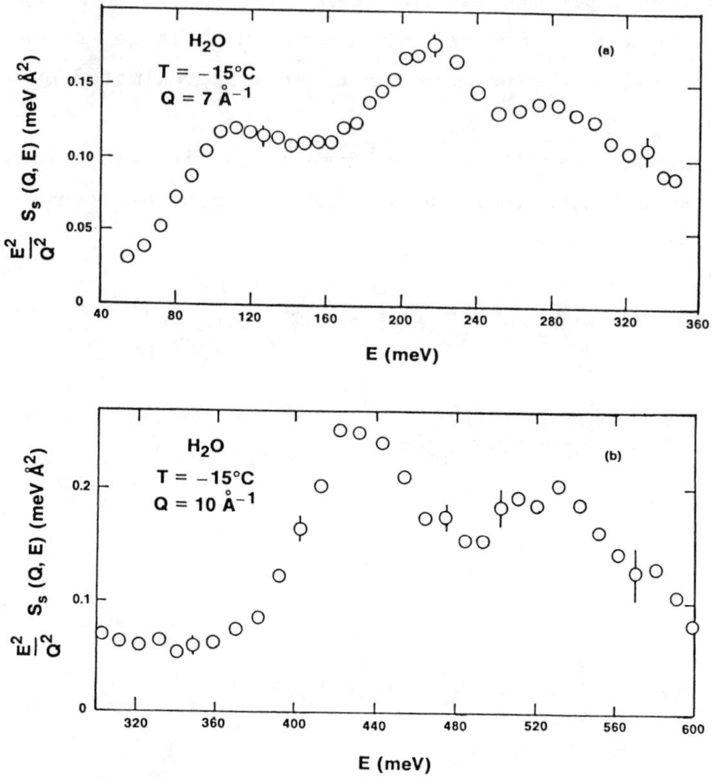

Fig. 11: onstant Q plots of $G_s(Q,\omega)$ for water at $T = 258^\circ K$ in two energy ranges. (a) Energy range covering L and BEN-modes. Notice the clearly visible satelite peak at 300 meV. (b) Energy range covering the STR-mode and its satelite peak at 530 meV.

From Fig. 11(b), $G_s(Q,\omega)|_{EXP} = 0.25$ and $G_s^{cl}(Q,\omega)|_{CMD} = 0.14$, thus the relation (36) is approximately satisfied. In practice the detailed balance factor is very temperature sensitive and the CMD temperature of water may not correspond to the real temperature of water. So this level of agreement is satisfactory. The formula (33), however, is not valid in other region of ω because the spectral densities in other region cannot be represented by those of harmonic oscillators.

3.2 Temperature Dependence of the Satelite Peak

Both Fig. 11(a) and 11(b) show pronounced satelite peaks occurring at the high energy side of the main vibrational bands. In Raman spectroscopy literature the satelite peaks are referred to as combination bands [41]. For example the satelite peak at 530 meV has been assigned to the overtone of the librational and the stretch bands on the basis of frequency shift analysis. From a quantum mechanical point of view it is more meaningful to regard this extra peak as arising from a perturbation of the stretch vibrational mode by the H-bond. Ricci and Chen [42] and Ricci et al. [43] have proposed a specific mechanism, an excitation of the fundamental stretch vibraion of a bound water dimmer with a simultaneous breaking of an H-bond, which allowed them to calculate both the frequency shift and the intensity of the satelite peak as a function of temperature. I shall briefly describe the calculation in the following.

Consider a hydrogen atom which is bonded to an oxygen atom in a water molecule by a covalent bond V_{OH} and is also involved in an H-bond V_{HB} with another oxygen atom, belonging to another water molecules. For simplicity we shall assume the two bonds are strictly co-linear as depicted in Fig. 13(A). Let r^* be the equilibrium position of the hydrogen atom and r_0 the coordinate of the oxygen atom involved in the hydrogen bond. In a first approximation, we shall take a Morse potential to represent both V_{OH} and V_{HB}. Thus

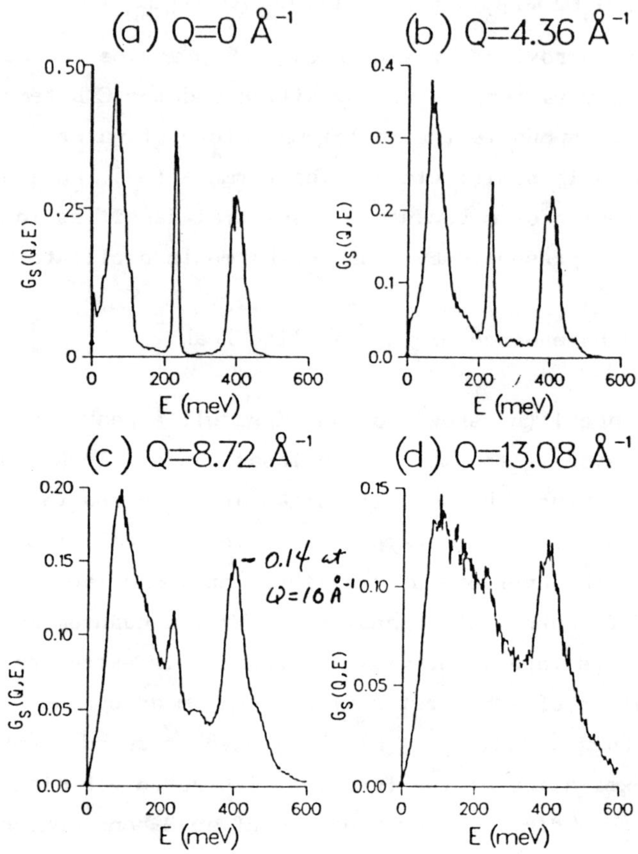

Fig. 12: CMD generated Q-dependent density-of-states for SPC flexible-molecule water at T = 250°K. Note the striking Q dependence of the L-mode and relatively Q-independence of the STR-mode.

$$V_{OH} = D_1\{\exp[-2a_1r] - 2\exp[-a_1r]\} \tag{37}$$

$$V_{HB} = D_1\{\exp[-2a_2(d-r)] - 2\exp[-a_2(d-r)] \tag{38}$$

where r is the position of the hydrogen atom measured from the equilibrium position r^* and $d = r_0 - 2r^*$. A typical set of parameters chosen by Ricci and Chen [42] were as follows:

$$\begin{aligned} &D_1 = 3.757 \text{ eV} && a_1 = 2.566\text{Å}^{-1} \\ &D_2 = 0.130 \text{ eV} && a_2 = 1.24\text{Å}^{-1}, \\ &r^* = 1\text{Å}, \quad r_0 = 2.8\text{Å}, && d = 0.8\text{Å} . \end{aligned} \tag{39}$$

We note that the OH-bond strength D_1 is 29 times larger than D_2, the average strength of the hydrogen bond, which was taken to be 130 meV, thus V_{HB} is a much weaker potential compared to V_{OH}.

The Hamiltonian of the hydrogen atom is

$$H = \frac{\hbar^2}{2m_p}\nabla^2 + V_{OH} + V_{HB} = H_0 + V_{HB} \tag{40}$$

Fig. 13(B) shows the profile of the total potential, $V = V_{OH} + V_{HB}$, in a solid line and that of V_{OH} in open circles. We know analytical forms of eigen-functions $\psi_n^{(0)}$ and eigen-values $E_n^{(0)}$ of H_0 [44]. The eigenvalues are given by the expression,

$$E_n^{(0)} = -D_1 + (n + \tfrac{1}{2})4D_1/k - (n + \tfrac{1}{2})^2 4D_1/k^2 . \tag{41}$$

where the constant $k = (8m_pD_1/\hbar\, a_1)^{1/2} = 32.8$. In particular $E_0^{(0)} =$ -3531.41 meV and $E_1^{(0)} = 3101.17$ meV, so that the energy of the stretch

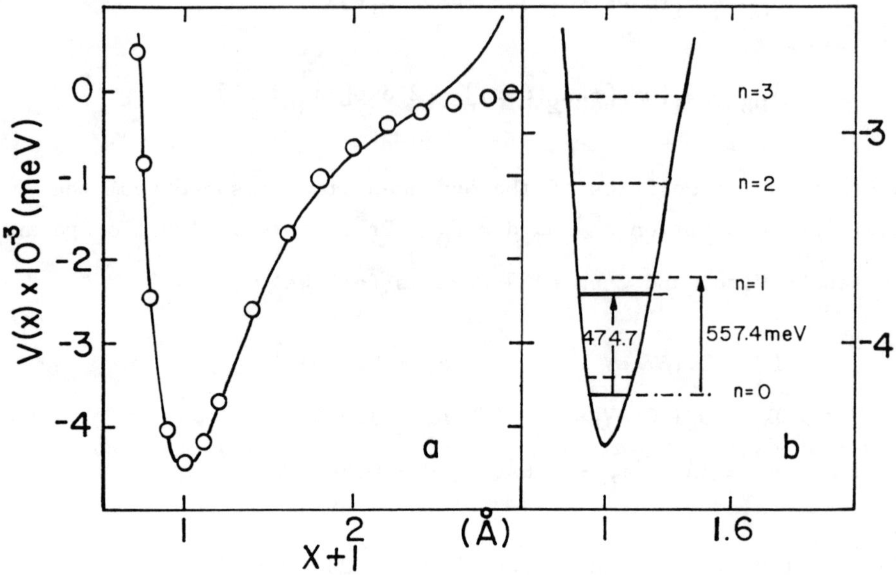

Fig. 13: Quantum mechanical calcultion of the linear hydrogen-bond. (A) Geometry of the linear H-bond. (B) The potential well, and (C) the energy diagram.

vibration of a free water molecule is $\hbar\omega = E_1^{(0)} - E_0^{(0)} = 430.24$ meV.

We next treat V_{HB} as a perturbation to H_0 and calculate the eigen functions and eigen-values of the Hamiltonian H. For the energies of the ground state and the first excited state, we have

$$E_0 = E_0^{(0)} + \langle \psi_0^{(0)} |V_{HB}|\psi_0^{(0)} \rangle = -3611.39 \text{ meV} \quad (42)$$

$$E_1 = E_1^{(0)} + \langle \psi_1^{(0)} |V_{HB}|\psi_1^{(0)} \rangle = -3184.24 \text{ meV} \quad (43)$$

From these values, we obtain the fundamental stretching vibration of a H-bonded hydrogen atom as $\hbar\omega_1 = E_1 - E_0$.

Referring now to Fig. 13 (c) which shows the energy level diagram. We designate the eigenvalues of H_0 by dashed lines and that of H by solid lines. The short arrow sign indicate the transition from $|\psi_0\rangle$ to $|\psi_1\rangle$ which has the transition energy $h\omega_1 = 427.15$ meV mentioned above. This transition gives rise to the STR-mode one observes in INS. The long arrow sign indicates a transition from $|\psi_0\rangle$ to $|\psi_1^{(0)}\rangle$, which has a transition energy of $\hbar\omega_2 = E_1^{(0)} - E_0 = 510.22$ meV. This transition corresponds to an excitation of the fundamental stretch vibration with a simultaneous breaking of a hydrogen bond. This is the origin of the satelite peak experimentally observed on the higher energy side of the STR-mode. According to the energy level diagram, the satelite peak is shifted by 510.22-427.15 = 83.1 meV on the higher energy side.

In real water, the hydrogen bond strength has a certain spread around a mean value. One can choose for example a value, $\overline{D}_2$ = 3.0Kcal/mol = 130 meV, according to a previous spectroscopic estimate [45]. The satelite peak energy is then given by

$$E_2 = 430.24 + 79.98\ D_2/130 \text{ meV} \quad (44)$$

The spread of D_2 around the mean value 130 meV is taken to be Gaussian, with a width of ± 1.5 Kcal/mol (65meV), as can be estimated from a CMD result [46]. The intensity of the satelite peak is proportional to a matrix element, $|M_{10}|^2 = |\langle \psi_1^{(0)} | \exp(i\underline{Q}\cdot\underline{r}) | \psi_0 \rangle|^2$ which can be easily calculated from the wave functions. The calculated sateliet peak profile should be further convolved with an instrument resolution of ± 24 mev. The result is shown in Fig. 14. The solid line is the theory and the symbols denote the experiment. One sees that the agreement of the theory and experiment is good. Using Eq. (44) to fit the experimentally observed satelite peaks at different temperatures, one can estimate D_2 as a function temperature. From data given in Fig. 14, one gets D_2 = 3.2 ± 0.1 Kcal/mol in the supercooled water and D_2 = 2.5 ± 0.3 Kcal/mol for the above room temperature water. The experiment thus suggests a softening of the hydrogen bond strength as the temperature rises above room temperature.

4. SUMMARY AND PROSPECTS

The significant progress in neutron scattering research on the structure and dynamics of bulk water since 1980 that I have included in this article came from three sources: the availability of CMD simulations over a wide range of temperatures, exploration into the supercooled region by QENS and INS and, last but not least, improvements in neutron scattering instrumentation. CMD calculations allow one to visualize correlation functions which are not accessible by experiments. More importantly they give detailed molecular dynamics of different time scales which are difficult for experiments to isolate. The exploration of the supercooled region of water enables one to separate the different relaxation times scales between rotations and translations, resulting in a more accurate determination of each of these quantities. Improvements of neutron scattering instrumentation have resulted in a greatly increased neutron flux at the sample, thus shortened the data collection

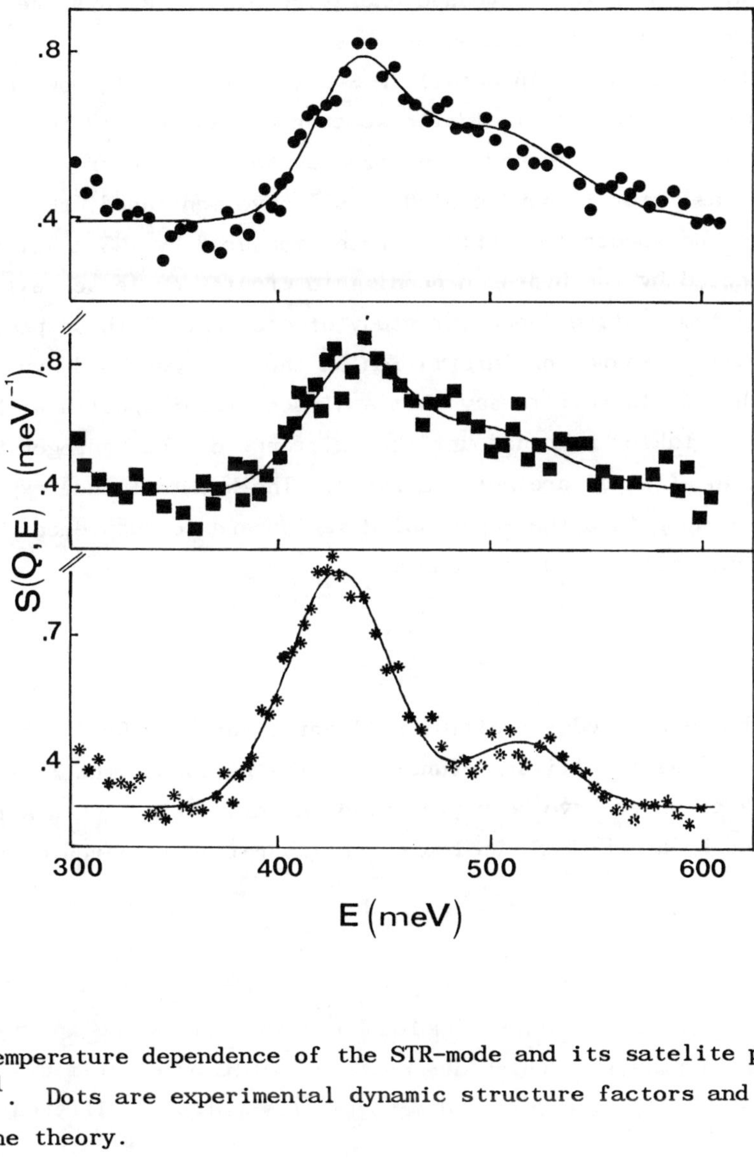

Fig. 14: Temperature dependence of the STR-mode and its satelite peak at Q = 11Å^{-1}. Dots are experimental dynamic structure factors and solid lines are the theory.

time. Furthermore the increased energy resolution allows one to measure the slower time scale when water is supercooled.

I have discussed in detail in Section II and III, how the quasi-elastic scattering and inelastic scattering can be used to obtain information on both the hydrogen-bond dynamics and hydrogen bond strength as water is supercooled. As I have mentioned, there are five peaks in the spectral density of water measured by INS. All five peaks are affected by the hydrogen bonding to greater or lesser extent. Thus detailed temperature dependent study of each one of these peaks is potentially a source of information on the hydrogen bond dynamics and strength. It is fair to say that analyses of INS spectra without explicitly taking into account the existence of the hydrogen bond network in of water are not realistic. The key to future experiments is to pursue deep into the supercooled region and to supplement INS experiments with concurrent CMD work.

The author acknowledge fruitful collaborations with Dr. Jose Teixeira and Marie Claire Bellissent-Funel over the period of the last ten years, the work of which have been presented in this article. This research is supported by a NSF-CNRS, US-France collaborative research grant.

APPENDIX A. Partial wave expansion of rotational intermediate scattering function $F_R(Q,t)$, Eq. (8).

Use the standard theorem for the expansion of the plane wave,

$$\exp[i\underline{Q}\cdot\underline{b}(t)] = 4\pi \sum_{\ell=1}^{\infty} \sum_{m=-\ell}^{\ell} i^{\ell} j_{\ell}(Qb) Y_{\ell m}^{*}(\hat{Q}) Y_{\ell m}(\hat{b}(t)) \tag{1A}$$

to get

$$\langle e^{-i\underline{Q}\cdot\underline{b}}\, e^{i\underline{Q}\cdot\underline{b}(t)}\rangle = (4\pi)^2 \sum_{\ell,m} \sum_{\ell',m'} (i)^{\ell}(-i)^{\ell'} j_{\ell}(Qb) j_{\ell'}(Qb) Y_{\ell m}^{*}(\hat{Q}) Y_{\ell' m'}(\hat{Q}) \times \langle Y_{\ell' m'}^{*}(\hat{b}) Y_{\ell m}(\hat{b}(t)) \rangle \tag{2A}$$

Average over the direction of $\hat{Q}$, and use (3A)

$$\langle Y_{\ell m}^{*}(\hat{Q}) Y_{\ell m}(\hat{Q}) \rangle_{\hat{Q}} = \frac{1}{4\pi} \delta_{\ell\ell'} \delta_{mm'} \tag{3A}$$

we get

$$\langle e^{-i\underline{Q}\cdot\underline{b}}\, e^{i\underline{Q}\cdot\underline{b}(t)}\rangle = 4\pi \sum_{\ell m} j_{\ell}^{2}(Qb) \langle Y_{\ell m}^{*}(\hat{b}) Y_{\ell m}(\hat{b}(t)) \rangle . \tag{4A}$$

Use the addition theorem of spherical harmonics (5A)

$$\sum_{m=-\ell}^{\ell} Y_{\ell m}^{*}(\hat{b}) Y_{\ell m}(\hat{b}(t)) = \frac{2\ell+1}{4\pi} P_{\ell}(\hat{b}\cdot\hat{b}(t)) , \tag{5A}$$

We finally obtain

$$\langle e^{-i\underline{Q}\cdot\underline{b}}\, e^{i\underline{Q}\cdot\underline{b}(t)} \rangle = \sum_{\ell=0}^{\infty} (2\ell + 1) j_{\ell}^{2}(Qb) \langle P_{\ell}(\hat{b}\cdot\hat{b}(t)) \rangle \tag{6A}$$

References

1. B. N. Brockhouse, Nuovo Cim. Suppl. 9, 45 (1958); Phys. Rev. Lett. 2, 287 (1959).

2. L. Van Hove, Phys. Rev. 95, 249 (1954).

3. M. Sakamoto, B. N. Brockhouse, R. G. Johnson and N. K. Pope, J. Phys. Soc. (Japan) 17, Suppl. B-II, 370 (1962).

4. J. H. Simpson and H. Y. Carr, Phys. Rev. 111, 1201 (1958)

5. P. Von Blanckenhagen, Ber. Bunsenges. Phys. Chem. 76, 891 (1972).

6. S. H. Chen and J. Teixeira, "Structure and Dynamics of Low Temperature Water as Studied by Scattering Techniques", p. 1-45, in Advances in Chemical Physics, Vol. LXIV, Edited by I. Prigogine and S. A. Rice, John Wiley and Sons (1986).

7. J. Teixeira, M. C. Bellissent-Funel, S. H. Chen and A. J. Dianoux, J. de Physique, Colloque C7, Suppl. 9, 45, C7-65 (1984).

8. W. Marshall and S. W. Lovesey, Theory of Thermal Neutron Scattering, Oxford, Clarendon Press (1971).

9. V F. Sears, Can. J. Phys. 44, 127 (1966); 44, 1299 (1966).

10. S. Yip, "Quasi-elastic Scattering in Neutron and Laser Spectroscopy", Chapt. 2, in Spectroscopy in Biology and Chemistry: Neutrons, X-rays, Laser, edited by S. H. Chen and S. Yip, Academic Press (1974).

11. R. G. Gordon, in Advances in Magnetic Resonance, J. S. Waugh, ed. (Academic Press, 1968). Vol. 3, p. 1.

12. R. W. Impey, P. A. Madden and I. R. McDonald, Mol. Phys. 4b, 513 (1982).

13. G. C. Lie and E. Clementi, Phys. Rev. A33, 2679 (1986).

14. D. Bertolini, M. Cassettari and G. Salvetti, J. Chem. Phys. 76, 3285 (1982).

15. O. A. Nabokov and Yu. A. Lubimov, Mol. Phys. 65, 1473 (1988).

16. E. Lang and H. D. Lüdemann, Ber. Bunsenges. Phys. Chem. 84, 462 (1980), 85, 603 (1981).

17. S. H. Chen, J. Teixeira and R. Nicklow, Phys. Rev. A26, 3477 (1982).

18. A. Sjölander, "Theory of Neutron Scattering by Liquids", Chapt. 7, in Thermal Neutron Scattering, edited by P. A. Egelstaff, Academic Press (1965).

19. H. R. Pruppacher, J. Chem. Phys. 5, 101 (1972).

20. K. T. Gillen, D. C. Douglass, M. J. R. Hoch, J. Chem. Phys. 57, 5117 (1972).

21. F. X. Prielmeier, E. W. Lang, R. J. Speedy and H. D. Lüdenann, Ber. Bensenges, Phys. chem. 92, 1111 (1988).

22. J. Teixeira, M. C. Bellissent-Funel, S. H. Chen, and J. Dianoux, Phys. Rev. A31, 1913 (1985).

23. F. Franks, J. Ravenhill, P. A. Egelstaff and D. I. Page, Proc. Roy. Soc. (London) A319, 189 (1970).

24. E. Fatusso and P. R. Mason, Proc. Phys. Soc. 90, 729 (1967).

25. G. Sposito, J. Chem. Phys. 74, 6943 (1981).

26. E. W. Lang, H. D. Lüdemann and L. Piculell, J. Chem. PHys. 81, 3820 (1984).

27. D. Kivelson and S. A. Kivelson, J. chem. Phys. 90, 4464 (1989).

28. H. E. Stanley and J. Teixeira, J. Chem. Phys. 73, 3404 (1980).

29. J. P. Boon and S. Yip, Molecular Hydrodynamics, McGraw Hill (1982).

30. S. H. Chen and S. Yip, Physics Today 29, 32 (1976).

31. S. H. Chen, K. Toukan, C. K. Loong, D. L. Price and J. Teixeira, Phys Rev. Lett. 53, 1360 (1984).

32. K. Toukan, M. A. Ricci, S. H. Chen, C. K. Loong, D. L. Price and J. Teixeira, PHys. Rev. A37, 2580 (1988).

33. K. Toukan and A. Rahman, Phys. Rev. B31, 2643 (1985).

34. R. Bansil, T. Berger, K. Toukan, M. A. Ricci and S. H. Chen, Chem. Phys. Lett. 132, 165 (1986).

35. G. E. Walrafen, J. Chem. Phys. 40, 3249 (1964).

36. G. E. Walrafen, in Water: A Comprehensive Treatise, edited by F. Franks (Plenum Press, 1972) Vol. 1, p. 151.

37. S. Krishnamurthy, R. Bansil and J. Wiafe-Akenten, J. Chem. Phys. 79, 5863 (1983).

38. J. Teixeira, M. C. Bellissent-Funel and S. H. Chen (unpublished data).

39. H. Boutin and S. Yip, Molecular Spectroscopy with Neutrons, MIT Press (1968).

40. J. Anderson, PhD thesis, Massachusetts Institute of Technology (1986).

41. G. E. Walrafen and L. A. Blatz, J. Chem. Phys. 59, 2646 (1973).

42. M. A. Ricci and S. H. Chen, Phys. Rev. A34, 1714 (1986).

43. M. A. Ricci, S. H. Chen, D. L. Price, C. K. Loong, K. Toukan, and J. Teixeira, Physica B 136, 190 (1986).

44. D. Rapp, Quantum Mechanics, (Holt, Reinhart and Winston, 1971), p. 101.

45. C. J. Montrose, J. A. Bucaro, J. Marshall-Coakley and T. A. Litovitz, J. Chem. Phys. 60, 5025 (1974).

46. F. H. Stillinger and A. Rahman, J. Chem. Phys. 60, 1545 (1974); F. H. Stillinger, Science 208, 451 (1980).

The properties of hydrogen bonded liquids studied by high pressure NMR

E.W. Lang and H.-D. Lüdemann

Institut für Biophysik und Physikalische Biochemie
Universität Regensburg, Postfach 397, D-8400 Regensburg

Abstract

The translational self diffusion coefficients D of a series of simple liquids and supercooled water obtained in a wide temperature range and at pressures up to 400 MPa are given. The model descriptions published for the ρ,T-dependence of D are tested with these data. With spin lattice relaxation time (T_1) measurements of the protons, deuterons and oxygen-17 nuclei in neat emulsified supercooled water as function of frequency models for the rotational mobility of water are evaluated. The anomalous temperature dependence of D and T_1 at $p \leq 150$ MPa is discussed. The deuteron-T_1 measurements have been extended to supercooled aqueous alkali and alkali earth halide emulsions. A simple two state model for the description of the temperature and concentration dependence of the deuteron-T_1 in these solutions is given.

Introduction

High pressure studies permit the separation of the influence of density variations from changes of thermal energy. For a typical liquid close to its atmospheric pressure boiling point, the application of a pressure of 100 MPa (≡ 1000 bar) causes a volume change of a few percent only. Since the compressibility even decreases with rising pressure, one has to design apparatus which can work under an internal load of several 100 MPa in order to obtain a temperature independent density variation of ≈ 10 %. In the following some simple NMR cells will be presented, which allow high resolution spectra to be obtained up to pressures of 400 MPa. The focal point of our work in this field was the study of the most abundant and most extraordinary terrestrial liquid: Water. The first figure gives a part of the phase diagram for light and heavy water. All unusual properties of water become more pronounced with falling temperature. It is thus indicated to study the liquid in the supercooled metastable range also. This range is described by the p,T region between the melting pressure curve and the homogeneous nucleation pressure curve. The homogeneous nucleation temperature was determined by Kanno, Speedy

J.C. Dore and J. Teixeira (eds.), Hydrogen-Bonded Liquids, 333–356.

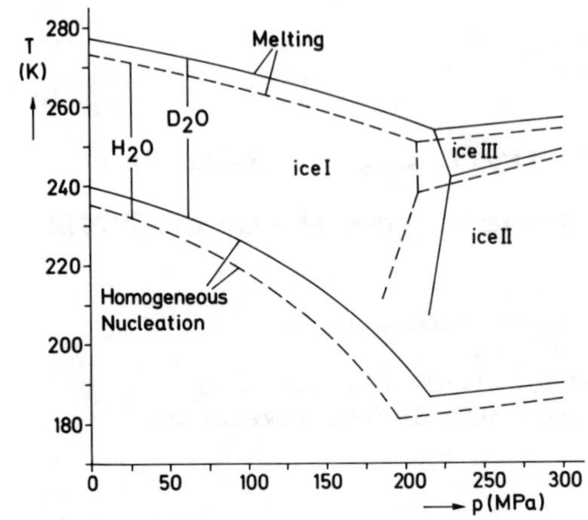

Figure 1. Phase diagram of light and heavy water.

and Angell (1,2) as function of pressure. It falls even faster with increasing pressure than the melting pressure curve and reaches in light water a minimum of approximately 180 K at a pressure of 200 MPa.

The crystallization of a supercooled liquid occurs most probably by heterogeneous nucleation on the surface of impurities within the liquid or at the walls of the container. The removal of such impurities by extreme purification is thus prerequisite to achieve reliable undercooling. Supercooled water becomes the easier to handle the smaller the metastable samples are, since the reduction in volume lowers the probability for the occurrence of heterogeneous nuclei. Therefore most investigations of supercooled water have dealt with water contained in narrow capillaries or with droplets of water ($d \approx 10$ μm) emulsified in a water-insoluble apolar carrier (3). Such emulsions are stabilized by the addition of a water insoluble emulsifying agent and can be kept in the supercooled metastable liquid state for periods of months. However, such emulsions are due to their multiple phase boundaries opaque white and thus most spectroscopic methods cannot be applied. Electromagnetic waves in the radiofequency range between 10 and 500 MHz, as used in NMR spectroscopy, readily penetrate the multiple phase boundaries of these emulsions and it was thus possible to study these emulsions by proton- and deuterium-NMR down to the close vicinity of the homogeneous nucleation pressure curve (4,5,6) and to include in subsequent studies aqueous solutions of simple salts (7,8,9).

In these experiments the spin lattice relaxation times T_1 of the protons, the deuterons and the oxygen-17 nuclei (10) were studied by the standard inversion recovery experiment (11). From these data the rotational correlation times τ_2 for single particle rotation can be derived by application of relaxation theory (12,13).

The deeply supercooled range is of special importance in these studies, since here the T_1-measurements become frequency dependent and thus allow a critical test of the various motional models (13). Between 300 K and 250 K T_1 could be measured both in bulk samples and in emulsions. In this broad temperature range the two approaches, yielded within the limits of accuracy, identical results, testifying, that the emulsification does not influence the T_1-data. Translational self diffusion coefficients D can be obtained from the classical Hahn spin echo experiment (14) performed in a defined linear magnetic field gradient (15). However, this experiment will yield reliable data only, if for the time of the experiment the dimension of the liquid phase is large compared to the mean squared displacement of the single molecules. Thus for these experiments the convenient emulsion technique cannot be applied and one has to supercool the bulk liquid in long thin capillaries (16,17,18). In the following we will present the results obtained in supercooled water with all its unique anomalies (18,19) and compare the data to similar studies on simple liquids and monoalcohols.

Experimental

Water in oil emulsions containing 50 to 60 % w/w of the aqueous phase are prepared by forcing the two liquids through several layers of a fine stainless steel net. For high pressure studies it proved most convenient to use a 1:1 mixture of methylcyclopentane/methylcyclohexane as a carrier phase (1). In order to stabilize the emulsions 3 % w/w of a water insoluble emulsifier (Span 65[R] ≡ Sorbitanetristearate, Serva, Heidelberg) are dissolved in the hydrocarbons. This emulsifier was first used by Rasmussen and McKenzie (3) and even long and thorough series of testing many other commercial nonionic emulsifiers returned in the end to this early choice (21). Because of the large magnetic moment of unpaired electrons, small amounts of paramagnetic impurities have a tremendous influence on the spin lattice relaxation times observed. The solutions have thus to be prepared and filled into the high pressure cells under perfect exclusion of oxygen and transition metal ions. Some possibilities for achieving this have been published (10,21).

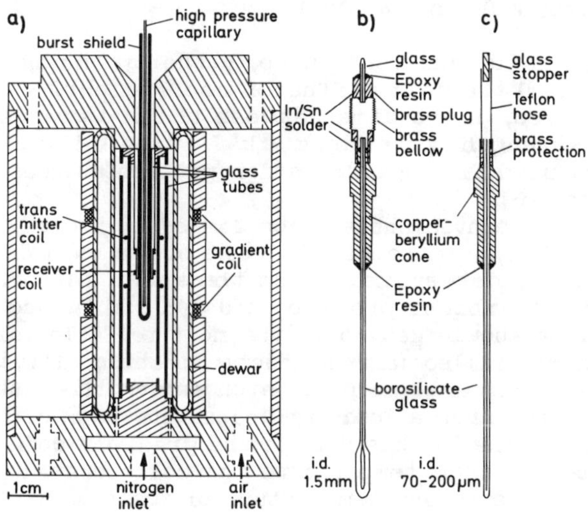

Figure 2. High pressure glass capillary system
a) Gradient NMR probe head
b) Glass cell for volatile substances
c) Narrow capillary for supercooling

For the study of high resolution NMR spectra at elevated pressure two different techniques are used. Starting with very early work by G.B. Benedek et al. (22) the groups of J. Jonas (23) and A. Merbach (24) have refined the design of nonmagnetic metallic autoclaves. In this approach the main constructive difficulties are the electrical lead-throughs for the RF-frequency and the achievement of a sufficiently high radial symmetry permitting the almost perfect field homogeneity of $\Delta B_0/B_0 \leq 10^{-8}$ necessary for high resolution studies. In these autoclaves pressures up to 1 GPa can

be obtained and the advanced constructions permit a rather large sample to be studied and thus have a relativly high sensitivity.
The second approach, the strengthened glass cell introduced by the pioneering work of H. Yamada (25) uses carefully annealed glass capillaries of i.d. $\leq$ 1,5 mm. A modified version of this technique is presented in fig. 2. Since glass is transparent to RF-fields this cell design can be readily used with the standard electronic configuration of commercial spectrometers, in all NMR-probeheads for sample tubes with d $\leq$ 10 mm. The whole arsenal of modern pulse NMR is thus immediately available.
The main drawback of this technique is the low filling factor of the high pressure glass cells, since the interior of the receiver coil is filled mainly with glass (c.f. fig. 2). In the standard capillaries given in fig. 2b pressures up to 300 MPa can be obtained (4) although the risk of a sudden breakage increases significantly at $p > 250$ MPa. In the smaller cells with i.d. $\approx$ 70 μm pressures of up to 400 MPa are reached regularly, but here the poor filling factor permits, even in modern high field spectrometers (B_0 = 7 Tesla), the study of the most sensitive nuclei in neat liquids only (17).
Fig. 2a gives the design of a 300 MHz probe head for high resolution proton NMR studies (17). In addition to a crossed coil design of the transmitter and receiver this probe features a gradient coil of the Helmholtz type permitting the measurement of self diffusion coefficients by the pulsed gradient technique (15,26). The echo amplitude A, in a Hahn spin-echo experiment, characterized by the spin-spin-relaxation time T_2, is reduced in the presence of a pulsed field gradient by

$$A(2\tau) = A(0)\exp(-(2\tau/T_2))\exp(-\gamma^2 D\delta^2(\Delta - \delta/3)g^2)$$

Here τ is the time between the 90° and 180° pulse, γ the gyromagnetic ratio of the nucleus studied, D the self diffusion coefficient, δ the duration of the gradient pulses, Δ the time between the two gradient pulses, and g the gradient strength, given by $g = k \cdot I$ with k the coil constant, that has to be obtained from a calibration with substances of known diffusion coefficients (26).
The determination of D was most conveniently accomplished by increasing the current I and thus raising g while keeping all other variables constant. The coil constant was determined using the known self diffusion coefficient of water at ambient pressure and 298 K obtained by Mills (27) which is generally acknowledged to be the most reliable value (28). The coil constant was controlled with a redetermination of D for benzene (15). In most cases it appeared preferable to derive the intensity as function of the gradient from a Fourier-transformation of the second half of the spin echo. Since in this design the gradient coil is kept at atmospheric pressure and room temperature the coil constant is perfectly independent of the pressure and temperature of the sample. The small diameter of the glass tube, though responsible for the low sensitivity, effectively eliminates thermal convection of the liquid under study. This convection is a real danger in diffusion studies since it will contribute to the molecular mobility, and thus produce erroneously large self diffusion coefficients. The complete descriptions of the cells and the pressure generating systems have been published (29,30).

THEORY

1) Diffusion

In the absence of strong, anisotropic attractive interactions, like hydrogen bonds, short-ranged repulsive interactions dominate the structure and dynamics of simple liquids. Modern perturbation theories of liquids successfully relate the thermodynamic properties of real fluids to those of a purely repulsive system (31). This approach is also applied in the hard sphere interpretation of the dynamic properties of fluids which assumes that diffusion is not influenced by attractive forces (15,32). In the following it will be discussed how hard sphere theories may be extended to include the retarding effect of attractive forces upon single particle dynamics. The resulting expression for the diffusion coefficient $D(T,\rho)$ describes the T,ρ-dependence of D successfully in molecular liquids with weakly attractive interactions but fails in those liquids where strongly anisotropic attractive interactions, like hydrogen bonds, dominate structure and dynamics.

1.1) The Hard Sphere Diffusion Coefficient D_{HS}

For dilute and moderately dense gases of hard spheres the diffusion coefficient can be deduced from the Chapman-Enskog solution of the Boltzmann equation as

$$D_{HS} = (3/8)(kT/\pi m)^{1/2}(\sigma/n)(1/g(\sigma)) = (D^o/n)(1/g(\sigma)) \tag{1}$$

with σ the hard sphere diameter, $n = \rho\sigma^3$ the packing fraction, ρ the number density, m the molecular mass and $g(\sigma)$ the radial distribution function when the hard spheres are in contact (15).

Computer simulations of hard sphere fluids revealed deviations for higher densities from the Chapman-Enskog expression of the diffusion coefficient D_{HS} due to correlated particle motions and backscattering events. To incorporate these effects in the dense fluid regions, the Chapman-Enskog expression (equ. 1) has been modified

$$D_{HS} = (D^o/n)\cdot P(n) \tag{2}$$

P(n) is a polynomial of the packing fraction n, the coefficients of which have been deduced from computer simulations of the hard sphere fluid (15). Values of D_{HS} from molecular dynamics simulations have been reassessed recently (33) and represented by

$$D_{HS} = (D^o/n)(1 - n/1.09)(1 + n^2(0.4-0.83n^2)) \tag{3}$$

In real liquids the density dependence of D is often well reproduced by equ. 2. However, the experimental D are always found lower than predicted by equ. 2. This has been ascribed by Chandler in the rough hard sphere model (RHS) to rotation-translation coupling (34)

$$D_{RHS} \approx A\cdot D_{HS}(T,n) \tag{4}$$

The rotation-translation coupling parameter $A \leq 1$ should be independent of temperature and density. The computer fitting of the experimental results to equ. 2 and 4 however shows, that A strongly increases with rising temperature while σ decreases slightly. Obviously the hitherto neglected attractive interactions must have a significant influence upon D_{exp}. Speedy (35) could incorporate these two observations into the hard sphere treatment by describing the temperature dependence of the diameter σ_B and the attractive interaction by a Lennard-Jones treatment with

$$\sigma_B = \sigma_{LJ} \; (2/(1 + (1+2kT/\varepsilon)^{1/2}))^{1/6} \tag{5}$$

$$D_{LJ}(T,n) = D_{HS}(T,n,\sigma_B) \exp(-\varepsilon/2kT) \tag{6}$$

with σ_{LJ} the Lennard-Jones diameter and ε the well depth of the potential. Both parameters are obtained from fitting the experimental results to equ. 6.

The treatment of diffusion data given above fails for hydrogen bonded liquids, since their dynamic properties vary much more rapidly with temperature than predicted in this formalism. Also log D versus 1/T plots show a curvature, the slope increasing with falling temperature. The dynamic properties of these liquids can often be described by the empirical VTF equation:

$$D = D_o \cdot \exp(-B/(T-T_o)) \tag{7}$$

with T_o the ideal glass transition temperature (36). For low pressure supercooled water this description fails. In this liquid a dynamic scaling law of the form

$$D = D_o \; (T/T_s-1)^{\gamma} \tag{8}$$

gives a quantitative description of the data (17). T_s is interpreted as the low temperature stability limit of the liquid (51).

2) Spin-Lattice Relaxation Rates

The relaxation of a spin system in a static field B_o is characterized by two relaxation rates, the longitudinal or spin-lattice relaxation rate $R_1 = 1/T_1$ and the transverse or spin-spin relaxation rate $R_2 = 1/T_2$. During the spin-lattice relaxation process energy is transferred from the spin system to the lattice whereas the spin-spin relaxation process effects a redistribution of the energy within the spin system (11). In both situations the relaxation is caused by dynamical processes in the liquid leading to local time dependent fluctuations of electric and/or magnetic fields. In liquid water two different spin interactions will be of importance - the magnetic dipole-dipole interaction and the electric quadrupole interaction.

2.1) The quadrupole relaxation of the deuterons (D) in D_2O

The deuterium spin relaxation is dominated by the interaction of the nuclear quadrupole moment eQ with the electric field gradient (efg) eq at the nucleus. To a good approximation the efg is generated in liquid water exclusively by intramolecular charges. It is oriented along the OD-bond and is almost axially symmetric. The interaction fluctuates in time because of the reorientation of the OD-bond of the diffusing molecules. The spin-lattice ($R_1 = 1/T_1$) and spin-spin ($R_2 = 1/T_2$) relaxation rates of the deuterium nucleus (I = 1) can be expressed as

$$R_1 = (3\pi^2/20)\ \chi_{eff}^2\ [2g(\omega_D\tau) + 8g(2\omega_D\tau)] \tag{9}$$

$$R_2 = (3\pi^2/20)\ \chi_{eff}^2\ [3g(o) + 5g(\omega_D\tau) + 2g(2\omega_D\tau)] \tag{10}$$

with the effective quadrupole coupling constant $\chi_{eff} = e^2\langle q_{OD}\rangle Q/h$, where Q means the nuclear quadrupole moment and $\langle q_{OD}\rangle$ represents the component q_{OD} of the efg along the OD-bond averaged over the fast librations of the water molecules. The slower rotational diffusion of the molecules provides the main contribution to the spectral intensity $g(\omega)$ at the Larmor frequency ω_D. If the tumbling of the molecules is isotropic and characterized by a single correlation time τ, the spectral intensity is given by

$$g(\omega\tau) = \tau/(1 + (\omega\tau)^2) \tag{11}$$

For low molecular weight liquids of normal viscosity the correlation time τ is much shorter than ω^{-1}. In this region T_1 and T_2 are independent of frequency.

2.2) <u>The quadrupole relaxation of the deuterons (D) in electrolyte/D_2O solutions</u>

Dissolved ions induce a motional anisotropy of the reorientational fluctuations of coordinated water molecules due to strong ion-dipole forces. The orientational fluctuations of the hydration water molecules can be decomposed into fast librations superimposed onto a slower anisotropic reorientation (τ_i) around the local director (ion-oxygen line). In addition the molecules tumble around the ion (τ_r) and chemical exchange (τ_{ex}) occurs. The spin-lattice relaxation rate of the deuterons on the hydration water molecules is then given by

$$(R_1)_{hyd} = (3\pi^2/20)\ \chi_{eff}^2\ \{(1/4)(3\cos^2\Theta-1)^2\cdot F(\omega_D\tau_o) + 3\sin^2\Theta\cos^2\Theta\cdot F(\omega_D\tau_1) + (3/4)\sin^4\Theta\cdot F(\omega_D\tau_2)\} \tag{12}$$

with $F(\omega\tau)$ and $g(\omega\tau)$ defined by equations 9 and 11 and

$$\frac{1}{\tau_o} = \frac{1}{\tau_r} + \frac{1}{\tau_{ex}},\quad \frac{1}{\tau_1} = \frac{1}{\tau_o} + \frac{1}{\tau_i},\quad \frac{1}{\tau_2} = \frac{1}{\tau_o} + \frac{4}{\tau_i}$$

The angle Θ measures the orientation of the OD-bond relative to the local director. The temperature dependence of the cooperative tumbling and exchange processes with correlation time τ_o is given by the VTF-equation (equ. 7) with D_o replaced by τ_{oo} whereas the thermally activated anisotropic mode with correlation time τ_i possesses an Arrhenius T-dependence.

$$\tau_i = \tau_{io} \exp(E_a/kT)$$

Because the chemical exchange is fast ($1/\tau_{ex} >> R_1$) only an average relaxation rate can be measured. It can be given within a two-site approximation (hydration water of the cation, bulk water) as

$$R_1 = (n_h/R)(R_1)_{hyd} + ((R-n_h)/R)(R_1)_{bulk} \qquad (13)$$

R = moles D_2O/mole salt $\qquad$ n_h = hydration number of the cation

2.3) <u>The magnetic dipole relaxation of the protons</u> (^1H) <u>in</u> $H_2{}^{16}O$

The magnetic dipole interaction between two protons (I,S) depends on their relative distance $r_{I,S}$ and on the orientation of the molecule in a laboratory fixed frame. Both vary in time due to rotational and translational diffusion depending on whether the two spins reside on the same molecule (intra) or on different molecules (inter).

The intramolecular spin-lattice relaxation rate of the protons (I = 1/2) may be written as

$$(R_1)_{intra} = (3/20)\ \gamma_H^4 h^2\ \langle R_{HH}^{-3}\rangle^2 F(\omega_H\tau) \qquad (14)$$

$\langle R_{HH}\rangle$ is the H-H distance within a water molecule averaged over the fast librations of the molecules. γ_H is the gyromagnetic ratio of the proton and h Plancks constant divided by 2π.

Hence $(R_1)_{intra}$ monitors the reorientation of the H-H vector within a water molecule.

The intermolecular spin-lattice relaxation rate of the protons will be written as

$$(R_1)_{inter} = (6\pi/5)\gamma_H^4 h^2 (N_H/2\cdot D_H\cdot d_{HH})\cdot F(\omega_H\tau_d) \qquad (15)$$

The spectral intensity $g(\omega_H\tau)$ is given for a translational diffusion process by (13)

$$g(\omega_H\tau_d) = [(3/2)u^2+(15/2)u+12] \cdot[(1/8)u^6+u^5+4u^4+(27/2)u^3+(81/2)u^2+81u+81]^{-1} \qquad (16)$$

$$u^2 = 2\omega_H\tau_d, \qquad \tau_d = d_{HH}^2/2\ D_H$$

d_{HH} gives the average distance of closest approach of two protons and D_H the self-diffusion coefficient of the protons.

2.4) The magnetic dipole interaction of the proton (^1H) and the oxygen (^{17}O) nucleus in $H_2{}^{17}O$

The dipolar coupling to the oxygen 17 nucleus (I = 5/2) provides an additional relaxation path for the protons. This interaction is to a good approximation of purely intramolecular origin and it fluctuates

because of reorientational motions of the OH-bond. Hence this additional interaction monitors exactly the same rotational motions as the deuterium relaxation.
The additional $^1H-^{17}O$ dipolar relaxation rate is given by

$$R_1(^1H-^{17}O) = (7/6)(\gamma_H\gamma_O h)^2 \langle R_{OH}^{-3}\rangle^2 \cdot F(\omega_H,\omega_O,\tau) \qquad (17)$$

with $F(\omega_H,\omega_O,\tau) = g(\omega_H-\omega_O,\tau) + 3g(\omega_H\tau) + 6g(\omega_H+\omega_O,\tau)$ and the spectral intensity for rotational diffusion given by equ. 11. R_{OH} is the OH-bond length in a water molecule. γ_H and γ_O are the gyromagnetic ratios of the proton and the oxygen-17 nucleus and h Plancks constant divided by 2π.

Results and Discussion

1. Neat Liquids

1.1) Self Diffusion

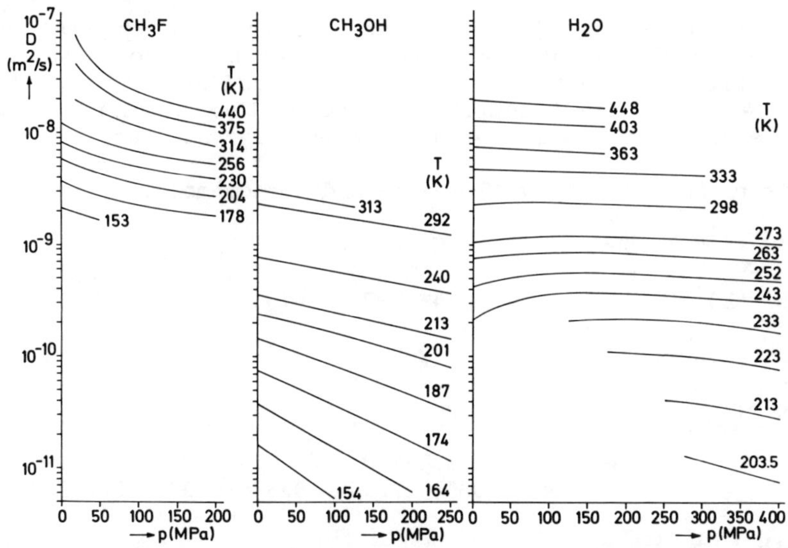

Figure 3. Self diffusion coefficients of methylfluoride, methanol and light water

Fig. 3 compiles the pressure dependence of the self diffusion coefficients of three pure liquids for a wide range of temperatures. The compounds have similar masses, molar volumes and dipole moments (CH_3F μ = 1.85 Debye; CH_3OH μ = 1.70 Debye; H_2O μ = 1.85 Debye) but differ in their ability to form hydrogen bonds. The p,T-dependence of D seen for methylfluoride is typical for a whole series of polar simple liquids composed of rigid methane derivatives that have been studied in our group (37,38,39,40). All isotherms of these compounds in the log D versus p plots do show a pronounced curvature, D decreases with pres-

sure. In methanol and all monoalcohols studied hitherto the curvature of the isotherms is less developed. The influence of the hydrogen bonding upon translational mobility is most clearly seen in the strong temperature dependence of D. The formation of hydrogen bonded chains lowers the mobility especially at low temperatures and in the supercooled state. The data for methanol given in fig. 3 are compiled from several sources (41,42,43). The data for light water (16,17,28,44,45) reveal the most interesting p,T-dependence. At high and very low temperatures one observes a slight decrease of D with increasing pressure. The isotherms between 300 and 240 K however, show on initial application of pressure an enhancement of diffusion, which amounts in the 243 K isotherm to a factor of 1.6. The data in the supercooled range were obtained in the thin glass capillaries given in fig. 2.
The pronounced curvature of the isotherm for D in the simplest liquid CH_3F makes it obvious, that the characterization of the pressure dependence by activation volumes defined by

$$\frac{d \ln D}{dp} = -\frac{\Delta V^{\ddagger}}{RT}$$

is of limited use, since $\Delta V^{\ddagger}$ must depend strongly on temperature and pressure. The description of D with the Chandler RHS model (34) (equ. 4) gives a good fit of the existing data, as it does for most similar liquids (38,39,40,46). Table 1 compiles the parameters obtained for a series of these substances. As stated in the theory section, the strong

TABLE 1. Fit parameters of the Chandler- and Speedy-model

Substance	CH_3F	CF_3H	CH_3Cl	$CHCl_3$
References	(37)	(38)	(39)	(39)
Temperature (K)	150-375	142-322	190-440	220-397
Max. Pressure (MPa)	200	200	200	150
CHANDLER model				
A	0.4-1.0	0.3-1.0	0.4-0.7	0.4-0.8
d (nm)	0.357±0.04	0.395±0.05	0.398±0.02	0.482±0.05
SPEEDY model				
ε/k_B (K)	56	86	129	107
σ_{LJ} (nm)	0.405	0.434	0.436	0.534
% Std. dev.	8.5	6.6	8.2	4.5

temperature dependence of the A-parameter is in contradiction to the statements of the rough hard sphere theory. In the last columns of tab. 1 the data obtained from the fit to the Speedy model (35) are given. Equ. 5 and 6 introduce a clear functional dependence of the parameters σ_{LJ} and ε and thus restrict the flexibility of the fitting procedure. Considering this, the standard deviation of the fit is better than expected. Especially if one takes into account, that the density data were ob-

tained for many substances by long extrapolations of experimental results. Because of the strong density dependence of D_{HS} in equ. 3 a variation of the density by 2 % may change D_{HS} by more than 20 %.
When simple monoalcohols and water (47,48,49) were subjected to an analysis by the RHS model (equ. 4) it became obvious, that A varied strongly with temperature, but this was ascribed to the change of the extent of hydrogen bonding with temperature only. Extension of the methanol data into the deeply supercooled range (43), as shown in fig. 2, still permits the description of the single isotherm by equ. 4. The diameter σ is increased for methanol between 300 and 150 K from 0.360 to 0.385 nm, which is very reasonable and not unexpected, when this result is compared to the data of Table 1. However in the same temperature range the A-parameter falls from 0.240 at 300 K to 0.016 at 150 K. Obviously the fit of D to the single isotherms is still possible but leads to physically unacceptable A-parameters. The almost constant diameter σ seems to indicate that even at this low temperatures individual single molecules are the mobile entities in the translational mass transport.
The description of the alcohol data by the Speedy formalism fails completely. This was to be expected, since the strong temperature dependence observable in the dynamic properties of hydrogen bonded liquids cannot be accounted for by a single Lennard Jones type interaction which only allows a relatively weak temperature dependence of D.
At the present state of knowledge, the best description of the isobaric temperature dependence of D in simple hydrogen bonded liquids is offered by the empirical VTF-equation (36). All self diffusion data for methanol given in fig. 3 as well as comparably extensive data tables for ethanol (43) and 2-methylpropanol-2 (50) can be described within the precision of the data by equ. 7. Table 2 compiles the results of the fitting process. The preexponential factor D_o appears independent of pressure while

TABLE 2. Fit parameter obtained for the description of the self diffusion in methanol by the VTF-equation (7).

Pressure (MPa)	T_o (K)	B (K)	D_o (m^2/s)
SVP	63	700	$2,8\cdot10^{-9}$
50	66	720	$2,8\cdot10^{-9}$
100	69	740	$2,8\cdot10^{-9}$
150	73	750	$2,8\cdot10^{-9}$
200	76	760	$2,8\cdot10^{-9}$
250	78,5	770	$2,8\cdot10^{-9}$

both the ideal glass transition temperature and the exponent B increase significantly with rising pressure. It must however be emphasized, that the data for all alcohols end far above the respective T_o and that the fitting demands a long extrapolation over the low temperature range close to the ideal glass transition. In this p,T region, which has not been covered by experiments hitherto, deviations from the behaviour predicted by the VTF-equation would become most obvious.

The self diffusion data for water at temperatures smaller than ≈ 300 K can because of the anomalous pressure dependence of D obviously not be fitted by any hard sphere treatment. At high pressures ($p > 150$ MPa) the VTF equation yields a good representation of the data, while at $p \leq 150$ MPa the fitting process leads to T_o values that deviate significantly from the accepted ambient pressure ideal glass transition temperature $T_o \simeq 125$ K. For pressures below 150 MPa the fit to the dynamic scaling law (equ. 8) appears superior, although one cannot decide, with the self diffusion data collected hitherto, which of the alternative forms of data analysis gives the better description. The parameters derived for light water at some integral pressures are collected in Table 3.

TABLE 3. Best fit parameters for the description of the isobaric temperature dependence of D in H_2O. In the p,T region above the melting pressure curve, the accepted published self diffusion coefficients (13) were used for the fits. T_o and T_s values in parentheses were derived from τ_2-data (10).

		$D = D_o \exp(-B/(T-T_o))$				$D = D_o(T/T_s-1)^{\gamma}$			
P (MPa)	T (K)	$D_o \cdot 10^8$ (m^2/s)	B (K)	T_o (K)	Std. dev.%	$D_o \cdot 10^8$ (m^2/s)	T_s (K)	γ	Std. dev.%
0.1	242-333	4.14	347	177	3.1	1.67	223(223)	1.82	2.4
10	252-333	6.46	455	161	1.3	1.61	219	1.92	0.8
100	238-333	10.1	622	133	0.7	1.20	203(207)	2.19	1.0
200	218-333	8.93	614	131	1.5	0.93	191	2.48	1.8
300	204-333	5.78	514	143	3.6	0.90	189	2.60	2.7
400	204-273	3.24	410	155	2.1	1.06	189	2.82	2.8

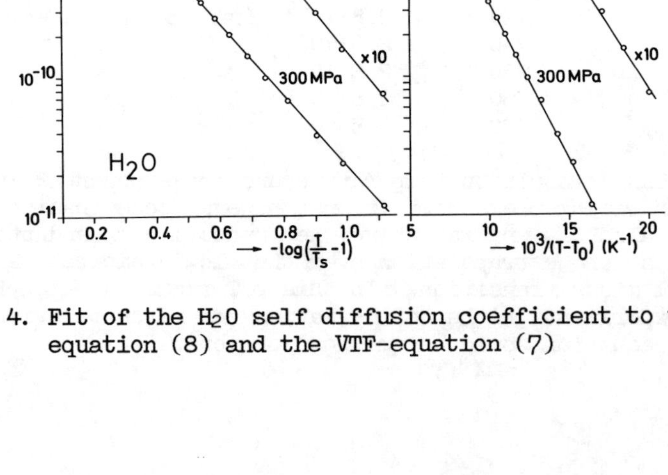

Figure 4. Fit of the H_2O self diffusion coefficient to the Speedy-Angell equation (8) and the VTF-equation (7)

Fig. 4 compares for 300 and 400 MPa the fit of the data to equ. 7 resp. 8. A more detailed discussion will be given together with the results of the analysis of the spin lattice relaxation time measurements.

1.2) Spin lattice relaxation, quadrupolar

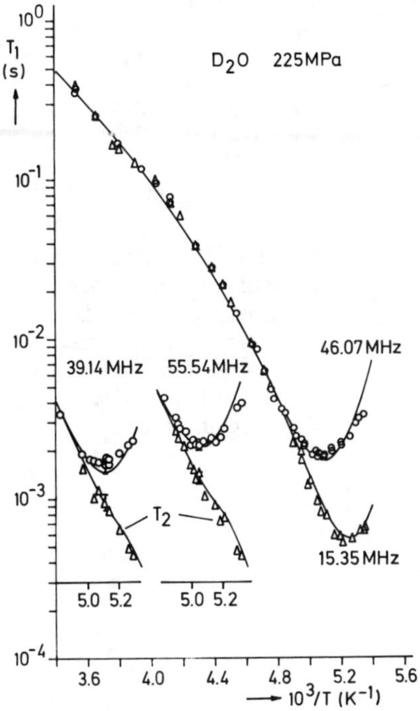

Figure 5. Deuteron spin lattice relaxation times. T_1 for heavy water at 225 MPa as function of frequency

In Fig. 5 the 225 MPa isobar of the deuteron spin lattice relaxation times T_1 is given (4,6,52). For some observe frequencies the spin-spin relaxation times T_2 could be obtained in addition. At temperatures below 220 K the T_1-data become frequency dependent. In this range the removal of thermal energy has slowed down the rotational motion to an extent that the correlation time τ_2 becomes comparable to the reciprocal observe frequency ω^{-1}. Differentiation of equ. 9 yields for the minimum of the T_1 curves the condition $\omega\tau = 0.6158$. The low temperature data allow thus a critical test of the motional model used for the description of the spectral density function and permit in addition an unambiguous determination of the quadrupole coupling constant (6). The quadrupole coupling constant for D_2O was derived to 200 ± 10 kHz. All deuteron relaxation rates collected hitherto can be described quantitativly by isotropic rotational diffusion with a single correlation time τ. For temperatures above 240 K this result is further corroborated by oxygen-17 T_1 relaxation times in oxygen-17 enriched heavy water (10) which reveal up to 460 K a pressure and temperature dependence identical to the deuteron T_1. Since the main field gradient of the oxygen-17 nucleus is directed perpendicular to the plane formed by the three atoms of the water molecule, it monitors the mobility of an axis perpendicular to the OD bond vector. This experimental finding thus gives additional support to the description of the rotational mobility by equ. 11.

The ratio of the self diffusion coefficients obtained for light and heavy water is given in fig. 6 together with the ratio of the rotational correlation times τ_2 obtained from oxygen-17 spin lattice relaxation time measurements (10). These two ratios are identical within the limits of experimental error. From simple liquid dynamics, however, one would expect D to depend on the square root of the mass ratio (= 1.05) while the τ_2 ratios should correspond to the square root of the ratio of the

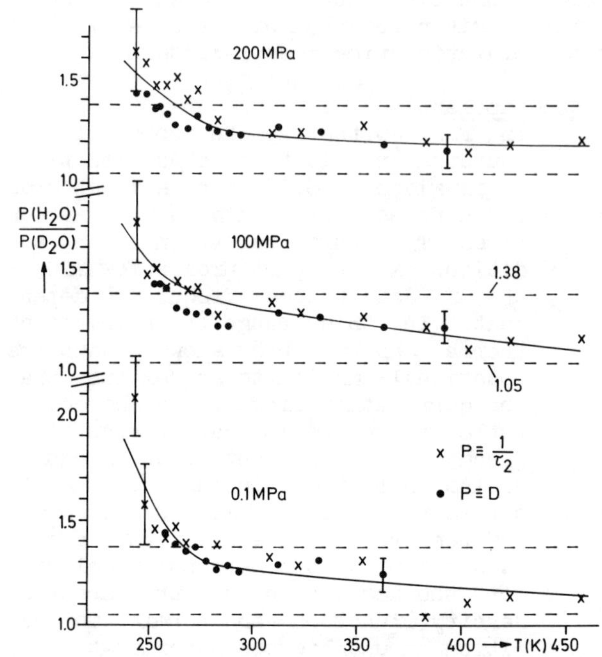

Figure 6. Isobars of the dynamical isotope effect in light and heavy water as derived from the ratios of the rotational correlation times τ_2 and the self diffusion coefficients D.

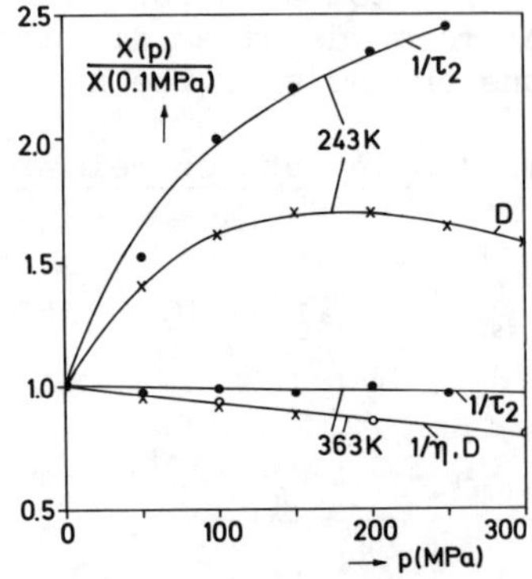

Figure 7. Reduced isotherms of the self diffusion coefficient D, the reciprocal viscosity $1/\eta$ and the reciprocal correlation time $1/\tau_2$ for light water.

moments of inertia (= 1.38). Especially in the supercooled range both ratios reveal a pronounced temperature dependence. The p,T-dependence of the mobility of the water molecules is obviously not determined by the mechanics of the single free water molecule but is a collective property reflecting the dynamics of the hydrogen bond network. Fig. 7 compares the pressure dependence of the self diffusion coefficient, characterizing translational motion, with the pressure dependence of $1/\tau_2$, the inverse rotational correlation times, as obtained from oxygen-17 spin lattice relaxation times. The 363 K-isotherms reveal the pressure effects as observed for all normal liquids: Increasing the density reduces the translational mobility, as described by the viscosity η or the self diffusion coefficient D, significantly stronger than the rotational diffusion characterized by $1/\tau_2$. This normal behaviour is strikingly reversed in the supercooled region, where the translational mobility is enhanced along the 243 K-isotherm of H_2O by a factor of 1.6 while the inverse rotational correlation times increase by approximately a factor of 2.5. Also the pressure where the maximal mobility is observed along each isotherm appears to be significantly lower for D ($\approx$ 150 MPa).
All thermodynamic and dynamic anomalies of water originate from the open

network structure of this liquid. The ability of the water molecule to participate with its two lone pairs and its two hydrogens in four hydrogen bonds enforces an approximately tetrahedral local arrangement with rather poor packing efficiency. Removal of thermal energy forces the molecules into low energy configurations with optimal relative orientations and distances which resemble local structures observed in low pressure polymorphs of ice and clathrates. The quasitetrahedral force field presented by the surrounding networks constrains a molecule to reside in one of a few well-defined orientational states which are separated by potential barriers. Because of reduced thermal excitations upon cooling tho lifetime of these fully H-bonded local structures must increase with a consequent slowing down of diffusive modes. Compression of the sample brings next nearest neighbours closer to a central molecule with nearest neighbour correlations being largely unchanged. The potential energy surface for rotation is then less sharply partitioned and there is a greater variety of orientational states, separated by smaller energy barriers, so that rotational diffusion is enhanced.

In all ice and clathrate networks each molecule is hydrogen bonded to four neighbours with a bond length of 0.28 ± 0.01 nm. The significant difference is that whereas in the clathrates and in ice I, the distance of closest approach of the nonbonded neighbours is ≃ 0.45 nm in ices II, III, IV and V they lie in the range 0.32 - 0.35 nm. Because rotational correlation times are shorter than the corresponding translation correlation times ($\tau_t \simeq \langle d^2 \rangle / 6\,D$, $d \simeq 0.3$ nm), orientational fluctuations initiate structural rearrangements within the transient H-bond network. Consequently, the enhancement of local molecular reorientations upon initial compression must facilitate also translational motions. However, an increasing number of next neighbours must form obstacles for translational motions. Thus self-diffusion starts to decrease already at pressures where rotation is still enhanced by further compression.

1.3) Spin lattice relaxation, dipolar

In fig. 8 the isobars of the proton spin lattice relaxation times of light water are given as function of frequency. Protons relax by dipole-dipole interaction and T_1 is thus determined by intra- and intermolecular contributions. The study of proton relaxation in oxygen-17 enriched light water permits the separation of these two contributions. The analysis of all data collected hitherto provides additional support for the description of the rotational diffusion by isotropic motion characterized by a single Debye correlation time (equ. 11).

The intermolecular contribution to the proton T_1 amounts at room temperature to ≈ 30 % of the total rate. In the temperature range of frequency dispersion between 190 and 230 K it contributes only between 5 and 15%. Considering the experimental uncertainties, the description of the total proton relaxation rate is relatively insensitive to the particular model chosen. The T_1-inter lines given in fig. 9 were calculated from equ. 15 and 16 with the experimental diffusion coefficients presented in fig. 3. The full lines in fig. 8 give the result obtained from summing up all contributions. The temperature dependence of τ_2 and D is described by the VTF-equation. The parameters used in the fit are collected in Table 4.

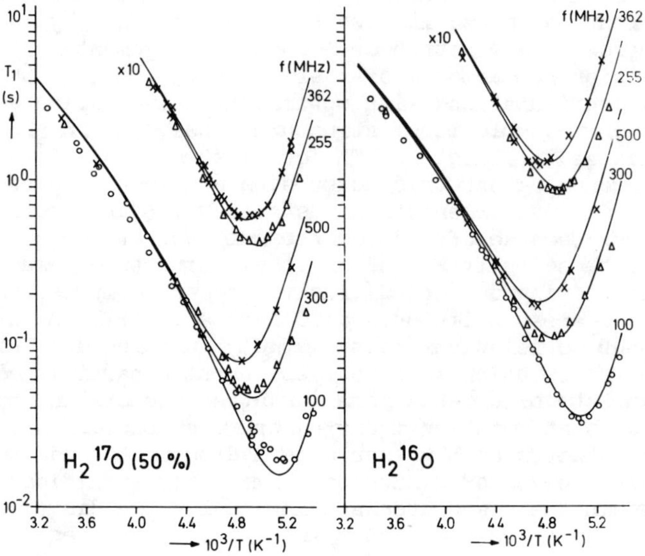

Figure 8. 200 MPa isobars of the proton spin lattice relaxation times T_1 in light water and water enriched with 50 % oxygen-17 as function of frequency

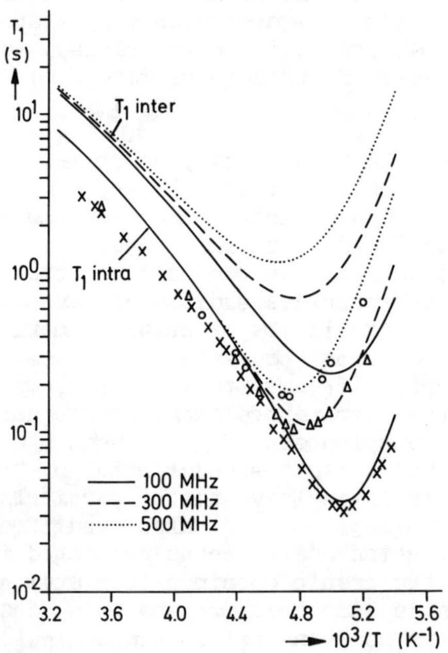

Figure 9. Separation of the 200 MPa isobars from fig. 8 into the intra- and intermolecular contributions with equations 14-17

TABLE 4. Molecular dimensions of the water molecule and fit parameter used in the data analysis

	τ_0 (ps)	B (K)	T_0 (K)	χ_{eff} (kHz)[a]	$(R_{HH})_{eff}$ (nm)[b]	$(R_{OH})_{eff}$ (nm)[b]	Γ_{OH}	Γ_{OD}	Γ_{HH}	d_{HH} (nm)
H_2O	.036	735	125	---	.159	.1055	.82	.87	.94	.280
D_2O	.050	687.6	132	201						

a) $\chi_{eff} = \Gamma_{OD} \cdot \chi$ b) $(R_{AB})_{eff} = \langle R_{AB}^3 \rangle^{1/3} = (R_{AB}^3/\Gamma_{AB})^{1/3}$

At temperatures between 250 and 300 K most T_1 data were determined in bulk water and in the emulsions described in the experimental sections. No significant differences were found between these two sets of data. This indicates that the dispersion of the water into droplets with diameters of a few micrometer does not influence the correct determination of the relaxation rates.

The deuteron and all proton T_1-measurements of the emulsion could be extended in the pressure range very close to the homogeneous nucleation temperature. The correlation times τ_2 derived from these data do thus provide the most complete set of data for a test of the two descriptions of the isobaric temperature dependence of the dynamic properties observed in supercooled liquid water. Speedy and Angell proposed for the low pressure data the description by the dynamic scaling law given in equ. 8 (51). This description was extended by Speedy (53) to negative and high pressure characterizing $T_s(p)$ as a continuous spinodal line or locus of the mechanical stability limit. Analysis of the τ_2 and self diffusion data at pressures $\leq$ 150 MPa give full support to this conclusion. The high pressure isobars can be described equally well by the dynamic scaling law and the VTF equation. It must thus remain an open question, whether the mechanical stability limit conjecture is still valid in this range or whether water turns into a normal hydrogen bonded liquid at pressures

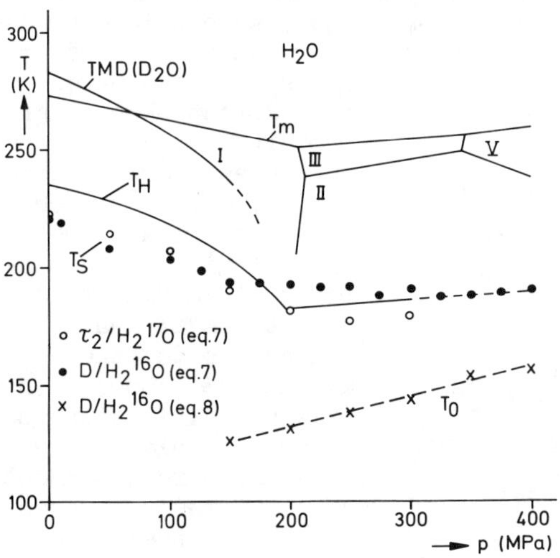

Figure 10. Phase diagram of light water T_s=singularity temperatures, T_0=ideal glass transition temperatures. TMD (D_2O)=temperature of maximum density in D_2O.

sufficiently high to suppress the temperature of maximum density (TMD) below the homogeneous nucleation temperature. This latter possibility has recently been favoured by Angell (54).
Fig. 10 gives the phase diagram of light water with the temperatures of maximum density (55) and the T_s and T_0 values derived from the fit of the self diffusion coefficients and rotational correlation times to equ. 7 and 8.

2.) Water dynamics in salt solutions.

Many highly concentrated salt solutions can be easily supercooled by slow cooling processes and finally form homogeneous glasses (55-58). These bulk solutions have often been analyzed in the temperature range where the relaxation times are frequency dependent (59-63). Dilute solutions crystallize more easily and thus it becomes impossible to study the concentration dependence of the relaxation rates in the dispersion region. The combined application of high pressure (≈ 200 MPa) and the emulsion technique, however, permit the investigation over the complete range of concentrations from the pure water to the saturated or even supersaturated salt solutions. In a first set of experiments the p,T,c dependence of the deuteron-T_1 in solutions of simple alkali and alkali earth halides in heavy water have been studied (7,8,9,64,65,66,67). Fig. 11 compiles some of the deuteron T_1-isotherms. The influence of the ions upon the mobility of the deuterons becomes much more pronounced in the supercooled region. At 283 K most salts shorten the deuteron spin lattice relaxation time of the heavy water, these substances are classified as structure forming salts. In the 3 molal aqueous solutions of cesium chloride the deuteron T_1 is longer than in pure water. Salts that influence the water in the direction of higher mobility of the water molecules are termed structure breaker (68, 69). The broken lines in fig. 11 give the deuteron spin lattice relaxation times of neat D_2O. Inspection of the 238 K isotherms shows that this classification becomes at this temperature pressure dependent. At low pressures even a 3 molal magnesium chloride solution has a longer T_1 than the bulk liquid. The tetrahedral network structure of pure water is here obviously well developed and the addition of any

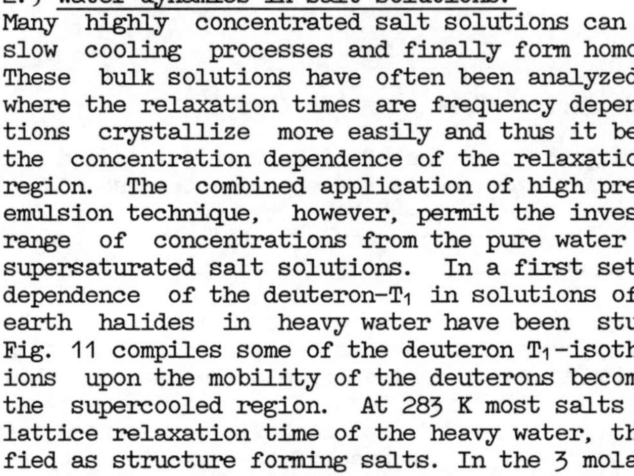

Figure 11. Isotherms of the deuteron-T_1 in some 3 molal supercooled aqueous salt solutions. The broken line gives the deuteron T_1 of neat D_2O

ion destroys the almost ideal arrangement and thus leads to increased mobility. At pressures above ≈ 100 MPa the normal sequence of structure forming and structure breaking is restored. The anomalous increase of the deuteron T_1 with pressure which is typical of neat water is around 240 K found, although less pronounced, even in the 3 molal solutions of most salts studied hitherto. For 1:1 electrolytes 3 molal solutions contain approximately 8 water molecules per ion. Remnants of the tetrahedral water structure are obviously found even at this high concentration and their deformation by pressure application is still observable.

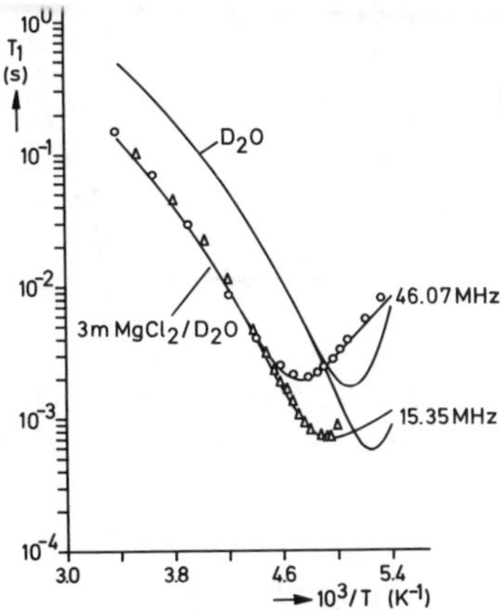

Figure 12. 225 MPa isobar of the deuteron-T_1 in neat D_2O and 3 molal $MgCl_2$ in D_2O at two frequencies

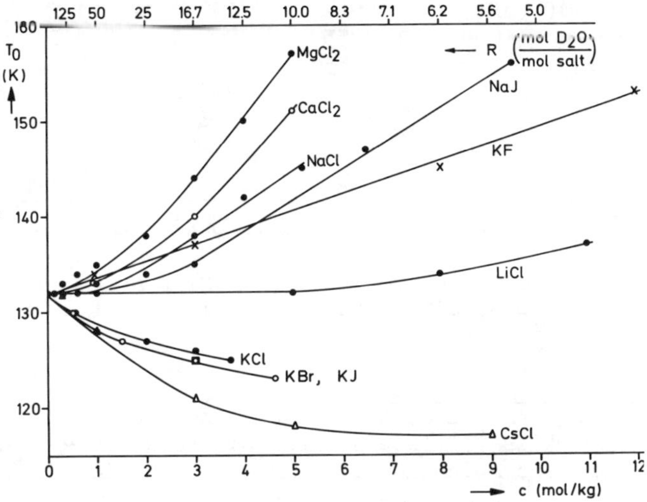

Figure 13. Ideal glass transition temperature T_0 at 225 MPa as obtained from fitting of the deuteron-T_1 data in the aqueous salt solutions to equation 13.

Fig. 12 shows the 225 MPa isobars of the deuteron T_1 in neat D_2O and in 3 molal magnesium chloride. The effects seen here are typical for most salts with a well developed hydration shell. Addition of salt raises the minimum of the T_1 and the minimum becomes flattened and broadened. The two site model proposed in the theory section can describe quantitatively the influence of the salt upon the water mobility. Inspection of fig. 11 shows that the influence of the chloride, bromide and jodide ion upon the water dynamics is small compared to the changes induced by varying the metal ion. For these halides it suffices to characterize the dynamical changes observed for the whole water by the hydration of the cation. Within the model described by equ. 13 the temperature and concentration dependence of the T_1 isobars can be described by the parameters given in Table 5. Fig. 13 collects the concentration dependence of the ideal glass transition temperature T_0. Ions with a well developed hydration shell (structure making salts) raise T_0 while larger cations and anions lead to a lowering of T_0. A more involved discussion of the model

TABLE 5. Parameters for the description of the cation in the two state model given by equ. 13.

	LiCl	NaCl	NaJ	KCl/KBr/KJ	CsCl/CsBr
τ_{00}±20 (fs)	200	120	85	65	65
τ_{i0} (fs)	80	62	69	62	62
E_a±0.8 (kJ/mol)	19.8	18.7	18.7	17.9	17.0
B (K)	687.6	687.6	687.6	687.6	687.6
χ_{eff}±10 (kHz)	192	192	201.2	201.2	201.2
	R=4.55	R=9.5	R=4.2	R≥13.5[a]	R=5.6
n_h ±1	4.55	5	3	7	2
	R≥6.25	R≥12.5	R=8.6	R≥11.0[b]	R=10
	6	6	4	6	5
			R≥16.7		R≥16.7
			6		8

R(moles D_2O/mol salt), a: KCl, b: KBr, KJ

described here and the application of multinuclear NMR to the dynamics in aqueous salt solutions is given in the article written by one of us (E.W. Lang) in this volume.

Concluding Remarks

During the last two decades the general understanding of the dynamic properties of liquids has come to a state where simple models derived from a combination of molecular dynamics simulation and theoretical gas kinetic approaches together with experimental data in a wide region of the ρ,T-space allow a quantitative description of the existing experimental D and τ_2 data of liquids consisting of rigid and flexible dipolar molecules. It remains to be seen, whether these concepts permit an extrapolation into the supercooled and the high density regions not covered by experiments hitherto.

For hydrogen bonded liquids this situation is less satisfactory. Here the isobaric temperature dependence of the dynamic properties can often be fitted to the empirical VTF-equation. However, it appears impossible to assign a basic physical concept to all parameters used in this fits.

The situation is even more obscure in the case of the threedimensional hydrogen bonded network liquid water. It has become obvious that even experiments covering the whole region of the metastable liquid down to the immediate vicinity of the homogeneous nucleation temperature curve cannot distinguish between the various descriptions offered in the literature. Also the many computer simulations of liquid water have not converged to a consistent model permitting a quantitative description of the influence of ρ and T variations upon thermodynamic and dynamic properties of water. The concepts and problems have become clearer during the last decade, the solutions to these nontrivial difficulties are not apparent.

Acknowledgement

Our own contributions have been sponsored generously by the Deutsche Forschungsgemeinschaft and the Fonds der Chemischen Industrie. The contributions of our students, who did most of the work presented here may be learned from the reference list. Experimental work depends on the reliable support by the technicians in the various labs and workshops. It is a pleasure to thank Mrs. A. Birkmann, Mrs. I. Cuno and Mr. E. Treml for their competent and dedicated help in the laboratories. In the faculty workshops Mrs. S. Heyn, R. Knott, D. Niesner and G. Wührl supported us far beyond their duty.

References

1. H. Kanno, R.J. Speedy, C.A. Angell (1975), Supercooling of Water to -92°C Under Pressure, Science 189, 880-881
2. C.A. Angell, H. Kanno (1976), Density Maxima in High-Pressure Supercooled Water and Liquid Silicon Dioxide, Science 193, 1121-1122
3. D.H. Rasmussen, A.P. McKenzie (1972), Effects of Solute on Ice-Solution Interfacial Free Energy; Calculation from Measured Homogeneous Nucleation Temperatures. In: H.H. Jellinek (ed.) 'Water Structure and the Water Polymer Interface', Plenum Press New York, 126-145
4. E.W. Lang, H.-D. Lüdemann (1980), Pressure and Temperature Dependence of the Longitudinal Deuterium Relaxation Rates in Supercooled Heavy Water to 300 MPa and 190 K, Ber. Bunsenges. Phys. Chem. 84, 462-470
5. E.W. Lang, H.-D. Lüdemann (1977), Pressure and Temperature Dependence of the Longitudinal Proton Relaxation Times in Supercooled Water to -87°C and 2500 bar, J. Chem. Phys. 67, 718-723
6. E.W. Lang, H.-D. Lüdemann, L. Piculell (1984), Nuclear Magnetic Relaxation Rate Dispersion in Supercooled Heavy Water under High Pressure, J. Chem. Phys. 81, 3820-3827
7. E.W. Lang, H.-D. Lüdemann (1985), p,T,c-Dependence of ^2H Spin-Lattice-Relaxation Rates in Supercooled LiCl-D_2O Solutions, 508-516
8. W. Fink, E.W. Lang (1988), The p,T,c-Dependence of Deuterium Spin-Lattice Relaxation Times in Undercooled $MgCl_2/D_2O$ Solution, J. Phys. Chem., 92, 6440-6445
9. W. Fink, H. Radkowitsch, E.W. Lang (1988), The p,T,c-Dependence of Deuterium Spin-Lattice Relaxation Times in Undercooled NaCl/D_2O Solutions, Chem. Phys., 124, 239-249
10. E. Lang, H.-D. Lüdemann (1981), High Pressure 0-17 Longitudinal Relaxation Time Studies in Supercooled H_2O and D_2O, Ber. Bunsenges. Phys. Chem. 85, 603-611
11. T.C. Farrar, E.D. Becker (1971), Pulse and FTNMR-Introduction to Theory and Methods, Academic Press, New York
12. H.W. Spiess (1978), Rotation of Molecules and Nuclear Spin Relaxation. In: P. Diehl, E. Fluck, R. Kosfeld eds. 'NMR-Basic Principles and Progress', Springer, Berlin 15, 55-214

13. J. McConnell (1987), The Theory of Nuclear Magnetic Relaxation in Liquids, Cambridge University Press, Cambridge
14. E.L. Hahn (1950), Spin Echoes, Phys. Rev. 80, 580-594
15. H.J.V. Tyrrell, K.R. Harris (1984), Diffusion in Liquids, Butterworth, London
16. F.X. Prielmeier, E.W. Lang, R. J. Speedy, H.-D. Lüdemann (1987), Diffusion in Supercooled Water to 300 MPa, Phys. Rev. Lett. 59, 1128-1131
17. F.X. Prielmeier, E.W. Lang, R.J. Speedy, H.-D. Lüdemann (1988), The Pressure Dependence of Self-Diffusion in Supercooled Light and Heavy Water, Ber. Bunsenges. Phys. Chem., 92 1111-1117
18. C.A. Angell (1982), Supercooled Water. In: F. Franks ed. 'Water - A Comprehensive Treatise', Plenum Press, New York 7, 1-81
19. E.W. Lang, H.-D. Lüdemann (1982), Anomalien des flüssigen Wassers, Angew. Chemie 94, 351-365, I. Ed. 21, 315-329
20. C.A. Angell (1983), Supercooled Water, Ann. Rev. Phys. Chem., 34, 593-630
21. W. Fink (1987), Dissertation, Universität Regensburg
22. G.B. Benedek (1963), Magnetic Resonance at High Pressure, Wiley Interscience, New York
23. J. Jonas (1978), Magnetic Resonance Spectroscopy at High Pressure. In: H. Kelm ed. 'High Pressure Chemistry', D. Reidel, Dordrecht, 65-110
24. W.L. Earl, H. Vanni, A.E. Merbach (1978), A High Pressure Probe for High Resolution Nuclear Magnetic Resonance, J. Magnet. Res. 30, 571-576
25. H. Yamada (1974), Pressure Resisting Glass Cell for High Pressure High Resolution NMR Measurements, Rev. Sci. Instr. 45, 640-642
26. E.O. Stejskal, J.E. Tanner (1965), Spin Diffusion Measurements: Spin Echoes in the Presence of a Time-dependent Field Gradient, J. Chem. Phys. 42, 288-292
27. R. Mills (1973), Self-Diffusion in Normal and Heavy Water in the Range 1-45°, J. Phys. Chem., 77, 685-689
28. M. Weingärtner (1982), Self Diffusion in Liquid Water. A Reassessment. Z. Phys. Chem., Neue Folge 132, 129-149
29. U. Gaarz, H.-D. Lüdemann (1976), Pressure Dependence of the Self Association of 9-Methylpurine in Aqueous Solution, Ber. Bunsenges. Phys. Chem., 80, 607-614
30. G. Völkel, E.W. Lang, H.-D. Lüdemann (1979), High Pressure High Resolution NMR III: Concentration Dependence of ΔV^* and ΔG^* for the Rotation of the Dimethylaminogroup in Aqueous Solutions of some N,N-Dimethylamides, Ber. Bunsenges. Phys. Chem., 83, 722-729
31. J.D. Weeks, D. Chandler, H.C. Andersen (1971), Role of Repulsive Forces in Determining the Equilibrium Structure of Simple Liquids, J. Chem. Phys. 54, 5237-5247
32. J. H. Dymond (1985), Hard-sphere Theories of Transport Properties, Chem. Soc. Rev. 14, 317-356
33. R.J. Speedy (1987), Diffusion in the Hard Sphere Fluid, Mol. Phys. 62, 509-515
34. a. D. Chandler (1974), Translational and Rotational Diffusion in Liquids I. Translational single-particle correlation functions. J.

Chem. Phys. 60, 3500-3507
b. D. Chandler (1975), Rough Hard Sphere Theory of the Self-Diffusion Constant for Molecular Liquids, J. Chem. Phys. 62, 1358-1363

35. R.J. Speedy, F.X. Prielmeier, T. Vardag, E.W. Lang, H.-D. Lüdemann (1989), Diffusion in Simple Fluids, Mol. Phys., 66, 577-590
36. C.A. Angell (1988), Perspectives on the Glass Transition, J. Phys. Chem. Solids 49, 863-871
37. E.W. Lang, F.X. Prielmeier, H. Radkowitsch, H.-D. Lüdemann (1987), High Pressure NMR Study of the Molecular Dynamics of Liquid Methylfluoride and Deutero-Methylfluoride, Ber. Bunsenges. Phys. Chem., 91, 1017 1025
38. E.W. Lang, F.X. Prielmeier, H. Radkowitsch, H.-D. Lüdemann (1987), High Pressure NMR Study of the Molecular Dynamics of Liquid Fluoroform and Deutero-Fluoroform, Ber. Bunsenges. Phys. Chem., 91, 1025-1033
39. F.X. Prielmeier, H.-D. Lüdemann (1986), Density Dependence of the Molecular Dynamics of CH_3Cl, CH_2Cl_2 and $CHCl_3$, Mol. Phys. 58, 593-604
40. T.M. Vardag, H.-D. Lüdemann (1988), High Pressure NMR Study of the Molecular Dynamics of Liquid Chlorodifluoromethane, Chem. Phys. 128, 527-535
41. R.L. Hurle, L.A. Woolf (1982), Tracer Diffusion in Methanol and Acetonitrile under Pressure. J. Chem. Soc. Faraday Trans. I. 78, 2921-2928.
42. H.S. Sandhu (1975), Coefficient of Self Diffusion in Liquids Using Pulsed NMR Techniques, J. Magn. Res. 17, 34-40
43. N. Karger, T. Vardag, Regensburg, unpublished results
44. K.R. Harris, L.A. Woolf (1980), Pressure and Temperature Dependence of the Self-Diffusion Coefficient of Water and Oxygen-18 Water, J. Chem. Soc. Faraday Trans I, 76, 377-385
45. K. Krynicki, C.D. Green, D.W. Sawyer (1980), Pressure and Temperature Dependence of Self Diffusion in Water, Faraday Discussion Chem. Soc, 66, 199-208
46. M. Has (1988), Diplomarbeit, Universität Regensburg
47. D.J.Wilbur, J.De Fries, J. Jonas (1976), Self-Diffusion in Compressed Liquid Heavy Water, J. Chem. Phys. 65, 1783-1786
48. R.L. Hurle, A.J. Esteal, L.A. Woolf (1985), Self Diffusion in Monohydric Alcohols under Pressure, J. Chem. Soc. Faraday Trans. I, 81, 769-779
49. J. Jonas, A. Akai (1977), Transport Processes in Compressed Liquid Methanol, J. Chem. Phys. 66, 4946-4950
50. M. Woznyj, F.X. Prielmeier, H.-D. Lüdemann (1984), Pressure Dependence of the Melting and Self Diffusion in 2,2-Dimethylpropane, 2,2-Dimethylpropionitrile and 2-Methylpropanol-2, Z. Naturforsch. 39a, 800-806
51. J. Speedy, C.A. Angell (1976), Isothermal Compressibility of Supercooled Water and Evidence for a Thermodynamic Singularity at -45°C, J. Chem. Phys. 65, 851-858
52. D. Girlich (1988), Diplomarbeit, Universität Regensburg
53. R.J. Speedy (1982), Stability Limit Conjecture. An Interpretation of the Properties of Water, J. Phys. Chem. 86, 982-991

54. C.A. Angell (1988), Supercooled Water - Approaching the Limits, Nature 331, 206-207
55. C.A. Angell, E.J. Sare (1970), Glass-Forming Composition Regions and Glass Transition Temperatures for Aqueous Electrolyte Solutions, J. Chem. Phys. 52, 1058-1068.
56. H. Kanno, C.A. Angell (1977), Homogeneous Nucleation and Glass Formation in Aqueous Alkali Halide Solutions at High Pressure, J. Phys. Chem., 81, 2639-2643
57. H. Kanno, J. Shirotani, S. Minomura (1981), Pressure Dependence and Cationic Radius Effect of the Glass Transition Temperature in Aqueous Alkali Acetate Solutions, Bull. Chem. Soc. Jpn, 54, 2607-2609
58. C.A. Angell, E.J. Sare, J. Donnella, D.R. MacFarlane (1981), Homogeneous Nucleation and Glass Transition Temperatures in Solutions of Li-Salts in D_2O and H_2O. Doubly Unstable Glass Regions, J. Phys. Chem., 85, 1461-1464
59. E.J. Sutter, J.F. Harmon (1975), Molecular Motion in Supercooled Liquids. I. Pulsed Nuclear Magnetic Resonance of Lithium-7 in 11 M Aqueous Lithium Chloride, J. Phys. Chem., 79, 1958-1961
60. J.F. Harmon, E.J. Sutter (1978), Molecular Motion in Supercooled Liquids. II. Nuclear Magnetic Resonance Relaxation of Deuterons and Protons in 11 M Aqueous Lithium Chloride, J. Phys. Chem., 82, 1938-1942
61. N. Boden, M. Mortimer (1978), Reorientation of D_2O in Concentrated Aqueous Solutions of Lithium Chloride Studied by Nuclear Magnetic Relaxation, J. Chem. Soc. Faraday Trans. II, 74, 353-366
62. A. Geiger, H.G. Hertz (1976), Proton Magnetic Relaxation Study of Water Orientation Around I^- and Li^+, J. Solution Chem. 5, 365-388
63. H. Weingärtner (1980), The Mechanism of ^7Li Relaxation in a Supercooled Aqueous LiI Solution, J. Magn. Res. 41, 74-87.
64. E.W.Lang, W. Fink, H.-D. Lüdemann (1984), The p,T,c-Dependence of Deuterium Spin-Lattice Relaxation Rates in Supercooled LiCl-, NaCl- and $MgCl_2$-D_2O Solutions, J. Physique 45, C7-173-177
65. E.W. Lang, L. Piculell (1986). New Results of High-Pressure NMR Studies on Supercooled Water and Aqueous Electrolyte Solutions. In: 'Water and Aqueous Solutions', G.W. Neilson, J.E. Enderby eds., Hilger, Bristol 31-40
66. E.W. Lang, F.X. Prielmeier (1988), Multinuclear Spin-Lattice Relaxation Time Studies of Supercooled Aqueous LiCl-Solutions, Ber. Bunsenges. Phys. Chem., 92, 717-724
67. W. Fink, H. Radkowitsch, E.W. Lang (1988), Deuteron Spin-Lattice Relaxation Times in Undercooled Aqueous Potassium- and Cesium-Halide Solutions, Z. Naturforsch. 43a, 538-546
68. M.N. Buslaeva, O.Ya. Samoilov (1986), Microdynamics of Solvation in E. Kalman, A.A. Kornyschev, J. Ulstrup eds. 'Studies in Physical and Theoretical Chemistry', Elsevier, Amsterdam, 391-414
69. G. Engel, H.G. Hertz (1968), On the Negative Hydration. A Nuclear Magnetic Relaxation Study, Ber. Bunsenges. Phys. Chem., 72, 808-834

LOW FREQUENCY LIGHT SCATTERING IN HYDROGEN BONDED LIQUIDS

Benassi P., Mazzacurati V., Nucara A.
Dipartimento di Fisica - Universita' di Roma "La Sapienza" -
P.le Aldo Moro, 2 - 00185 Roma - Italy
Ruocco G.
Dipartimento di Fisica - Universita' di L'Aquila -
P.zza Annunziata, 1 - 67100 L'Aquila - Italy

ABSTRACT. A discussion of polarized and depolarized low frequency light scattering spectra in liquids is presented. The shapes of the anisotropic and isotropic contributions are discussed for both liquid water and heavy water as well as for same water-alchools mixtures. The peculiarities of these spectra will be shown by comparison to those of simple liquids, and will be related to the translational and orientational dynamics, which in this frequency range are strongly influenced by the exsistence of H-bonds.

1. INTRODUCTION

Light scattering spectroscopy is a very powerful technique to study dynamics and structural properties in condensed matter and in the last years has become increasingly important because of the fast development of high intensity mono-mode CW ion lasers, detectors and spectrum analyzers.

In particular the new class of gratings monochromators recently developed [1] makes possible the quantitative analysis of the entire low frequency spectral region, which is typical of acoustic waves and relaxational dynamics. This analysis was indeed previously limited either by the low finesse of Fabry-Perot interferometers or by the low resolving power of the available gratings spectrometers[1]. To introduce notations and relevant quantities for our discussion, in the §2 we briefly treat the basic light scattering theory with particular emphasis on the low frequency region in molecular liquids. In §3 we discuss recent experimental results in H-bonded liquids and their interpretations.

2. LIGHT SCATTERING THEORY IN MOLECULAR LIQUIDS

In the framework of the linear response theory [2], the light scattering differential cross-section can be written as

$$\partial^2\sigma\left(\vec{q}, \omega, \vec{n}_i, \vec{n}_s\right)\Big/\partial\Omega\partial\omega = \Sigma_{\alpha\beta\gamma\delta}\left(n_s\right)_\alpha \left(n_i\right)_\beta \left(n_s\right)_\gamma \left(n_i\right)_\delta\, \partial^2\sigma_{\alpha\beta\gamma\delta}\left(\vec{q},\omega\right)\Big/\partial\Omega\partial\omega \qquad (1)$$

J.C. Dore and J. Teixeira (eds.), Hydrogen-Bonded Liquids, 357–368.

with

$$\partial^2\sigma_{\alpha\beta\gamma\delta}(\vec{q},\omega)\Big/\partial\Omega\partial\omega = \frac{\omega_i\,\omega_s^3}{2\pi c^4}\frac{1}{N}\int dt\, e^{i\omega t}\int d\vec{r}_1\, d\vec{r}_2\, e^{-i\vec{q}(\vec{r}_1-\vec{r}_2)} < P^+_{\alpha\beta}(\vec{r}_1,t)\,P_{\gamma\delta}(\vec{r}_2,0) > \qquad (2)$$

being ($\vec{k}_i$, ω_i, $\vec{n}_i$) and ($\vec{k}_s$, ω_s, $\vec{n}_s$) the wave-vectors, frequencies and polarizations unit vectors of incident (i) and scattered (s) photons respectively, $\vec{q} = \vec{k}_s - \vec{k}_i$ and $\omega = 2\pi\tilde{\nu}c = \omega_s - \omega_i$ represent the exchanged quantities.

In these equations the indeces $\alpha,\beta,\gamma,\delta$ run over x,y,z (laboratory frame coordinates), c is the light velocity in vacuum and $P_{\alpha\beta}(\vec{r},t)$ are the components of the macroscopic polarizability density tensor of the medium. The < > brackets in equations (2) indicate the usual thermodynamical-quantum average. Thank to the macroscopic isotropy of liquid phase, it may be easily shown that the only relevant quantities in equation (1) are those for which [4]

$$\alpha=\gamma=\beta=\delta \qquad \text{or} \qquad (\alpha=\gamma)\neq(\beta=\delta) \qquad (3)$$

Therefore only two quantities are needed to describe light scattering experiments in liquids i.e.

$$\frac{\partial^2\sigma_{zzzz}(\vec{q},\omega)}{\partial\omega\partial\Omega} \quad \text{and} \quad \frac{\partial^2\sigma_{xzxz}(\vec{q},\omega)}{\partial\Omega\partial\omega} \qquad (4)$$

In particular using 90° scattering geometry (i.e. $\vec{k}_i \perp \vec{k}_s$), separate measurements of polarized ($\vec{n}_s // \vec{n}_i$) and depolarized ($\vec{n}_s \perp \vec{n}_i$) scattering directly give these two quantities, i.e.

$$\frac{\partial^2\sigma_{VV}(\vec{q},\omega)}{\partial\Omega\partial\omega} = \frac{\partial^2\sigma_{zzzz}(\vec{q},\omega)}{\partial\Omega\partial\omega} \qquad (5)$$

$$\frac{\partial^2\sigma_{VH}(\vec{q},\omega)}{\partial\Omega\partial\omega} = \frac{\partial^2\sigma_{HV}(\vec{q},\omega)}{\partial\Omega\partial\omega} = \frac{\partial^2\sigma_{HH}(\vec{q},\omega)}{\partial\Omega\partial\omega} = \frac{\partial^2\sigma_{xzxz}(\vec{q},\omega)}{\partial\Omega\partial\omega} \qquad (6)$$

Here the subscripts V(vertical) and H(Horizontal) indicate the polarization of the incoming and scattered photons with respect to the scattering plane.

It is immediate to see that equation (5) and (6) are related to the correlation functions of diagonal (VV) and off-diagonal (VH, HV or HH) elements of the polarizability density tensor.

The light-scattering cross -section is therefore entirely described in terms of the space-time Fourier transform of the correlation function of the tensor $\vec{P}(r,t)$. This macroscopic quantitiy in a molecular system can be expressed throughout the "effective polarizability" of the l-th molecule $\pi^l_{\alpha\beta}(t)$ as

$$P_{\alpha\beta}(\vec{r},t) = \sum_{l=1}^{N} \pi^{l}_{\alpha\beta}(t)\, \delta\left(\vec{r} - \vec{R}^{l}(t)\right) \qquad (7)$$

being N the total number of molecules and $\vec{R}^l(t)$ the instantaneous position of the center of mass of the l-th molecule.

In very low density systems the effective polarizability may be identified with the "bare molecular polarizability" and depends on the electronic states of the isolated molecule. In dense systems intermolecular interactions perturbate the electronic states thus changing the value of the molecular polarizability. In non-covalent media, where these interactions do not imply a severe rearrengement of the electronic states, the effect on the polarizability of the l-th molecule due to the sorroundings ones, can be treated as a perturbation and we may write [5]

$$\pi^{l}_{\alpha\beta}(t) = \alpha^{l}_{\alpha\beta}(t) + \Delta\pi^{l}_{\alpha\beta}(t) \qquad (8)$$

where $\alpha^{l}_{\alpha\beta}(t)$ is the bare molecular polarizability and $\Delta\pi^{l}_{\alpha\beta}$ is an "incremental" or "induced polarizability" which, in principle, depends on the entire set of the nuclear coordinates of the system.

In molecular liquids the most relevant contribution to the induced polarizability is the so called Dipole-Induced-Dipole effect (DID): this takes into account the change in the polarizability of each molecule due to the dipoles "induced" on all the sorroundings by the exciting field. In rare gas the DID contribution is the only detectable induced scattering up to very high density, where another mechanism due to short range collisions becomes evident. This last contribution comes from the perturbation of the atomic states due to the direct overlap of the electronic clouds (EO, i.e. electronic overlap) and may be taken satisfactorily into account only by quantum-mechanical calculations. DID and EO both contribute also in molecular liquids, toghether with other mechanisms arising from the presence of charges, dipoles, higher order polarizabilities and molecular anisotropies. A detailed treatment can be found in current literature for many molecular systems [6].

Because we will deal with low frequency light scattering (i.e. the frequency range with $|\omega| < 100\ \mathrm{cm}^{-1}$) we may neglect the molecular polarizability modulation coming from internal degree of freedom (intramolecular vibrations) and looks at the molecule as a rigid body. To specify scattering equations in a useful form, it is convenient to exctract from $\alpha^{l}_{\alpha\beta}(t)$ and $\Delta\pi^{l}_{\alpha\beta}(t)$ their isotropic and anisotropic components writing:

$$\alpha^{l}_{\alpha\beta}(t) = \alpha\delta_{\alpha\beta} + \beta Q^{l}_{\alpha\beta}(t) + \gamma R^{l}_{\alpha\beta}(t) \qquad (9)$$

$$\Delta\pi^{l}_{\alpha\beta}(t) = {}^{0}\pi^{l}(t)\, \delta_{\alpha\beta} + {}^{2}\pi^{l}_{\alpha\beta}(t) \qquad (10)$$

where $\alpha = \frac{1}{3}\mathrm{Tr}\,\{\overleftrightarrow{\alpha}(t)\}$, β and γ are the molecular polarizability anisotropies, $Q^{l}_{\alpha\beta}(t)$ and $R^{l}_{\alpha\beta}(t)$ are functions of time throughout the set of the angular variables of the l-th molecule [16] and ${}^{0}\pi^{l}(t) = \frac{1}{3}\mathrm{Tr}\,\{\overleftrightarrow{\pi}^{l}(t)\}$. By definition we have

$$\mathrm{Tr}\left\{\overleftrightarrow{Q}^{l}(t)\right\} = \mathrm{Tr}\left\{\overleftrightarrow{R}^{l}(t)\right\} = \mathrm{Tr}\left\{{}^{2}\overleftrightarrow{\pi}^{l}(t)\right\} = 0 \qquad (11)$$

and consequentely in macroscopically isotropic systems the following relations will hold:

$$\Sigma_{ll'}\left\langle {}^{2}\pi^{l}_{zz}(t)\ {}^{2}\pi^{l'}_{zz}(0)\right\rangle = \frac{4}{3}\Sigma_{ll'}\left\langle {}^{2}\pi^{l}_{zx}(t)\ {}^{2}\pi^{l'}_{zx}(0)\right\rangle \qquad (12a)$$

$$\Sigma_{ll'}\left\langle Q^{l}_{zz}(t)\ Q^{l'}_{zz}(0)\right\rangle = \frac{4}{3}\Sigma_{ll'}\left\langle Q^{l}_{zx}(t)\ Q^{l'}_{zx}(0)\right\rangle \qquad (12b)$$

$$\Sigma_{ll'}\left\langle R^{l}_{zz}(t)\ R^{l'}_{zz}(0)\right\rangle = \frac{4}{3}\Sigma_{ll'}\left\langle R^{l}_{zx}(t)\ R^{l'}_{zx}(0)\right\rangle \qquad (12c)$$

Thank to the properties (12) it is useful introduce the new quantities $\frac{\partial^2\sigma_{ISO}}{\partial\omega\partial\Omega}$ and $\frac{\partial^2\sigma_{ANISO}}{\partial\omega\partial\Omega}$, defined as

$$\frac{\partial^2\sigma_{ISO}(\vec{q},\omega)}{\partial\Omega\partial\omega} = \frac{\partial^2\sigma_{VV}}{\partial\omega\partial\Omega} - \frac{4}{3}\frac{\partial^2\sigma_{HV}}{\partial\omega\partial\Omega} \qquad (13a)$$

$$\frac{\partial^2\sigma_{ANISO}}{\partial\omega\partial\Omega} = \frac{\partial^2\sigma_{HV}}{\partial\Omega\partial\omega} \qquad (13b)$$

These will give us directly the differential cross-section of isotropic and anisotropic scattering connected to the trace and anysotropies of the molecular effective polarizabilities. Making use of equations (2)(5)(6)(7)(8)(9)(10) and (13) we will have for these quantities:

$$\partial^2\sigma_{ISO}(\vec{q},\omega)\Big/\partial\Omega\partial\omega = \omega_i\omega_s^3\Big/c^4\left[\alpha^2 S(\vec{q},\omega) + \frac{1}{2\pi N}\Sigma_{ll'}\int dt\, e^{i\omega t}\left\langle {}^{0}\pi^{l}(t)\ {}^{0}\pi^{l'}(0)\right\rangle\right] \qquad (14)$$

$$\partial^2\sigma_{ANISO}(\vec{q},\omega)\Big/\partial\Omega\partial\omega = \omega_i\omega_s^3\Big/c^4\left[\frac{\beta^2}{2\pi N}\Sigma_{ll'}\int dt\, e^{i\omega t}\left\langle Q^{l}_{zx}(t)\ Q^{l'}_{zx}(0)\right\rangle + \right.$$

$$\left.\frac{\gamma^2}{2\pi N}\Sigma_{ll'}\int dt\, e^{i\omega t}\left\langle R^{l}_{zx}(t)\ R^{l'}_{zx}(0)\right\rangle + \frac{1}{2\pi N}\Sigma_{ll'}\int dt\, e^{i\omega t}\left\langle {}^{2}\pi^{l}_{zx}(t)\ {}^{2}\pi^{l'}_{zx}(0)\right\rangle\right] \qquad (15)$$

where $S(\vec{q},\omega)$ is the center of mass dynamic structure factor.

The isotropic cross-section is thus dominated by the dynamic structure factor which, in the $|\vec{q}|$ range explored by visible light scattering, is well represented by generalized hydrodynamics and gives rise to the well-known Rayleigh-Brillouin triplet

For highly anisotropic molecule, the anisotropic component will be dominated by the β and γ terms, which gives rise to the reorientational quasi-elastic contribution in the low frequency region of light scattering spectra. The integrated intensity of this contribution (relative to the intensity of the Brillouin doublet) is of the order of $\frac{4}{45}\frac{\beta^2+3\gamma^2}{\alpha^2\, S(\vec{q})}$ and its frequency extension ranges from 0 to ≈ 500 cm^{-1}.

In the case of nearly isotropic molecules first order DID contribution may dominate the depolarized spectrum. It can be written as

$$\Delta\, \overset{1}{\pi}_{\alpha\beta}(t) = \alpha^2 \sum_{l'}{}' \, T^{(2)}_{\alpha\beta}\left(\vec{R}^{\,ll'}(t)\right) \qquad (16)$$

where $\overleftrightarrow{T}^{(2)}(\vec{R}_{ll'})$ tensor is the usual dipole propagator i.e.

$$T^{(2)}_{\alpha\beta}(\vec{R}) = \frac{\left[3\,\hat{R}_\alpha \hat{R}_\beta - \delta_{\alpha\beta}\right]}{|R|^3} \qquad (17)$$

and $R^{ll'}(t) = R^l(t) - R^{l'}(t)$, $\hat{R} = \vec{R}/|\vec{R}|$. In this case therefore

$$^{o}\pi^{l}(t) = 0 \qquad (18)$$

and the induced contribution coming from this term does not appear in the isotropic spectrum. The DID spectrum in liquids has been evaluated by Bucaro and Litovitz [7], and it results of nearly exponential shape with a decaying frequency ω_o. This quantity has been calculated and measured for many different liquids, the observed values ranging from 7 cm^{-1} ($SnBr_4$) to 88 cm^{-1} (NH_3) and it is found to be around 50 cm^{-1} in liquid water.

In liquids of anisotropic molecules the DID low frequency shape is intimately mixed to the reorientational contribution. In this case it is difficult to distinguish these contributions one from the other and evaluations of intensities and comparison with different experimental data (i.e. NMR reorientational times, temperatures and density behaviour, neutron scattering data etc. etc.) are necessary to make a reliable assignment.

The formalism we have introduced allows us to discuss experimental measurements of low frequency light scattering data and to extract informations on the dynamics of H-bonded systems in the time scale we are looking at.

2 - Low frequency light scattering spectra of H-bonded liquids.

Some typical depolarized spectra are shown in figures 1,2 and 3 for high purity H_2O and D_2O liquid samples at different temperatures. An entire set of similar measurements have been performed at many different temperatures ranging from +80 down to - 20°C in supercooled fluid [8]. All spectra show to a different extent a sharp peak centered at the laser frequency, the width of which is given by the instrumental resolution (typically δ = 0.035 cm^{-1} FWHM). In the depolarized spectra this feature can be totally assigned to an elastic spourious contribution probably coming from the sample container, being its intensity quite independent from the collection optics alignement.

In polarized spectra this phenomenon is still present, even if it merges within the Rayleigh peak, the natural width of which may also be predicted well below the instrumental resolution. Therefore data are assumed to be reliable down to a frequency shift of 0.06 cm^{-1}. In the depolarized spectrum, furthermore, can be also observed a small leakege of the polarized Brillouin doublet (see figure 1), about 10^{-4} of its intensity in the VV configuration. This has been taken into account by measuring the polarization rejection η of our analyzer, so that clean HV spectra has been obtained (see figure 1, dotted line) from experimental data by the subtraction

$$I_{HV}(\omega) = I_{HV}^{exp}(\omega) - \eta\, I_{VV}^{exp}(\omega) \qquad (19)$$

A very accurate measurements of depolarization ratio [9] shows that the isotropic cross-section both in water and heavy water is given, in this frequency region, by the Brillouin doublet plus a very small and nearly frequency independent contribution. This last is the low frequency part of an isotropic contribution which extends its tail up to 500 cm^{-1} and the integrated intensity of which is about 10^{-4} the Brillouin integrated intensity.

This contribution has been previously assigned to long range induction mechanisms other than DID and it has been shown that it is not a peculiarity of H-bonded liquid being also present in liquid H_2S [10].

As far as the anisotropic spectrum is concerned, the identification of its high frequency part with a first order DID contribution is straightforward, because of its nearly exponential shape (up to 50 cm^{-1}) and of its absolute integrated intensity. This part of the spectrum is identical for water and heavy water and almost temperature independent.

The peculiar characteristic of the low frequency spectrum of water, observed and studied by many authors [7,11,12,13], is the strong temperature dependence of the width of a quasi-elastic central component.

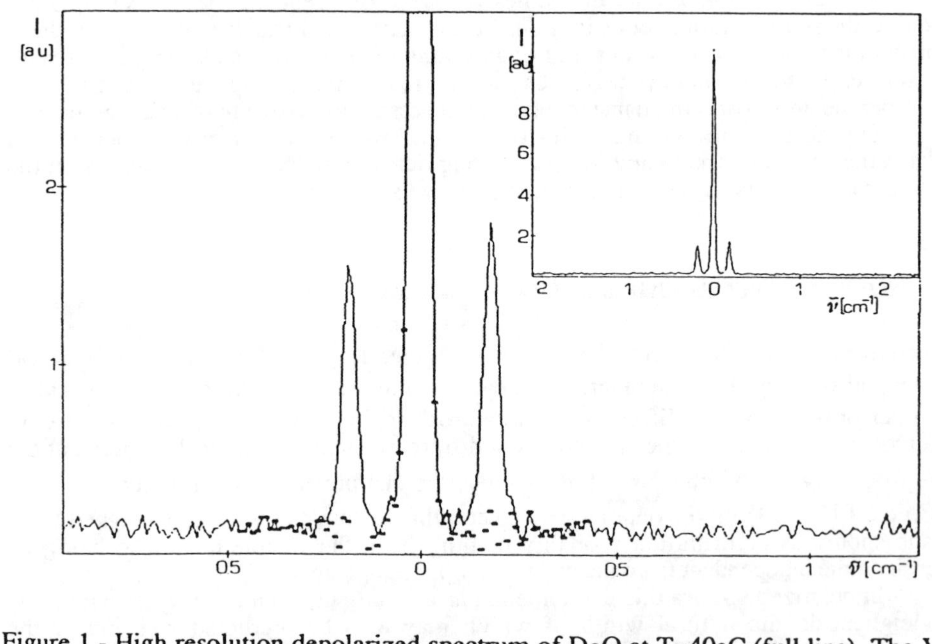

Figure 1 - High resolution depolarized spectrum of D_2O at T=40°C (full line). The VV filtering is subtractred of and a "clean" HV spectrum is obtained as explained in the text (dotted line). In the inset the same spectrum is shown in a full scale.

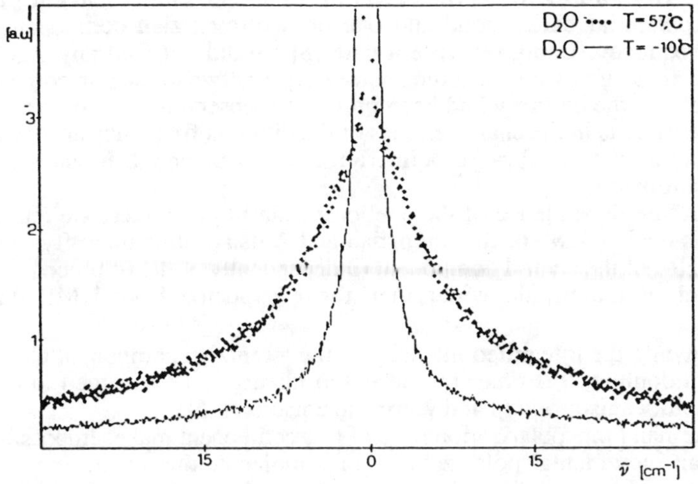

Figure 2 - Low frequency depolarized (HV) spectra of high purity D_2O at two different temperatures. The strong temperature dependent contribution is evident in this frequency range.

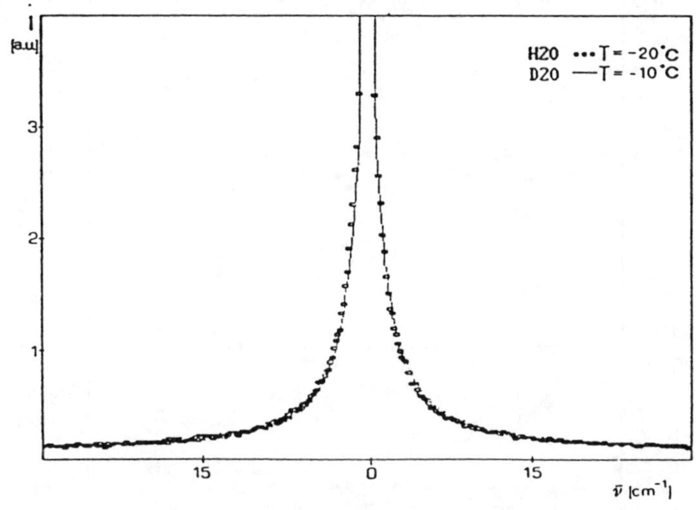

Figure 3 - Low frequency depolarized spectra of H_2O at T=-20°C and D_2O at T=-10°C. The comparison shows that the spectral shape of H_2O at a given temperature is very close to the D_2O one at same higher temperature

An accurate fit of all the observed spectrum from 0.06 cm^{-1} to 50 cm^{-1} can be made using an exponential background and one or two lorentzian components both centered at zero frequency. Using our recent data [8] we did not find any significant improvement in the fit accuracy between the choice of one or two lorentzian components, improvement which, on the contrary had been previously observed. We believe that the origin of this discrepancy is in the characteristics of the different frequency analyzer used, which joins the resolution of a Fabry Perot interferometer with the F.S.R. and a contrast of a grating monochromator.

The temperature dependence of the one lorentzian fit parameters are reported in figure 4, where Γ is the HWHM of the lorentzian and A its absolute intensity. It can be noted that the HWHM of the central component (independently of the fit procedure used) is about one order of magnitude wider then those expected from NMR data for reorientational time.

Moreover, being the integrated intensity of the lorentzian component about 10^{-2} times the Brillouin doublet, this contribution is too intense to be assigned to the bare molecular polarizability anisotropy (β and γ terms in equation (15)).

On the other hand the polarization ratio of this component makes impossible it to be originated by an incremental polarizability of a molecule due to the formation or breaking of hydrogen bonds. It is indeed very easy to show [13] that this kind of mechanisms should have a polarization ratio $\rho = I_{VV}/I_{VH} \geq 3$ which largely exceed the measured ρ=4/3.

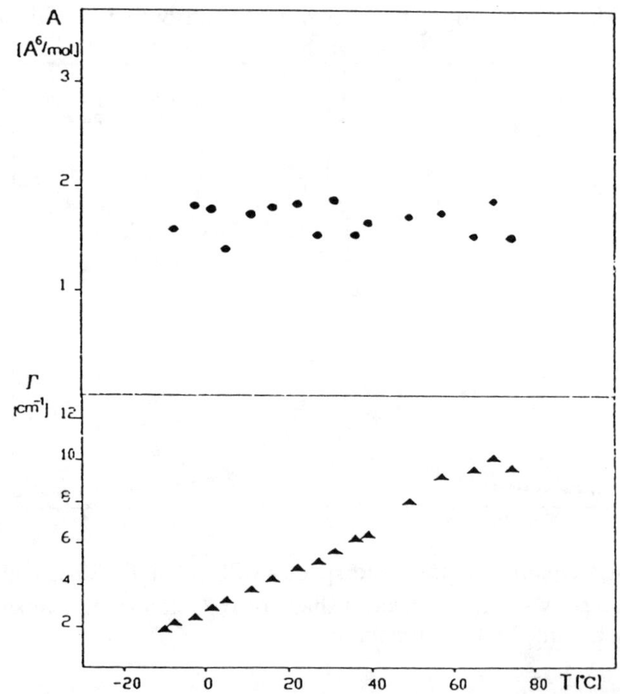

Figure 4 - Temperature behaviour of the fit parameters for D_2O low frequency depolarized spectra. A is the absolute intensity of the Lorentzian component (•) and Γ its HWHM (▲) (see text).

On such basis, this peculiar temperature dependent contribution has been temptatively assigned [13] to a characteristic shape of the DID spectrum, which could be typical of hydrogen-bonded liquids. If this hypothesis is correct, the observed spectrum should be the Fourier transform of a correlation function involving pure translational degrees of freedom. To test this hypothesis a Molecular Dynamics (MD) calculation has been performed using a TIP4P potential and the low frequency light scattering spectrum has been calculated at some different temperatures [14]. This potential was been already used to simulate successfully the isotropic components and the characteristic temperature evolution of the shoulder which appear in water spectrum around 180 cm $^{-1}$ [15,16]. In [14] both DID and rotational contribution have been included in the simulation as well as their cross-terms, and the good agreement with experimental data is shown in figure 5.

In figure 6 the pure orientational spectrum, the DID spectrum and their cross term, are presented in order to emphasize the relative relevance of these contributions. It can be observed that the time scale of orientational and translational correlation functions is very similar and that the value of their cross-correlation function is very high. This result leads to the conclusion that the strong directionality of H-bonds intimately mixes, on this time scale, translational and orientational variables, creating a high degree of correlation between the angular positions of molecules and the relative motions of their centers of mass.

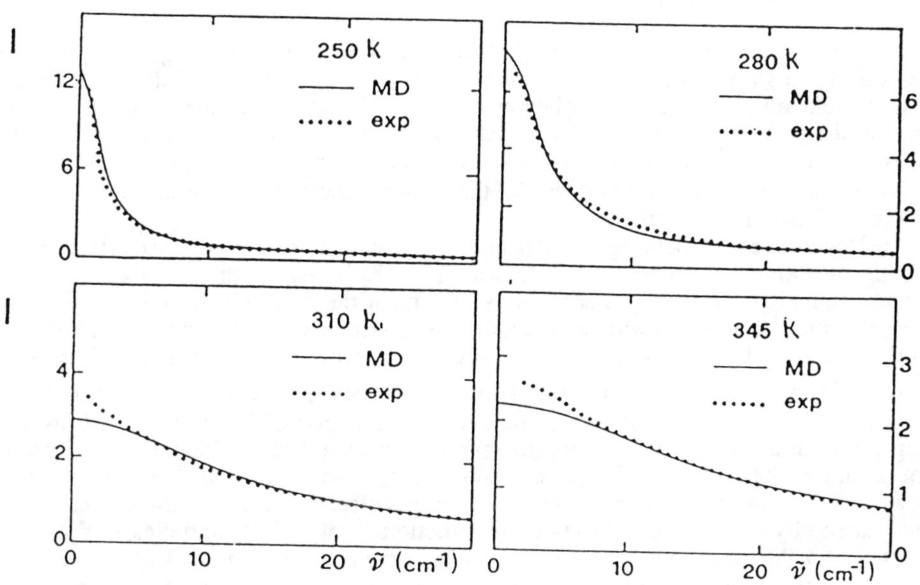

Figure 5 - Comparison between experimental data (dotted the line) and Molecular Dynamics results (full line) for H_2O spectra at four different temperatures (see reference [14]).

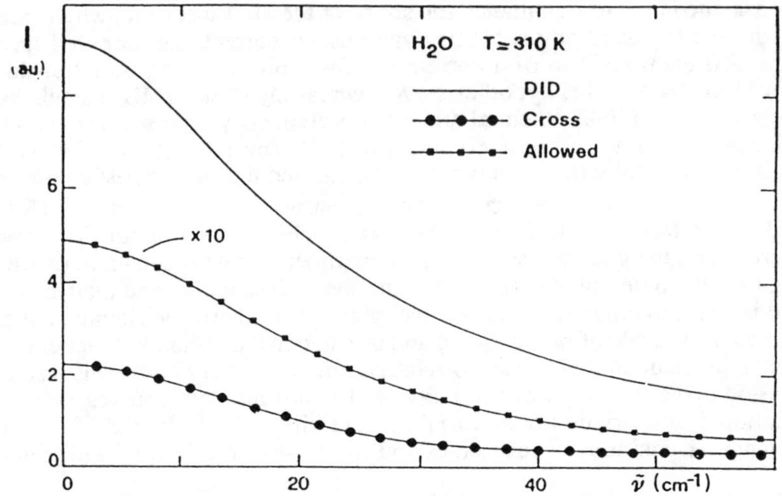

Figure 6 - Molecular Dynamics results for H_2O spectrum at T=310K. The pure orientational spectrum (-■-■-), the DID spectrum (——) and their cross term (●—●) are shown as discussed in the text.

The conclusion we have drawn allows us to predict a coupling between sound waves and translational and/or orientational relaxation, if the characteristics time is comparable. Unluckely this cannot be observed in light scattering spectra of pure water, even at the lower temperature at which experiments can be performed (i.e. the supercooled region).In fact at T = -20° the central component of the depolarized spectrum reachs its minimum value (1.5 cm^{-1})while the acoustic wave frequency still remains about one order of magnitude smaller.

Nevertheless somethings similar to what we expect for pure water has been found recently by Benassi, D'Arrigo, and Nardone [17-18] in water-ethanol mixtures. These authors studied polarized light scattering spectra in the Brillouin frequency region of water-ethanol mixtures in a wide temperature and concentration range using a Fabry-Perot interferometer. The aim was to obtain sound velocities and longitudinal kinematic viscosity fitting data with the hydrodynamics theoretical prediction for $S(\vec{q},\omega)$, and to compare such results with the ultrasonic data of D'Arrigo and Paparelli [19]. Lowering temperature of a 0.23 molar mixture, the authors found a direct evidence of a relaxation process in the GHz region which involves both shear and bulk viscosities. The spectrum shown in fig.7 (taken at the temperature where this effect is most remarkable) cannot be fitted indeed by the simple hydrodynamic function (broken line) and clearly shows the presence of a "Mountain peak", which can be predicted in the framework of a generalized hydrodynamic theory, assuming that the kinematic viscosity coefficient can relax following the relation

$$D_V(\omega) = D_V(\infty) + \frac{b}{1 + i\omega\tau} \qquad (20)$$

The same authors, looking at the depolarized spectrum in the same experimental condition, observe a central peak the frequency width of which falls in the same frequency region of acoustic waves and closely resembles the HWHM of the central structure generated in the $S(\vec{q},\omega)$ by viscosity relaxation. From this similarity the authors argued that the microscopic variables responsible for the central component in the depolarized spectrum must be strongly related to those responsible for the anomalous shape of the $S(\vec{q},\omega)$. If all the conclusion we have drawn for pure water may be extended to the water-ethanol mixture as far as the depolarized spectrum is concerned, than the observed "Mountain peak" can be assigned to the coupling of sound waves with microscopic translational and/or rotational degree of freedom.

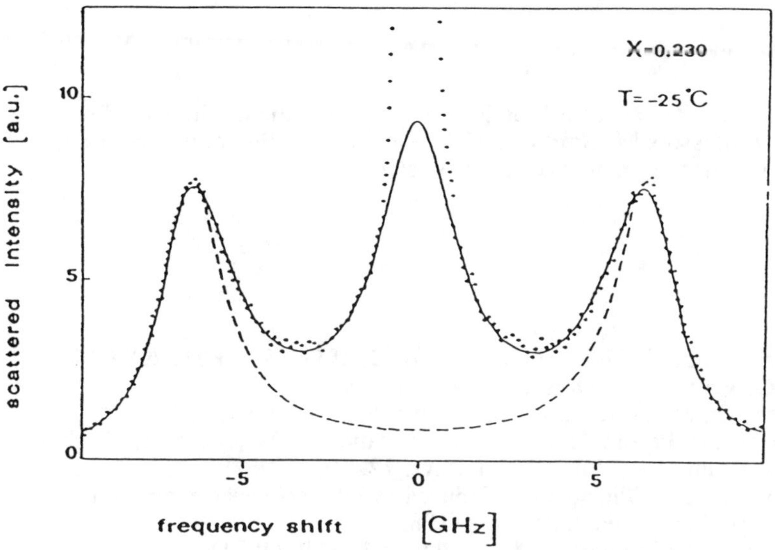

Figure 7 - VV spectrum of a 23% molar mixture of water-ethanol at T= - 25°C (dotted line) fitted using the simple hydrodynamic function (broken line) and the generalized one, including the "Mountain peak" (full line).

Due to the time scale of the relaxation process in the water-alchool mixture the sound velocity in this system undergoes to a positive dispersion in the $\vec{q}$ region explored by light scattering experiment ($|q| \sim 10^{-3}$ Å^{-1}), changing its value (for 0.23 molar mixture at T=-30°C) from $c(0) = 1.70 \times 10^5$ cm/sec. (ultrasonic measurements) to $c(\infty) = 2.25 \times 10^5$ cm/sec for $\omega >> \Gamma_{relax}$. In the same way, positive sound dispersion could be present in pure water at in the $|\vec{q}|$ region of 0.2 Å^{-1} room temperature, (where $\omega = c(0)q \sim \Gamma_{relax} \sim 8$ cm^{-1}). Positive sound velocity dispersion has been also predicted by Clementi et al. from M.D. data [20] in the same q range.

This picture can be an explanation for the experimental findings made by Teixeira et al. [21]. These authors measured the coherent anelastic neutron cross-section in heavy water at room temperature for exchanged $|\vec{q}| \geq 0.5$ Å^{-1}. They observe a clear evidence of a collective excitation, the propagation velocity of which is about 3300 m/sec, two times higher than the measured sound velocity in light scattering experiments. One of the possible assignement suggested by the authors themselves is a positive dispersion of the acoustic velocity due to an interaction between longitudinal sound waves and a relaxation process. This assignement was ruled out from the authors because the dielectric relaxation times are too long to be at the origin of the phenomenon. If this time will be derived from the width of the depolarized central component observed, the argument need to be reconsidered.

Other experiments are in progress in water-alchool mixtures as well as in pure alchools to gain better insight on this problem.

Acknowledgements

The authors are particularly indebted, for their patient criticism, discussion and comments, to Professors M. Nardone, M.A. Ricci and G. Signorelli and are grateful to Mrs.S. Maiolo for her substantial editorial assistence.

References

[1] V. Mazzacurati, P. Benassi, G. Ruocco - J. of Physics **E21**, 798 (1988).
[2] R.A. Cowley - Proc. Phys. Soc **84**, 281 (1964).
[3] V. Mazzacurati, G. Ruocco - Mol. Phys. **61**, 1391 (1987).
[4] B.J. Berne, R. Pecora, "Dynamics light scattering" Wiley (1976).
[5] A.D. Buckingham - Adv. Chem. Phys. **12**, 107, (1967).
[6] See for examples "Phenomnena induced by intermolecular interactions" Ed. G. Birnbaum - Plenum Press (1985).
[7] J. Bucaro, T. Litovitz - J. Chem. Phys. **54**, 3846 (1971).
[8] A. Nucara - Thesis, University of Rome (1989).
[9] P. Benassi, V. Mazzacurati, M. Nardone, M.A. Ricci, G. Ruocco, A. De Santis R. Frattini, M. Sampoli - Mol. Phys. **62**, 1467, (1987).
[10] A. De Lorenzi, A. De Santis, R. Frattini, M. Sampoli - Phys. Rev.**A33**, 3900 (1986).
[11] O. Conde, J. Teixeira - J. Physique **44**, 525 (1983).
[12] W. Denninger, G. Zundel - J. Chem. Phys. **74**, 2769 (1981).
[13] V. Mazzacurati, P. Benassi - Chem. Phys. **112**, 147 (1987).
[14] V. Mazzacurati, M.A. Ricci, G. Ruocco, M. Sampoli - Submitted to Chem. Phys. Lett.
[15] R. Frattini, M. Sampoli, M.A. Ricci - Chem Phys. Lett, **141**, 297 (1987).
[16] M.A. Ricci, G. Ruocco, M. Sampoli - Mol. Phys. **00**, 0000.
[17] P. Benassi, G. D'Arrigo, M. Nardone - J. Chem. Phys, **89**, 4469 (1988).
[18] M. Nardone, P. Benassi - Chem. Phys, Lett. **145**, 575 (1988).
[19] G. D'Arrigo, A. Paparelli - J. Chem. Phys. **88**, 405, (1988).
[20] M. Wojcik, E. Clementi - J. Chem. Phys. **85**, 6085 (1986).
[21] J. Teixeira, M.C. Bellissent-Funel, S.H. Chen, B. Dorner - Phys. Rew. Lett. **54**, 2681 (1985).

DIELECTRIC MEASUREMENTS OF WATER, ALCOHOLS AND THEIR MIXTURES

G. Salvetti,
Institute of Atomic and Molecular Physics (IFAM-CNR)
Via del Giardino 7, 56100 Pisa (I)

SUMMARY. This paper will outline the role of dielectrometry in the study of the hydrogen bond and its effects on the structure and dynamics of associated liquids. After a brief description of the experimental techniques used, data referring to water, alcohols and their mixtures are reported. A review of the models suggested by various authors over the past fifty years to explain the dielectric relaxation of these liquids is also presented.

The last section is devoted to the illustration of a recent model developed by the Liquid Research Group at IFAM; this model traces the dynamic behaviour of H-bonded liquids back to the stochastic formation and breaking of their hydrogen bonds. The results predicted by the model and the experimental results for water are shown to be in good agreement.

1. INTRODUCTION

Dielectrometry must be considered to be one of the most effective experimental techniques available for detecting the existence of hydrogen bonding [1] and for tracing its role in setting up the static and dynamic properties of H-bonded liquids. Indeed, the extent of hydrogen bonding in these liquids determines both the static dielectric constant, generally much higher than that observed in normal liquids, and the dielectric relaxation spectrum. In 1939 J.G. Kirkwood [2] developed his theory of the static dielectric constant; this was subsequently modified by H. Frohlich [3], who introduced an adimensional correlation factor, G_K, to take into account the dipolar correlations arising from the presence of hydrogen bonding, a factor not previously considered by L. Onsager [4].

The dielectric relaxation of H-bonded liquids shows characteristic features at low and intermediate frequencies which are not present in normal liquids. Indeed, the relaxation spectra (here only liquids with $G_K > 1$ will be considered) are dominated by a large band; other bands at higher frequencies with much smaller amplitudes are also observed.

Generally the main relaxation bands can be fitted using a single characteristic time τ_D. The close relationship that exists between the main relaxation band and hydrogen bonding became evident when it was observed that the amplitude of the band, Δ, decreased when the level of alcohol dilution was increased in alcohol-nonpolar solvent mixtures [5]. Moreover, Δ decreases when the pressure or the temperature of the sample increases. This behaviour and the features of the other relaxation bands of monoalcohols suggest that the main relaxation band is linked to the presence of large structures, whereas the higher frequency relaxation processes are due to the

J.C. Dore and J. Teixeira (eds.), Hydrogen-Bonded Liquids, 369–380.

relaxation of the mononers and/or dimers. The relaxation spectrum of water shows only one large band at approximately 17 GHz at room temperature; it seems quite probable that a second band at much higher frequencies also exists, but this has not yet been sufficiently studied.

Water-alcohol mixtures are systems of increasing interest, particularly in relation to the water-rich region. However, further advances in both experiment and theory must be achieved if we are to gain a better understanding of the perturbation effects provoked by the presence of alcohol molecules on the structure and dynamics of water. I will not deal with this complex subject here, limiting myself instead to a few general considerations of the properties of these mixtures.

2. EXPERIMENTAL TECHNIQUES

Dielectric techniques can be gathered together under two main headings: frequency domain and time domain techniques. Historically, frequency domain experiments were the first to be carried out; using a single experimental apparatus, relevant data can be obtained, but only within a small frequency range. The maximum sensitivity is given by resonance spectrometers working in a differential configuration. These instruments are generally "home-made" by the individual laboratory for their specific experimental purposes. Collecting data by means of this technique in order to study the dielectric relaxation of an H-bonded liquid is extremely time consuming, however, since it is necessary to utilise different apparatuses to cover the entire frequency range of interest. When one considers that the measurements must also be performed across a wide temperature range in order to gather information regarding the activation energy and the structural evolution of the sample, it becomes clear that a systematic study of the problem involves considerable difficulties.

The time domain technique has been developed over the past twenty years to overcome these problems; it has proved to be a powerful tool which is being used more and more widely by those studying H-bonded liquids. In fact, with this technique the entire relaxation spectrum can be obtained over a wide frequency range and within a very short amount of time. A detailed description of the experimental apparatuses for both frequency domain and time domain dielectrometry can be found in refs. [6] and [7], respectively and in the paper by E. Tombari in this book.

3. EXPERIMENTAL RESULTS

The dielectric spectra for water, normal alcohols and their mixtures are reported here. The analysis will be devoted particularly to those features which researchers have been utilising to develop theoretical models for dielectric relaxation.

3.1 <u>Water</u>

Fig. 1 shows the real, $\varepsilon'(\nu)$, and the imaginary, $\varepsilon''(\nu)$, parts of the complex dielectric constant of H_2O at 25° C. The maximum of the $\varepsilon''(\nu)$ curve occurs at $\nu_0 \cong 17$ GHz; that is, the characteristic time turns out to be $\tau_D = 1/2\pi\nu_0 \cong 8$ ps. The fitting curves (where ϖ is the angular frequency) may be obtained from Eq. 1.

$$\varepsilon'(\omega) = \varepsilon_\infty + \frac{\varepsilon_0 - \varepsilon_\infty}{1 + (\omega\tau)^2} \quad \text{and} \quad \varepsilon''(\omega) = \frac{(\varepsilon_0 - \varepsilon_\infty)\,\omega\tau}{1 + (\omega\tau)^2} \tag{1}$$

The points (ε',ε'') lie on a semicircle centered on the ε' axis and intersecting this axis at points $\varepsilon' = \varepsilon_0$ and $\varepsilon' = \varepsilon_\infty$ [8].

This graphic representation, known as the Cole-Cole plot, immediately gives clear evidence as to whether the experimental points refer to a single relaxation time process or if a distribution of relaxation times will be necessary. Water follows eq. 1 very well (see Fig. 2); in the metastable region as well a single dielectric relaxation time has been found [9].

It is only at frequencies much higher than ν_0 that the Cole-Cole plot does not fit the experimental points, owing to the presence of a second band [10]. The presence of a much higher frequency band is also suggested by the value of $\varepsilon_\infty \cong 5$, which is much larger than the ε_∞ value of ice (~3.1) and the square of the refractive index of water (~1.8).

The relaxation time τ_D decreases as the sample temperature increases and its behaviour cannot be described by an Arrhenius type equation, as would be expected for a Debye relaxation process. This is particularly evident in the metastable region [9]. Therefore, the dielectric relaxation of water cannot be considered to be a typical activated process with an energy barrier of height E. An activation energy can be calculated only for small temperature intervals; the energy changes with temperature, increasing as the temperature decreases (Fig. 3). On the other hand, the unicity of the relaxation time supports the hypothesis that a single microscopic mechanism exists, despite the structural complexity of liquid water.

Furthermore, the Debye relationship $\tau T/\eta$ = cost (where η is the viscosity) is well verified, at least at T > 0°C, and can thus be considered as further evidence that water relaxation behaves in the same way as a simple dipole relaxation.

These contradictions are one example of the numerous challenges which water presents to researchers. Another quantity used in analysing the dielectric behaviour of water is the mean "lifetime," τ_{HB}, of the hydrogen bond. This value was obtained for the first time by C.J. Montrose et al. [11] and then by O. Conde and J. Teixeira [12]. Figure 4 shows τ_{HB} and the related quantity τ_f (which will be defined below) as a function of 1/T.

3.2 Mono-alcohols

For the sake of simplicity, the discussion here will be limited to a consideration of methanol and n-pentanol. Methanol shows a high degree of structural complexity, 3 being the maximum number of HBs (hydrogen bonds) that the molecule can form with its neighbors [13], whereas n-pentanol is typical of the normal superior alcohols in that its molecular structure is characterised by linear chains of different lengths. Indeed, 2 is the maximum number of HBs per molecule.

Figure 5 shows the dielectric relaxation time τ_D for methanol vs 1/T down to the supercooled region [14]. The experimental points from ~160°K up to room temperature are fitted by a straight line very well, deviations being observable only near the boiling point. The τ_D of n-pentanol is presented in

Figure 6, where once again a straight line fits all of the experimental points [15], but the temperature interval explored was too narrow.

One general characteristic of the H-bonded liquids under examination here seems to be the non-Arrhenius behaviour of their relaxation processes. To a greater or lesser degree, depending on the individual case, the activation energy is a function of temperature - that is, the liquid structure changes with temperature, thus affecting the height of the energy barrier for dipole relaxation. For a detailed analysis of the dielectric properties of other H-bonded liquids, see ref. [8].

3.3 Alcohol-Water Mixtures

Alcohol-water mixtures, like alcohol-alcohol mixtures, show a dielectric spectrum with a main relaxation band well fitted by a single relaxation time. The characteristic parameters for the relaxation of these liquids cannot be obtained from the values measured in the pure components. In fact they behave like pure H-bonded liquids! This observation is amply supported by the curves in Fig. 7, which show the dielectric relaxation time of the water-methanol mixture at various temperatures as a function of the molar fraction of water, x_W [14]. Thus, from the experimental results it follows that:

i) The microscopic mechanism underlying dielectric relaxation is generally the same, at least for all of the H-bonded liquids being considered here. Polyalcohols, however, do exhibit peculiar features due to the presence of more than one OH group, whereas in the tertiary alcohols steric effects and a distribution of relaxation times can be observed [15].

ii) Mixtures of H-bonded liquids behave like pure H-bonded liquids. In a mixture of an H-bonded liquid and a non-polar solvent, as the solvent concentration increases the amplitude of the main relaxation band decreases.

iii) The observed temperature and pressure dependence of the main relaxation band suggests the existence of a correlation between the parameters of hydrogen bonding, microscopic structure and dielectric relaxation.

4. A REVIEW OF SOME MODELS FOR DIELECTRIC RELAXATION

4.1 Dielectric Relaxation of Mono-Alcohols

Many attempts have been made to explain the dielectric behaviour observed in H-bonded liquids in terms of the microscopic properties of the molecules involved, and different models of liquid structure and dynamics have emerged. Here a few of the most important are reviewed, in chronological order.

P. Debye in 1929 considered dipolar relaxation to be an angular Brownian motion of the dipole and obtained a single relaxation time [16]. Subsequently, W. Kauzmann described the dielectric relaxation as an activated process, the dipole being embedded in a microcrystalline structure. According to this model, the fluctuations of the structure produce the environmental condition for dipole relaxation - that is, an "activated structure" - by absorbing the activation energy E. This model predicted an Arrhenius-type behaviour for τ_D as a function of temperature [17].

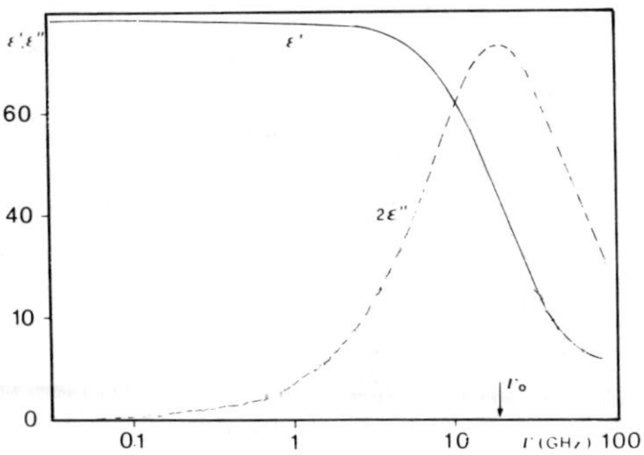

Fig. 1 The ε'(v) and ε''(v) of H_20 at 25°C.

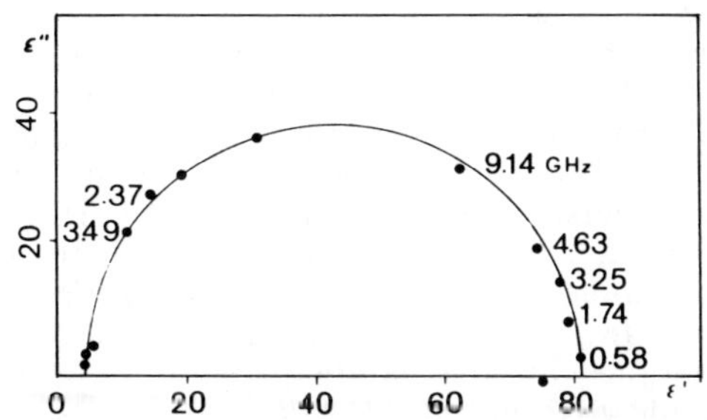

Fig. 2 The Cole-Cole plot for H_20 at 20°C.

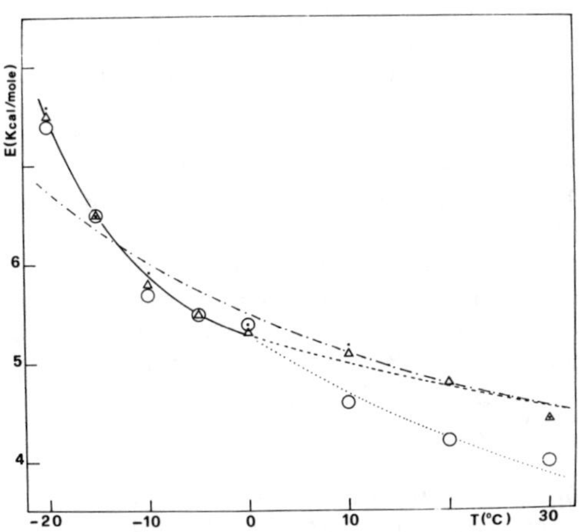

Fig. 3 The activation energy of H_20 as a function of temperature (ref.9).

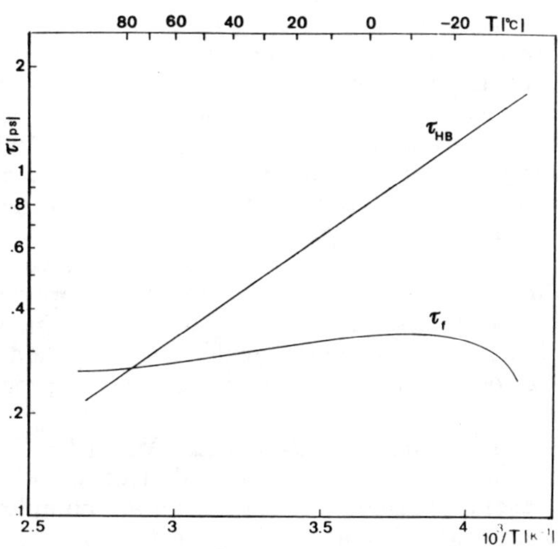

Fig. 4 The τ_{HB} and τ_f of H_20 as a function of 1/T (ref.29).

In 1955 D.J. Denney and R.H. Cole tried to explain the observed unicity of the relaxation time by assuming that the process was cooperative, a large number of molecules being simultaneously involved [18].

E. Bauer, M. Magat and C. Brot (1963) suggested that the actual reorientation responsible for the main relaxation band of monoalcohols takes place in single molecules which are temporarily unbonded to the linear multimers. The breaking of the HBs determines the dipole reorientation; the H-bond "lifetime" is thus equivalent to the dielectric relaxation time and the activation energy is the energy of the HBs [19].

In 1966 J. Middelhock and C.J.F. Böttcher showed that the activation energies of the straight chain heptanol isomers may differ widely [20]. P. Bordewijk et al. (1968) then observed that a mixture of heptanol-1 and heptanol-4 can be described extremely well by a single relaxation time [21], whereas a distribution of relaxation times would be expected if one followed the model of Bauer et al. and the experimental results of ref. [20]. Bordewijk suggested that the multimers in liquids are identical and that the relaxation is thus due to the reorientation of the multimer as a whole. The energetically favoured multimer should then be the cyclic tetramer.

W. Dannhauser et al. (1970) retained the assumption of linear association suggested by Bauer et al., but did not take the breaking of the HB to be the rate-determining step. According to their model, this step was only a prerequisite for the occurrence of dipolar reorientation, which they considered to be a cooperative process between the unbonded molecule and its environment. The activation energy would then be the H-bond energy plus the reorientation energy [22]. In 1976 B.M. Fung and T.W. McGaughy suggested that the reorienting units are not monomers and that dielectric relaxation is the result of an intermolecular anisotropic rotation and the reorientation of the cluster as a whole [23].

R. Minami et al. (1980) followed Dannhauser's model and assumed that dipole relaxation is due to the cooperative rotation of the OH groups in the linear multimers. The rotation originates at the terminal molecules of the chain (the molecules with one HB) and then propagates from one extremity to the other, following a random walk process with two absorption walls [24].

4.2 Dielectric Relaxation of Water

If the description of the dielectric relaxation of alcohols has been shown to present difficulties, even more complexities are involved in describing the behaviour of water. Only a few of the models reported in the literature will be mentioned here.

J.B. Hasted et al. (1952) suggested that the dielectric relaxation in water may be due to "asymmetrically two-bonded" molecules. The molecules with 0 and 1 HB would then be responsible for the higher frequency band [25] and the main relaxation band should be related to the breaking of the dymers. This model accounts for the ε_∞ value of water.

J.E. Anderson in 1967 proposed a two-state model in which the water molecule is imagined to be interacting with two different environments. He succeeded in writing the dipole correlation function using a stochastic approach [26]. H.E. Stanley and J. Teixeira (1980) then described the static properties of water in terms of the probability p_B that an HB would be intact at a given temperature, T. They were able to demonstrate that the fraction f_i of molecules with i number of intact HBs is given by the binomial distribution:

$$f_i = \binom{4}{i} p_B^{\,i} (1 - p_B)^{4-i} \tag{2}$$

and that the water structure is a percolated network of HBs [27]. However, these authors only offer a few qualitative hypotheses concerning the dynamics of water.

Recently a model for the dynamics of liquids has been proposed by the Liquid Group at IFAM in Pisa. This will be reported in some detail here and the predictions which it generates will be compared with experimental data.

5. A "MICROSCOPIC" MODEL FOR THE DYNAMICS OF WATER

A first version of this model was published some years ago [28, 29] and a revised version will appear shortly in the Journal of Chemical Physics [30].

In this model we assume that the microscopic mechanisms which underlie the dynamics of water are the break-up and reformation of the HBs. The definition of the hydrogen bond adopted here is essentially an on-off definition; i.e., the partially established HBs are not taken into account. This means that below a certain threshold the HB is regarded as being intact, whatever its definition (geometric or energetic) may be [31,32,33].

The HBs are considered to be equivalent and the maximum possible number per molecule, given our definition of the HB, is four. The H-bond dynamics are described by introducing a stochastic variable, η, which simulates the random change over time of the number of HBs established by the water molecule. The water molecule can explore a total of five states as the number of HBs changes from 0 to 4 and, using a "quantum-like formalism," the corresponding statistical picture is given by the state:

$$|\psi(t)\rangle = \sum_{i=0}^{4} |\eta_i\rangle \, p_i(t) \tag{3}$$

where $p_i(t)$ is the probability that the molecule is in the state $|\eta_i\rangle$ at time t.

Assuming that the simultaneous break-up or formation of two or more HBs per molecule are statistically negligible events, it is possible to write a master equation for the time evolution of the probability $p_i(t)$, in abbreviated notation:

$$\frac{dp_i(t)}{dt} = \sum_{J=0}^{4} p_J(t) \, (K_\eta)_{iJ} \tag{4}$$

where $(K_\eta)_{iJ}$ are simple expressions containing the rate of formation, A, and the rate of break-up, B, of one HB (see diagram).

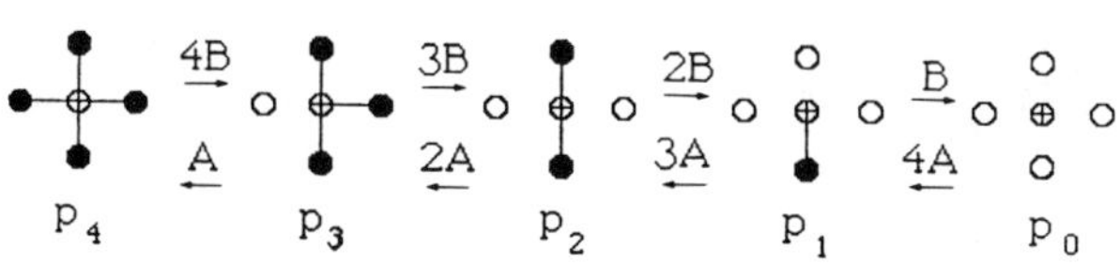

If the probability $p_{HB}(t)$ that a randomly considered HB is intact at time t is defined as:

$$p_{HB}(t) = \frac{4p_4(t)+3p_3(t)+2p_2(t)+p_1(t)}{4} = \frac{n_{HB}}{4} \qquad (5)$$

where $n_{HB}(t)$ is the average number of HBs per molecule at time t, one obtains: $A = 1/\tau_f$; $B = 1/\tau_{HB}$, where τ_f is the mean "deathtime" and τ_{HB} the mean "life-time" of the HB.

The equilibrium distribution turns out to be the binomial distribution of the Stanely-Teixeira model of eq. 2 with $p_{HB}(\infty) = p_B$. The equilibrium probability, p_B, is related to τ_f and τ_{HB} by the equation:

$$p_B / (1-p_B) = \tau_{HB}/\tau_f \qquad (6)$$

The probability p_B is temperature dependent so, if the sample temperature changes, the populations f_i will arrive at a new binomial distribution in a time which is the inverse of the first non-vanishing eigen value of eq.(s) 4.

To describe the orientational relaxation of water and to compare the predictions of the model with the experimental data it is necessary to calculate the correlation functions:

$$\Phi_{ni}(t) = \langle P_n(0), P_n(t) ; \eta_i \rangle \qquad (7)$$

where $P_n(t) = P_n [\cos \theta(t)]$ donotes the nth Legendre polynomial.

The time evolution of $\Phi_{ni}(t)$ is given by the equation:

$$\frac{d\Phi_{ni}(t)}{dt} = - D_{Rn}(\eta_i)\Phi_{ni}(t) + \sum_{J=0}^{4} (K_\eta)_{iJ}\, \Phi_{nJ}(t) \qquad (8)$$

where $D_{Rn}(\eta_i)$ is the orientational diffusion coefficient in the state $| \eta_i \rangle$ and the coefficients $(K\eta)_{iJ}$ are those of eq. 4.

The total correlation function is:

$$\Phi_{Tn}(t) = \sum_{i}^{4} \Phi_{ni}(t)\, f_i \qquad (9)$$

where f_i is the statistical weight of the state $| \eta_i \rangle$ given by the binomial distribution.

By solving eq.(s) 9 with the initial conditions $\Phi_{nJ}(0) = 1$ for J=i and $\Phi_{nJ}(0) = 0$ for $J \neq i$ we find:

$$\Phi_{Tn}(t) = \sum_{J=0}^{4} \Delta_{nJ} \exp(-\lambda_{nJ} t) \qquad (10)$$

where λ_{nJ} are the eigen values of eq.(s) 8. Eq.(s) 8 were derived within the framework of a general theory (see ref. 29 and the references therein) which is able to give information on the dynamics of these systems, particularly within the long-time regime. Indeed, not only the orientational but also the translational dynamics can be studied in this way and, for example, the self-diffusion coefficient of water can be derived [28,29].

Let us proceed by analysing the rotational relaxation of water. To compare theory and experiment, I will utilise the reduced version of the model, obtained by assuming that the relaxation is essentially due to the water molecules with a maximum of two HBs. This means that we are considering two states, one in which the molecules are "mobile" and the other in which they are "immobile."

The populations of the two states are:

$$F_M = f_0 + f_1 + f_2 \ ; \quad F_I = f_3 + f_4 \tag{11}$$

where the subscripts M and I refer to the "mobile" and the "immobile" states, respectively.

The correlation functions of interest are now given by the eq.(s):

$$\begin{aligned} d\Phi_M(t)/dt &= -(D_{RM} + \frac{1}{\tau_M})\, \Phi_M(t) + \frac{1}{\tau_I}\Phi_I(t) \\ d\Phi_I(t)/dt &= -(D_{RI} + \frac{1}{\tau_I})\, \Phi_I(t) + \frac{1}{\tau_M}\Phi_M(t) \end{aligned} \tag{12}$$

where τ_I and τ_f are the "residence times" in the two states.

By solving eq.(s) 12 with the initial conditions given for eq.(s) 8 one obtains for the total correlation function:

$$\Phi(t) = \Delta_- \exp(\lambda_- t) + \Delta_+ \exp(\lambda_+ t) \tag{13}$$

Eq.(s) 13 are formally equal to those obtained by Anderson [26].

The lower eigen value λ_- is related to the higher amplitude Δ_- of the relaxation spectrum; that is, it represents the main relaxation band of water, whereas the second exponential refers to higher frequency processes. The parameter λ_-^{-1} is the characteristic time of the correlation function and it must be compared with the measured τ_D values.

From theory the following may be derived (assuming $D_{RI} = 0$):

$$\lambda_- \cong D_{RM} F_M \,/\, (1 + D_{RM} F_M \tau_I) \tag{14}$$

It is now necessary to give values to the parameters introduced in the model, that is, $\tau_{HB}(T)$ and $p_B(T)$ and $D_{RM}(T)$. For $\tau_{HB}(T)$, values in Fig. 4 will be utilised, whereas the $p_B(T)$ values can be evaluated following Stanley and Teixeira's suggestions [27]. The values for the rotational diffusion coefficient $D_{RM}(T)$ may also be obtained from the experimental data.

In fact, an analysis of the QENS spectroscopy data for water [31] gives two characteristic times t_0 and τ_1, which can be interpreted as the "residence times" of the water molecule in the "immobile" and "mobile" states, respectively. Consequently, the rotational diffusion coefficient of the second Legendre polynomial correlation function in eq.(s) 8 can be assumed to be $D_{R2}(\eta_M) = 1/\tau_1$ and the value of $D_{R1}(\eta_M)$, that is, the diffusion coefficient D_{RM} of the correlation function of the first Legendre polynomial in eq. 14 turns out to be $D_{RM} = 1/3\ \tau_1$ [30].

From the experimental data in ref. [34] the temperature dependence of τ_1 is given by the equation: $\tau_1(T) = 0.0485 \exp(925/T)$.

Figure 10 shows the experimental points τ_D vs $1/T$; the shaded region represents the theoretical uncertainty band related to the scattering of the existing $\tau_{HB}(T)$ and $p_B(T)$ determinations. The values of the characteristic time τ_{NMR} coming from NMR experiments are also shown together with the uncertainty belt of our theoretical predictions.

This model can also be applied to normal alcohols. For example in n-pentanol, which is composed only of linear chains of varying lengths, molecules with 0, 1 or 2 HBs may be found; this means that a three-state model will describes the behaviour of this alcohol. We have used the existing dielectric data to estimate the values of τ_{HB} for some alcohols. In this way we obtained τ_{HB} values which increase with the length of the alkylic chain (~2 ps for methanol, up to ~150 ps for decanol) [29].

6. CONCLUSIONS

The main dielectric relaxation band of H-bonded liquids is intimately related to the presence of a structure in which the molecular dipole rotates owing to the fluctuating character of the HBs which the molecule can establish. Attempts to trace the dynamics of these molecules back to the properties of particular molecular aggregates have not proved satisfactory.

The model presented here is capable of explaining not only dielectric relaxation, but also other dynamic properties as revealed by NMR and SANS studies [34]. Moreover, the theoretical framework [35] on which the model is based permits us to interpret the results emerging from the latest experimental work and computer simulations of molecular dynamics, from recent studies of the effects arising out of different HB definitions, and from the assumptions for each state $|\eta_i\rangle$ of different τ_{HB} values [29].

Physically the picture which emerges from our theoretical approach is a mesoscopic one in which the molecule is considered together with its environment. Its dynamics are equivalent to the oscillatory motion of a particle contained in a cage consisting of the neighbouring molecules; this cage undergoes a fluctuation-dissipation process which mimics the influence of the remaining molecules of the sample.

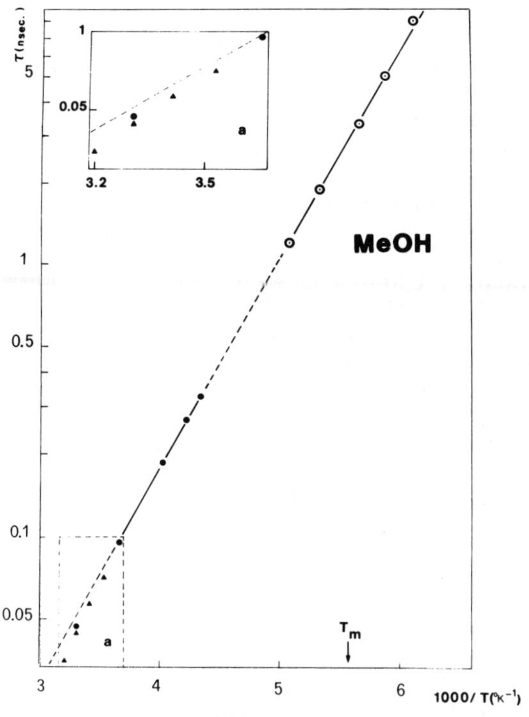

Fig. 5

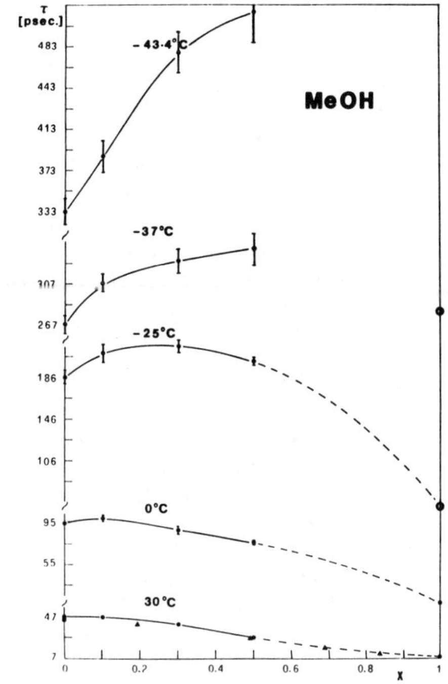

Fig. 7

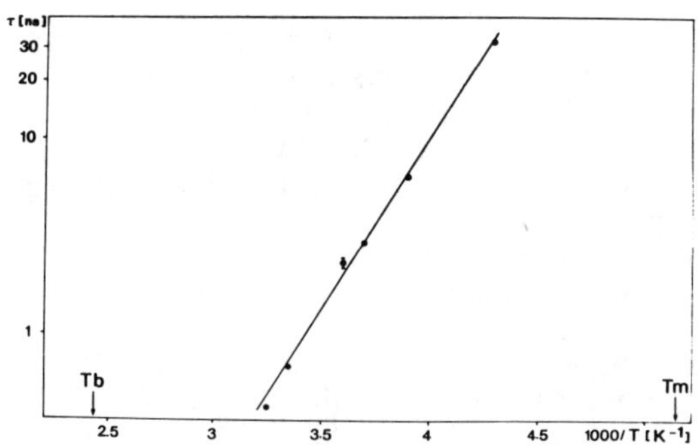

Fig. 6

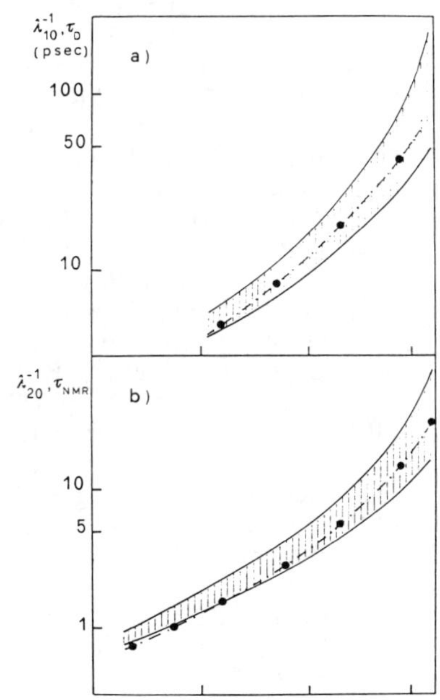

Fig. 8

Fig. 5: τ vs 1/T of MeOH (ref.14).
Fig. 6: τ vs 1/T of n-POH (ref. 15).
Fig. 7: τ vs x of MeOH-H_2O mixture (ref.14).
Fig. 8: a) τ_D vs 1/T of H_2O;
b) τ_{NMR} vs 1/T of H_2O (ref.30).

REFERENCES

[1] Rao, C.N.R. (1972) in Water. A Comprehensive Treatise. Vol.1, F. Franks (ed.), Plenum, NY.

[2] Kirkwood, J.G. (1932) J.Chem.Phys., 7:911.

[3] Frohlich, H. (1949) Theory of Dielectrics, Clarendon Press, Oxford.

[4] Onsager, L. (1936), J.Am.Chem.Soc. 58:1486.

[5] Glasser, L., Crossley, J. and Smyth, C.P. (1972) J.Chem.Phys. 57:3977.

[6] Grout, E.H., Sheppard, R.J. and South, G.P. (1978) Dielectric Behaviour of Biological Molecules in Solution, Oxford Univ. Press.

[7] Cole,R.H., Berberian,J.G., Mashimo,S., Ghryssikos,G., Burns,A. & Tombari, E. (in press) J.Appl.Phys.; Cole, R.H.(1977) Ann.Rev.Phys.Chem. 28:283.

[8] Böttcher, C.J.F. and Bordewijk, P. (1978) Theory of Electric Polarization. Vol.II, Elsevier Scientific Publ. Co., Amsterdam.

[9] Bertolini, D., Cassettari, M. and Salvetti, G., (1982) J.Chem.Phys. 76:3285.

[10] Afsar, M.N. and Hasted, J.B. (1978) Infrared Phys. 18:835.

[11] Montrose, C.J., Bucaro, J.A., Marchall-Coakley, J. and Litovitz, T.A. (1974) J.Chem.Phys. 60:5025.

[12] Conde, O. and Teixeira, J. (1983) J.Phys.(Paris) 44:525.

[13] Jorgensen, W.L. (1980) Chem.Phys.Lett. 70:326.

[14] Bertolini, D., Cassettari, M. and Salvetti, G., (1983) J.Chem.Phys. 78:365.

[15] D'Aprano, A., Donato, I.D., Bertolini, D., Cassettari, M. and Salvetti, G. (1985) Mol.Phys. 55:475.

[16] Debye, P. (1929) Polar Molecules, Dover, NY.

[17] Kauzmann, W. (1942) Rev.Mod.Phys. 14:12.

[18] Denney, D.J. and Cole, R.H. (1955) J.Chem.Phys. 23:1762.

[19] Bauer, E. and Magat, M. (1963) Bull.Soc.Chim.F.D., 341.

[20] Middelhock,J. and Bottcher, C.J.F. (1966) Mol.Relax.Processes, Chem.Soc.Spec.Publ. 20:69.

[21] Bordewijk, P., Gransch, F. & Böttcher, C.J.F. (1969) J.Phys.Chem. 73:3255.

[22] Dannhauser, W. and Flueckinger, A.F. (1970) Phys.Chem.Liquids 2:37.

[23] Fung, B.M. and McGaughy, T.W. (1976) J.Chem.Phys. 65:2970.

[24] Minami,R. Itoh,K., Takahashi, H. & Higasi, K. (1980) J.Chem.Phys. 73:3396.

[25] Hasted, J.B., Haggis, C.H. and Buchanan, J.J. (1952) J.Chem.Phys. 20:1452.

[26] Anderson, J.E. (1967) J.Chem.Phys. 47:4879.

[27] Stanley, H.E. and Teixeira, J. (1980)J.Chem.Phys. 73:3404.

[28] Bertolini, D., Cassettari, M., Ferrario, M., Salvetti, G. and Grigolini, P. (1984) J.Chem.Phys. 81:6214.

[29] Bertolini, D., Cassettari, M., Ferrario, M., Grigolini, P. and Salvetti, G. (1985) in Memory Function Approaches to Stochastic Problems in Condensed Matter, Evans, M.W., Grigolini, P. and Pastori-Parravicini, G. (eds.), Wiley, NY, p.277.

[30] Bertolini, D., Cassettari, M., Ferrario, M., Grigolini, P., Salvetti, G. and Tani, A. (1985) J.Chem.Phys. (in press).

[31] Geiger, A., Stillinger, F.H. and Rahaman, A. (1979) J.Chem.Phys. 70:4184.

[32] Mezei, M. and Beveridge, D.L. (1981) J.Chem.Phys. 74:622.

[33] Blumberg, R.H., Stanley, H.E., Geiger, A. and Mausbach, P. (1984) J.Chem.Phys. 80:5230.

[34] Chen, S.H. and Teixeira, J. (1985) Adv.Chem.Phys. 64:1; and Toukan, K., Ricci, M.A., Chen, S.H., Loog, C.K. and Price, D.L. (1988) Phys.Rev.A 37:2580.

ULTRASONIC PROPERTIES OF ALCOHOLIC AQUEOUS SOLUTIONS

G.D'Arrigo
Department of Physics - University of Rome "La Sapienza"
P.le A.Moro, 2 - 00185 Rome, Italy
and
Gruppo Nazionale di Struttura della Materia del CNR
Rome,Italy

ABSTRACT. The ultrasonic properties of many aqueous solutions, and in particular those with monohydric alcohols, exhibit anusual behaviors in the water-rich region of concentration. These anomalies can be related to the unique structural properties of water and to the ability of the solute molecules to promote a positive structuring on the solvent. After a short summary of the experimental and theoretical backgrounds underlying the u.s. propagation in liquid systems, we will analyse the u.s. behavior in water- alcohol mixtures. A particular attention will be devoted to the solutions with ethanol, ethoxyethanol and butoxyethanol where the temperature range investigated was extended down to the freezing temperatures (T_F). The critical-like behavior observed near T_F is discussed.

1. INTRODUCTION

Binary mixtures of water with alcohols are typical examples of solutions where the solute molecules exhibit opposite interaction effects giving rise to solution properties strongly dependent on concentration. The alcohol molecule contains both hydrophilic and hydrophobic groups. The former can hydrogen bond with water and so tend to hold it in solution while the apolar groups tend to force the solute out of solution. Depending on the relative importance of the two groups and on the values of physical parameters such a pressure (p), temperature (T) and composition (x= alcohol mole fraction), the solubility conditions can largely vary from situations where the solute is miscible in all proportions to those where a phase separation occurs. In some cases micelle formation results, a compromise where the polar heads remain in solution while apolar groups are removed from the solution and restricted in the micellar core.

The peculiar concentration dependence of thermodynamic properties of water-alcohol mixtures has been extensively reviewed by Franks and Read [1]. Similar anomalies in the water-rich region of composition are also found in many other physical properties such as viscosity, Rayleigh intensity of scattered light, dielectric, NMR and ultrasonic

J.C. Dore and J. Teixeira (eds.), Hydrogen-Bonded Liquids, 381–391.

relaxations. The origin of such behaviors is usually attributed to the enhancement of the water structure around the solute hydrophobic groups.

Ultrasonic (u.s.) propagation is quite sensitive to structural changes in fluid systems. The low frequency sound velocity (c_o) is related to an equilibrium property, the isentropic compressibility β_S, while the u.s. absorption coefficient (α) as a function of frequency allows to explore the dynamical behavior of the system. We will show here the general behavior of the u.s. properties in alcohol aqueous solutions. Particular references will be made to ethanol and ethoxyethanol solutions (which are miscible in all proportions) and to butoxyethanol solutions (which exhibit a miscibility loop with a LCST at 50°C) where the investigations were extended to temperatures near the T-x freezing curve.

2. ULTRASONIC PROPAGATION IN LIQUIDS

2.1. Experimental

The frequency (f) range covered by conventional u.s. techniques is between about 1 to 250 MHz. Molecular processes with characteristic times in the range 10^{-6}-10^{-9} s can be then probed by such techniques. Lower frequency ranges (down to kHz) requires special methods (resonances, reverberation) while the GHz region can be investigated by Brillouin light scattering.

$c(\omega)$ and $\alpha(\omega)$ measurements are usually performed by pulse techniques where short u.s. pulses, generated by a transducer (e.g. a piezoelectric plate), travel through the liquid sample and are reflected forth and back by a receiving transducer. In a typical variable-path arrangement the superposition with a reference c.w. signal upon changing the emitter- receiver distance allows to get the wavelength while the measurement of the amplitude of the reflected pulses as a function of the distance gives the absorption coefficient. Details of the experimental techniques and methods can be found in Ref.[2].

2.2. Theoretical backgrounds

According to standard hydrodynamics laws, the phase velocity and the absortpion coefficient of a longitudinal elastic wave of frequency ω (=2πf) travelling in a fluid should be given by [3]

$$c^2 = (\rho \beta_S)^{-1} \tag{1}$$

$$\alpha(\omega) = \frac{\omega^2}{2c^3}\left[D_\ell + (\gamma-1)\, D_T\right] \tag{2}$$

where

$$D_\ell = \rho^{-1}(\eta_B + \frac{4}{3}\ \eta_s) \tag{3}$$

is the longitudinal kinematic viscosity and $D_T = \lambda / \rho C_p$ the thermal diffusivitiy. D_T is usually ignored in fluids since it is negligible compared to D_ℓ.

From an experimental point of view [4,5] one finds that most of fluid system do not follow the hydrodynamic behavior, i.e. c and α/f^2 depend on the frequency. This observation means that for an u.s. wave (wavenumber k, frequency ω) travelling in a fluid, the conditions for local equilibrium, i.e. $k\xi<<1$; $\omega\tau_\xi<<1$, are not attained (ξ and τ_ξ are the correlation length and the correlation time, respectively, characterizing fluctuations of microscopic properties (e.g. density, energy)). In order to account for such divergences, one could generalize hydrodynamics laws by considering frequency dependent transport coefficients [$\eta_B(\omega)$; $\eta_s(\omega)$] by means of integrals of time-correlation functions of microscopic fluxes (Green-Kubo relations). However this approach, if conceptually satisfying, requires ad hoc dynamic variables coupled to viscosities. The most common approach to the non hydrodynamic behavior of u.s. wave propagation is then made through thermodynamic relaxation theories. In such case one considers specific relaxation processes coupled to sound waves. Temperature and pressure changes produced by u.s. waves can modify thermal and/or mechanical equilibria in the medium giving rise to relaxation processes where elastic energy is dissipated.

In the relaxation theories one obtains frequency dependent specific heats and/or compressibility moduli which are put in the hydrodynamics laws. It is assumed that such approach is equivalent to the generalized hydrodynamics laws (this equivalence was demonstrated by Zwanzig [6] in the case of thermal relaxation processes).

The relaxation processes which can be studied by ultrasonics depend on the nature of the fluid and on the characteristic times (relaxation times) of these processes compared to the u.s. frequency range. Examples of equilibria which can be perturbed by u.s. waves in a fluid are:

a) equilibria between external (translational and rotational) and internal (vibrational) degrees of freedom in weakly interacting polyatomic fluids;
b) rotational isomerism;
c) chemical reactions in chemically active media;
d) configurational or structural equilibria among different molecular arrangements in strongly interacting media, e.g. associated hydrogen bonded liquids, polar liquids, and so on;
e) concentration fluctuations in binary mixtures.

As an example of concepts and results involved in relaxational theories, let us briefly consider the <u>structural relaxation</u> [4][5] occurring in water, alcohols and other associated liquids. As a consequence of pressure changes accompanying the u.s. waves, it is assumed that the volume change consists of two terms: a) a change

(δV_∞) due to an uniform reduction of the molecular distances. This contribution, which is similar as in crystals, is very rapid and can be considered almost istantaneous; b) a delayed change ($\delta V_r(t)$) due to molecular rearrangements. This process (configurational or structural) takes a finite time because it needs the breaking of molecular bonds and/or steps to overcome potential barriers. By generalizing the statistical thermodynamics relation for the isothermal compressibility β_T one obtains a complex frequency dependent $\tilde{\beta}_T(\omega)$

$$\tilde{\beta}_T(\omega) = \beta_T(o) - \frac{i\omega}{k_B TV} \int_o^\infty dt\, e^{-i\omega t} <\delta V(o)\ \delta V(t)> \tag{4}$$

in terms of an integral of time correlation functions of volume fluctuations. In the case of an exponential decay of fluctuations $\delta V_r(t) = \delta V_r(o)\exp(-t/\tau_V)$ (single time structural relaxation), one obtains

$$\tilde{\beta}_T(\omega) = \beta_T(o) - i\omega \frac{\tau_V \beta_r}{1+\omega\tau_V} \tag{4'}$$

Putting this expression in the hydrodynamics laws one gets $c(\omega)$ and $\alpha(\omega)$. An equivalent procedure is to consider a complex sound velocity

$$\tilde{c}(\omega) = \frac{\omega}{\tilde{k}} = \frac{\omega}{k - i\alpha} \simeq \frac{\omega}{k} + i\,\frac{\alpha\omega}{k^2} \tag{5}$$

and to use the relation $\tilde{c}^2(\omega) = \gamma \tilde{K}_T(\omega)/\rho$ where the complex frequency dependent bulk modulus $\tilde{K}_T = (1/\tilde{\beta}_T)$ in the case of single relaxation process takes the form

$$\tilde{K}_T(\omega) = (K_o + \frac{K_r \omega^2\ \tau_V^2}{1+\omega^2\ \tau_V^2}) + i\ K_r\, \frac{\omega\ \tau_V}{1+\omega^2\ \tau_V^2} \tag{6}$$

One then obtains:

$$c^2(\omega) = \rho^{-1}(K_{o,s} + K_{r,s}\, \frac{\omega^2\ \tau_V^2}{1+\omega^2\tau_V^2}) \tag{7}$$

$$\alpha(\omega) \simeq \frac{\omega^2}{2\rho c^3}\ \frac{K_{r,s}\ \tau_V}{1+\omega^2\tau_V^2} \tag{8}$$

The corresponding behaviors are shown in Fig.1. Comparing with the hydrodynamics predictions (Eqs.1-2), one can see that this approach is formally equivalent to a bulk viscosity $\eta_B(\omega)$ given by

$$\eta_B(\omega) = \frac{K_{r,s}\tau_V}{1+\omega^2\tau_V^2} \quad (9)$$

This expression is analogous to the frequency dependent shear viscosity $\eta_s(\omega)$

$$\eta_s(\omega) = \frac{G_\infty \tau_s}{1+\omega^2\tau_s^2} \quad (10)$$

as obtained from viscoelastic theories (G_∞= shear modulus; τ_s= Maxwell or shear relaxation time).

From Eqs.7-8 we can see that for $\omega\tau_V<<1$ (hydrodynamic conditions) it is $c^2(o)=K_{o,s}/\rho$ and $(\alpha/f^2)_{\omega\to o} =(2\rho c_o^3)^{-1}\eta_B(o)$ while in the limit $\omega\tau_V>>1$ (solid-like conditions) one obtains $c_\infty^2 =(K_{o,s}+K_r)/\rho=K_{\infty,s}/\rho$ and $(\alpha/f^2)_{\omega\to o}\to 0$.

3. SOUND PROPAGATION IN ALCOHOLIC SOLUTIONS

3.1. Experimental results

3.1.1. The low frequency sound velocity behavior[7].

A typical example of the sound velocity anomalous composition dependence in these systems is shown in Fig.2 which refers to ethanol solutions. Correspondingly, the isentropic compressibility isotherms display well defined minima which disappear at high T. This eccentric behavior, according to which the compressibility of the system becomes lower than in pure solvent in spite of the higher value in the added solute, is a clear indication of a structural change produced in water. Other interesting aspects shown from the figure are: i) the noticeable increase of $c_o(x)$ in diluted solutions, expecially at low T; ii) the negative temperature coefficient of the c_o peak values and the slight shift of the corresponding concentrations (x_{PSV}) toward higher values as T decreases.

A comparison with similar isotherms in different alcohol solutions shows, in addition, other characteristic features. For example, the c_o peak values (at fixed T) are slightly dependent on the type of alcohol while the corresponding x_{PSV} strongly depend on the "size" of the hydrophobic groups. Characteristic values in monohydric alcohols at 25°C are shown in the Table.

TABLE 1 - Characteristic concentrations of the sound velocity (x_{PSV}) and absorption (x_{PSA}) peak values in aqueous solutions with monohydric alcohols at 25°C. c_{peak} is the sound velocity peak value (c_{water} = 1498 m/s)

Alcohol	c_{peak}(m/s)	x_{PSV}	x_{PSA}
Methanol	1570	0.16 - 0.19	0.35
Ethanol	1610	0.10 - 0.12	0.24
Iso-Propanol	1610	0.085	0.18
n-Propanol	1590	0.05	0.14
t-Butanol	1600	0.04	0.11

3.1.2. The u.s. absorption behavior [12,13,14].

Even the u.s. attenuation is quite anomalous in alcohol aqueous solutions. It is characterized by the following features:

a) α/f^2 vs x (at fixed T and f) displays peak values at compositions x_{PSA} which are always higher than x_{PSV} (see Fig.3 and Table). Peak values noticeably increase as T decreases with slight changes in the corresponding x_{PSA}. These characteristic concentrations depend on the added alcohol: in solutions with monohydric alcohols they regularly decrease on going from methanol to t-butanol (see Table). In critical solutions (e.g. butoxyethanol) x_{PSA} is near (but not at) the critical concentration x_c;
b) the u.s. absorption (at fixed x and T) exhibits a non-hydrodynamic behavior around x_{PSA}. Such effect noticeably increases as T decreases toward the freezing point temperature T_F. (Fig.4);
c) at fixed f and T (e.g. at room temperature) the absorption peak values largely depend on the type of alcohol. However, as $T \rightarrow T_F$ the attenuation shows, as a rule, a diverging-like behavior whatever the alcoholic solution (see Fig.5);
d) in critical solutions (e.g. butoxyethanols) the u.s. attenuation is much larger near T_F rather than at the critical point;
e) although less pronounced, dispersion effects are also evident in the sound velocity. However, these effects are larger at x_{PSA} rather than at x_{PSV} and become considerable as $T \rightarrow T_F$.

As a conclusive remark, we observe that in water-alcohol solutions the overall behavior of c and α/f^2 as a function of x,T and ω shows an extraordinary resemblance with that observed in critical binary mixtures (either with LCST or UCST), x_{PSA} and T_F taking the places of the critical parameters x_c and T_c.

3.2. Discussion

The existence and the different location of the u.s. velocity and absorption peaks strongly suggest distinct structural conditions at x_{PSV} and x_{PSA}. This conclusion is corroborated by other physical properties which exhibit extrema near these two concentrations. We will attempt to rationalize the rather complex u.s. experimental behavior according to the following picture.

3.2.1. The sound velocity [7]. The sudden increase of the sound velocity upon addition of small amount of alcohol (Fig.2) can be explained on the grounds of hydrophobic interactions. According to this picture, the apolar groups of the added solute stimulate or enhance the formation of tetraedrically coordinated H-bonded water structures around the solute molecule. These structures could correspond, for example, to short-living clathrate-like arrangements as suggested by Glew [8] and Franks [9] or to unstrained H-bonded cages or polyedra structures as proposed by Stillinger [10] for supercooled water. However, these filled cage-like structures in the solutions should be denser and more rigid than the empty structures in pure water thus decreasing the overall compressibility. A similar conclusion might be driven by using the Stanley-Teixeira [11] arguments which relate the anomalous compressibility of low T pure water to large density fluctuations.

Hydrophobic interaction can qualitatively explain most of the sound velocity characteristic features listed in Sec.3.1.1. In particular, it explains the apolar group "size" dependence of x_{PSV} and the almost independent c_o peak values on the type of alcohol (see Table). The existence of a minimum compressibility at x_{PSV} suggests that the encagement of the solute molecules reach a maximum at this concentration and that a new type of water alcohol structures should be formed for $x > x_{PSV}$. A crude two-state model [7] show indeed that the volumetric properties (i.e. densities and compressibilities) of ethanol solutions in the range $0<x<x_{PSV}$ can be explained by means of cage-like structures whose properties are similar as in the stable solid clathrate hydrates. However, in the range $x_{PSV}<x<x_{PSAC}$ this model requires volumetric properties associated to water-alcohol structures where the solute molecules are no more filling cavities but rather collaborate as constitutive parts in forming small mixed aggregates. The concentration x_{PSA} should then characterize a limit for the stability of such new aggregates: for $x > x_{PSAC}$ the amount of water is not enough (and/or the alcohol-alcohol interaction so strong) to substain aggregation so that the new structures should collapse giving rise to "regular" solutions where more and more "free" alcohol molecules are present.
As we'll show in the next Section, the dynamics of u.s. attenuation support this picture.

3.2.2. The u.s. absorption. Early attempts to explain the peculiar behavior of the u.s. absorption in water-alcohol solutions were based on the so-called quasichemical models where the u.s. losses and the non-hydrodynamic behavior were linked to the perturbation of chemical equilibria among specific water-alcohol aggregates [12]. These models predict discrete relaxation processes (mainly one or two) whose characteristic times are related to the reaction rates of molecular aggregations. Although with some limitations (strengths and relaxation times cannot be probed independently), the experimental frequency dependence apparently conformed to such models. However, the good agreement of u.s. spectra to discrete relaxation times equations was a consequence either of the narrow frequency range investigated or of the small relaxation strenghts such as observed at room temperature. Investigations over extended frequency and temperature ranges in ethanol [13,14], propanol [15], ethoxyethanol [16] and butoxyethanol [16] solutions show, in fact, that as temperature decreases toward T_F, the u.s. dynamics shift from a discrete to a continuum distribution of relaxation times. This shift is accompanied by a very large u.s. attenuation and by a slowing-down of the characteristic times. On accounting of these findings and of the previously suggested presence of mixed aggregates around x_{PSAC}, concentration fluctuations models have been considered. These models relates the u.s. dynamics to a sound wave-diffusion coupling mechanisms: changes of local equilibrium concentration fluctuations produced by the pressure and temperature variations accompanying the sound wave decay according to a diffusion equation giving rise to energy dissipation. Since every component (of wavevector k) of the Fourier expansion of the spatial concentration fluctuations decays with a characteristic time $\tau_k = 2Dk^2$ (D = mutual diffusion coefficient), a continuum spectrum of relaxation times is expected.

We compared [13,16] the low temperature u.s. spectra in our solutions against two approaches of this model, namely the Romanov-Soloviev (R-S) [17] and the Ferrell-Bhattacharjee (F-B) [18] theories. In the first case the calculation of the u.s. absorption and velocity is accomplished by evaluating the contribution to the volume of small scale concentration fluctuations and then extracting a complex frequency dependent bulk modulus. The explicit ω and T dependence of u.s. attenuation predicted by the R-S approach can be put in the form [13]

$$\frac{\alpha}{f^2}(\omega,T) = A_{RS}(T)\ F_{RS}(\omega^*) \qquad (11)$$

where: $A_{RS}(T)$ is a quantity related to the diffusion coefficient D and to the volume and enthalpy changes due to concentration fluctuations; ω^* is a reduced frequency given by

$$\omega^*(\omega,T) = \frac{\omega}{D(T)} = \frac{\omega}{2D\ \xi_{min}^{-2}} \qquad (12)$$

and $F_{RS}(\omega^*)$ is a scaling function which comes out from the assumption of a Debye distribution for the k-Fourier components of concentration fluctuations.

On accounting of the noticeable resemblance of the u.s. behavior near T_F with a critical behavior, we also compared our results with the F-B theory which is tailored for binary critical mixtures. Even if the predicted u.s. behavior takes a form similar to the Eq.(3.1), the F-B treatment differs from the R-S one in that it uses a complex frequency dependent specific heat and takes into account for dynamic scaling requirements in critical phenomena by considering an appropriate scaling function $F_{FB}(\omega^*)$. The F-B theory gives, in addition, a very simple relation among the u.s. absorption at fixed frequency and $\Delta T = T-T_c$, where T_c is the critical temperature.

Comparisons in different alcohol solutions near x_{PSA} show that as temperature decreases toward T_F, the frequency dependence is well accounted by either R-S or F-B predictions. However, very near to T_F the F-B theory gives the best quality fit to the experimental results but the temperature trend does not conform to its predictions. Such findings show that the u.s. dynamics near the freezing point is similar, but not the same, as in critical mixtures. We are then tempted to link the u.s. anomalies in alcohol aqueous solutions to peculiar first-order phase transition phenomena near x_{PSA}. The occurrence of critical-like or "pseudocritical" phenomena near a first-order phase transition of alcohol solutions has been in the past invoked on both experimental and theoretical grounds.

From a structural point of view these results might be rationalized by supposing that the water-alcohol aggregates previously envisaged near x_{PSA} will tend both to increase in number or to clump as temperature decreases giving rise to long-range microstructures and then to large concentration fluctuations. This structural picture, which looks quite similar to that proposed by Stillinger to explain singularities in pure supercooled water, needs to be substanciated and characterized by further experimental and theoretical investigations.

REFERENCES

[1] Franks F. and Reid D.S. (1973) in Water: A Comprehensive Treatise edited by F.Franks (Plenum, N.Y.) Vol.II, Chap.5

[2] Mc Skimin H.J. (1964), "Ultrasonic Methods for Measuring the Mechanical Properties of Liquids and Solids", in Physical Acoustics - Principles and Methods Vol:I Part A Ed. by W.P.Mason - Academic Press; Beyer R.T. and Letcher S.V. (1969), Physical Ultrasonics, Acad. Press N.Y.-London; Matheson A.T. (1971), Molecular Acoustics , Wiley Intersc., London

[3] Landau L.D. and Lifshitz E.M. (1971) Mechanique des Fluides Editions MIR Moscow; Berne B.J. and Pecora R. (1976), Dynamics Light Scattering Wiley and Sons N.Y.

[4] Hertzfeld K.F. and Litovitz T.A. (1959), "Absorption and Dispersion of Ultrasonic Water" Academic Press N.Y., London

[5] Blandamer M.I. (1973), Introduction to chemical ultrasonics Academic Press, London

[6] Zwanzig R. (1965), J.Chem.Phys. 43, 714

[7] see, for example, D'Arrigo G. and Paparelli A. (1988), J.Chem.Phys.88, 405 and references cited therein.

[8] Glew D.N. (1962), J.Phys.Chem. 66 605

[9] Franks F. and Johnson H.H. (1962), Trans.Faraday Soc 58, 656; ibid. (1968) 64, 2962

[10] Stillinger F.H. (1980), Science 209, 451

[11] Stanley H.E. and Teixeira J. (1980), J.Chem.Phys. 73,, 3404

[12] see, for example, Blandamer M.I., in "Water; A Comprehensive Treatise" edited by F.Franks (Plenum, N.Y.) Vol.II, Chap.3 and references cited therein
[13] D'Arrigo G. and Paparelli A. (1988), J.Chem.Phys. 88, 7687
[14] Benassi P., D'Arrigo G. and Nardone M. (1988), J.Chem.Phys. 89, 4469
[15] Madigosky W.M. and Warfield R.W., (1987), J.Chem.Phys. 86, 1491
[16] D'Arrigo G. and Paparelli A. (to be published)
[17] Romanov V.P. and Soloviev V.A. (1965), Sov.Phys.Acoust.11, 68 269
[18] Farrell R.A. and Bhattacharjee J., (1985) Phys.Rev.A31, 1788

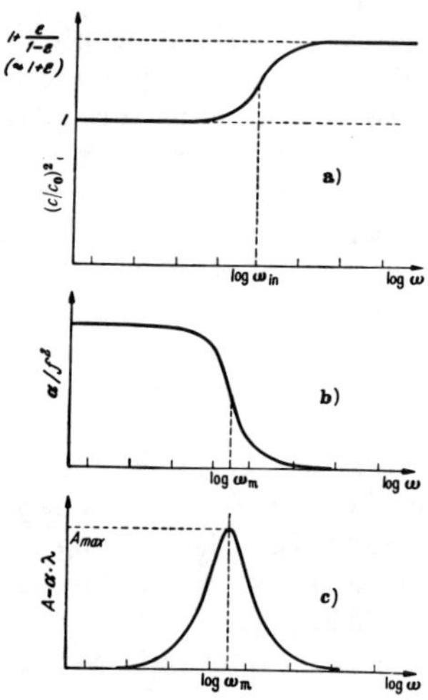

Fig.1. c^2/c_o^2, α/f^2 and $\alpha\lambda$ vs f for a single relaxation process.

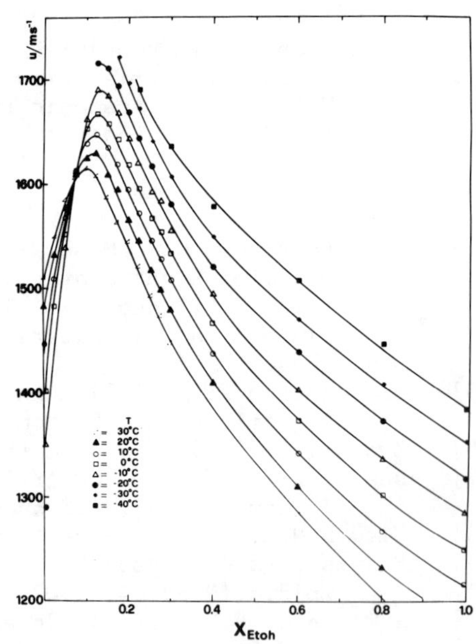

Fig.2. c_o vs x isotherms in ethanol solutions.

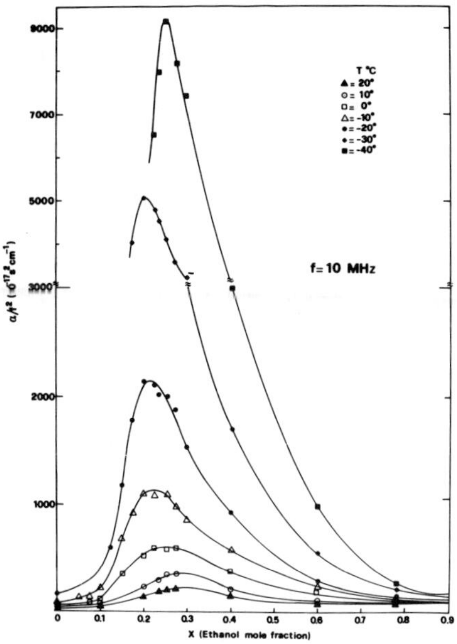

Fig.3. α/f^2 vs x isotherms in ethanol solutions

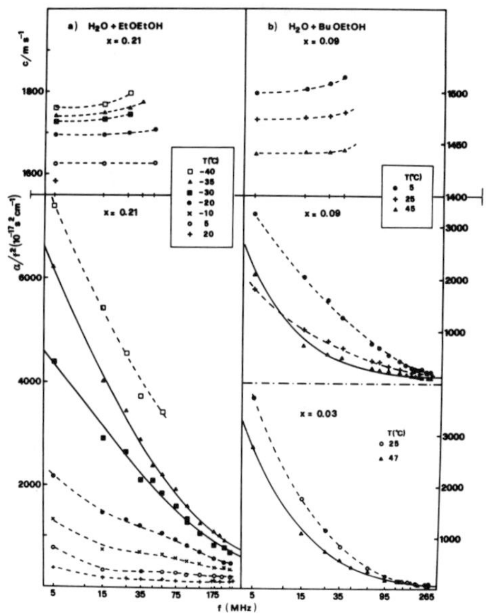

Fig.4. α/f^2 vs f in ethoxyethanol and butoxyethanol solutions at x_{PSA}

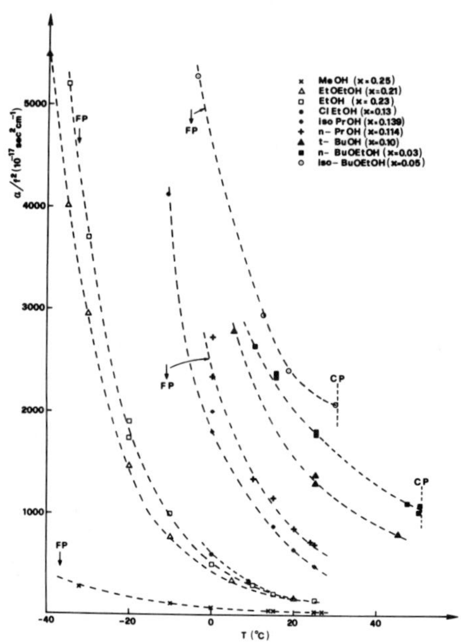

Fig.5. α/f^2 vs T in some alcohol solutions at x_{PSA} (15 MHz)

MULTI-NUCLEAR RELAXATION TIME STUDIES IN UNDERCOOLED AQUEOUS ELECTROLYTES

E. W. Lang, W. Fink and H. Radkowitsch
Institut für Biophysik
Universität Regensburg, POB 397
D-8400 REGENSBURG, FRG

ABSTRACT

The dynamic properties of water molecules coordinated to the ions Li^+ and F^- in undercooled LiCl and KF solutions have been investigated with NMR. Relaxation time studies at low temperatures provide details of molecular motions as well as certain features of the local arrangement in the coordination sphere of these ions.

1. INTRODUCTION

NMR relaxation rate studies of various nuclei are well suited to study the rotational and translational diffusion of molecules in liquids. These diffusional processes are fast on the NMR time scale (≈ns). Thus details can only be revealed if molecular motions can be slowed down sufficiently. This can often be accomplished in undercooled liquids.

Hydrostatic pressure and/or ionic solutes disturb the random, transient hydrogen bonded network of liquid water. Crystallization is inhibited strongly and water behaves in many ways like a normal hydrogen bonded liquid (1). At high solute concentrations aqueous LiCl solutions are glassforming (2), whereas aqueous KF solutions do not seem to form glasses readily.

Both solutions are ideal systems for NMR studies because in addition to water nuclei ^6Li, ^7Li and ^{19}F can also be studied easily.

Whereas molecular motions are isotropic in pure water, dissolved ions induce some anisotropy in the reorientational motions of coordinated water molecules (3). A simple motional model is discussed which, together with a two-site approximation, is used to interpret relaxation rate curves $R_1(p,T,c,\omega)$ of the nuclei ^2H, ^1H, ^6Li, ^{19}F in aqueous solutions of LiCl and KF.

J.C. Dore and J. Teixeira (eds.), Hydrogen-Bonded Liquids, 393–404.

2. NUCLEAR SPIN RELAXATION and MOLECULAR MOTIONS

In liquid water and aqueous electrolyte solutions molecular motions occur on two largely separated time scales. Hindered translations and rotations can be considered as thermal excitations of the quasi-static H-bond network (4). At high temperatures these quasi-lattice vibrations (QLV) are strongly damped. At lower temperatures they are much faster than translational and rotational diffusive motions. Because of time-scale separation QLV and diffusive motions can be considered to be statistically independent. They may be characterized by appropriate correlation functions G(t) of the fluctuating orientational and/or spatial variables. The spin-lattice relaxation rate R_1 is related to the Fourier-Laplace transform of G(t), the spectral density

$$g(\omega_I,\tau) = \int G(t)\exp(i\omega_I t)dt \tag{1}$$

via(5)

$$R_1(\omega_I,T,p,c) = (1/T_1) = \langle\chi\rangle_{QLV} \cdot F(\omega_I,\tau(T,p,c)) \tag{2}$$

The factor $\langle\chi\rangle$ - averaged over the QLV - reflects the coupling of the spin system to the lattice degrees of freedom via the perturbing, time-dependent spin-lattice interaction providing the relaxation mechanism. Only electric quadrupole - electric field gradient (efg) and magnetic dipole - dipole couplings will be of relevance in the following. $F(\omega,\tau)$ represents a combination of spectral density functions $g(\omega\tau)$. Explicit expressions are given in the paper by Lang and Lüdemann in this volume (12), hence they will not be reproduced here. For practical reasons $g(\omega\tau)$ can be sampled at a few resonance frequencies ω_I of the nucleus at most, thus the entire spectrum of diffusive motions cannot be obtained. A motional model is therefore necessary to provide the spectral density of states $g(\omega\tau)$.

2.1 A MOTIONAL MODEL OF THE HYDRATION WATER DYNAMICS

Neutron (6) and X-ray (7) scattering as well as computer simulations (8) provide details of the average local structure of the coordination sphere of ions. Water molecules always orient their hydrogen atoms away from cations, their dipole moment vector being tilted relative to the radial direction. Towards anions water molecules orient one of the OH-bonds in an almost linear fashion. The width of the peaks in the corresponding radial distribution functions indicate the range of angles and distances allowed by the given interaction potentials. Consistent with this average local structure reorientational fluctuations of hydration waters are decomposed as follows. The molecules execute small amplitude librations about their mean orientation. These fast torsional oscillations are superimposed onto anisotropic fluctuations around the local director (ion - oxygen distance) with correlation time τ_i. In addition the molecules tumble around the ion with correlation time τ_r or leave the coordination sphere after a mean residence time τ_{ex}. Both modes are highly cooperative and will be combined to an effective tumbling mode $1/\tau_0 = 1/\tau_r + 1/\tau_{ex}$.

2.2 CORRELATION FUNCTION AND RELAXATION RATE OF HYDRATION WATER

With this motional model in mind an appropriate time correlation function G(t) of the orientational fluctuations of hydration water molecules has been derived (5)

$$G(t) = \langle G(0) \rangle_{QLV} \cdot \Sigma |D^2_{m0}(\Omega_{DP})|^2 \exp(-m^2 t/\tau_i) \exp(-t/\tau_0) \quad (3)$$

$D^2_{m0}(\Omega_{DP})$ are elements of the Wigner rotation matrix with Ω_{DP} representing the orientation of the principal axis system (P) of the spin-lattice coupling relative to the local director frame(D) This expression rests on the following assumptions:

a) time-scale separation pertains between QLV and diffusive modes
b) the anisotropic local mode and the tumbling modes are statistically independent
c) the spin-lattice coupling is completely randomized on exchange
d) an exponential decay of the partial orientational correlation functions prevails for times $t > \tau$

The corresponding spectral intensities are obtained via equ. 1 and have to be inserted into equ. 2 to obtain the hydration water relaxation rate $(R_1)_{hyd}$(12).

2.3 TEMPERATURE DEPENDENCE OF THE CORRELATION TIMES

In undercooled liquids cooperative motions slow down strongly causing, in fragile liquids (11), a characteristic non-Arrhenius temperature dependence of transport and relaxation coefficients. This behaviour is well represented by the empirical VTF-equation (12)

$$\tau_0 = \tau_{00} \exp(B/(T - T_0)) \quad (4)$$

It predicts a slowing down of cooperative molecular motions at the ideal glass transition T_0 and can be deduced from approximate theories (13) which either consider the accessible free volume per particle or the configurational entropy of the metastable liquid as the factors which control diffusion and flow.

The local anisotropic reorientation of the water molecules around the local director can be considered a thermally activated mode with a characteristic Arrhenius behaviour

$$\tau_i = \langle 2\pi/\omega_{lib} \rangle \cdot \exp(E_a/kT) = \tau_{io} \cdot \exp(E_a/kT) \qquad (5)$$

The preexponential factor corresponds to an attempt frequency for barrier crossing, hence is given by an average librational frequency of hydration water molecules (14).

3. MULTI-NUCLEAR RELAXATION EXPERIMENTS

Deuterium relaxation rates $R_1(^2H)$ have been measured in various undercooled aqueous alkali and alkali earth halide solutions (5,15-17). These rates monitor reorientational fluctuations of the OD-bond only. Hence these investigations provide suitable tests of the motional model proposed. The two unknowns (E_a, τ_{00}) have been adjusted to the experimental data sampled at high field (B_0=7 Tesla, $\omega_D/2\pi$ = 46 MHz). Then relaxation rate curves obtained at different resonance frequencies ω_D and/or in more dilute solutions have been calculated. It could be demonstrated, that the model can reproduce experimental relaxation rate curves sampled over the whole concentration range and over a large temperature range rather well (5,15-17). These investigations have been extended to other nuclei recently (18).

3.1 UNDERCOOLED AQUEOUS KF SOLUTIONS

Besides the proton (^1H) and deuteron (^2H) relaxation, the ^{19}F relaxation can also be studied conveniently in this system. Proton and fluorine nuclei relax via magnetic dipole-dipole interactions. Their relaxation rate thus monitors reorientational and translational diffusive motions. Furthermore the principal axis systems of these spin-lattice couplings have different orientations relative to the local director frame. The aim of these studies was to see how well the relaxation rate curves of these nuclei may be reproduced with the model parameters adjusted to the deuteron relaxation data only. These investigations should also try to answer the recent controversy concerning the orientation of water molecules relative to an F^- anion (19-21). The fluoride anion interacts with water much stronger than the potassium cation. The latter disturbs the water dynamics only marginally. Hence a 12 molal KF solution, corresponding to a composition R(moles water/mole salt) $\simeq$ 4, may be considered to contain only hydration water of the F^- anion. The model parameters have been adjusted to the deuteron relaxation data sampled at 46 MHz and are compiled in table 1. An almost linear orientation of the water molecules relative to the F^- anion has been assumed. The resulting relaxation rate curve is, however, not very sensitive to the average orientation. Thus deuteron relaxation rate curves cannot distinguish between a linear and an almost symmetric arrangement.

TABLE 1
Compilation of model parameters entering the relaxation rate expressions (see ref.12 for details)

	τ_{00} (ps)	τ_{i0} (ps)	E_a (kJ/mol)	B (kJ/mol)	χ_{eff} (kHz)	R_{HH} (nm)	R_{HF} (nm)
H_2O	0.200	0.047	21	5.46		0.16	0.18
D_2O	0.200	0.065	21	5.09	185		
			T_0 (K)			d_{HH}	d_{HF}
C(molal)		1	3	8	12	(nm)	(nm)
H_2O		127	130	137	144	0.28	0.35
D_2O		134	137	144	152		

Relaxation rate curves for more dilute solutions have been calculated then within the two-site approximation (equ.13 of ref.12) and are shown in figure 1. The pronounced curvature of the $\ln R_1$ versus $1/T$ diagram for $T>T_{max}$ is due to the dominance of the cooperative tumbling motions. At low temperatures ($T<T_{max}$) the strong slowing down of co-operative motions is hidden behind the weaker temperature dependence of the local anisotropic mode, which takes over to determine the relaxation rate for $T<T_{max}$ (see equ.12 of ref.12 in the limit $\tau_i<<\tau_0$ for $T<T_{max}$ where $\omega\tau>>1$ holds)
The relaxation of the protons is governed by dipolar interactions. Within the coordination sphere these couplings are modulated by reorientational fluctuations of the intramolecular ^1H-^1H vectors and intermolecular ^1H-^{19}F vectors. The corresponding spectral density functions are given by equ.14 and equ.17 of ref.12. They are calculated with the model parameters as deduced from the deuterium relaxation suitably scaled according to the dynamic isotope effect (18,23). Positional fluctuations also modulate the dipolar interaction with the protons outside the coordination sphere. This intermolecular ^1H-^1H contribution to the proton relaxation rate is calculated within a translational diffusion model with reflecting boundary conditions. The appropriate spectral density function is given in equ.16 of ref.12. The diffusion coefficients of ^1H and ^{19}F are known at ambient temperature and pressure only (19). They have been assumed to follow a VTF-equation also. The distances of closest approach have been taken from radial distribution functions calculated with molecular dynamics techniques (21,22). All parameters are compiled in table 1. Figure 2 shows the proton relaxation rate curve as calculated with an almost linear orientation of the water molecules (full line) or with a symmetric arrangement (broken line). Only the ^1H-^{19}F distances and

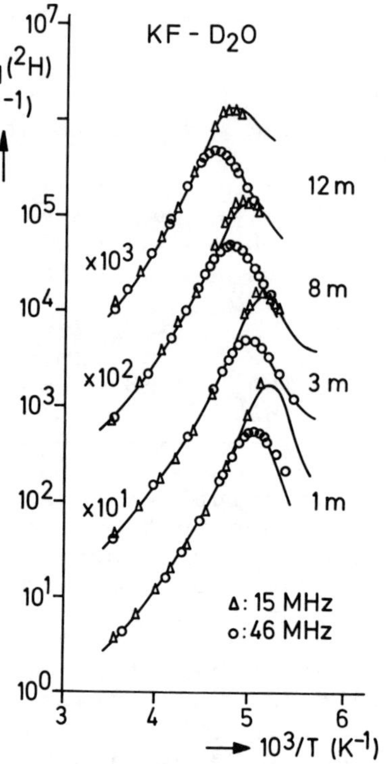

Figure 1. ^2H-relaxation rates in undercooled KF/D_2O solutions (p=225 MPa). Full lines are calculated as explained in the text.

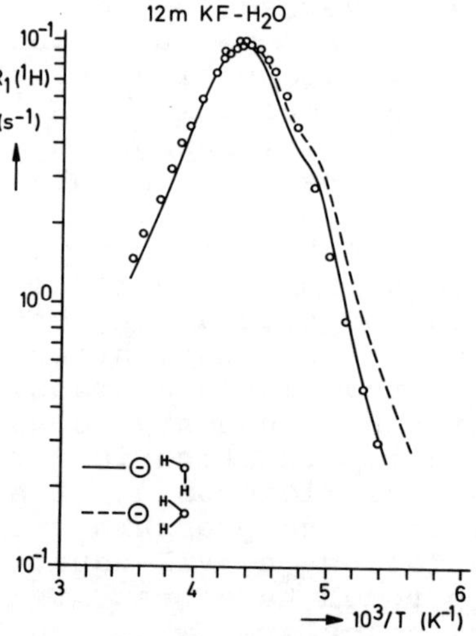

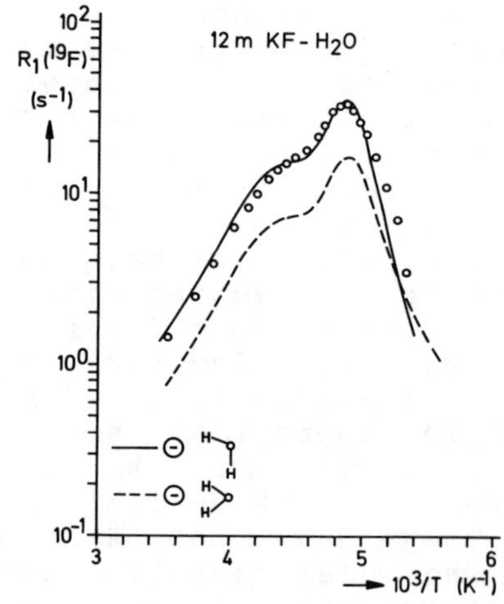

Figure 2. ^1H-relaxation rate in a 12m undercooled KF/H_2O solution.

Figure 3. ^{19}F-relaxation rate in a 12m undercooled KF/H_2O solution.

Full (broken) line in both figures correspond to a linear (symmetric) arrangement of the hydration waters of F^-.

the angles entering the amplitude factors $|D_{m0}(\Omega_{DP})|^2$ in equ. 2 have been changed accordingly. A comparison with the experimental data favours the linear configuration, although the resulting relaxation rate curves do not differ much. Any difference is apparent at low temperatures only and is due to the ^1H-^{19}F interaction, which causes a little shoulder in the relaxation rate curve and appears to be more pronounced in the symmetric arrangement. Obviously the model needs further improvement because the subsidiary rate maximum due to ^1H-^{19}F interactions is less clearly developed in the experimental data. Within the two-site approximation the relaxation rate curves $R_1(^1H)$ for the more dilute solutions can be reproduced also satisfactorily (23).

Within a reasonably long-lived ($\tau_{ex} \geq \tau_r, \tau_i$) hydration sphere the ^{19}F nucleus is relaxed mainly via magnetic dipole couplings to the protons of the hydration waters. This coupling is modulated by reorientational fluctuations of the ^{19}F-^1H vectors and depends sensitively (via $1/R_{FH}$**6) on the average ^{19}F-^1H distance. This rotational contribution $R_1(^{19}F$-$^1H)$ is identical to the rate $R_1(^1H$-$^{19}F)$ contributing to $R_1(^1H)$ except that one has to sum over all protons in the hydration sphere. In addition a translational contribution due to dipolar couplings of the ^{19}F nucleus to protons outside the hydration sphere has been taken into account. All parameters are compiled also in table 1. Again, figure 3 compares the relaxation rate curve $R_1(^{19}F)$ for the 12 molal solution calculated with an almost linear configuration (full line) and with a symmetric orientation of the hydration water (broken line). The experimental data clearly favour the almost linear arrangement. The maximal rate $R_1(T_{max})$ is proportional to $(1/R_{FH})$**6 and is underestimated considerably with the symmetric arrangement due to a larger ^{19}F-^1H distance (0.23 nm versus 0.18 nm). It is of interest also that the calculation with the linear configuration is not sensitive to the local anisotropic mode. The latter modulates only the coupling of the ^{19}F nucleus to the more distant proton. Because of the large distance (0.286 nm) this contribution may be neglected. The two maxima in the relaxation rate curve correspond to spectral density functions $g(\omega_F+\omega_H)$, $g(\omega_F)$ and $g(\omega_F-\omega_H)$ respectively with $\omega_F/2\pi$=282 MHz and $\omega_H/2\pi$=300 MHz (note that at every T_{max} $\omega\tau(T_{max})\simeq 1$ holds). Again relaxation rate curves $R_1(^{19}F)$ corresponding to more dilute solutions can be reproduced within the two-site approximation fairly well (23).

3.2 UNDERCOOLED LiCl SOLUTIONS

Aqueous LiCl solutions provide the opportunity to study the dynamic behaviour of undercooled liquids, which range

from very difficult to vitrify to excellent glass formers by increasing the solute concentration. Outside the glass-forming composition range ($R \geq 5$) hydrostatic pressure ($p \simeq 200$ MPa) must be applied to provide the maximum possible undercooling. The system has been thoroughly studied over the whole p,T,c-range with deuterium NMR and the results have been discussed in terms of the current motional model (5,9). The relevant parameters for the calculation of the relaxation rates $R_1(^2H)$ are compiled in table 5 of ref.12.
These investigations have been extended to other nuclei in a 11 molal solution, corresponding to a composition $R \simeq 4$ recently (18). All water molecules in this solution may be considered to belong to the hydration sphere of the Li^+ cations. Hence this system is well suited to deduce the parameters of the rotational model and to test the ability of the model to reproduce the relaxation rate curves of various nuclei. The relaxation rates of ^1H, ^6Li, ^7Li and ^{35}Cl have been measured at ambient pressure at temperatures $T \geq T_g$ and are shown in figure 4. The results have been interpreted in terms of the current model thoroughly (5,9,18), hence only a short discussion may suffice.
The relaxation of ^1H and ^6Li nuclei is governed by dipolar couplings despite of the fact that ^6Li (I=1) possesses an, albeit exeedingly small, nuclear quadrupole moment. This can be deduced directly from the two minima in the relaxation time curve $T_1(T)$ of ^6Li dissolved in D_2O. These minima arise from spectral density functions $g(\omega_{Li})$, $g(\omega_{Li}+\omega_D)$ and $g(\omega_{Li}-\omega_D)$ respectively ($\omega_{Li}/2\pi$=44.17MHz and $\omega_D/2\pi$=46.07MHz) entering the relaxation rate expression due to a dipolar coupling of spins I,S (see equ.17 of ref.12). ^2H, ^{35}Cl and ^7Li nuclei interact exclusively via electric quadrupole electric field gradient (efg) couplings. The latter is of intramolecular origin for the ^2H nuclei, but is due to the surrounding water dipoles for the ^7Li nuclei (24). Note that for a highly symmetric arrangement of the water dipoles around the Li-ion the efg would almost vanish. The relaxation time curves $T_1(^7Li)$ in light and heavy water superimpose if the temperature is scaled by $T(H_2O)=T(D_2O)-\Delta T_g$, the difference in calorimetric glass temperatures (25) to account for the isotope effect. This is a direct proof of the absence of any contribution from dipolar couplings to the ^7Li relaxation, which ,however, is known to contribute in more dilute solutions (3,24). The observation that all minima in the relaxation time curves of the various nuclei occur at similar temperatures proofs that the same molecular motions are responsible for the various spin-lattice interactions.
The local structure of the coordination sphere of Li^+ is well known (6-8,26). Water molecules show a tetragonal

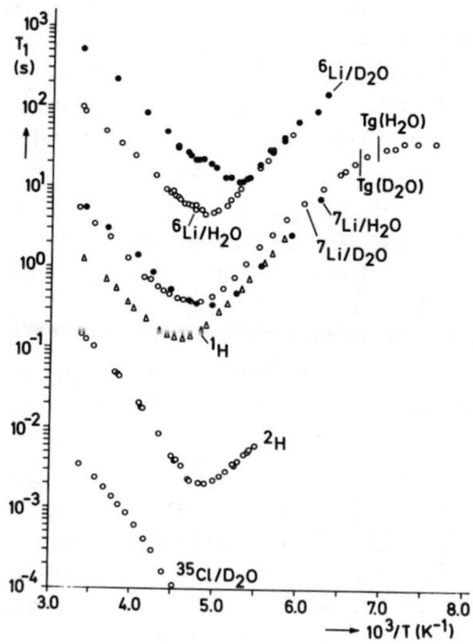

Figure 4. Multi-nuclear spin-lattice relaxation times in undercooled aqueous LiCl solutions.

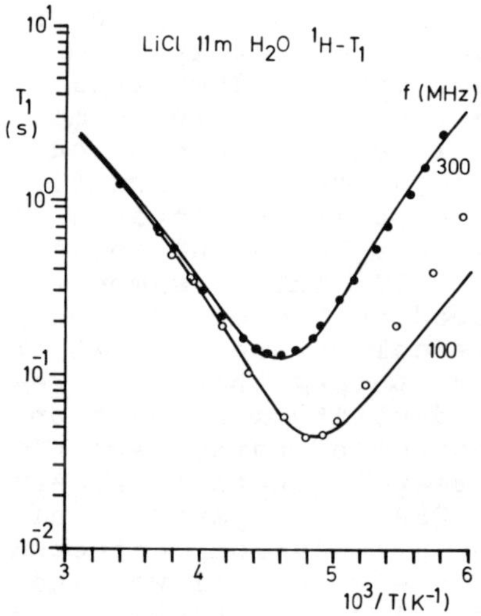

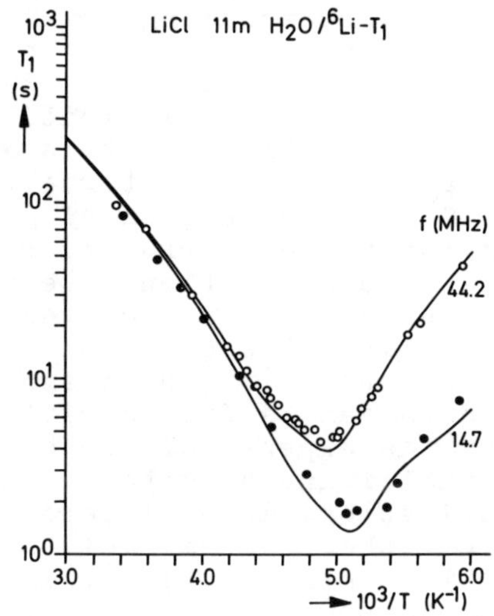

Figure 5. ^1H-T_1 in undercooled 11m $LiCl/H_2O$

Figure 6. ^6Li-T_1 in undercooled 11m $LiCl/H_2O$

Full lines in both figures are calculated according to motional model discussed in the text.

orientation in the hydration shell with the protons pointing away from the cation and the dipole moment vector tilted away from the radial direction.
The diffusion coefficients of the protons and lithium nuclei, entering the intermolecular relaxation rate contribution (equ.16 of ref.12), have also been measured to very low temperatures. The diffusion coefficient D(H) is larger than D(Li) indicating an extra proton mobility over that of the water molecule due to a short mean exchange lifetime of the protons in accord with conductivity measurements.
Taking the model parameters as deduced from $R_1(^2H)$ and with all distances and angles taken from scattering and simulation data, the relaxation rate curves $R_1(^1H)$ and $R_1(^6Li)$ could be calculated. The relevant parameters are compiled in table 1 of ref.18. The experimental relaxation rate curves obtained at $\omega_H/2\pi$=300 MHz and $\omega_{Li}/2\pi$=44.17 MHz (corresponding to B_0=7 Tesla) could be reproduced fairly well as may be seen from figures 5 and 6. Meanwhile these relaxation rates have been determined at $\omega_H/2\pi$=100 MHz and $\omega_{Li}/2\pi$=14.72 MHz (B_0=2.35 Tesla) also. These preliminary results are included in figures 5 and 6, together with the relaxation rate curves as calculated with the current model. At least the proton $R_1(^1H)$ curves show that the low temperature data do not scale with ω^{-2}. This behaviour would be expected for a Lorentzian spectral density function corresponding to an exponentially decaying correlation function G(t). As the relaxation rate at low temperatures is, according to the current model, dominated by the local anisotropic mode, these results ask for a more elaborate treatment of this motion. It should be mentioned that an earlier investigation of the deuteron relaxation in LiCl/D_2O solutions of similar composition invoked a distribution of correlation times (Cole-Davidson) to describe the relaxation rate curves (27). This asymmetric distribution function serves to introduce higher frequency motions into the problem, the physical nature of which remains, however, unclear. Needless to say that the present data can be fitted with such a distribution function, but at the expense of one consistent set of parameters for all nuclei investigated. In this respect another recent investigation is of interest here (28). Heitjans et al. studied the system LiCl·7D_2O with β-radiation detected NMR. The spin-lattice relaxation rate R_1 of ^8Li was measured to temperatures far below the glass temperature of the system. They could find clear indications for two dynamic processes and suggest that the local anisotropic modes are responsible for the weak relaxation rate maximum observed below the glass transition. Hence these local modes should be still activated in the glassy state.

Acknowledgement
We are grateful to Prof. Lüdemann for supporting this work and for many stimulating discussions. We also thank Prof. Heitjans for sending us a preprint of ref. 28. The skillful technical help of R. Knott, S. Heyn and E. Treml made this work feasible.

REFERENCES

[1] C.A. Angell (1982), Supercooled Water, in F. Franks ed. 'Water-A Comprehensive Treatise', Vol. 7, p. 1-76, Plenum, New York
[2] C.A. Angell, E.J. Sare, J. Donnella, D.R. MacFarlane (1981), 'Homogeneous Nucleation and Glass Transition Temperatures in Solutions of Li Salts in D_2O and H_2O. Doubly Unstable Glass Region', J. Phys. Chem. **85**, 1461-64
[3] H.G. Hertz, R. Tutsch, H. Versmold (1971), 'Molecular Motion and Structure around the Hydrated Ions Li^+ and Al^{3+}', Ber. Bunsenges. phys. Chem. **75**, 1177-1191
[4] S.A. Rice, M. Sceats (1981), 'A Random Network Model for Water', J. Phys. Chem. **85**, 1108-1119
[5] E.W. Lang, H.-D. Lüdemann (1985), 'p, T, c-Dependence of ^2H Spin-Lattice Relaxation Rates in Supercooled LiCl-D_2O Solutions', Ber. Bunsenges. Phys. Chem. **89**, 508-516
[6] J.E. Enderby (1987), 'Diffraction Studies of Aqueous Ionic Solutions', in M.-C. Bellissant-Funel, G.W. Neilson eds., The Physics and Chemistry of Aqueous Ionic Solutions, Nato ASI Series C, Vol. 208, p. 129-145, Reidel, Dordrecht, Holland
[7] A.H. Narten, F. Vaslow, H.A. Levy (1973), 'Diffraction Pattern and Structure of Aqueous Lithium Chloride Solutions', J. Chem. Phys. **58**, 5017-5023
[8] K. Heinzinger (1985), 'Computer Simulations of Aqueous Electrolyte Solutions', Physica **131B**, 196-216
[9] E.W. Lang, H.-D. Lüdemann, L. Piculell (1984), 'Nuclear Magnetic Relaxation Rate Dispersion in Supercooled Heavy Water Under High Pressure', J. Chem. Phys. **81**, 3820-3827
[10] P.L. Cummins, G.B. Bacskay, N.S. Hush, S. Engström, B. Halle (1985), 'The Effect of Intermolecular Interactions on the ^2H and ^{17}O Quadrupolar Coupling Constants in Ice and Liquid Water', J. Chem. Phys. **82**, 2002-2013
[11] C.A. Angell (1988), 'Perspectives on the Glass Transition', J. Phys. Chem. Solids **49**, 863-871
[12] see paper by E.W. Lang, H.-D. Lüdemann, this volume
[13] J. Jäckle (1986), 'Models of the Glass Transition', Rep. Prog. Phys. **49**, 171-231
[14] D.W. James, R.F. Armishaw (1975), 'Structure of Aqueous Solutions of Monovalent Halides', Aust. J. Chem. **28**, 1179-1186

[15] W.Fink, E.W.Lang (1988),'Pressure, Temperature and Composition Dependence of Deuterium Spin-Lattice Relaxation Times in Undercooled $MgCl_2/D_2O$ Solutions', J.Phys.Chem.**92**,6440-6445
[16] W.Fink, H.Radkowitsch, E.W.Lang (1988),'p,T,c-Dependence of Deuterium Spin-Lattice Relaxation Times in Undercooled NaCl/D_2O and NaJ/D_2O Solutions', Chem.Phys. **124**,239-249
[17] W.Fink, H.Radkowitsch, E.W.Lang (1988),'Deuterium Spin-Lattice Relaxation Times in Undercooled Aqueous Potassium and Cesium Halide Solutions',Z.Naturforsch. **43a**,538-546
[18] E.W.Lang, F.X.Prielmeier (1988),'Multi-Nuclear Spin-Lattice Relaxation Time Studies of Supercooled Aqueous LiCl Solutions', Ber.Bunsenges.Phys.Chem. **92**,717-724
[19] H.G.Hertz, G.Keller, H.Versmold (1969),'Molecular Motions in the Hydration Sphere of the F^- Ion in Aqueous Solution',Ber.Bunsenges.Phys.Chem.**73**,549-62
[20] K.J.Müller, H.G.Hertz (1984),'Experimental Proof that Water Arrangement in the Hydration Sphere of F^- is Symmetric', Z.Phys.Chem.Neue Folge **140**,31-54
[21] R.Ahlrichs, H.-J.Böhm, H.G.Hertz, K.J.Müller (1984), 'The Hydration of F^-: An Explicit Demonstration of a Basic Discrepancy Between NMR Results and Those Obtained From Various Computations and Neutron Diffraction Experiments', Z,Phys.Chem.Neue Folge **142**,67-80
[22] R.W.Impey, P.A.Madden, I.R.McDonald (1983),'Hydration and Mobility of Ions in Solution',J.Phys.Chem. **87**, 5071-5083
[23] H.Radkowitsch, E.W.Lang,'Hydration Water Dynamics in Undercooled Aqueous KF Solutions', in preparation
[24] H.Weingärtner (1980),'The Mechanism of ^{7}Li Relaxation in a Supercooled Aqueous LiI Solution',J.Magn.Reson. **41**,74-87
[25] H.Kanno, I.Shirotani, Sh.Minomura (1980),'Isotope Effect of the Glass Transition Temperature of Aqueous Solutions. LiCl and $ZnCl_2$ Solutions in Water and D_2O' Bull.Chem.Soc.Jpn.**53**,2079-2080
[26] I.C.Baianu, N.Boden, D.Lightowler, M.Mortimer (1978), 'A New Approach to the Structure of Concentrated Aqueous Electrolyte Solutions Using Pulsed NMR Methods', Chem.Phys.Lett. **54**,169-175
[27] N.Boden, M.Mortimer (1978),'Reorientation of D_2O in Concentrated Aqueous Solutions of Lithium Chloride Studied by Nuclear Magnetic Relaxation', J.Chem.Soc. Faraday Trans.II **74**,353-366
[28] W.Faber, P.Heitjans, A.Schirmer (1988),'β-NMR on $LiCl \cdot 7D_2O$ in the Glassy and the Liquid State', Proceedings of the ILL-Workshop 'Dynamics of Disordered Materials', in print

VIBRATIONAL DYNAMICS OF LIQUID AND GLASSY ELECTROLYTES

J. DUPUY, J.F. JAL, A. AOUIZERAT-ELARBY, P. CARMONA
J. DIANOUX* and P.CHIEUX*
Département de Physique des Matériaux - U.A. 172 CNRS - U.C.B LYON - 69622 Villeurbanne - France
* INSTITUT LAUE LANGEVIN - 156X 38042 - GRENOBLE CEDEX - France

ABSTRACT. The interest of neutron inelastic scattering to study collective excitations in disordered systems is illustrated by the modifications of the frequency distribution of the coordinated water in LiCl, RH_2O according to the liquid or vitreous thermodynamic state of this electrolyte.

1. GENERAL CONSIDERATIONS

Neutrons of very low energy are a unique probe for the study of molecular dynamics in solids and liquids since the neutron energy is then comparable to that of lattice vibrational molecular motions.

The relevant energies are of the order of 10^{-3} to 10^{-1} eV with wavelength in the range of 10^{-8} cm, so inelastic neutron scattering gives information on short distance order and short time.

The time scale between 10^{-14} to 10^{-11} seconds is typically the range of vibrational motions and of diffusive motion in "microscopic diffusive processes", the N.M.R. giving the average behaviour (10^{-9} seconds).

Therefore inelastic neutron scattering (INS) has become a powerful tool for studying the microscopic environment in liquids and solids.

For hydrogen-bounded liquids this technic is unique because hydrogen scattering is essentially incoherent (allowing us to study self diffusion) and its cross-section is at least one order of magnitude greater than that of other elements. From the dynamical structure factor S(q.E), the frequency distribution g(E) can be derived which gives the density of vibrational frequencies. It is related to the partition function of the system and therefore permits the determination of thermodynamic properties like C_V. g(E) is represented by a continuous spectrum (because all frequencies are active), with peaks corresponding to characteristic frequencies. The intensity of the peaks is function of the amplitude of vibration of normal modes of the system. Their sharpness depends on several factors :

i) in the case of a crystal, the dispersion of the branch corresponding to the normal mode.

ii) the strength of the bond holding a molecule in its environment

iii) the amount of long range ordering in the system

J.C. Dore and J. Teixeira (eds.), Hydrogen-Bonded Liquids, 405–412.

Particularly interesting for a disordered solid system, is the low energy part of the frequency distribution which allows us to reach the behaviour of the temperature dependence of the "lattice" vibrational part of the specific heat (see [1]).

A complete derivation leading to generalized frequency distribution and related moments has been presented in ref. [2]

For a polyatomic topologically disordered system containing hydrogen the phonon expansion of the inelastic part depends only on incoherent scattering

We start from the relation

$$S_{inc}(q, E) = \frac{4\pi}{\sigma_{incoh.}} \frac{k_o}{k} \frac{d^2\sigma_{inc}}{d\Omega\, dE}\Bigg|_{inél.} = e^{-2W(q)} \sum_{n=1}^{\infty} \frac{f_n(E)}{n!} \left(\frac{h^2 q^2}{2M}\right)^n \quad (1)$$

with $\hbar q = \hbar\,(k_o - k)$, the momentum transfer (k_o et k being the wave vectors of the neutrons before and after scattering).

$$\hbar E = \frac{\hbar}{2\,m}\,(k_o^2 - k^2),$$ the energy transfer (m = the neutron mass)

$e^{-2W(q)}$ is the Debye-waller factor, M the mass of the scattering unit

$$f_n(E) = \int_{-\infty}^{+\infty} f_{n-1}(E)\, f_1(E - E')\, dE \qquad (n > 1)$$

with $$f_1(E) = \frac{g(E)}{\hbar E\,(1 - e^{-\beta})}$$ and $$\left(\beta = \frac{\hbar E}{kT}\right)$$

g(E) is the phonon density of state

The measured intensity contains the peak of elastically scattered neutrons and some parasitic intensity due to multiphonon processes. These must be removed in order to obtain the one-phonon cross section which is directly related to the phonon density of states.

Until now, very few analysis of frequency distributions for hydrogen-bounded systems have been made, this is particularly true for aqueous systems. H. Boutin and S. Yip [3] have reviewed the influence of ions on the molecular motion of water. More recently P. Madden and R.W. Impey [4] compared experimental results with simulation of the dynamics of coordinated water. M.C. Bellissent-Funel et al. [5] has compared the vibrational informations on H_2O and aqueous $ZnCl_2$ solutions obtained by inelastic neutron scattering with Raman scattering.

All of these studies indicate that some H_2O-ion interaction occurs, but that, for the different concentrations studied, the presence of ions alters the structure only very locally.

However at concentrations around 12.5 M for $ZnCl_2$ in water, M.C. Bellissent-Funel assigns the peak at 11 meV to the bending deformation of $ZnCl_4^{2-}$ complexes [M is the molarity which corresponds here to 4.5 moles of water for 1 mole of salt, a very high concentration indeed]. Concentrated aqueous electrolytes being good glass formers at some well defined concentrations, we can therefore study the differences between frequency distributions in vitreous, supercooled and liquid state and analyse the elastic response in the vicinity of the glass transition. A careful examination of the various energy ranges will lead us to discuss the questionable interpretation of the additional modes due to long wavelength acoustic phonons. In this paper we will point out the *principal characteristics of the frequency distributions in the glassy and liquid state with a special interest in the dynamics of the coordinated water and in its possible network forming ability as a function of temperature and ionic concentration.*

2. CHARACTERIZATION OF THE SYSTEM AND OF ITS RESPONSE TO INELASTIC SCATTERING

We have analyzed the metastable states of different salts diluted in water [6]. LiCl, RH_2O is as far as we know, the thermodynamically as well as kinetically best defined one [7] [8].

The non equilibrium phase diagramme is given in figure 1.

figure . 1

Metastable equilibrium domains in the phase diagram of LiCl, RH_2O determined on warming the quenched glass samples with a heating rate v =3.10-2 K.sec-1

- ◆ Tg, glass transition
- – o liquidus line
- ◻ ice crystallization
- ◊ pentahydrate crystallization
- x dihydrate crystallization
- ◆ trihydrate crystallization

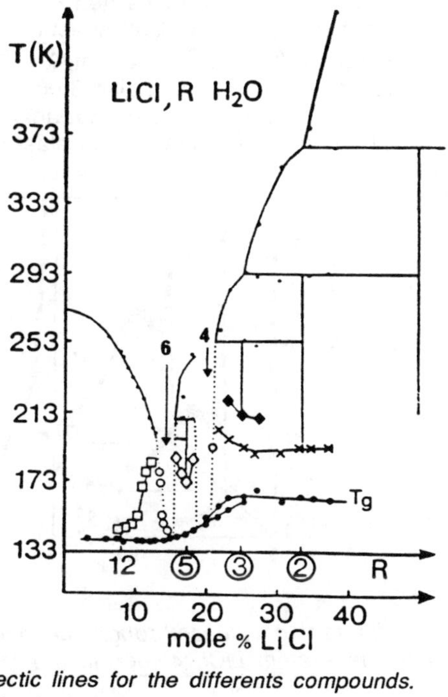

The horizontal continuous lines are the peritectic lines for the differents compounds. The vertical continuous lines correspond to the hydrates
The discontinuous lines are postulated metastable extensions of the different compound branches of the liquidus line.

The definition of the concentration is given in terms of $x = \frac{\text{moles of salt}}{100 \text{ moles of water}}$

or in terms of R = number of moles of water per mole of salt ($x = \frac{100}{1+R}$)

The important property shown in this diagram is the ability around the total hydration values R = 6 and R = 4 of LiCl to maintain a disordered structure from very low temperature up to the melting point although Tg is significantly lower than $T_{m.p.}$ The low temperature limit of the undercooled state is T_0 = 145 K for R=6 and T_0 = 157 K for R = 4 [8].

In figure 2, an example of the dynamical structure factor S_{inc} (q, E) is given for R = 4 at several q values and two temperatures . The shape remains approximately the same, but the intensity of the inelastic spectra increases with increasing q value, as in the incoherent approximation. The two temperatures correspond to the glass transition domain and the liquid state. The large quasi elastic signal seen in the liquid is due to diffusive processes of hydrogen containing molecules.

The experiments were carried out at the Institut Laue Langevin in Grenoble using the focussing time-of flight spectrometer IN6. The flux at the sample was

$\phi \approx 8.9 \cdot 10^4 \text{ n.cm}^{-2} \cdot \text{sec}^{-1}$ at $\lambda = 5.1$ Å

the dimensions of the sample being 3 cm x 5 cm x 0,02 cm

In the present paper, we do not analyse the diffusive motion of water molecules nor the low energy excitations in the glassy state.

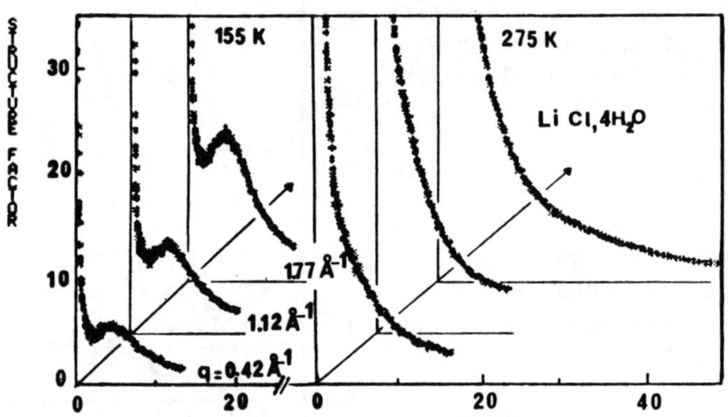

Figure 2
Cuts at different q values through the energy loss part of the dynamical structure factor of the system LiCl, $4H_2O$, in the glassy (155K) and liquid state (275K).

3. FREQUENCY DISTRIBUTION

As seen briefly from the relation (1), some corrections have to be done in order to obtain the frequency distribution.

The elastic part is corrected using a purely incoherent scatterer (vanadium) or better using our own sample at 20K. We excluded a range of energies from -150 μeV to 150 μeV corresponding to 3 times the instrumental resolution.

Since we are in the incoherent approximation the spectra can be summed over all q values in order to increase the statistics of the data.

Corrections for background and multiphonon are applied following available programs [9] and the spectra are normalized such as $\int g(E)\, dE = 1$.

The multiphonon connection is not unique and a self consistency check of the results over a large temperature range has to be performed. For the time being we present only the characteristics of the frequency distributions in the glassy state and in the liquid state.

The results obtained for compositions R=6 and R=4 are given in the figure 3 in comparison with the frequency distribution measured for pure water kindly provided by M.C. Bellissent - Funel (these last data being not corrected multiphonons and not normalized).

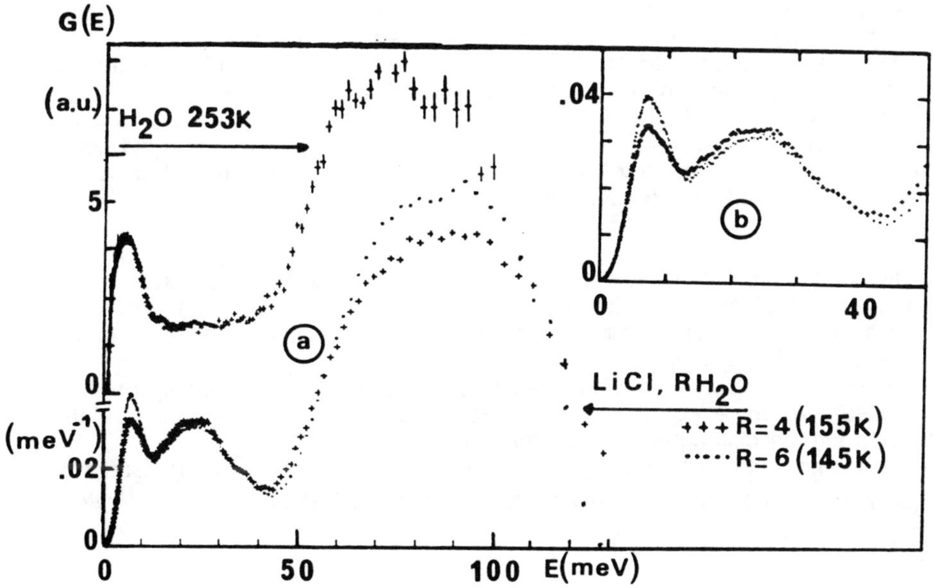

Figure 3

Frequency distributions measured in the glass transition domain for LiCl, 6H_2O and LiCl, 4 H_2O. A comparison is made with supercooled H_2O

This figure exhibits the principal features of the dynamics of non coordinated and coordinated water.

The characteristics frequencies of the hydrogen network correspond to intermolecular modes (THz domain). For the two R values we observe three maxima in the distribution, respectively centered around 7 meV (56 cm^{-1}), 25 meV (200 cm^{-1}) and 75 meV (600 cm^{-1}). The first one is assigned to the center of mass motion, the second to the stretching and the third to the rotational motion. The last assignation has been verified by substituting D_20 to H_20. (Due to change of the inertial moment we note a large shift of this last maximum, to lower energies).

The difference between the two concentrations is noticeable in the bands near 7 meV and 75 meV. The origin of the variation at 7 meV seems to be related to the variation in the number of water molecules. The change in the intensity of the rotational motion could be related to its amplitude. Indeed, due to absence of direct interactions between Li^+ and Cl^- for R = 6 [10], the rotation of the surrounding water molecules is less hindered than for R = 4 where we expect a Cl^- ion participating to the first coordination shell around Li^+.

If we compare the glassy electrolyte to supercooled water, the intensity change for frequencies near 200 cm^{-1} is certainly connected with the strong ion-water interaction.

However *the most spectracular change in the frequency distribution is induced by melting.* This is seen in Figures 4 and 5 where we compare the frequency distributions in glassy and liquid state for the two R values. The characteristic frequencies for the center of mass motion and the rotational motion are the same in the two thermodynamic states, the intensity being strongly reduced.

The *frequency shift to lower energy* observed for the stretching vibrational mode in the liquid state may be related to the reduction of the order range.

We know [10] that in the glassy state significant order can be found at distance as large as 10 to 15 Å. This medium range order disappears in the liquid structure where it is smeared out by the density fluctuations with an effect on the corresponding vibrational frequencies.

In conclusion, the ability for aqueous electrolytes to form glasses gives an interesting physical base to discuss the vibrational density of states. We have seen that it is possible for condensed hydrogenated systems to obtain frequencies distributions with a good statistical accuracy.

However in order to efficiently analyse the distribution of density of states it is necessary to have good structural informations. In particular the partial ion-water and ion-ion distributions should be known.

This is possible for LiCl, RD_2O which makes this system as a good candidate for the study of hydrogen-bounded disordered materials with the emphasis on water.

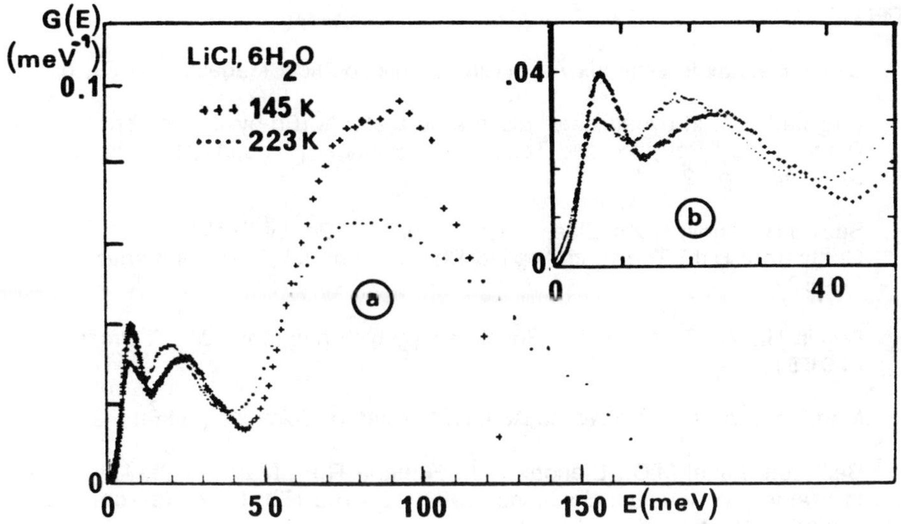

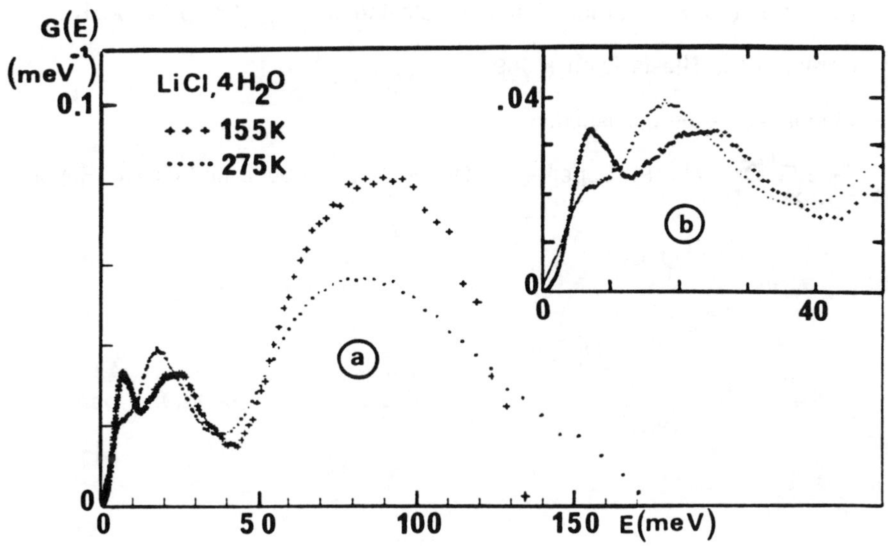

Figures 4 et 5
Frequency distributions for R=6 and R=4 in the glassy and liquid state.

REFERENCE

[1] See for example Dianoux A.J. (1989), Philosophical Magazine, B 59, p. 17.

[2] Egelstaff P.A. in dynamics of disordered material (Ed by D. Richter, Dianoux A.J., Petry W. and Texeira J.) Springer proceedings in Physics, 37 (1989) p. 2.

Suck J.B., Rudin H., in Glass Metals II, p. 217 (Ed. by Beck H., Guntherodt H.J.) Topics in Applied Physics, Vol. 53, Springer Verlag (1983).

[3] Boutin H., Yip S., Molecular Spectroscopy with neutrons, M.I.T. Press (1968).

[4] Madden, P.A., Impen R.W., in Ann. N.Y. Acad. of Sciences (1986).

[5] Bellissent-Funel M.C., Dianoux A.J., Fontana R.P., Maisano C.,Migliardo P., in Water and Aqueous Solutions (1986) 199-106 (Ed. G.W. Neilson, J.E. Enderby, Bristol Adam. Hilger).

[6] Aouizerat-Elarby A., Thesis Lyon (1988).

[7] Elarby-Aouizerat A., Jal J.F., Chieux P., Letoffé J.M., Claudy P., Dupuy J. (1988) Journal of non crystalline solids, 104, p. 203-210.

[8] Carmona P., Thesis Lyon (1989).

[9] Dianoux J., to be published.

[10] Jal J.F., Carmona P., Soper A.K. , Dupuy J., to be published (R.A.L. Report 1989).

THE PROGRAMME

a) Lecture presentations (see "Contents")

b) Additional Lectures:-

H.J.C.Berendsen:	Hydration and hydrogen-bonding in biological systems.
M.Klein:	Interaction potentials and molecular dynamics.
D.Middendorf:	Collective excitations in biopolymers.
G.Palinkas:	Electron diffraction studies and computer simulation.

c) Other Sessions

i) Poster session (including some short presentations) see following pages

ii) Panel discussion:
"What are the new developments and what implications do they have for future work?"

S.Bratos, S-H.Chen, H-D.Lüdemann, G.Palkinkas and G.Walrafen with J.C.Dore (Chairman).

iii) General discussion:
"What do we really know/want to know about liquid water?"
Introduced and led by J.Teixeira.

iv) An overview of the meeting presented by G.Palinkas.

POSTERS

1. "Proton Dynamics in Formic Acid"
   C.K.Loong*, F.Trouw, D.J.Leloir (Argonne National Laboratory) and S.H.Chen, (MIT).

2. "Structural Studies of Liquid Formic Acid by Neutron Scattering",
   G.Swan*, J.C.Dore (University of Kent) and M-C. Bellissent-Funel (CEN Saclay).

3. "Plastization of Water's Hydrogen-Bonded Network and the Glass Transition Temperature Minimum"
   K.Hoffer*, A.Halbrucker, E.Mayer (Innsbruck), P.Johari (Ontario).

4. "Reverse Monte Carlo Simulation Technique and Applications from LJ System to the possibility of Simulating disordered Molecular Systems"
   L.Pusztai* and R.L.McGreevy (Oxford).

5. "Expansivity of m-Fluoro Aniline: a Model of a Thermodynamic Property for the Whole Liquid State"
   C.Alba Simionesco* (CNRS Paris).

6. "First Approach of Pg along an Isothermal Path"
   C.Alba Simionesco*(CNRS Paris).

7. "Inelastic Neutron Scattering Studies of the Dynamics of ice Ih"
   J.C.Li* and D.K.Ross (University of Birmingham)

8. "MD Simulation of $MgCl_2$ in Methanol"
   Y.Tamura*, E.Spohr and K.Heinzinger (Max-Planck Institute for Chemistry

9. "1H NMR Studies of Hydrogen Bonded Systems at Low Temperatures: Phenol-Amine Complexes"
   M.Ilczyszyn*, H.Ratajczak and J.A.Hadd (Ladd)

10. "Ice Formed in Supercooled Electrolytic Solutions"
    J.Dupuy* (Lyon)

11. "Dielectric Relaxation Measurements by Time Domain Reflectometry on Propylene Glycol/Water Solutions"
    E.Tombari*, G.Chryssikos, B.Gestblom and R.H.Cole (Brown Univ)

---

* Designates an oral presentation as a paper or a short contribution

12. "Structure in Urea Solutions"
    J.Turner* and J.L.Finney (Birkbeck College, University of London)

13. "Conformation Studies of n-Propylamine by Raman Spectroscopy"
    A.M.Amorim da Costa (Coimbra Portugal)

14. a) "On the Electrostatic Directionality of N-H...O = (Hydrogen Bonding)
    b) "Hayes-Stone Intermolecular Perturbation Theory Calculations on the Formamide/Formaldehyde Complex"
    J.B.O.Mitchell and S.L.Price (Cambridge)

15. "A Search for Structural Anomalies in Temperature Studies of Water in Porous Silicas"
    E.Davies, J.C.Dore (Kent), W.Drost-Hansen (Miami) and P.Chieux (ILL).

16. "Neutron Diffraction Studies of t-Butanol"
    D.Huxley, J.C.Dore (Kent), P.Chieux (ILL).

17. "Molecular Dynamics Simulation of Single Ions in Water"
    E.Guardia, R.Rey and J.A.Padro (Universitat de Barcelona, Spain)

18. "Slow Counterion Dynamics in Polyelectrolyte Solutions Studied with NMR"
    R.H.Tromp, J. de Bleysey and J.C.Leyte (Dept. of Physical Chemistry, University of Leiden, Netherlands)

19. "Viscometric and Spectrometic Investigation of Poly Vinyl Pyrrolidone Solutions"
    A.Guner, F.Yigit and O.Guyne (Hacettepe University, Turkey)

20. "Thermodynamic Results for a Solute at the Interface between two liquids"
    R.Bonaccorsi, F.Floris and J.Tomasi (Dip di Chimica, Univ di Pisa)

21. "Multifractal Nature of Brownian Motion"
    M.A.Lopez-Quintela, C.Tojo, M.C.Bujan-Nunez, (Univ. of Santiago, Spain)

22. "Inelastic Neutron Scattering Results of Deeply Supercooled H_2O in Porous Glass"
    K.Bradley, S.H.Chen, C.K.Loong (Argonne), J.Teixeira, M.C.Bellissent-Funel (CEN Saclay)

Postscript

The most extensive coverage of hydrogen-bonding phenomena has been the three volume compilation by Schuster, Zundel and Sandorfy published in 1976. Specific studies of water have been covered in the series 'Water: a comprehensive treatise' edited by Franks and extending to seven volumes. The last volume was published in 1982 and there has been a change to an annual production with a new series entitled 'Water Science Reviews'. The present book of the papers presented at Cargèse provides a somewhat different focus from these preceding volumes and by gathering together individuals working in different areas of "H-bond science", there has been a deliberate attempt to fuse together the current trends and developing ideas. The contents emphasise work that is in progress and include some techniques that are novel, rather than the established consensus and resolved outlook which frequently emerges from a critical review. Although some papers include supporting background, there is a strong emphasis on new features to be further studied. This factor has also been apparent in the compilation of the manuscripts since several participants needed extra time to complete the analysis of their data and it was therefore impossible to assemble the contributions immediately after the meeting.

By bringing together experts in different fields it became apparent that there has been a diversity of interest which has resulted in a loss of communication between active groups. The well-established spectroscopic techniques have provided a framework for the discussion of hydrogen-bonding which has not always been sufficiently well recognised by groups engaged in new and emerging areas of study. This viewpoint was clearly expounded in the panel discussion where the overlap of research methods was considered and the need to make appropriate connections between different techniques was strongly emphasised. There was an implicit criticism underlying this discussion that some groups are "discovering old truths" but not everyone was prepared to accept this view. It was, however, clear that the potential for misunderstanding was still present in transferring nomenclature across different subjects, i.e. what does one actually mean by "strong and weak hydrogen-bonds"? This type of difficulty will only be avoided if a satisfactory dialogue is established across the subject boundaries and, in that sense, the meeting itself had helped to clarify the issues.

Inevitably, much of the discussion centered on the properties of water. In the past there have been many controversial exchanges surrounding this complex material but these seem now to play a less important role. The provocative title "What do we really know about water?" for a discussion topic on the last day failed to create any major disagreements! What has emerged is that the structural characteristics of water depend on a delicate balance between the ordering of molecular positions due to the orientational effects of hydrogen-bonding and the disordering effects of thermal fluctuations. The special properties of super-cooled water arise from the space-filling properties of the tetrahedral network and involve a dynamic balance which can be treated by normal statistical physics principles. There is no need to invoke the vague ideas of flickering-cluster or quasi-crystalline models nor to engage in heated debate about the relative merits of continuum or mixture models. The dynamic properties of water and their relation to transport properties were also discussed. The basic phenomena are understood but essential concepts such as the hydrogen bond lifetime are still controversial both in simulation and real experiment. The coupling between vibrational, rotational and diffusive motions and their temperature dependence also remains a largely unsolved problem but original approaches to these questions have been presented in several lectures.

Although the principles required for a complete understanding of water are now defined there still remains a need for much more detailed work to improve the quantitative agreement between experimental observation, computer simulation and theoretical prediction. This programme necessarily involves the study of water under extreme conditions of temperature and pressure which will obviously be easier on a computer than for the real material! The issue then becomes one of finding a suitable interaction potential that satisfactorily represents the known data. So far, the systems have been based on two-body forces (sometimes with polarizability effects added) and these are known to be incomplete. There are then two approaches based on either an empirical treatment using a simplified form of potential with adjustable parameters or a sophisticated approach using *ab initio* methods to provide a basis for the interaction formalism.

Although water has a dominating influence there are many other systems that have particular characteristics of interest, which help to establish the more general phenomenon of hydrogen-bonding. These systems range from relatively simple molecules such as the hydrogen halides, through small organic molecules to the complex macro-molecules of biological importance. Experimental measurements are now providing valuable results for comparison with computer simulations and some spectacular successes have been achieved. However, the increased complexity leads directly to difficulties in detailed interpretation of available data and the increased speed and storage requirements for computation. A closer link will be required in which simulation studies with realistic potentials can be compared directly with experimental data of high precison and reliability.

Ab initio calculations are already playing a significant role in the description of energy surfaces for simple hydrogen-bonded molecules and have been used to establish effective two-body interaction potentials. The use of more sophisticated versions in simulation studies is currently constrained by available computer power. The continuing improvement in computational facilities is expected to lead to significant improvements but will raise further questions of a fundamental nature such as:-

i) how important are three-body (and higher order) terms in the description of the interaction?

ii) can "co-operativity" manifest itself in large scale structures?

iii) do quantum corrections cause significant change from predictions based on a classical treatment.

The results of any simulation or theory are only as good as the potentials used in the computations and the ability to distinguish between different formulations has not yet been possible.

The meeting has also seen the emergence of several new approaches to other research areas based on the presence of a second component in the form of ions, other liquids or solids. The importance of aqueous solutions in this context is paramount but the inclusion of this topic in the meeting was deliberately restricted due to the highly successful ASI meeting on 'Aqueous Solutions' presently held at Cargèse [M-C. Bellissent-Funel and G.Neilson (eds) Kluwer (pub), 1988]. The investigation of liquid mixtures has previously been centered on thermodynamics and vibrational spectroscopy but it now seems that a much wider interest is developing in which new experimental methods are being used and the data are interpreted in terms of microscopic theories and computer simulation. Within this class, the particular case of water/alcohol mixtures seems likely to be a fruitful field for detailed investigation since it concerns an effective perturbation of the hydrogen-bond structures formed in the pure liquids. Another type of perturbation in hydrogen-bonding occurs at the proximity to a solid interface. This is primarily a sub-branch of work on water and has obvious implications for many areas of science. It is to be expected that there will be a continuing interest in this fundamental work which is essentially multi-disciplinary and provides a foundation for the whole area of colloid and interface science. Several papers involving preliminary investigations have shown that valuable information can be obtained despite difficulties in sample preparation and interpretation of the observation.

In conclusion, we may recognise the importance of hydrogen-bonding to an understanding of the physical, chemical and biological world which surrounds us. It has always been a fundamental but difficult area of the natural sciences and has rarely been depicted as exciting or epoch-making in its new discoveries. However, the practitioners of the gentle art of "hydrogen-bond investigations" will know that slow progress does not deter and that the intellectual challenge remains as strong today as it always has been, and probably always will be!

Author Index

Chemical Index

Subject Index